Conservation for the Anthropocene Ocean

Conservation for the Anthropocene Ocean

Interdisciplinary Science in Support

of Nature and People

Editors

Phillip S. Levin
The Nature Conservancy & University of Washington
School of Environmental and Forest Sciences, Seattle, WA, USA

Melissa R. Poe
University of Washington, Washington Sea Grant, Seattle, WA, USA

ACADEMIC PRESS

An imprint of Elsevier

Library of Congress Cataloging-in-Publication Data
A catalog record for this book is available from the Library of Congress

British Library Cataloguing-in-Publication Data
A catalogue record for this book is available from the British Library

ISBN: 978-0-12-805375-1

For information on all Academic Press publications visit our website at
https://www.elsevier.com/books-and-journals

  Working together
to grow libraries in
developing countries

www.elsevier.com • www.bookaid.org

Publisher: Sara Tenney
Acquisition Editor: Kristi Gomez
Editorial Project Manager: Pat Gonzalez
Production Project Manager: Edward Taylor
Designer: Christian Bilbow

Typeset by TNQ Books and Journals

Contents

Section II
Principles for Conservation in the Anthropocene

Section III
Conservation in the Anthropocene in Practice

15.   Marine Conservation as Complex Cooperative and Competitive Human Interactions

Xavier Basurto, Esther Blanco, Mateja Nenadović and Björn Vollan

16.   Transdisciplinary Research for Conservation and Sustainable Development Planning in the Caribbean

Katie K. Arkema and Mary Ruckelshaus

Section IV
Looking Forward

20. The Future of Modeling to Support Conservation Decisions in the Anthropocene Ocean

Éva E. Plagányi and Elizabeth A. Fulton

21. The Big Role of Coastal Communities and Small-Scale Fishers in Ocean Conservation

Anthony Charles

List of Contributors

Aurora Alifano FishWise, Santa Cruz, CA, United States

Edward H. Allison University of Washington, Seattle, WA, United States

Katie K. Arkema Stanford University, Stanford, CA, United States; University of Washington, Seattle, WA, United States

Natalie C. Ban University of Victoria, Victoria, BC, Canada

Xavier Basurto Duke University, Beaufort, NC, United States

Nathan Bennett University of British Columbia, Vancouver, BC, Canada

Esther Blanco Indiana University, Bloomington, IN, United States; Innsbruck University, Innsbruck, Austria

Mariah Boyle FishWise, Santa Cruz, CA, United States

Elena Buscher University of Victoria, Victoria, BC, Canada

Courtney Carothers University of Alaska Fairbanks, Anchorage, AK, United States

Elizabeth B. Cerny-Chipman Oregon State University, Corvallis, OR, United States

Kai M.A. Chan University of British Columbia, Vancouver, BC, Canada

Anthony Charles Saint Mary's University, Halifax, NS, Canada

Isabelle M. Côté Simon Fraser University, Burnaby, BC, Canada

Rachel Donkersloot Alaska Marine Conservation Council, Anchorage, AK, United States

Nicholas K. Dulvy Simon Fraser University, Burnaby, BC, Canada

Tammy E. Davies BirdLife International, Cambridge, United Kingdom

Lauren Eckert University of Victoria, Victoria, BC, Canada

Elena M. Finkbeiner Stanford University, Monterey, CA, United States

Nicole Franz Food and Agriculture Organization of the United Nations, Rome, Italy

Elizabeth A. Fulton CSIRO Oceans & Atmosphere, Hobart, TAS, Australia; Centre for Marine Socioecology, Hobart, TAS, Australia

David Goldsborough VHL University of Applied Sciences, Leeuwarden, The Netherlands

Steven Gray Michigan State University, East Lansing, MI, United States

Benjamin S. Halpern UC Santa Barbara, Santa Barbara, CA, United States; Imperial College London, Ascot, United Kingdom

Julia Hoffmann Christian Albrechts Universität zu Kiel, Kiel, Germany

Aerin L. Jacob University of Victoria, Victoria, BC, Canada

Holly K. Kindsvater Rutgers University, New Brunswick, NJ, United States

John N. Kittinger Conservation International, Honolulu, HI, United States

Sarah C. Klain University of British Columbia, Vancouver, BC, Canada

John Z. Koehn University of Washington, Seattle, WA, United States

Marloes Kraan Wageningen Marine Research, Den Helder, The Netherlands

Darienne Lancaster University of Victoria, Victoria, BC, Canada

Heather M. Leslie University of Maine, Walpole, ME, United States

Phillip Levin University of Washington, School of Environment and Forest Sciences, Seattle, WA, United States

Sara G. Lewis FishWise, Santa Cruz, CA, United States

Jane Lubchenco Oregon State University, Corvallis, OR, United States

Marc Mangel University of California Santa Cruz, Santa Cruz, CA, United States

Darcy L. Mathews University of Victoria, Victoria, BC, Canada

Mateja Nenadović Duke University, Beaufort, NC, United States

Kirsten L.L. Oleson University of Hawai'i at Mānoa, Honolulu, HI, United States

Paige Olmsted University of British Columbia, Vancouver, BC, Canada

Julia Olson NOAA Fisheries, Woods Hole, MA, United States

Malin L. Pinsky Rutgers University, New Brunswick, NJ, United States

Anton Pitts University of British Columbia, Vancouver, BC, Canada

Éva E. Plagányi CSIRO Oceans & Atmosphere, St Lucia, QLD, Australia

Melissa R. Poe University of Washington, Seattle, WA, United States

Martin F. Quaas Christian Albrechts Universität zu Kiel, Kiel, Germany

Chris Rhodes University of Victoria, Victoria, BC, Canada

Leslie Robertson University of British Columbia, Vancouver, BC, Canada

Christine Röckmann Wageningen Marine Research, Den Helder, The Netherlands

Andrew A. Rosenberg Union of Concerned Scientists, Cambridge, MA, United States

Mary Ruckelshaus Stanford University, Stanford, CA, United States; University of Washington, Seattle, WA, United States

Terre Satterfield University of British Columbia, Vancouver, BC, Canada

Jörn O. Schmidt Christian Albrechts Universität zu Kiel, Kiel, Germany

Steven Scyphers Northeastern University, Boston, MA, United States

Rebecca L. Selden Rutgers University, New Brunswick, NJ, United States

Kevin St. Martin Rutgers University, New Brunswick, NJ, United States; University of Tromsø, Tromsø, Norway

Jenna M. Sullivan Oregon State University, Corvallis, OR, United States

Nancy J. Turner University of Victoria, Victoria, BC, Canada

Nathan Vadeboncoeur University of British Columbia, Vancouver, BC, Canada

Luc van Hoof Wageningen Marine Research, Den Helder, The Netherlands

Björn Vollan Innsbruck University, Innsbruck, Austria; University of Marburg, Marburg, Germany

Rüdiger Voss Christian Albrechts Universität zu Kiel, Kiel, Germany

Charlotte Whitney University of Victoria, Victoria, BC, Canada

Esther S. Wiegers Food and Agriculture Organization of the United Nations, Rome, Italy

Elizabeth A. Williams University of British Columbia, Vancouver, BC, Canada

The Name of Science

In a world of logical inquiry
even blasphemous theories can realize redemption

and while the law of evolution
would say we get smarter with time,
we remain confounded by the accomplishments
of the ancients

for the curious mind
query is followed by technology
letting us see the unseen
and measure the unknown

the scientific pursuit of better crops
has yielded time honored accomplishments
as wheat and corn
hybrid over the millennia from unassuming grasses.

through science we feed the multitudes
though great and natural world is displaced

and as we celebrate science
let us not forget
"better living through chemistry"
and let's not forget the destructive powers
unleashed by science
and technologies that distract the multitudes
commercial enterprise is the usual attendant factor
in the "management" of resources and life

while the compromises have leaned on the natural world
A person of the ocean learns to fear terms as
"sustainable", "precautionary principle"
which have become the terms accompanying the demise
of species and habitats across the seas
It is not the wild, but rather the humans
and their affects that need management

the unbridled curiosities and infatuation
with trying to understand the great mysteries of life
our place in the universe or maybe multiverse
and the intricacies of nature
bid us to probe ever deeper
piling even more questions upon our answers
help us know that life is precious

should knowledge or the intelligence of humanity
be measured by the health of our planet
we realize the great failing of humanity

With the failings and the folly,
it would be a sorry day to end science
and the pursuit of knowledge and understanding

Science is a discipline worthy of our respect
and is usually right
… at least for a while

By Chief Gidansta (Guujaaw)
Haida Nation

Preface

Humanity as a whole is thriving. Over the last two centuries, average life spans have more than doubled. We have conquered diseases and are more resilient to natural disasters. We have developed technologies that allow us to live in formerly inhospitable climates, and to be more food and water secure.

The earth is our home, but the way we use our planet has taken its toll (Ellis et al., 2013). We humans have intentionally and unintentionally shaped and reshaped the earth's environments for thousands of years (Marris, 2013). And as the largest biome, the oceans have long been a major focus of human use (Thurstan et al., 2015). For at least 200,000 years people have fished, gathered shellfish, and hunted marine mammals (Erlandson and Rick, 2010). The ocean has long been central to the culture and identity of those dependent on it (Poe et al., 2014). We have used the sea as a means of transportation, and as a source of energy. We are connected to the ocean to such a degree that our brains are hardwired to react positively to the ocean, decreasing stress and increasing innovation and insight (e.g., White et al., 2010). It is not surprising, then, that globally, nearly 3 billion people live within 60 miles of the coast, and population growth along the coasts is 6 times greater than inland areas (Feist and Levin, 2016).

Humanity's extraordinary success has long impacted ocean ecosystems. Indeed, even before the existence of written records, the impact of humans on the animals they exploited was significant. For instance, the giant clam is found on a number of islands in the South Pacific, but not in Fiji. While no records exist revealing its presence in Fiji, archeological data suggests that clams occurred on Fijian reefs, but were driven to extinction there around 750 BC (Seeto et al., 2012). Moreover, large-scale pollution of the ocean by industrial activities has been occurring for many thousands of years. Roman and Carthaginian mining, for example, led to substantial heavy metal pollution throughout the North Atlantic (Davis Jr. et al., 2000).

While humans have certainly interacted with and impacted the oceans for eons, the scale and scope of human influence upon the oceans today is unprecedented (Fig. 1; Duarte, 2014). For example, coral reef fishes in Hawaii are now severely depleted by commercial fishing (McClenachan and Kittinger, 2013), and since European contact, they no longer support robust fisheries (Van Houtan et al., 2013) that were historically fished by native Hawaiians for millennia (Kittinger et al., 2011). In addition, the rapid, postindustrial revolution rise of human populations and their preferential settlement along coastlines (Feist and Levin, 2016) has led to the urbanization and development of shorelines with concomitant

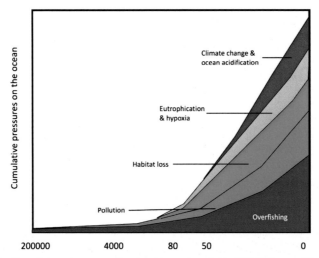

FIGURE 1 Cumulative pressures on the global ocean over the last 200,000 years. The size of the polygons corresponding to each pressure is heuristic and is not meant to reflect their quantitative impacts. *Redrawn from Duarte, C.M., 2014. Global change and the future ocean: a grand challenge for marine sciences. Frontiers in Marine Science 1, 63.*

habitat and ecosystem effects resulting from coastal erosion, pollution, and habitat modification (e.g. Levin et al., 2015). Similarly dramatic, global-scale impacts are associated with shipping, agriculture, energy development, and modifications to freshwater flow, among many other human activities. Furthermore, in human-dominated oceans, our activities are strongly interacting (Fig. 2) often in ways we do not fully understand. Indeed, human activities in and near the ocean have become so prevalent (Halpern et al., 2008) that alarms regarding the future of ocean ecosystems have been sounded, leading to a number of high-profile initiatives to better manage and conserve our seas (e.g., European Union, 2008; Obama, 2010).

Importantly, the impacts of people on marine ecosystems are uneven—not all people impact oceans to the same degree everywhere, and the benefits and risks to people from altered ecosystems are not evenly shared. Such asymmetrical impacts and benefits are easily illustrated by shrimp farming. Globally, aquaculture produces some 4.5 million metric tons of shrimp annually, with nearly 90% of the production occurring in Asia (National Marine Fisheries Service, 2014). Shrimp farming has had a devastating impact on ecologically sensitive and important habitats, most notably mangroves—a critical habitat for fish, invertebrates, and birds. The demise of coastal mangrove habitats has adverse impacts on coastal fisheries and wildlife, and has destabilized coastal zones with negative effects on benthic communities (Hatje et al., 2016). Shrimp farming has also resulted in a variety of negative impacts to local communities and their access to mangrove resources for local livelihoods and food security (Belton, 2016; Stonich and Bailey, 2000). While the ecological and social justice impacts of shrimp farming are experienced disproportionately in Asia,

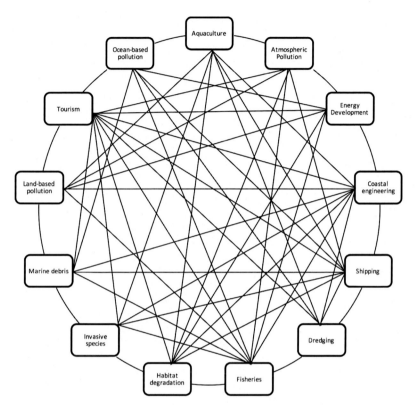

FIGURE 2 Human activities affecting the ocean are highly linked and interacting.

the bulk of shrimp are consumed in the US and Europe. Thus, it is consumer demand in the West that is driving social and environmental degradation in Asia.

To be sure, we are now living in the Anthropocene—an age reflecting the cumulative history of anthropogenic changes to our planet, and an epoch that has yielded fundamental changes in biophysical, social, and cultural systems that depend on and interact with the ocean (Brondizio et al., 2016). The term Anthropocene was coined less than 20 years ago (Crutzen and Stoermer., 2000); yet, it is permeating both popular and academic discourse (Dalby, 2016). It has provided a foundation for transcending the human–nature dichotomy that has afflicted recent discussions in conservation (Levin, 2014; Mace, 2014), and forces us to think beyond the discrete categorization of human and nature, driving us toward a focus on a future where both human and nature thrive. Indeed, the Anthropocene challenges the ideological strictures of modern conservation and demands attention to values, power relationships, and politics (Dalby, 2016).

Kueffer and Kaiser-Bunbury (2014) elegantly describe three axes that must be considered for conservation in human-dominated seascapes (Fig. 3A). They argue that we must actively conserve remnants of historical habitat that would cease to exist without human action. Historical wild habitats may be the preference of traditional conservation biology, but in human-dominated seas, the maintenance of biodiversity will, in some cases, depend on the design and creation of novel habitats that can withstand human impacts. Indeed, we are inadvertently creating novel ecosystems in the world's oceans, and these, in some cases, could be coopted as part of a conservation strategy. Finally, Kueffer and Kaiser-Bunbury (2014) make the case that a dichotomy between production-dominated and conservation-focused seascapes is outmoded. Rather, through the use of biodiversity-friendly and sustainable ocean-use practices, we could view all ocean areas as opportunities for conservation.

While the ecologically focused scheme of Kueffer and Kaiser-Bunbury (2014) provides new practical perspectives for conservation, it still does not fully integrate biophysical and human systems. Fig. 3B highlights that conservation in the Anthropocene requires that we move away from a dichotomous world view. Here we highlight that conservation must not only maintain ecological attributes of the system, it must also concern itself with constituents of social systems (Hicks et al., 2016). Thus, conservation must address connections of people to each other and the ocean, the ability to act meaningfully to determine one's future, and ensure that human needs are met.

The stakes are high in this era of unparalleled transformation for our oceans and the people and communities that depend on them. While ecology, oceanography, and other natural sciences have increased our knowledge of the threats and impacts to ecological integrity, the unique scale and scope of changes

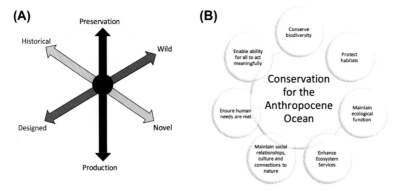

FIGURE 3 (A) Three continua across which modern conservation in the occurs, from historical to novel habitats, from wild to designed ecosystems, and from landscapes primarily used for production to those used for biodiversity preservation. (B) Conservation in the Anthropocene Ocean requires that we move away from a dichotomous world view. Conservation must not only maintain ecological attributes of the system, but it must also concern itself with constituents of social systems. *(A) Redrawn from Kueffer, C., Kaiser-Bunbury, C.N., 2014. Reconciling conflicting perspectives for biodiversity conservation in the Anthropocene. Frontiers in Ecology and the Environment 12, 131–137.*

increases uncertainty about responses of dynamic social-ecological systems. Thus, to understand and protect the biodiversity of the ocean, and to ameliorate the negative impacts of ocean change on people, we need to expand the tools we use to understand and predict the complexity of changes, including knowledge and science that help understand human beliefs, values, behaviors, and vulnerabilities. Conversely, on a human-dominated planet, it is simply impossible to understand and address human well-being, and chart a course for sustainable use of the oceans, without understanding the implications of environmental change for human societies that depend on marine ecosystems and resources. There is no longer an option—now is the time for an interdisciplinary approach to the conservation of our oceans.

Broadly speaking, this book asks how the diverse social and natural science disciplines can provide principles and tools for conserving marine social–ecological systems. The complexity of this problem is daunting, requiring diverse expertise. We have thus assembled an appropriately eclectic group of contributors to this volume. Some of the authors are experts in biophysical aspects of marine ecosystems; their knowledge is essential for understanding the biological, ecological, and physical changes occurring in the world's oceans. Other authors are authorities in the social sciences. These individuals provide a rich understanding of governance and the relationships of people and the ocean, and highlight how changes in biophysical and human well-being are linked. Some authors were invited because they transcend disciplines and demonstrate how biophysical and human dimensions can be merged to yield new insights on appropriate paths for conservation. And finally, many authors have direct experience working with stakeholders, implementing management actions, and developing conservation policies.

The volume is divided into four sections. The first (Setting the Stage) provides a backdrop by highlighting fundamental topics in contemporary marine conservation. These chapters tackle the foundational issue of conservation—navigating the science–policy interface. They also explore key ecological issues associated with climate change and extinctions, as well as issues of food security and social resilience. The second section (Principles of Conservation in the Anthropocene), presents a variety of essential considerations for marine conservation. These range from issues of scale, equity, and cultural practices to what constitutes "natural" or "community." This section of the book also provides social–ecological takes on privatization of ocean resources, ecosystem services, and ocean tipping points. The third section (Conservation in the Anthropocene in practice) provides a series of case studies where principles of contemporary conservation have been applied. Some of these chapters are place-based, focusing on specific systems, while others tackle issues that occur across broad areas, such as human rights in fisheries and stakeholder engagement. The fourth section of the book (Looking Forward) considers the cutting-edge of ocean conservation. It asks how we can protect human communities and marine ecosystems in the face of a changing climate, and also how we can use emerging tools and deep engagement with stakeholders to provide the best available information to support conservation. Finally, in the conclusion we highlight a guiding theme of *interconnectedness* that binds us to the ocean and links the chapters of this book.

This book reflects the collective wisdom of authors who bring a remarkable range of knowledge and perspective. It is our hope that these efforts will stimulate readers to consider how we might best approach marine conservation and management in the Anthropocene. We are living in an increasingly dysfunctional age when the oceans and the people they support are at risk. Even so, we remain optimistic. Conservation thinking has evolved from a view of "nature in spite of people" through "nature for people" to "nature and people" (Mace, 2014). Yes, we live in a time of accelerated change and enormous uncertainty that impacts everything in the ocean from seafood production to shipping and coastal zone development to species continuance; but collectively the chapters in this volume chart a conservation path committed to both thriving ocean ecosystems and human communities. Effective paths to conservation depend on bringing together diverse knowledge and perspectives—from indigenous peoples, scientists, public and industry stakeholders, and policy makers. This challenging work has begun. We ask you, the reader, to ensure that it continues.

Phillip S. Levin
Melissa R. Poe
Seattle, WA, United States

REFERENCES

Belton, B., 2016. Shrimp, prawn and the political economy of social wellbeing in rural Bangladesh. Journal of Rural Studies 45, 230–242.

Brondizio, E.S., O'brien, K., Bai, X., Biermann, F., Steffen, W., Berkhout, F., Cudennec, C., Lemos, M.C., Wolfe, A., Palma-Oliveira, J., Chen, C.T.A., 2016. Re-conceptualizing the Anthropocene: a call for collaboration. Global Environmental Change 39, 318–327.

Crutzen, P.J., Stoermer, E.F., 2000. The Anthropocene. Global change Newsletter. International Geosphere–Biosphere Programme (IGBP) 41, 17–18.

Dalby, S., 2016. Framing the Anthropocene: the good, the bad and the ugly. The Anthropocene Review 3, 33–51.

Davis Jr., R.A., Welty, A.T., Borrego, J., Morales, J.A., Pendon, J.G., Ryan, J.G., 2000. Rio Tinto estuary (Spain): 5000 years of pollution. Environmental Geology 39, 1107–1116.

Duarte, C.M., 2014. Global change and the future ocean: a grand challenge for marine sciences. Frontiers in Marine Science 1, 63.

Ellis, E.C., Kaplan, J.O., Fuller, D.Q., Vavrus, S., Klein Goldewijk, K., Verburg, P.H., 2013. Used planet: a global history. Proceedings of the National Academy of Sciences 110, 7978–7985.

Erlandson, J.M., Rick, T.C., 2010. Archaeology meets marine ecology: the antiquity of maritime cultures and human impacts on marine fisheries and ecosystems. Annual Review of Marine Science 2, 231–251.

European Union, 2008. Directive, marine strategy framework. Directive 2008/56/EC of the European Parliament and of the Council of 17 June 2008 establishing a framework for community action in the field of marine environmental policy. Official Journal of the European Union 164, 19–40.

Feist, B.E., Levin, P.S., 2016. Novel indicators of anthropogenic influence on marine and coastal ecosystems. Frontiers in Marine Science 3, 113.

Halpern, B.S., Walbridge, S., Selkoe, K.A., Kappel, C.V., Micheli, F., D'Agrosa, C., Bruno, J.F., Casey, K.S., Ebert, C., Fox, H.E., Fujita, R., Heinemann, D., Lenihan, H.S., Madin, E.M.P., Perry, M.T., Selig, E.R., Spalding, M., Steneck, R., Watson, R., 2008. A global map of human impact on marine ecosystems. Science 319, 948–952.

Hatje, V., de Souza, M.M., Ribeiro, L.F., Eça, G.F., Barros, F., 2016. Detection of environmental impacts of shrimp farming through multiple lines of evidence. Environmental Pollution 219, 672–684.

Hicks, C.C., Levine, A., Agrawal, A., Basurto, X., Breslow, S.J., Carothers, C., Charnley, S., Coulthard, S., Dolsak, N., Donatuto, J., Garcia-Quijano, C., Mascia, M.B., Norman, K., Poe, M.R., Satterfield, T., St Martin, K., Levin, P.S., 2016. Engage key social concepts for sustainability. Science 352, 38–40.

Kittinger, J.N., Pandolfi, J.M., Blodgett, J.H., Hunt, T.L., Jiang, H., Maly, K., McClenachan, L.E., Schultz, J.K., Wilcox, B.A., 2011. Historical reconstruction reveals recovery in Hawaiian coral reefs. PLoS One 6, e25460.

Kueffer, C., Kaiser-Bunbury, C.N., 2014. Reconciling conflicting perspectives for biodiversity conservation in the Anthropocene. Frontiers in Ecology and the Environment 12, 131–137.

Levin, P.S., 2014. New conservation for the Anthropocene Ocean. Conservation Letters 7, 339–340.

Levin, P.S., Williams, G.D., Rehr, A., Norman, K.C., Harvey, C.J., 2015. Developing conservation targets in social-ecological systems. Ecology and Society 20.

Mace, G.M., 2014. Whose conservation? Science 345.

Marris, E., 2013. Rambunctious Garden. Bloomsbury, New York.

McClenachan, L., Kittinger, J.N., 2013. Multicentury trends and the sustainability of coral reef fisheries in Hawaii and Florida. Fish and Fisheries 14, 239–255.

National Marine Fisheries Service, 2014. Fisheries of the United States: 2014. National Marine Fisheries Service, Silver Spring, MD.

Obama, B., 2010. Executive Order 13547: Stewardship of the Ocean, Our Coasts, and the Great Lakes (Washington, DC).

Poe, M.R., Norman, K.C., Levin, P.S., 2014. Cultural dimensions of socioecological systems: key connections and guiding principles for conservation in coastal environments. Conservation Letters 7, 166–175.

Seeto, J., Nunn, P.D., Sanjana, S., 2012. Human-mediated prehistoric marine extinction in the Tropical Pacific? Understanding the presence of hippopus hippopus (Linn. 1758) in Ancient Shell Middens on the Rove Peninsula, Southwest Viti Levu island, Fiji. Geoarchaeology 27, 2–17.

Stonich, S., Bailey, C., 2000. Resisting the Blue Revolution: contending coalitions surrounding industrial shrimp farming. Human Organization 59, 23–36.

Thurstan, R.H., McClenachan, L., Crowder, L.B., Drew, J.A., Kittinger, J.N., Levin, P.S., Roberts, C.M., Pandolfi, J.M., 2015. Filling historical data gaps to foster solutions in marine conservation. Ocean and Coastal Management 115, 31–40.

Van Houtan, K.S., McClenachan, L., Kittinger, J.N., 2013. Seafood menus reflect long-term ocean changes. Frontiers in Ecology and the Environment 11, 289–290.

White, M., Smith, A., Humphryes, K., Pahl, S., Snelling, D., Depledge, M., 2010. Blue space: the importance of water for preference, affect, and restorativeness ratings of natural and built scenes. Journal of Environmental Psychology 30, 482–493.

Acknowledgments

We are grateful to each of the authors who contributed their expertise and vision to this collection.

Our thinking on and exposure to interdisciplinary social-ecological science has benefited from collaborators with whom we participate in the Ocean Tipping Points project, the Social Wellbeing Indicators for Marine Management working group, the Ocean Modeling Forum, the California Current Integrated Ecosystem Assessment, and the Lenfest Ocean Task Force. We appreciate the support of the Gordon and Betty Moore Foundation, the David and Lucille Packard Foundation, Pew Charitable Trusts, the Lenfest Foundation, Washington Sea Grant, and NOAA for supporting these collaborations.

We also thank Hillary Thorpe, Penny Dalton, Elliott Hazen, Adrian Stier, Patricia Clay, Patricia Pinto da Silva, Christina Hicks, Davin Holen, Anne Beaudreau, Rebecca McLain, Steve Alexander, Joyce LeCompte, Tom Thornton, Blake Feist, Beth Sanderson, Kelly Biedenweg, Dale Blana, Karma Norman, Reade Davis, Thorsten Bleckner, Sara Breslow, Tom Koontz, Jennifer Silver, Patrick Christie, Maciej Tomczak, Ryan Kelly, Dave Fluharty, Dan Okamoto, Kristin Marshall, Leila Sievanen, P. Sean McDonald, and Emily Howe for their critical reviews. We appreciate the guidance and assistance of Pat Gonzalez and Kristi Gomez.

We extend deep gratitude to the Haida Nation and our partners with the Archipelago Management Board and Gwaii Haanas. Thank you for sharing with us and teaching us that life in the sea around us is the essence of our own well-being.

Section I

Setting the Stage

Chapter 1

Bridging the Science–Policy Interface: Adaptive Solutions in the Anthropocene

Jenna M. Sullivan[1], Elizabeth B. Cerny-Chipman[1], Andrew A. Rosenberg[2], Jane Lubchenco[1]
[1]Oregon State University, Corvallis, OR, United States; [2]Union of Concerned Scientists, Cambridge, MA, United States

INTRODUCTION: OUR VISION AND CHALLENGES TO ACHIEVING IT

Conservation success in the Anthropocene will require learning from the past, responding to and preparing for accelerating environmental and informational changes, and expecting surprises (Inniss et al., 2016; Lubchenco, 1998; Vitousek et al., 1997). Scientific insights can help inform a simple but overarching question: How can people use the planet equitably without using it up? Doing so requires understanding and balancing short-term versus long-term priorities, competing uses, and conflicting information. None of those challenges are new. But they are now exacerbated by emerging trends of information overload and confusion, mistrust of science (Gauchat, 2012), and the no-analogue state of environmental conditions. As a result, to be helpful to society, scientists must understand and respond effectively to these collective challenges.

Scientists, policymakers, resource managers, civil society, and industry all have a stake in ocean resources and ecosystems, and each has a key role to play in achieving sustainable use. In this chapter, we focus on the role of scientists in not only providing information, but also engaging with society, becoming trusted partners with other stakeholders, helping create flexible institutions, and working toward innovative solutions. Our vision recognizes the integrated nature of ecological and social systems and the changing environmental policy landscape. It acknowledges the need for scientists at all levels to facilitate collective movement toward an integrated culture of conservation. We envision a community of conservation professionals where insights from social and biophysical sciences inform public understanding and evidence-based policy actions, and in which the needs of society and resource managers inspire new

Conservation for the Anthropocene Ocean. http://dx.doi.org/10.1016/B978-0-12-805375-1.00001-5

research that advances scientific knowledge and provides useful understanding and solutions.

Achieving this vision is difficult. Scientists often fail to understand the vital importance of strong and enduring relationships with policymakers and stakeholders, built and sustained through time with frequent communication and mutual respect. In addition, natural scientists must heed the advice emerging from the social science community that to be trusted, scientists need to be seen as not only competent, but also warm (Fiske and Dupree, 2014), caring, and having shared values. Currently, the public tends to perceive scientists as competent but cold and, too often, set apart (Baron, 2010; Branscomb and Rosenberg, 2012; Rosenberg et al., 2013).

Scientists often have outdated notions of how to engage effectively with society, e.g., assuming that simply providing information will ensure accurate interpretation of scientific findings, a clear understanding of their implications, and improved policy outcomes. Furthermore, many scientists might not realize that they can take a proactive role in engaging stakeholders rather than waiting to be approached. A sustained, respectful two-way dialogue is critical for building trust and understanding between scientists and policymakers. As Admiral Thad Allen, former Commandant of the US Coast Guard and National Incident Commander for the Deepwater Horizon oil spill disaster, has said, "a crisis is not the time to start exchanging business cards." Scientists need to be increasingly adept at engaging with users of scientific information early and often, listening and responding to their needs, communicating scientific evidence and its uncertainty effectively, and adhering to best practices to ensure integrity.

Policymakers and the public can find interpretation of the science particularly challenging when evidence from various scientific reports is in conflict. This conflict may represent true discrepancies in evidence gathered from independent, legitimate scientific sources, or it may reflect deliberate attempts to achieve particular policy outcomes by distorting, suppressing, or cherry-picking scientific evidence, or simply sowing doubt and confusion (Krimsky, 2005; Oreskes and Conway, 2011). As a result, trust in scientific information varies from topic to topic and is complicated by personal beliefs and affected by political positioning (McCright et al., 2013; McCright and Dunlap, 2011). Appropriate roles of science include framing the issues, describing status and trends, projecting the likely consequences of different policy options, being candid about uncertainty, and offering potential solutions when appropriate (or relevant) (e.g., Costello et al., 2008). Scientists need to understand that policymakers will also take into account other kinds of information, including politics, legal constraints such as tribal and international treaties, economics, and values, in making decisions.

Scientific understanding evolves as environmental conditions and needs for management action change and new knowledge emerges. In fact, the implementation of management actions can result in new findings that inform ongoing scientific work. That new information can in turn motivate revised, "adaptive" management. Although this may be ideal for improving the efficacy of management, it is easier said than done. Management and research timeframes do not always align well. The very idea of revising management measures (i.e., "moving the goal posts" in the eyes of some stakeholders) may be politically and legally difficult (Nie and Schultz, 2012). New scientific results may also point out new uncertainties. Furthermore, reexamining issues when new information appears can be akin to reopening a can of worms—careful negotiations with stakeholders may be at risk. Hence, what may appear reasonable to scientists may be difficult or impossible for management in practice.

This chapter discusses these and other impediments to and opportunities for effective interactions between scientific research and policy action. In the following sections, we highlight some important differences in the cultures of scientists and policymakers, discuss how scientific information can be made usable for a policy context, and describe how to develop policy that is flexible enough to incorporate new scientific information. Using case studies, we highlight examples of challenges and successes at the science–policy interface and identify specific avenues for scientists to become more effectively engaged in the policy and management process.

SUCCESS AT THE SCIENCE–POLICY INTERFACE

Understanding Differences Between Scientists and Policymakers

Although some policymakers are scientists or have scientific background, most do not. Even scientists in a policymaking position must, by necessity, have a different perspective than a "working" scientist, for example, focusing on the mission of his or her agency. Scientists and policymakers have different perspectives and often conflicting demands, motivations, rewards, and constraints. The jobs are different and the cultures and timeframes are distinctive. As with any activity that crosses cultural boundaries, improved integration of scientific research and policy will require well-defined processes and scientists and policymakers who are willing to make concerted efforts to understand each other. Having more scientists in policy agencies and offices can assist greatly in harmonizing these differences.

Based on our observations of decades of interactions between policy and science, some patterns emerge in the ways in which scientific information is most usefully conveyed to policymakers. Likewise, certain kinds of policies are more responsive to new scientific information, such as those that require continuous monitoring, periodic assessment, and iterative policy modification.

Science for Policy

Policymakers need scientific information that is (in part after Clark and Majone, 1985):

1. *Clear and succinct.* Scientific knowledge must be communicated in ways that are understandable to nonscientists but preserve the meaning and relative certainty of different findings.

2. *Timely and relevant.* Scientific information is most helpful if it addresses issues faced by policymakers now; is relevant to the policymaker's constituents, mandates, or interests; and is not prescriptive of a particular policy outcome. Science-based findings that describe the likely consequences of different policy options are particularly helpful.

3. *Legitimate and credible.* Scientific findings are especially useful if they represent consensus findings across the relevant scientific communities and disciplines (Cash et al., 2003; Cook et al., 2013). Credibility is enhanced when multiple perspectives are synthesized and presented with clearly described evidence, when research findings come from individuals and institutions perceived as credible and legitimate, and when transparent processes are involved in research and policy integration. For example, peer-reviewed findings are generally deemed more credible than non-peer-reviewed results. A scientific assessment from the National Academy of Sciences often carries more weight than one conducted by a single group of scientists.

4. *Actionable.* Findings are most helpful when scientists describe options for specific actions that a policymaker can take to address an issue; this is more useful than simply diagnosing the problem (Palmer, 2012).

Effective communication tools that can satisfy the aforementioned criteria include scientific assessments or consensus statements. Formal scientific assessments (e.g., reports from the Intergovernmental Panel on Climate Change, the US National Climate Assessment, or the US National Research Council) are particularly helpful because they emerge from a strong, transparent process including rigorous reviews, balanced perspectives, and credible institutions. Scientific consensus statements from ad hoc communities of scientists (e.g., the Scientific Consensus Statement on Marine Reserves[1] or on Marine Ecosystem-Based Management[2]) can also be timely, useful, and credible. For those, special attention should be paid to presenting a balanced synthesis of the information.

Policy That Anticipates Evolving Knowledge

Both scientists and policymakers recognize that environmental conditions and scientific understanding of an issue may change through time. Forward-looking policies will incorporate explicit mechanisms to monitor conditions and update policies on a periodic basis. Policies that are responsive to changing conditions have specific shared characteristics. They are:

1. *Adaptive.* Policy implementation should include monitoring plans and mechanisms by which they can be reviewed and revised based on new

---

1. www.nceas.ucsb.edu/consensus/consensus.pdf.
2. www.compassonline.org/science/EBM_CMSP/EBMconsensus.

information. Frequently, the policy or legislation provides broad goals and leaves the specifics for implementation by executive branch agencies through rule-making regulations. It is good practice to ensure that monitoring and periodic review and revisions to regulations, as well as overarching policies, occur on a regular basis.

2. *Holistic*. Traditionally, conservation policies and management actions have focused on individual environmental issues [endangered species, marine mammals, water quality, invasive species, ocean acidification (OA), climate change] or sectors (fishing, mineral extraction, oil and gas, transportation, renewable energy, aquaculture). Interactions across issues or sectors can derail even the best-formulated policies. For example, management of most renewable resources, such as fisheries, should now take climate change into account (e.g., see the National Oceanic and Atmospheric Administration's (NOAA) Climate Science Strategy Regional Action Plans[3]).

   Many of the activities associated with each issue or sector can trigger the need for an Environmental Impact Assessment (EIA). Policies and management actions should be crafted with the larger framework in mind and take into account the impacts, trade-offs, and synergies that can occur across different uses or issues. For example, considerable savings might be achieved in performing a single EIA rather than a series of piecemeal and incomplete EIAs (CEQ, 2010; US Commission on Ocean Policy, 2004; NOAA Integrated Ecosystem Assessment Program[4]).

3. *Logical*. Scientists can assist policymakers in understanding scientific analyses sufficiently so that data are correctly interpreted and used. Specific policies and management actions should clearly and logically flow from diagnoses and analyses of relevant scientific evidence. Although science is not the only input into choosing appropriate policies, it is a key foundation for conservation action.

CASE STUDIES: EFFECTIVE APPROACHES TO BRIDGING THE INTERFACE

The following case studies highlight ways in which science can support policymaking and policy can be responsive to science across a range of problems and issues. Each case study illustrates a different type of science–policy interaction or different challenge. We chose examples with which at least one of us was intimately familiar to enable a deeper understanding of the context, players, and outcomes.

Knowledge-to-Action Success Story: Washington State Blue Ribbon Panel on Ocean Acidification

Washington State is famous for its oysters. The shellfish industry is an important part of Washington's economy, contributing $184 million to the economy

---

3. https://www.st.nmfs.noaa.gov/ecosystems/climate/rap/index.
4. https://www.integratedecosystemassessment.noaa.gov//

and supporting 2710 jobs in 2010 (Northern Economics, Inc, 2013). When oyster production at hatcheries failed dramatically for 5 years beginning in 2005, shellfish farmers looked to scientists for answers and policymakers for support. What transpired is a story of scientists, policymakers, and industry working together to understand, reduce, and adapt to the effects of an ongoing environmental disaster—ocean acidification (OA).

Scientists had been studying OA intensely, exemplified by a landmark scientific assessment report published by the UK Royal Society (Raven et al., 2005). Scientists had estimated that the ocean has absorbed approximately one-quarter of the anthropogenic CO_2 emitted into earth's atmosphere (Sabine et al., 2004). This CO_2 lowers the pH and availability of minerals that many organisms require to build shells and other hard structures and poses a threat to many marine species. Academic and agency scientists shared their concerns about OA with key members of the US Congress and state governments with whom they had productive working relationships. In 2009 Congress passed bipartisan legislation (Federal Ocean Acidification Research and Monitoring Act) to set up a federal OA research and monitoring program (see oceanacidification.noaa.gov). Meanwhile, university and government scientists determined that oyster die-offs in Washington and Oregon were largely caused by abnormally acidic waters, making shell building difficult (Barton et al., 2012); this finding in hatchery oysters was later expanded to include wild populations of other species in the broader California Current Large Marine Ecosystem along the US west coast (Bednaršek et al., 2014). A temporary work-around for hatcheries was devised to avoid using corrosive waters in oyster hatcheries, but concerns about the long-term fate of the industry remain.

OA was no longer a hypothetical threat. Impacts to economically important (shellfish in hatcheries) and ecologically important (pteropods in the coastal ocean) marine life had been demonstrated. Scientific discoveries, monitoring, and science–industry partnerships thus set the stage for the informed and innovative policy initiatives surrounding OA described later.

Recognizing the environmental and economic importance of OA to her state, Washington Governor Christine Gregoire worked with federal and state scientists, industry representatives, and governmental leaders to bring broader public attention to the issue and elicit policy recommendations that were grounded in stellar science. In 2012 she convened the Washington State Blue Ribbon Panel on Ocean Acidification, comprising governmental and academic scientists, industry and conservation representatives, public opinion leaders, and state, local, federal, and tribal policymakers. She selected two experienced statesmen with high public credibility and experience at the science–policy interface as cochairs and charged the Panel with creating a comprehensive strategy to address OA and its impacts in State waters. The Panel held hearings, listened to scientists, and delivered a report that synthesized scientific understanding of OA and its effects on marine ecosystems and assessed Washington's potential to mitigate, remediate, and adapt to changing ocean chemistry. The Panel recommended 42 specific action steps (Washington State Blue Ribbon Panel on Ocean Acidification, 2012), including the overarching need to reduce carbon emissions, initiate new monitoring and research, reduce run-off of nutrients into coastal areas and educate stakeholders, decision makers, and the public. Many of the suggestions were immediately codified by Governor Gregoire in Executive Order 12-07.

Two recommendations set the stage for ongoing monitoring, research, and policy formulation, and were designed to be responsive to new scientific findings. First, the Washington Ocean Acidification Center was established as a research hub at the University of Washington. Second, the Marine Resources Advisory Council standing committee was created within the Governor's office to advise state-level decisions relating to OA and broader marine resources. These efforts have focused on improving the understanding of OA in local waters through research and monitoring. Other efforts are seeking solutions at the local scale, including promoting the adoption of OA-resilient shellfish industry practices and expanded mitigation of local contributors to OA due to nutrient loading from nitrogen sources, including failing or inadequate septic systems and fertilizer run-off. Another outcome has been development of greater awareness of OA among stakeholders through education and outreach programs. Finally, these state efforts led to the establishment of the West Coast Ocean Acidification and Hypoxia Science Panel,[5] a broader regional effort under the auspices of the West Coast Governors' Alliance on Ocean Health.

This case study provides a clear example of legitimate, actionable, credible, and timely science informing actions by industry and immediate policy response to an environmental and economic threat. The Panel built upon and strengthened relationships among scientists and policymakers, resulting in policies informed by science and new scientific research targeting specific policy needs. Efforts in Washington State have provided a model for other states (including Maine, Maryland, and a consortium of West Coast States) and nations to address the science and policy of OA. Both Governor Gregoire and her successor Governor Jay Inslee have highlighted OA as yet another reason to aggressively reduce carbon emissions.

Developing Science Tools and Expertise to Support Policymaking: Marine Reserves and the California Marine Life Protection Act

Marine protected areas (MPAs), areas of the ocean where some human activities are restricted, are important tools for promoting biodiversity and habitat health. When well designed, marine reserves (MRs), a type of MPA where all extractive or destructive activities are prohibited, can be highly effective in protecting habitats and ecosystems. This case study focuses on ways in which

---

5. http://westcoastoah.org/.

scientists, conservation leaders, managers, and policymakers worked together to develop and implement legislation to create a network of MPAs and MRs within California State waters.

During the 1990s, many conservation groups and scientists around the world promoted MPAs to protect biodiversity (Ballantine, 1991; Kelleher and Kenchington, 1991). However, despite an abundance of MPAs and numerous scientific studies of them, a synthesis of scientific information was lacking. Moreover, scientists could not easily answer important practical questions posed by managers and conservation activists, such as: Are all MPAs the same? How big should an MPA be? How far apart should they be? How many are needed?

A 1997 symposium on MRs at the annual meeting of the American Association for the Advancement of Science (AAAS) concluded that there was a need for synthesis of existing research (Allison et al., 1998; Hill, 1997). A working group was convened in 1997 at the National Center for Ecological Analysis and Synthesis (NCEAS) with the goal of building actionable ecological theory for MRs to support policy and management decisions. The group included specialists in oceanography, ecology, larval biology, terrestrial conservation, modeling, population genetics, and more; it produced a useful synthesis about MRs published as a Special Feature in *Ecological Applications* (Lubchenco et al., 2003) and other peer-reviewed papers. Results articulated a range of benefits that MRs provide (Halpern and Warner, 2003; Lester et al., 2009), proposed the idea of networks of MRs, and encouraged the use of scientific tools to specify size and spacing guidelines (Leslie et al., 2003; Shanks et al., 2003).

Recognizing that the primary scientific literature was not accessible to most policymakers, resource users, and managers, and that misinformation and disinformation about MRs were rampant, the NCEAS authors teamed up with communication and policy specialists at COMPASS[6] and the Partnership for Interdisciplinary Studies of Coastal Oceans (PISCO)[7] to organize workshops where managers could share their needs and scientists could share new results. The workshops identified the need for products that would translate peer-reviewed findings into lay language in appealing formats, leading to a video about MRs and the first in what became a series of booklets (The Science of Marine Reserves, 2002). In addition, nongovernmental organizations (NGOs), managers, and agency scientists urged the scientific community to speak with one voice on the topic of MRs. In 2001 scientists drafted a Scientific Consensus Statement on Marine Reserves.[8] Both the new scientific findings and the Consensus Statement were presented at a special symposium at the AAAS meetings in 2001 and garnered significant press attention, thus spreading information to broader audiences. The products of the NCEAS working group were powerful tools for policymakers because they presented a credible, legitimate,

---

6. www.compassonline.org.
7. www.piscoweb.org.
8. www.nceas.ucsb.edu/consensus/consensus.pdf.

and unified scientific voice on reserve design and potential impacts and because they were embraced by broader communities of stakeholders.

As MR science was being synthesized, US policymakers and the public began to call for designation of formal protected areas. In 1999 California Lawmakers approved the Marine Life Protection Act (MLPA), which advocated a network of MPAs in state waters to be developed by the California Fish and Game Commission. Two initial attempts to design the MPA network failed because of limited public input, confusion surrounding interpretation of scientific information, and lack of funding. In 2004 following a significant input of philanthropic funding that enabled a renewed process and broader dissemination of the NCEAS results and PISCO booklets, California created the MLPA Initiative. It established regional Blue Ribbon Task Forces and several other groups, including Science Advisory Teams (SATs), to engage in a science-based and stakeholder-driven process to redesign the state's existing MPAs. A number of scientists from the NCEAS working group served on the regional SATs. To develop specific size and spacing guidelines for the MLPA, the SATs built on the NCEAS efforts, expanding and tailoring them to species of primary interest throughout California. An iterative process led to the approval of the state's first MPA network along the central coast in 2007 (Kirlin et al., 2013). Although the MPA networks were guided by science, the final designs represented a compromise to address trade-offs and concerns expressed by stakeholders and the public. By 2013, 16% of the state's waters was incorporated in 124 MPAs and 9.4% in 61 MRs (Gleason et al., 2013).

Science was central to the MLPA Initiative process. At the outset, the SATs in each region established scientific guidelines required for MPA network proposals. In many cases, proposals from stakeholders went through multiple revisions before they met the guidelines (Saarman et al., 2013). In this way, science defined the "decision-making space" in which the proposals could operate (Osmond et al., 2010), but stakeholders had a significant say about specific locations. In addition, scientists from the SATs developed educational presentations for the Task Force and stakeholder working groups to ensure the inclusion of basic scientific knowledge of MPAs (Kirlin et al., 2013). Following implementation of the state's MPA network, scientists have been essential for monitoring and adaptive management efforts through the MPA Monitoring Enterprise (Gleason et al., 2013).

The SATs developed "rules of thumb" as a simple and understandable format for communicating scientific guidelines (Carr et al., 2010). These rules of thumb were scientifically sound yet general enough to be easily incorporated into the planning process, focusing on critical decisions that the Task Force would need to make about MPA size, placement, and habitat. For example, in marine systems, spatial connectivity tends to be driven by movement of larvae rather than adults. As a result, the rule of thumb stated that the reserves should be spaced 50–100 km apart to encompass the travel range of most larvae based on a rigorous scientific review of the species of greatest interest (Carr et al., 2010).

Continued scientific research coupled with effective engagement with resource users and political leaders has resulted in a significant increase in global MR coverage. In the last decade, the fraction of strongly protected marine areas has gone from 0.1% of the surface area of the ocean to 1.6% (updated from Lubchenco and Grorud-Colvert, 2015). Because the design of MRs strongly influences whether they meet conservation objectives, scientific input has been critical to both the conceptual development of MRs and their implementation globally (Saarman et al., 2013). The MLPA process in California is a rare example of ocean governance at the subnational level that provides important lessons for other states and governments considering MPA and MR networks (Osmond et al., 2010).

Science Driving Solutions: Secure-Access Fisheries Management

In 1883 English biologist Thomas Henry Huxley made the (in)famous assertion that the ocean contains an inexhaustible bounty and therefore all fisheries regulation is useless. Faced with increasingly small yields of increasingly small fish, it became clear by the mid-20th century that this was not the case (Gulland, 1971). Creating scientifically, economically, and societally sound fisheries regulation within diverse biological and social contexts has been a central challenge for fishers, scientists, policymakers, and managers for many years. The need is increasingly important as human consumption of fish is predicted to increase dramatically as human populations and economies grow (FAO, 2012). Until recently, the vast majority of fisheries have been managed as common pool resources where fishers compete with one another until the quota for the fishery has been reached. The ensuing "tragedy of the commons" is a well-known consequence, often despite the best of intentions of fishery managers. Economists' diagnosis of the resulting overfishing focuses on the strong incentives for fishers to maximize catches today, despite the fact that doing so may compromise future fishing opportunities, simply because there is no guarantee that any individual will reap the benefits from future fisheries.

To be most effective, policies and management actions designed to end overfishing must thus consider both the ecological context of the fish and the societal context of the fishers. Emerging interdisciplinary, science-based approaches to fisheries management can provide strategic frameworks for aligning environmental and economic incentives to create robust and adaptive fisheries management policies (Beddington et al., 2007). In other words, changing economic incentives for fishers to focus on both the long and the short term is more likely to result in sustainable fisheries.

One strategy for optimizing both fishery and environmental stewardship goals is an approach in which fishers are granted secure access to the fishery, either through an assigned and exclusive place to fish or a guaranteed fraction of the scientifically determined total allowable catch. This approach goes by various names, but the most generic and least confusing is "secure access" program. Other terms for all or a subset of secure-access programs include "Limited Access Privilege Programs," "Individual Transferrable Quotas (ITQs)," "Individual Fishing Quotas (IFQs)," "rights-based fishery management (RBFM)," "catch shares," "transferrable fishery concessions," and "Territorial User Rights in Fisheries." Regardless of the name, all programs make clear that access to a fishery is not a "right" but a "privilege." (The term "right" has different connotations that are culturally dependent.) Secure-access management shifts the fishery from a common pool resource often characterized by a "race to fish" to an allocated harvest access resource that provides users greater incentive to employ sustainable fishing practices (Beddington et al., 2007; Costello et al., 2008; Hanna, 1997; National Research Council, 1999) because they have a stake in the future of the resource (either a place to fish or a fraction of the quota).

Although effects of secure-access fisheries are mixed, there are many examples in which a change to that approach successfully improved fisheries sustainability (Hilborn, 2007). Experience in the United States with ITQ and IFQ fisheries, including surf clam and ocean quahogs, Alaska halibut, and Alaska pollock, prompted legislation (e.g., the Magnuson-Stevens Act, or MSA, of 2006) to again allow consideration of catch share approaches in US federal fisheries (following a moratorium on such approaches). Researchers Costello et al. (2008) provided evidence that, across over 11,000 fisheries between 1950 and 2003, secure-access fisheries tended to be more sustainable than those under common pool management. These results accelerated adoption of secure-access approaches in many US federal fisheries and others around the world. In 2006, for example, there were 24 unique species in seven US federal catch share programs; by 2015, there were 107 species in 16 such programs. Today, around two-thirds of the seafood caught (by volume) in US federal waters is under a catch-share program.

The US Pacific groundfish trawl fishery illustrates a management change enabled by science. A history of overfishing and declining stocks led to this fishery of over 90 species being declared a "Federal Fishery Disaster" in 2000. With looming threats of fishery closures, fishers, scientists, managers, and environmental groups began working together to craft solutions. Significant changes in federal law resulted in new scientifically informed management policies that helped turn this fishery around. Two changes were particularly important. (1) The 2006 MSA provided a tough new mandate to end overfishing and recover stocks, including the required use of science-based annual catch limits and accountability measures. (2) The MSA allowed the option of using secure-access approaches in management. As a result, in 2011 an ITQ plan was adopted for the fishery. These policy advances built upon previous management changes to the fishery, including buybacks, closures, and quota reductions, and new funding that also helped jump-start the rebuilding process (Mamula and Collier, 2015).

The results have been dramatic. Many of the stocks have recovered substantially, and the vast majority of the species are now listed as "best choice" or "good alternative" by the independent Monterey Bay Aquarium Seafood Watch program on the basis of sustainability (Conway and Shaw, 2008). Fishery revenues increased as stocks rebounded, and bycatch has decreased significantly as fishers are now incentivized to minimize nontarget species and ensure the fishery is healthy. Similar economic, social, and environmental success stories have emerged across many fisheries of diverse scale, from small to industrial, and in both developed and developing countries (Barner et al., 2015; Lubchenco et al., 2016).

It is important to note that secure-access approaches do not constitute a panacea for overfishing (Allison et al., 2012; Branch, 2009; Chu, 2009; Hilborn et al., 2005). They do not always achieve every objective across economic, social, and environmental sectors and they must be well designed to be effective. It is important to not confound the usefulness of secure-access management as a tool for sustainable fisheries management with a failure due to inappropriate or poor design in a given context. A properly designed secure-access program requires attention not only to ecological and economic sustainability but also to social equity and impact on fishing communities.

Secure-access fisheries management in the United States is a case where insights provided by social and natural scientists have grounded a management tool and shaped the process of how it is used. Furthermore, policies informed by science at the national level enabled the use of a management tool for sustainability. The use of secure-access approaches to align incentives across sectors remains an area of active interdisciplinary research, the results of which will improve future use of this and other management tools.

In addition, like all fisheries management strategies, RBFMs cannot be static and must instead identify and adapt to environmental and social changes to be effective in the long term. For example, despite implementing an RBFM-based method in 2010 in the chronically overfished Gulf of Maine cod fishery and reducing quotas by 73% in 2013, biomass has continued to decline drastically and is now at only 4% of the value that would give maximum sustainable yield (Pershing et al., 2015). This failure of a cod stock to recover despite recent stringent fishing limits has been attributed to decades of overfishing, poor accountability, and enforcement in catch limits, and the recent rapid warming of Gulf of Maine waters, leading to increased mortality and reduced recruitment. This warming was not taken into account when the catch limit was set for the fishery. In contrast, Barents Sea cod stocks have responded positively to stronger management controls despite a warming climate (Kjesbu et al., 2014). Scientifically informed tools for sustainable resource management will become increasingly important in the Anthropocene, but must be adaptive and applied with attention to the current and future context.

Development of the US National Ocean Policy

Effective management of ocean resources can be context dependent and often involves mediation among stakeholders with conflicting priorities. Policy and management must be holistic and balance multiple uses with environmental, social, and economic objectives. This case study explores the scientific framework behind the creation in 2010 of US National Ocean Policy (NOP), its goal of streamlining and increasing effectiveness of the management of activities affecting US oceans and coasts, and the on-the-ground action that the NOP inspired. The NOP is an example of high-level national policy that aimed to incorporate scientific input into policy development, implementation, and assessment.

The impetus for the NOP was driven by recognition that ocean problems were increasing and by a large body of scientific work that emphasized the importance of taking a more integrated, ecosystem-based approach to management instead of the piecemeal, issue-by-issue, sector-by-sector approach that existed. The NOP was strongly influenced by reports from the two bodies set up to review ocean practices and policies in the United States [the Pew Oceans Commission (Pew Oceans Commission, 2003) and the US Commission on Ocean Policy (US Commission on Ocean Policy, 2004)] and the merger of the members of both into the Joint Ocean Commission Initiative.[9] Members included savvy politicians, business leaders, environmental champions, fishers, and scientists. Each commission received abundant scientific and expert testimony and produced recommendations for action solidly grounded in good science. The findings and recommendations of the two commissions overlapped substantially, sending a strong message to the nation that current management was ineffective in ensuring sustainable use of the ocean and that more integrated approaches were needed.

Managing resources across economic sectors and on an ecosystem basis requires involvement of multiple government agencies at both state and federal levels. Creation of an Ocean Policy Task Force (CEQ, 2010) brought together experts from federal agencies across the government to recommend to the President specific measures to improve ocean and coastal management. Based on the work of this Task Force and that of the previous commissions, President Obama issued Executive Order 13547 on stewardship of the oceans, coasts, and Great Lakes in July 2010. It directed agencies to work together and to develop new approaches for improving interactions with the states and with one another to achieve multiple goals. These included: (1) reduce conflict surrounding ocean use through integrated planning, (2) ensure the sum total of activities is not having a negative impact on ocean ecosystems, and (3) be more efficient and effective in discharging responsibilities (White House, 2010). The NOP is noteworthy for stating clearly the intention of achieving productive, healthy, and resilient ocean ecosystems to benefit coastal communities, economies, and the nation. Following this Executive Order, the federal government, under the auspices of a new National Ocean Council, developed a National Ocean Policy Implementation Plan to put into action the President's directive (National Ocean Council, 2013).

---

9. Joint Ocean Commission Initiative: www.jointoceancommission.org.

One outcome of the NOP was the creation of Regional Planning Bodies consisting of federal, state, and tribal authorities charged with working with the range of relevant stakeholders including industry and civil society, to create an integrated marine spatial plan for the region. Building on previous statewide efforts and embracing the opportunity to do comprehensive, smart planning for their regions, these two Regional Planning Bodies along the US East Coast assembled scientific analyses and massive scientific data bases and considered wide-ranging options for their regions. Both groups produced regional action plans[10] that were finalized by the National Ocean Council in December 2016 and are now moving into implementation. These plans represent cross-sectoral efforts to integrate data, agree upon a common vision, and develop plans to harmonize across possible uses while also taking ecosystem health into account.

Other regions have used the NOP for sector-specific action while the larger integrated plans are being designed. The WindFloat Pacific Offshore Wind Demonstration Project in Oregon provides an example. Led by the Department of the Interior and the Bureau of Ocean Energy Management, in collaboration with state and tribal governments and local and regional businesses and organizations, WindFloat Pacific represented the first offshore wind energy demonstration platform in the United States. It also serves as a case study for improving efficiency in future NOP projects that span federal agencies, state and local authorities, and the private sector.

As regional and federal ocean planning develops, scientists are mapping habitats and improving understanding economic trade-offs in response to the ongoing needs of policymakers. The formation of the US NOP highlights the potential for science to drive policy change and demonstrates that even when progress on large initiatives is difficult, they can motivate science and policy action at lower levels.

Summary of Case Studies

These four case studies provide examples of existing frameworks that have been used for integrating science into policy and highlight challenges and successes at the science–policy interface. They also highlight the need for "use-inspired" (*sensu* Stokes, 1997) and policy-relevant science. The Blue Ribbon Panel on Ocean Acidification case study illustrates a situation where policy and science worked in tandem from the outset. The California MLPA Initiative and the NCEAS working group on MRs are examples where scientists created clear and scientifically rigorous guidelines that helped move policies forward. Policies on secure-access fishery management were driven by the integration of environmental and economic information by scientists that evaluated benefits for fishers, the public, and the ocean. The NOP demonstrated an important policy framework that was created in response to scientific findings and incorporates

---

10. https://www.whitehouse.gov/blog/2016/12/07/nations-first-ocean-plans.

science as a clear priority in ongoing policymaking. It provides a mechanism to connect and streamline national and regional ocean management and demonstrates the impact scientists can have on the policy process.

CONCLUSIONS

Science in Support of Resilient Institutions

Incorporating science into ocean policy and management is vital. A positive climate supported by strong legal structures, political will, funding, leadership, and public engagement are all necessary for science to be used successfully. Each of the examples described in the case studies took time—years to decades—for policies to be instigated. Science and knowledge generation is a smaller piece of a larger adaptive governance structure needed for sustainable resource management. Institutions, which include everything from organizations to rules, norms, and customs (see Hahn et al., 2008), must be flexible to support science and build resilient human and natural systems. Furthermore, systemic leaders who are able to see the big picture, are willing to step away from their own mental models, and can inspire others to articulate and cocreate the future are needed to promote a culture of sustainable resource use (see Senge et al., 2015).

Challenges in the Anthropocene will likely require creation of new tools and knowledge, with an ongoing need for iterative learning processes. Such adaptive structures change the role of scientists from unbiased observers to active participants in a larger system of knowledge generation and collective action (Folke et al., 2005). Scientists act to reinforce the adaptive capacity of institutions by collaborating with other actors such as citizens, NGOs, industry, policymakers, and managers. In adaptive governance, a constellation of polycentric institutions that span scales and incorporate diverse actors and knowledge sources provides a deep reservoir for finding alternative options and solutions (Norberg et al., 2008).

Looking Forward

As the Anthropocene continues to evolve, one challenge is becoming increasingly apparent: making science relevant in an era in which not everyone values evidence or science or data. We have presented examples where science has influenced policy and management directly, and in each of those examples the outcome reflected strong engagement of scientists with citizens and policymakers and effective communication of scientific knowledge in ways that were actionable and credible. We suggest that future successes will require a quantum leap in engagement of scientists with society. We do not mean simply communicating scientific information, but engaging in meaningful two-way exchanges with nonscientists. For citizens and policymakers to see science as useful and relevant, they must trust those who deliver that information and they must experience that using scientific information in fact produces useful outcomes. Many scientists, managers, and policymakers in the conservation community are already becoming conversant in the languages of science and policy as we collectively realize the vital importance of clear communication and trusted relationships to effectively incorporate science into solutions. But we must accelerate the rate at which this is happening if we hope to use the resources of our planet without using them up.

From Science to Solutions

Scientists at all stages of their careers have access to diverse opportunities for involvement in public, policy, and management endeavors. These opportunities span scales from local to international and draw upon different skills and expertise. Although some opportunities may not be open to every scientist, getting started in policy and management is often simply a matter of being proactive and willing to share time and knowledge with others. It also requires a willingness to listen and learn.

Get Involved: Opportunities for Scientists (At All Stages of Their Careers) to Engage in Policy and Management

1. Share scientific results with lay audiences through social and traditional media, and oral and written presentations.
2. Request and participate in communications training to learn to be more effective.
3. Invite seminar speakers from the worlds of policy and management to visit your institution.
4. Connect to the public with educational activities, including those that facilitate citizen science.
5. Participate in policy trainings, e.g., those run by a professional scientific society.
6. Volunteer to engage in a policy issue relevant to your scientific training, not your own research; learn to listen to what users need, not just promote the science that you know.
7. Volunteer, work for, or collaborate with a scientifically credible NGO with deep policy expertise.
8. Cultivate relationships with members of Congress or state or local policymakers.
9. Volunteer, work for, or collaborate with state or federal agencies to better understand their needs.
10. Work for or collaborate with a "boundary" organization such as COMPASS or the Union of Concerned Scientists.
11. Be willing to testify before Congress, State Legislature, or local boards or commissions.
12. Apply for policy internships and fellowships (AAAS Science and Technology Fellowship; Sea Grant Fellowship).

REFERENCES

Allison, E.H., Ratner, B.D., Åsgård, B., Willmann, R., Pomeroy, R., Kurien, J., 2012. Rights-based fisheries governance: from fishing rights to human rights. Fish and Fisheries 13, 14–29. http://dx.doi.org/10.1111/j.1467-2979.2011.00405.x.

Allison, G.W., Lubchenco, J., Carr, M.H., 1998. Marine reserves are necessary but not sufficient for marine conservation. Ecological Applications 8, S79–S92.

Ballantine, W.J., 1991. Marine reserves for New Zealand. Leigh Marine Laboratory Bulletin Number 25 University of Auckland. Auckland, New Zealand.

Barner, A., Lubchenco, J., Costello, C., Gaines, S., Leland, A., Jenks, B., Murawski, S., Schwaab, E., Spring, M., 2015. Solutions for recovering and sustaining the bounty of the ocean: combining fishery reforms, rights-based fisheries management, and marine reserves. Oceanography 25, 252–263. http://dx.doi.org/10.5670/oceanog.2015.51.

Baron, N., 2010. Escape From the Ivory Tower: A Guide to Making Your Science Matter. Island Press.

Barton, A., Hales, B., Waldbusser, G.G., Langdon, C., Feely, R.A., 2012. The Pacific oyster, *Crassostrea gigas*, shows negative correlation to naturally elevated carbon dioxide levels: implications for near-term ocean acidification effects. Limnology and Oceanography 57, 698–710. http://dx.doi.org/10.4319/lo.2012.57.3.0698.

Beddington, J.R., Agnew, D.J., Clark, C.W., 2007. Current problems in the management of marine fisheries. Science 316, 1713–1716. http://dx.doi.org/10.1126/science.1137362.

Bednaršek, N., Feely, R.A., Reum, J.C.P., Peterson, B., Menkel, J., Alin, S.R., Hales, B., 2014. *Limacina helicina* shell dissolution as an indicator of declining habitat suitability owing to ocean acidification in the California Current Ecosystem. In: Proc. R. Soc. B. The Royal Society, p. 20140123.

Branch, T.A., 2009. How do individual transferable quotas affect marine ecosystems? Fish and Fisheries 10, 39–57.

Branscomb, L.M., Rosenberg, A.A., 2012. Critic at large: science and democracy—researchers and conscientious citizens must unite against the partisan rancor in American politics and restore the role of scientific information in policymaking. Scientist 26, 23.

Carr, M.H., Saarman, E., Caldwell, M.R., 2010. The role of "rules of thumb" in science-based environmental policy: California's marine life protection act as a case study. Stanford Journal of Law, Science & Policy 2, 1.

Cash, D.W., Clark, W.C., Alcock, F., Dickson, N.M., Eckley, N., Guston, D.H., Jäger, J., Mitchell, R.B., 2003. Knowledge systems for sustainable development. Proceedings of the National Academy of Sciences of United States of America 100, 8086–8091. http://dx.doi.org/10.1073/pnas.1231332100.

CEQ, 2010. Final Recommendations of the Interagency Ocean Policy Task Force. White House Council Environmental Quality, Washington, DC.

Chu, C., 2009. Thirty years later: the global growth of ITQs and their influence on stock status in marine fisheries. Fish and Fisheries 10, 217–230.

Clark, W.C., Majone, G., 1985. The critical appraisal of scientific inquiries with policy implications. Science, Technology & Human Values 10, 6–19.

Conway, F., Shaw, W., 2008. Socioeconomic lessons learned from the response to the federally-declared West Coast groundfish disaster. Fisheries 33, 269–277. http://dx.doi.org/10.1577/1548-8446-33.6.269.

Cook, C.N., Mascia, M.B., Schwartz, M.W., Possingham, H.P., Fuller, R.A., 2013. Achieving conservation science that bridges the knowledge–action boundary. Conservation Biology 27, 669–678. http://dx.doi.org/10.1111/cobi.12050.

Costello, C., Gaines, S.D., Lynham, J., 2008. Can catch shares prevent fisheries collapse? Science 321, 1678–1681. http://dx.doi.org/10.1126/science.1159478.

FAO, 2012. The State of the World Fisheries and Aquaculture 2012 (SOFIA). FAO, Rome.

Fiske, S.T., Dupree, C., 2014. Gaining trust as well as respect in communicating to motivated audiences about science topics. Proceedings of the National Academy of Sciences of United States of America 111, 13593–13597. http://dx.doi.org/10.1073/pnas.1317505111.

Folke, C., Hahn, T., Olsson, P., Norberg, J., 2005. Adaptive governance of social-ecological systems. Annual Review of Environment and Resources 30, 441–473. http://dx.doi.org/10.1146/annurev.energy.30.050504.144511.

Gauchat, G., 2012. Politicization of science in the public sphere a study of public trust in the United States, 1974 to 2010. American Sociological Review 77, 167–187. http://dx.doi.org/10.1177/0003122412438225.

Gleason, M., Fox, E., Ashcraft, S., Vasques, J., Whiteman, E., Serpa, P., Saarman, E., Caldwell, M., Frimodig, A., Miller-Henson, M., Kirlin, J., Ota, B., Pope, E., Weber, M., Wiseman, K., 2013. Designing a network of marine protected areas in California: achievements, costs, lessons learned, and challenges ahead. Ocean & Coastal Management 74, 90–101. http://dx.doi.org/10.1016/j.ocecoaman.2012.08.013 (Special Issue on California's Marine Protected Area Network Planning Process).

Gulland, J.A., 1971. The Fish Resources of the Ocean. Fishing News Books, West Byfleet, UK.

Hahn, T., Schultz, L., Folke, C., Olsson, P., 2008. Social networks as sources of resilience in social-ecological systems. Complex Theory Sustainable Future 119–148.

Halpern, B.S., Warner, R.R., 2003. Matching marine reserve design to reserve objectives. Proceedings of Royal Society of London B Biological Sciences 270, 1871–1878. http://dx.doi.org/10.1098/rspb.2003.2405.

Hanna, S.S., 1997. The new frontier of American fisheries governance. Ecological Economics 20, 221–233. http://dx.doi.org/10.1016/S0921-8009(96)00082-1.

Hilborn, R., 2007. Moving to sustainability by learning from successful fisheries. AMBIO Journal of Human Environment 36, 296–303. http://dx.doi.org/10.1579/0044-7447(2007)36[296:MTSBLF]2.0.CO;2.

Hilborn, R., Parrish, J.K., Litle, K., 2005. Fishing rights or fishing wrongs? Reviews in Fish Biology and Fisheries 15, 191–199.

Hill, R., 1997. Scientists Call for Protection Oceans. The Oregonian.

Inniss, L., Simcock, A., Ajawin, A.Y., Alcala, A.C., Bernal, P., Calumpong, H.P., Araghi, P.E., Green, S.O., Harris, P., Kamara, O.K., Kohata, K., Marschoff, E., Martin, G., Ferreira, B.P., Park, C., Payet, R.A., Rice, J., Rosenberg, A., Ruwa, R., Tuhumwire, J.T., Van Gaever, S., Wang, J., Westawski, J.M., 2016. The First Global Integrated Marine Assessment: World Ocean Assessment I.

Kelleher, G., Kenchington, R.A., 1991. Guidelines for Establishing Marine Protected Areas. A Marine Conservation and Development Report. IUCN, Gland, Switzerland.

Kirlin, J., Caldwell, M., Gleason, M., Weber, M., Ugoretz, J., Fox, E., Miller-Henson, M., 2013. California's Marine Life Protection Act Initiative: supporting implementation of legislation establishing a statewide network of marine protected areas. Ocean & Coastal Management 74, 3–13. http://dx.doi.org/10.1016/j.ocecoaman.2012.08.015 (Special Issue on California's Marine Protected Area Network Planning Process).

Kjsebu, O.S., Bogstad, B., Devine, J.A., Gjøsæter, H., Howell, D., Ingvaldsen, R.B., Nash, R.D.M., Skjæraasen, J.E., 2014. Synergies between climate and management for Atlantic cod fisheries at high latitudes. Proceedings of the National Academy of Sciences of the United States of America 111, 3478–3483. http://dx.doi.org/10.1073/pnas. 1316342111.

Krimsky, S., 2005. The funding effect in science and its implications for the judiciary. Journal of Law & Policy 13, 43.

Leslie, H., Ruckelshaus, M., Ball, I.R., Andelman, S., Possingham, H.P., 2003. Using siting algorithms in the design of marine reserve networks. Ecological Applications 13, 185–198.

Lester, S.E., Halpern, B.S., Grorud-Colvert, K., Lubchenco, J., Ruttenberg, B.I., Gaines, S.D., Airamé, S., Warner, R.R., 2009. Biological effects within no-take marine reserves: a global synthesis. Marine Ecology Progress Series 384, 33–46.

Lubchenco, J., 1998. Entering the century of the environment: a new social contract for science. Science 279, 491–497. http://dx.doi.org/10.1126/science.279.5350.491.

Lubchenco, J., Cerny-Chipman, E.B., Reimer, J.N., Levin, S.A., 2016. The right incentives enable ocean sustainability successes and provide hope for the future. Proceedings of the National Academy of Sciences of United States of America. http://dx.doi.org/10.1073/pnas.1604982113 201604982.

Lubchenco, J., Grorud-Colvert, K., 2015. Making waves: the science and politics of ocean protection. Science 350, 382–383. http://dx.doi.org/10.1126/science.aad5443.

Lubchenco, J., Palumbi, S.R., Gaines, S.D., Andelman, S., 2003. The science of marine reserves. Special Issue Ecological Applications 13, S1–S228.

Mamula, A., Collier, T., 2015. Multifactor productivity, environmental change, and regulatory impacts in the U.S. West Coast groundfish trawl fishery, 1994–2013. Marine Policy 62, 326–336. http://dx.doi.org/10.1016/j.marpol.2015.06.002.

McCright, A.M., Dentzman, K., Charters, M., Dietz, T., 2013. The influence of political ideology on trust in science. Environmental Research Letters 8, 044029.

McCright, A.M., Dunlap, R.E., 2011. The politicization of climate change and polarization in the American Public's Views of global warming, 2001–2010. Sociological Quarterly 52, 155–194. http://dx.doi.org/10.1111/j.1533-8525.2011.01198.x.

National Ocean Council, 2013. National Ocean Policy Implementation Plan. Washington DC, USA.

National Research Council, 1999. Sharing the Fish: Toward a National Policy on Individual Fishing Quotas. National Academies Press, Washington DC, USA.

Nie, M.A., Schultz, C.A., 2012. Decision-making triggers in adaptive management. Conservation Biology 26, 1137–1144. http://dx.doi.org/10.1111/j.1523-1739.2012.01915.x.

Norberg, J., Wilson, J., Walker, B., Ostrom, E., 2008. Diversity and resilience of social-ecological systems. Complex Theory Sustainable Future 46–79.

Northern Economics, Inc, 2013. The economic impact of shellfish aquaculture in Washington, Oregon, and California. Prepared for Pacific Shellfish Aquaculture.

Oreskes, N., Conway, E.M., 2011. Merchants of Doubt: How a Handful of Scientists Obscured the Truth on Issues From Tobacco Smoke to Global Warming. Bloomsbury Publishing USA, New York, NY.

Osmond, M., Airamé, S., Caldwell, M., Day, J., 2010. Lessons for marine conservation planning: a comparison of three marine protected area planning processes. Ocean & Coast. Management 53, 41–51. http://dx.doi.org/10.1016/j.ocecoaman.2010.01.002.

Palmer, M.A., 2012. Socioenvironmental sustainability and actionable science. BioScience 62, 5–6. http://dx.doi.org/10.1525/bio.2012.62.1.2.

Pershing, A.J., Alexander, M.A., Hernandez, C.M., Kerr, L.A., Bris, A.L., Mills, K.E., Nye, J.A., Record, N.R., Scannell, H.A., Scott, J.D., Sherwood, G.D., Thomas, A.C., 2015. Slow adaptation in the face of rapid warming leads to collapse of the Gulf of Maine cod fishery. Science 350, 809–812. http://dx.doi.org/10.1126/science.aac9819.

Pew Oceans Commission, 2003. America's Living Oceans: Charting a Course for Sea Change.

Raven, J., Caldeira, K., Elderfield, H., Hoegh-Guldberg, O., Liss, P., Riebesell, U., Shepherd, J., Turley, C., Watson, A., 2005. Ocean Acidification Due to Increasing Atmospheric Carbon Dioxide. The Royal Society.

Rosenberg, A.A., Halpern, M., Shulman, S., Wexler, C., Phartiyal, P., 2013. Reinvigorating the role of science in democracy. PLoS Biology 11, e1001553. http://dx.doi.org/10.1371/journal.pbio.1001553.

Saarman, E., Gleason, M., Ugoretz, J., Airamé, S., Carr, M., Fox, E., Frimodig, A., Mason, T., Vasques, J., 2013. The role of science in supporting marine protected area network planning and design in California. Ocean & Coastal Management 74, 45–56. http://dx.doi.org/10.1016/j.ocecoaman.2012.08.021 (Special Issue on California's Marine Protected Area Network Planning Process).

Sabine, C.L., Feely, R.A., Gruber, N., Key, R.M., Lee, K., Bullister, J.L., Wanninkhof, R., Wong, C., Wallace, D.W.R., Tilbrook, B., et al., 2004. The oceanic sink for anthropogenic CO_2. Science 305, 367.

Senge, P., Hamilton, H., Kania, J., 2015. The dawn of system leadership. Stanford Social Innovation Review 27–33 Winter 2015.

Shanks, A.L., Grantham, B.A., Carr, M.H., 2003. Propagule dispersal distance and the size and spacing of marine reserves. Ecological Application 13, 159–169.

Stokes, D.E., 1997. Pasteur's Quadrant: Basic Science and Technological Innovation. Brookings Institution Press, Washington, DC.

The Science of Marine Reserves, 2002. Partnership for Interdisciplinary Studies of Coastal Oceans.

US Commission on Ocean Policy, 2004. An Ocean Blueprint for the 21st Century.

Vitousek, P.M., Mooney, H.A., Lubchenco, J., Melillo, J.M., 1997. Human domination of Earth's ecosystems. Science 277, 494–499. http://dx.doi.org/10.1126/science.277.5325.494.

Washington State Blue Ribbon Panel on Ocean Acidification, 2012. Ocean Acidification: From Knowledge to Action, Washington State's Strategic Response (No. 12-01-015). Washington Department of Ecology, Olympia, WA.

White House, 2010. Executive Order 13457, Stewardship of the Ocean, Our Coasts and the Great Lakes.

Chapter 2

Climate Variability, Climate Change, and Conservation in a Dynamic Ocean

Malin L. Pinsky, Rebecca L. Selden
Rutgers University, New Brunswick, NJ, United States

INTRODUCTION

Oceanography and climate science have revealed that the ocean is a highly dynamic environment and that ocean climate will change dramatically in the coming decades and centuries. Although traditional efforts at marine conservation and management of natural resources have tended to largely overlook these changes (Hilborn and Walters, 1992; McLeod and Leslie, 2009), new developments in science and practice are revealing that both conservation and fisheries management can often be more effective by considering ocean dynamics. Similarly, an appreciation for the complex dynamics of human behavior is also becoming clear, such that changes in the ocean affect human behavior, which in turn feeds back to affect ocean ecosystems. How these coupled dynamics will play out in the face of climate change remains an important question. This chapter lays out key features of climate, ecosystem, and human dynamics; how they interact; and emerging adaptation, conservation, and management approaches.

THE PACE OF CLIMATE CHANGE IN THE OCEAN

Physical conditions in the ocean vary across a range of temporal and spatial scales, from days to centuries and beyond. On top of this natural variability, anthropogenic greenhouse gas emissions are driving directional changes in climate, the main focus of this chapter. The oceans have absorbed about 93% of the Earth's increase in heat since 1971 (Rhein et al., 2013), and average surface temperatures increased by 0.44°C over that time frame. Particular areas like the North Sea, East China Sea, Gulf of Maine, and Newfoundland-Labrador Shelf, however, have warmed much more rapidly over the last few decades (Belkin, 2009), while areas like the northwest Atlantic, northern Mediterranean, and

Conservation for the Anthropocene Ocean. http://dx.doi.org/10.1016/B978-0-12-805375-1.00002-7

western Australia have experienced extreme but short-lived marine heat waves (Hobday et al., 2016).

Particularly when considering ecological consequences, it is useful to measure how fast isotherms (lines of constant temperature) move across the globe with global warming. This "climate velocity" indicates the speed at which an organism would need to move to maintain a stable temperature. Spatial gradients in ocean temperature are much shallower than on land, and so climate velocities in the ocean are as fast or faster than on land (up to 200 km/decade from 1960 to 2009), even though absolute rates of warming are lower (Burrows et al., 2011). In addition, the ocean has relatively few microclimates compared with those on land (Robinson et al., 2011), so short distance movements are less likely to encounter dramatically different conditions.

The "other" CO_2 problem, aside from global warming, is ocean acidification. The ocean takes up carbon dioxide from the atmosphere, which dissolves to form a weak acid. In turn, the pH, carbonate ion concentration, and calcium carbonate saturation states of the ocean decline. Open ocean acidification has already resulted in a 26% increase in hydrogen ion concentration (drop in pH of 0.1 units) (Rhein et al., 2013), although coastal carbonate chemistry is highly variable and trends remain more poorly understood. Other important changes include a decline in oxygen concentrations and rising sea levels (Rhein et al., 2013).

OBSERVED RESPONSES OF MARINE FISH AND INVERTEBRATES

Ocean conditions are important for the physiological performance and fitness of marine species (Metcalfe et al., 2012), and climate change has impacted (1) individual physiology, (2) population-level demography and abundance, (3) distributions of species across space or time, and (4) species interactions and marine communities. At each scale, evolutionary adaptation may allow species and communities to buffer climate impacts.

Although physiological performance can initially improve with warming, continued warming, oxygen limitation, and acidification increase physiological stress (Kroeker et al., 2013; Pörtner and Knust, 2007; Seebacher et al., 2015). Mean conditions, variability, and extremes are all important to consider here. Sustained adverse physiological conditions can lead to changes in population demographic rates, including birth and death rates, which influence the capacity of marine species to sustain harvesting. For example, reduced recruitment and increased cod mortality due to high temperatures in the Gulf of Maine appear likely to inhibit the rebuilding potential of the stock (Pershing et al., 2015).

These changes in demographic rates can lead to alteration of relative species abundances. A meta-analysis found that many communities around the world show consistent increases in the abundance of warm-water species and declines in cold-water species (Poloczanska et al., 2013). In the South Atlantic Bight, northern species are less abundant in years following mild winters, whereas

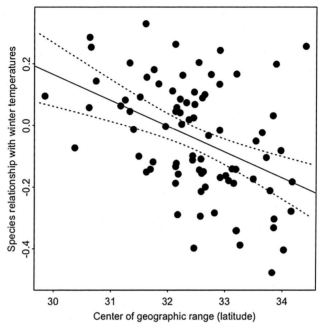

FIGURE 2.1 Summer biomass responses of 83 taxa to winter temperature anomalies, organized from left to right by the mean latitude of each species. Positive values indicate taxa whose summer biomass is higher following warm winters.

southern species are more abundant (Fig. 2.1) (Morley et al., 2017). Some of these shifts relate to changes in species distribution (Pinsky et al., 2013), and species biomass can shift toward new suitable habitat based on three nonexclusive mechanisms: (1) larval dispersal, (2) adult movement, and (3) differential changes in demographic rates. Empirically, high mobility and fast growth rates are associated with faster rates of range shifts (Pinsky et al., 2013; Sunday et al., 2015). Sessile organisms that rely on larval dispersal to move out of adverse conditions may be particularly vulnerable where climate velocities are moving in opposition to major currents. The degree to which different climate-related stressors covary will also influence the utility of shifting distributions for coping with these stressors.

In addition to shifts in space, climate change alters the timing of important life history processes. Spring phytoplankton blooms shift little in their timing due to their dependence on light availability (Edwards and Richardson, 2004). However, organisms that are dependent on temperature for physiological development have seasonal cycles that shift earlier in warm years (Edwards and Richardson, 2004). The asynchrony in timing of seasonal cycles among trophic levels affects the productivity and demography of higher trophic level species, including fish (Beaugrand et al., 2003) and seabirds (Hipfner, 2008).

More generally, differential responses of species within an assemblage to changes in ocean conditions can alter key species interactions. In the Eastern Bering Sea, warming-induced range expansion and increased abundance of a predatory flatfish has resulted in increased overlap with its pollock prey, leading to reduced recruitment of a prey species that supports one of the largest fisheries in the world (Hunsicker et al., 2013). In addition to affecting pairwise species interactions, changes in ocean conditions have led to reorganization of some marine food webs. Arctic marine food webs are particularly vulnerable to warming-induced alteration because invading boreal taxa tend to have a greater fraction of top predators (Mueter and Litzow, 2008) that are also more generalist (Kortsch et al., 2015) than their arctic counterparts.

The degree to which evolution mediates these effects is dependent on genetic variation in populations, as well as the rate and consistency in the direction of change. Rapid rates of anthropogenic change and fluctuations in climate can limit the capacity for directional selection (Seebacher et al., 2015). In corals, both short-term physiological acclimatization and long-term evolution are important mechanisms by which species copes with warm temperatures (Palumbi et al., 2014). The relative importance of these processes across species remains an important question.

FUTURE RESPONSES

Best practices and methods for projecting future marine ecosystem states continue to evolve (Cheung et al., 2016), but the broad outlines of the effects of climate change are becoming clear. Projections of future changes in many ways mirror the observed changes to date, while also being more extreme. Species distributions will continue to shift, for example, and fish are projected to shift an average of 45–49 km/decade under moderate climate warming (Cheung et al., 2009). An important constraint, however, is the availability of other habitats. Coral reef fish arriving poleward of the furthest coral reefs will often find it hard to survive, while the decreasing availability of light at higher latitudes may limit the extension of at least some coral species (Muir et al., 2015). Geographic constraints induce important variation across regions as well, and not all species shifts will be poleward at a regional scale. In the Gulf of Mexico or Gulf of Maine, for example, cooler waters are deeper and south, not north, and species are therefore expected to move in those directions (Kleisner et al., 2016; Pinsky et al., 2013). This regional variation has particularly important implications for regional conservation or management efforts.

As species distributions change, new communities of species are likely to be created, and major questions still surround how these novel assemblages will function. Species richness is generally projected to decline in the tropics, but warmer temperatures in other locations may actually attract high-diversity communities from low latitudes, driving local increases in richness (Cheung et al., 2009). Colonizations that occur faster than extirpations could also drive local increases in richness, although as a transient effect.

As a fundamental constraint, the primary productivity of the ocean limits the energy available for all species higher in the food chain. Current projections from Earth System Models suggest that primary productivity will decline globally over the 21st century, up to a $9\% \pm 8\%$ decline with a business as usual emissions scenario (Boyd et al., 2014). This global decline is composed of strong declines in the tropics and temperate latitudes, where warming increases ocean stratification and reduces the supply of nutrients to the photic zone, and increases near the poles, where warming reduces thermal limitation and light limitation from ice cover.

CONSEQUENCES FOR FISHERIES AND FISHERIES MANAGEMENT

Changes in primary productivity are predicted to have strong indirect effects on fish production and fisheries yield. Fish productivity is limited not only by low density of phytoplankton and zooplankton (Blanchard et al., 2012), but also by mismatches in the seasonal availability of prey for juveniles (Beaugrand et al., 2003). As a result of changes in temperature, productivity, and other conditions, climate change is projected to lead to a large-scale redistribution of global catch potential with large increases in high latitude regions and declines in the tropics (Blanchard et al., 2012; Cheung et al., 2010).

For species with clear linkages between productivity and the environment, environmental conditions are beginning to be considered in management. Harvest rules for Pacific sardine are based on average sea surface temperatures (Lindegren and Checkley, 2013), although the rule has been controversial. Whether an environmental harvest control rule results in significant improvements over conventional management depends on the strength of the relationship between the environment and recruitment (Brunel et al., 2010). In cases where the relationship between environment and recruitment is uncertain or weak, precautionary approaches may better ensure management that increases yields and reduces risk of stocks falling to low levels (Punt et al., 2014). Species life history traits are increasingly being used as predictors of the relative biological sensitivity of species or functional groups to climate change (Hare et al., 2016), and could be used to prioritize where improved information on the link between the environment and fishery productivity may have the most benefit.

Distribution shifts can also complicate management. As species shift, their distributions may cross into new management jurisdictions or even cross national boundaries. These shifts make stock assessments more challenging and may require redefinition of what constitutes a particular stock (Link et al., 2011; Shackell et al., 2016). In addition, transboundary stocks are particularly prone to depletion when management is not coordinated, and shifting distributions due to climate only exacerbate the incentives to overharvest. The "mackerel wars" in Europe provide a dramatic example of this: as mackerel shifted into warming Icelandic and Faroese waters, both countries unilaterally decided to increase their quotas, which, in combination with the existing harvest from the European Union and other countries, threatened the stock (The Economist, 2010). Projections of future species distributions can highlight where stocks will straddle management boundaries, allowing stakeholders to make prearrangements for allocation of quotas that minimize conflict and decrease incentives to overharvest (Pinsky and Mantua, 2014).

Changes in ocean conditions can affect not only the productivity and distribution of target stocks but also the likelihood that nontarget species are caught (Hazen et al., 2013). As quotas are increasingly limited by bycatch of so-called "weak stocks," decisions about where to fish are increasingly made to avoid these species. Where distributions of these bycatch species are related to ocean conditions, environmental data can be used to predict where these species are likely to occur, allowing managers to make changes in spatial zoning in near real time. Dynamic management has the potential to improve the efficiency of closures for bycatch and to forego less target catch by allowing fishing more precisely where and when the bycatch species is not present (Maxwell et al., 2015). As a management measure to avoid catching juvenile cod in the northwest Atlantic, dynamic management has the potential to increase the value of fishery landings by $15–52 million (Dunn et al., 2016).

Despite its complexity, dynamic approaches are already being used to regulate fishing on short timescales. Information derived from satellite tags about the seasonal thermal habitat preferences of southern bluefin tuna has been used since 2003 to update the zoning for the management of the eastern Australia longline fishery in which bluefin tuna are caught as bycatch (Hobday et al., 2011) (Fig. 2.2). Zoning in this fishery was changed on average every 2 weeks during the season, providing evidence that managers and fishers can successfully implement complex and dynamic spatial zoning based on oceanographic conditions. The voluntary TurtleWatch program to reduce the number of turtles caught in longline fisheries is another example (Howell et al., 2008). These short-term approaches may provide a strong basis for long-term adaptation to changing oceanographic conditions as well.

CONSEQUENCES FOR MARINE CONSERVATION

As explained earlier, climate change has already and will continue to have dramatic consequences for marine species, marine communities, and the people that rely upon them. The only real way to limit or reverse these impacts is to reduce our emissions of greenhouse gases, or, perhaps, to sequester large amounts of carbon dioxide from the atmosphere. However, because our past emissions have already committed us to substantial global warming, there are also important questions about how to adapt marine conservation efforts to unavoidable climate change.

Many marine conservation efforts have focused on particular places, such as setting aside marine protected areas (MPAs) with limits to human activities

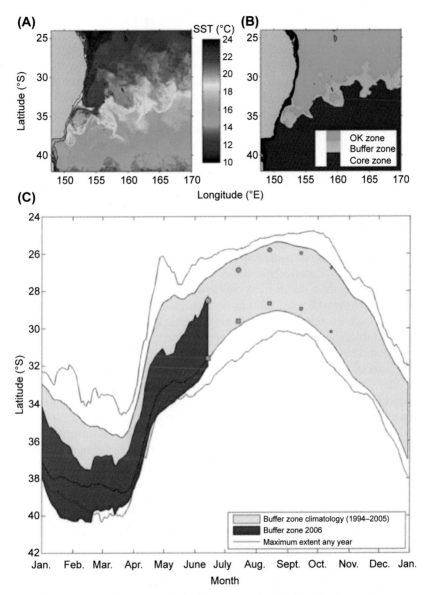

FIGURE 2.2   (A) Sea surface temperature (SST) maps, and (B) southern bluefin tuna habitat maps using a "nowcast" for June 15, 2006, from an oceanographic model for southeast Australia. (C) Daily climatology of bluefin tuna habitat zones. The mean latitude of the boundaries between core and buffer zones (lower line) and between buffer and OK zones (upper line) for 1994–2005 is indicated by the *yellow band*. The *thin blue lines* indicate the maximum northerly and southerly extent of these boundaries recorded during the period. The position of the habitat boundaries in 2006 up to June 15 is depicted by the *red band*. Tuna habitat maps and climatology are updated online approximately every 2 weeks during the fishing season. *Reprinted by permission from Hobday, A.J., Hartog, J.R., Spillman, C.M., Alves, O., Hilborn, R., 2011. Seasonal forecasting of tuna habitat for dynamic spatial management. Canadian Journal of Fisheries and Aquatic Sciences 68, 898–911, NRC Research Press, copyright 2011.*

like fishing. However, a lack of fishing would not prevent the local extirpation of a species pushed beyond its thermal tolerance by warming water. Whether or not turnover in the ecological community is of concern for marine conservation depends in part on the goals of a given MPA. This turnover may have relatively low management consequences if areas are managed for broad purposes, such as California's Marine Life Protected Act or the Great Barrier Reef Marine Park, although questions about baselines and what to consider "natural" are not straightforward (Hobday, 2011). On the other hand, protected areas with purposes tied to particular species may fail to meet their management objectives if distributions shift. The "Plaice Box" in the North Sea, for example, no longer adequately protects juvenile plaice (a flatfish) from fishing, and moving this area in a dynamic management approach may be necessary (van Keeken et al., 2007). Given the variety of responses and rates of response among species, there is unlikely to be a single adaptation method that works well in all cases.

The establishment of MPAs also has consequences for species' abilities to shift their distributions, particularly through a process called biotic resistance. Sea urchins from Australia have begun to colonize Tasmania and overgraze kelp forests, turning them into barrens with low species richness (Ling et al., 2009). However, MPAs with many lobsters (a key urchin predator) have a greater ability to exclude urchins and maintain kelp forests. This and other examples provide an intriguing contrast to the terrestrial experience, which has emphasized protected areas as sites of rapid colonization because of better habitat (Thomas et al., 2012).

More broadly than just protected areas, marine spatial planning efforts are underway to delineate areas of the ocean for distinct human uses, including fishing, energy development, transportation, recreation, mining, and conservation. Given the complexity of reaching an agreement among diverse stakeholders and the durability of certain ocean uses (mining and energy development often have 30+ year leases), revision of marine spatial plans is challenging except over long timescales. Current legal frameworks and guidelines for marine spatial planning, as well as current practice, however, generally overlook the impacts of climate change during plan development (Craig, 2012). A concern going forward is that shifts in species distributions may create unforeseen conflicts that could be avoided with a consideration of multiple ocean uses and both current and future species distributions.

For degraded ecosystems or particularly vulnerable species, conservation may also employ more active approaches. Coral reef restoration has been a particularly active field, including calls for choosing or even breeding stress-tolerant strains of coral (van Oppen et al., 2015). Similar considerations would apply to oyster reef or seagrass restoration. The goals of restoration efforts, however, are particularly sensitive to climate change, since historical ecosystems may not be achievable in future climates (Handel, 2015). Assisted colonization is a similarly active approach to conservation and involves moving a species to a new place that will be more suitable in the future. To many, this approach carries echoes of invasive species, and the risks may be great. On the other hand, some species may lack the ability to colonize suitable habitats or to evolve rapidly enough to withstand climate change. A carefully thought-out assisted migration program may be needed in these cases (Hoegh-Guldberg et al., 2008).

Planning and implementing marine conservation efforts in the face of climate change is difficult because of irreducible uncertainty about the future and because there will almost certainly be surprises in how future ecosystems function (Schindler and Hilborn, 2015). In light of these challenges, successful approaches will likely balance forward-looking strategies that broadly anticipate climate-driven changes against inherently robust and flexible strategies (Pinsky and Mantua, 2014). In the face of uncertainty, maintaining options and using monitoring to revaluate and adapt will be important (Schindler and Hilborn, 2015).

SOCIAL-ECOLOGICAL FEEDBACKS

The earlier sections of this chapter have set the stage in a couple important ways: climate is driving directional changes in ocean ecosystems, and these changes affect the success of the rules we use to govern human activities in the ocean. However, there are also important feedbacks between changes in the ocean and the ways in which humans behave. Fisheries and conservation are therefore part of a coupled social-ecological system (McCay et al., 2011). Coupled systems like this are often characterized by complex feedbacks, lags, cumulative impacts, and the potential for thresholds. How humans adapt to climate change, for example, is likely to have impacts on ecosystems to an extent as great as, if not greater than, the direct effects of climate change alone (Turner et al., 2010).

To illustrate these points, it is useful to consider how fisheries have adapted to past changes in species distributions. For a fishery to shift locations, individual fishermen can change the primary port they use for landing fish or travel further from their current ports. Alternatively, fishermen in some locations could catch more fish (e.g., at high latitudes), or others could catch fewer fish (e.g., at low latitudes). Regulations and business considerations may limit which of these options are feasible, while social processes and preferences may alter which options are most appealing or accessible (McCay et al., 2011).

In the northeast United States, many species have been shifting poleward over the past few decades, and many fisheries have also been moving poleward (Pinsky and Fogarty, 2012). However, these fisheries are not moving as fast as their target species (Fig. 2.3). For example, regulations, local ecological knowledge, and the availability of markets all appear to be slowing the red hake fishery (Pinsky and Fogarty, 2012). Because the fishery uses small-mesh nets, it is excluded from large regions of the Gulf of Maine and northern Georges Bank out of bycatch concerns. For the summer flounder fishery, a state-by-state allocation of fisheries quota based on the late 20th century distribution of the species has limited adaptation of the fishery, although boats

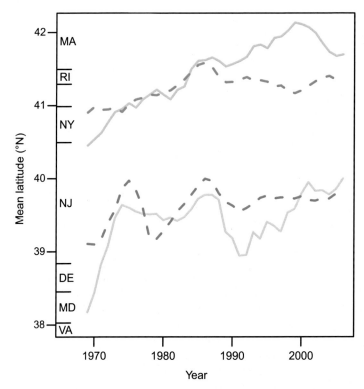

FIGURE 2.3 Mean latitude of red hake (*Urophycis chuss*) and summer flounder (*Paralichthys dentatus*) biomass over time on the northeast US continental shelf (*upper green solid line* and *lower blue solid line*, respectively). The slower poleward shifts of the red hake and summer flounder fisheries are plotted as *dashed lines* (*upper green dashed line* and *lower blue dashed line*, respectively). The latitudinal ranges of each state are shown on the left margin for reference. *See Pinsky, M.L., Fogarty, M., 2012. Lagged social-ecological responses to climate and range shifts in fisheries. Climatic Change 115, 883–891 for methodological details.*

from North Carolina now steam 500 km north to catch the species in New Jersey but land at home (NCPoliticalNews.com, 2013). These examples reveal how regulations and economics can mediate a fishery's response to shifting species in important ways.

Given the potential mismatch between fishing and climate, it becomes important to consider cumulative impacts. A lagging fishery, for example, may imply that low latitude fishermen are fishing more intensely for the remaining fish. When temperatures warmed in the Gulf of Maine, necessary reductions in the cod fishery quota were delayed because of political pressure from fishermen concerned about reduced profits (Pershing et al., 2015). The "creeping" enclosure of the commons and increased number of limited access fisheries may intensify this problem by reducing fishers' flexibility to exploit other species as replacements (Murray et al., 2010). Although rational, intensified fishing at the trailing edge may remove warm-adapted individuals that would otherwise provide the genetic material for populations that would do well in a warmer future.

However, a lagging fishery may also imply that a species is, in effect, outrunning the fishery. Fishing drives down abundance at the trailing edge, whereas less fishing on the leading edge would allow the population to colonize new territory more quickly, similar to a predator release phenomenon (Fuller et al., 2015). The combined effect could appear as if the social-ecological feedbacks were amplifying the climate effects, with the species biomass shifting poleward faster than with climate alone. Whether or not a fishery exploits a population's leading edge depends in part on the regulations in place. Species historically absent or at low abundance in a region may be unregulated, as was the case for blueline tilefish in the Mid-Atlantic Bight. This species is managed south of Cape Hatteras, North Carolina, but in 2015, a fishery quickly developed north of Cape Hatteras, before any management measures could be implemented (Menashes, 2015).

Other types of management may concentrate fishing on a species' leading edge. MPAs that exclude fishing have often been proposed as effective tools for climate adaptation, and yet they also concentrate fishing effort into narrower regions. The overall impacts of concentrated fishing in some regions and no fishing in other regions can actually be worse for species trying to colonize new territory, as compared with more even fishing across a wide region. However, how humans respond to MPAs determines the strength and even direction of this effect (Fuller et al., 2015). These kinds of feedbacks, from climate to marine species to human behavior to cumulative impacts, are only likely to become more important as the pace of environmental change increases.

NEW APPROACHES

The need for adaptation in a rapidly changing environment sets up a tension, as "flexibility" has sometimes been interpreted as leeway to relax sustainable fishing limits in favor of short-term economic gain. Responsible adaptation to a dynamic ocean, however, does not imply a weakening of sustainability goals. Because fisheries management, marine conservation, and even ecosystem-based management have traditionally avoided much focus on climate (Hilborn and Walters, 1992; McLeod and Leslie, 2009), efforts that create greater awareness of climate impacts can help move society toward effective adaptation. Examples moving in this direction include the Climate Change Web Portal (http://www.esrl.noaa.gov/psd/ipcc) and the OceanAdapt website (http://oceanadapt.rutgers.edu).

How to measure sustainability in a dynamic ocean also remains contentious, since traditional baselines are often no longer useful. Defining sustainability in terms of stable or increasing inclusive wealth (the sum of natural, human, and built capital) is becoming more widely accepted for this purpose

(UNU-IHDP and UNEP, 2014). Efforts are now underway to apply inclusive wealth to climate change questions, and initial results suggest that human institutions strongly affect the value of natural capital like fish, perhaps even more than direct climate effects (Fenichel et al., 2016).

Some adaptation steps are quite general across a range of climate impacts. One approach is simply to "address the basics" and reduce nonclimate stressors where possible, including overfishing, habitat damage, loss of prey, pollution, and bycatch. Populations and communities facing fewer stressors are more likely to be resilient to climate change. In the face of uncertainty, fisheries diversification may also be a useful strategy for fishers trying to adapt (Coulthard, 2009). Tools like tradable permits may help allow diversification, although they also tend to favor consolidation (Kasperski and Holland, 2013).

The observation that self-governance of common-pool resources can emerge, often in small communities, also suggests that sustainability will be easier to achieve if individuals have natural incentives to cooperate (Ostrom, 2009). However, a common attribute of successful local institutions is the existence of "clear boundaries," and climate-driven shifts of species across boundaries and among groups appear to violate this principle. More research is needed on the kinds of institutions that incentivize cooperation for sustainability in the face of climate-driven changes.

Overall, this is a critical moment for the integration of climate research across natural and social sciences and between theory and application. The oceans are already experiencing rapid, climate-driven changes, and there is growing awareness of these changes within the public and within the management community. The need now is for a better understanding of the dynamics and feedbacks within marine social-ecological systems, and the kinds of solutions that can allow adaptation and resilience over the coming decades and centuries.

ACKNOWLEDGMENTS

We thank Phil Levin, Melissa Poe, and Elliott Hazen for helpful comments on earlier drafts. We also thank the National Science Foundation (#OCE-1426891, #OCE-1430218, and #OCE-1521565), the Pew Charitable Trusts, and an Alfred P. Sloan Research Fellowship for support.

REFERENCES

Beaugrand, G., Brander, K.M., Alistair Lindley, J., Souissi, S., Reid, P.C., 2003. Plankton effect on cod recruitment in the North Sea. Nature 426, 661–664.

Belkin, I.M., 2009. Rapid warming of large marine ecosystems. Progress in Oceanography 81, 207–213.

Blanchard, J.L., Jennings, S., Holmes, R., Harle, J., Merino, G., Allen, J.I., Holt, J., Dulvy, N.K., Barange, M., 2012. Potential consequences of climate change for primary production and fish production in large marine ecosystems. Philosophical Transactions of the Royal Society B: Biological Sciences 367, 2979–2989.

Boyd, P.W., Sundby, S., Pörtner, H.O., 2014. Cross-chapter box on net primary productivity in the ocean. In: Field, C.B., Barros, V.R., Dokken, D.J., Mach, K.J., Mastrandrea, M.D., Bilir, T.E., Chatterjee, M., Ebi, K.L., Estrada, Y.O., Genova, R.C., Girma, B., Kissel, E.S., Levy, A.N., MacCracken, S., Mastrandrea, P.R., White, L.L. (Eds.), Climate Change 2014: Impacts, Adaptation, and Vulnerability. Part A: Global and Sectoral Aspects. Contribution of Working Group II to the Fifth Assessment Report of the Intergovernmental Panel on Climate Change. Cambridge University Press, Cambridge, UK and New York, NY, pp. 133–136.

Brunel, T., Piet, G.J., van Hal, R., Röckmann, C., 2010. Performance of harvest control rules in a variable environment. ICES Journal of Marine Science 67, 1051–1062.

Burrows, M.T., Schoeman, D.S., Buckley, L.B., Moore, P.J., Poloczanska, E.S., Brander, K.M., Brown, C.J., Bruno, J.F., Duarte, C.M., Halpern, B.S., Holding, J., Kappel, C.V., Kiessling, W., O'Connor, M.I., Pandolfi, J.M., Parmesan, C., Schwing, F.B., Sydeman, W.J., Richardson, A.J., 2011. The pace of shifting climate in marine and terrestrial ecosystems. Science 334, 652–655.

Cheung, W.W.L., Frölicher, T.L., Asch, R.G., Jones, M.C., Pinsky, M.L., Reygondeau, G., Rodgers, K.B., Rykaczewski, R.R., Sarmiento, J.L., Stock, C.A., Watson, J.R., 2016. Building confidence in projections of the responses of living marine resources to climate change. ICES Journal of Marine Science: Journal du Conseil 71, fsv250.

Cheung, W.W.L., Lam, V.W.Y., Sarmiento, J.L., Kearney, K., Watson, R., Pauly, D., 2009. Projecting global marine biodiversity impacts under climate change scenarios. Fish and Fisheries 10, 235–251.

Cheung, W.W.L., Lam, V.W.Y., Sarmiento, J.L., Kearney, K., Watson, R., Zeller, D., Pauly, D., 2010. Large-scale redistribution of maximum fisheries catch potential in the global ocean under climate change. Global Change Biology 16, 24–35.

Coulthard, S., 2009. Adaptation and conflict within fisheries: insights for living with climate change. In: Adger, W.N., Lorenzoni, I., O'Brien, K.L. (Eds.), Adapting to Climate Change: Thresholds, Values, Governance. Cambridge University Press, Cambridge, UK, pp. 255–268.

Craig, R.K., 2012. Ocean governance for the 21st century: making marine zoning climate change adaptable. Harvard Environmental Law Review 36, 305–350.

Dunn, D.C., Maxwell, S.M., Boustany, A.M., Halpin, P.N., 2016. Dynamic ocean management increases the efficiency and efficacy of fisheries management. Proceedings of the National Academy of Sciences of the United States of America 113, 201513626.

Edwards, M., Richardson, A.J., 2004. Impact of climate change on marine pelagic phenology and trophic mismatch. Nature 430, 881–884.

Fenichel, E.P., Levin, S.A., McCay, B.J., Martin, K.S., Abbott, J.K., Pinsky, M.L., 2016. Wealth reallocation and sustainability under climate change. Nature Climate Change 6, 237–244.

Fuller, E., Brush, E., Pinsky, M.L., 2015. The persistence of populations facing climate shifts and harvest. Ecosphere 6, 153.

Handel, S.N., 2015. Velocity of climate change and of restoration action: collision course? Ecological Restoration 33, 125–126.

Hare, J.A., Morrison, W.E., Nelson, M.W., Stachura, M.M., Teeters, E.J., Griffis, R.B., Alexander, M.A., Scott, J.D., Alade, L., Bell, R.J., Chute, A.S., Curti, K.L., Curtis, T.H., Kircheis, D., Kocik, J.F., Lucey, S.M., McCandless, C.T., Milke, L.M., Richardson, D.E., Robillard, E., Walsh, H.J., McManus, M.C., Marancik, K.E., Griswold, C.A., 2016. A vulnerability assessment of fish and invertebrates to climate change on the northeast U.S. continental shelf. PLoS One 11, e0146756.

Hazen, E.L., Jorgensen, S., Rykaczewski, R.R., Bograd, S.J., Foley, D.G., Jonsen, I.D., Shaffer, S.A., Dunne, J.P., Costa, D.P., Crowder, L.B., Block, B.A., 2013. Predicted habitat shifts of Pacific top predators in a changing climate. Nature Climate Change 3, 234–238.

Hilborn, R., Walters, C.J., 1992. Quantitative Fisheries Stock Assessment: Choice, Dynamics, and Uncertainty. Springer Science & Business.

Hipfner, J.M., 2008. Matches and mismatches: ocean climate, prey phenology and breeding success in a zooplanktivorous seabird. Marine Ecology Progress Series 368, 295–304.

Hobday, A.J., 2011. Sliding baselines and shuffling species: implications of climate change for marine conservation. Marine Ecology 32, 392–403.

Hobday, A.J., Alexander, L.V., Perkins, S.E., Smale, D.A., Straub, S.C., Oliver, E.C.J., Benthuysen, J.A., Burrows, M.T., Donat, M.G., Feng, M., Holbrook, N.J., Moore, P.J., Scannell, H.A., Sen Gupta, A., Wernberg, T., 2016. A hierarchical approach to defining marine heatwaves. Progress in Oceanography 141, 227–238.

Hobday, A.J., Hartog, J.R., Spillman, C.M., Alves, O., Hilborn, R., 2011. Seasonal forecasting of tuna habitat for dynamic spatial management. Canadian Journal of Fisheries and Aquatic Sciences 68, 898–911.

Hoegh-Guldberg, O., Hughes, L., McIntyre, S., Lindenmayer, D.B., Parmesan, C., Possingham, H.P., Thomas, C.D., 2008. Assisted colonization and rapid climate change. Science 321, 345–346.

Howell, E.A., Kobayashi, D.R., Parker, D.M., Balazs, G.H., Polovina, J.J., 2008. TurtleWatch: a tool to aid in the bycatch reduction of loggerhead turtles *Caretta caretta* in the Hawaii-based pelagic longline fishery. Endangered Species Research 5, 267–278.

Hunsicker, M.E., Ciannelli, L., Bailey, K.M., Zador, S., Stige, L.C., 2013. Climate and demography dictate the strength of predator-prey overlap in a subarctic marine ecosystem. PLoS One 8, e66025.

Kasperski, S., Holland, D.S., 2013. Income diversification and risk for fishermen. Proceedings of the National Academy of Sciences of the United States of America 110, 2076–2081.

Kleisner, K.M., Fogarty, M.J., McGee, S., Barnett, A., Fratantoni, P., Greene, J., Hare, J.A., Lucey, S.M., McGuire, C., Odell, J., Saba, V.S., Smith, L., Weaver, K.J., Pinsky, M.L., 2016. The effects of sub-regional climate velocity on the distribution and spatial extent of marine species assemblages. PLoS One 11, e0149220.

Kortsch, S., Primicerio, R., Fossheim, M., Dolgov, A.V., Aschan, M., 2015. Climate change alters the structure of arctic marine food webs due to poleward shifts of boreal generalists. Proceedings of the Royal Society B: Biological Sciences 282 20151546.

Kroeker, K.J., Kordas, R.L., Crim, R., Hendriks, I.E., Ramajo, L., Singh, G.S., Duarte, C.M., Gattuso, J.P., 2013. Impacts of ocean acidification on marine organisms: quantifying sensitivities and interaction with warming. Global Change Biology 19, 1884–1896.

Lindegren, M., Checkley, D.M., 2013. Temperature dependence of Pacific sardine (*Sardinops sagax*) recruitment in the California Current Ecosystem revisited and revised. Canadian Journal of Fisheries and Aquatic Sciences 70, 245–252.

Ling, S.D., Johnson, C.R., Frusher, S.D., Ridgway, K.R., 2009. Overfishing reduces resilience of kelp beds to climate-driven catastrophic phase shift. Proceedings of the National Academy of Sciences of the United States of America 106, 22341–22345.

Link, J.S., Nye, J.A., Hare, J.A., 2011. Guidelines for incorporating fish distribution shifts into a fisheries management context. Fish and Fisheries 12, 461–469.

Maxwell, S.M., Hazen, E.L., Lewison, R.L., Dunn, D.C., Bailey, H., Bograd, S.J., Briscoe, D.K., Fossette, S., Hobday, A.J., Bennett, M., Benson, S., Caldwell, M.R., Costa, D.P., Dewar, H., Eguchi, T., Hazen, L., Kohin, S., Sippel, T., Crowder, L.B., 2015. Dynamic ocean management: defining and conceptualizing real-time management of the ocean. Marine Policy 58, 42–50.

McCay, B.J., Weisman, W., Creed, C., 2011. Coping with environmental change: systemic responses and the roles of property and community in three fisheries. In: Ommer, R.E., Perry, R.I., Cochrane, K., Cury, P. (Eds.), World Fisheries: A Socio-Ecological Analysis. Blackwell Publishing, West Sussex, UK, pp. 381–400.

McLeod, K.L., Leslie, H.M., 2009. Ecosystem-Based Management for the Oceans. Island Press, Washington, DC.

Menashes, E.H., 2015. Mid-Atlantic Fishery Management Council (MAFMC); fisheries of the northeastern United States; scoping process. Federal Register 80, 29301–29302.

Metcalfe, J.D., Le Quesne, W.J.F., Cheung, W.W.L., Righton, D.A., 2012. Conservation physiology for applied management of marine fish: an overview with perspectives on the role and value of telemetry. Philosophical Transactions of the Royal Society B: Biological Sciences 367, 1746–1756.

Morley, J., Batt, R., Pinsky, M.L., 2017. Marine assemblages respond rapidly to winter climate variability, submitted.

Mueter, F.J., Litzow, M.A., 2008. Sea ice retreat alters the biogeography of the Bering Sea continental shelf. Ecological Applications 18, 309–320.

Muir, P.R., Wallace, C.C., Done, T., Aguirre, J.D., 2015. Limited scope for latitudinal extension of reef corals. Science 348, 1135–1138.

Murray, G., Johnson, T.R., McCay, B.J., St Martin, K., 2010. Creeping enclosure, cumulative effects and the marine commons of New Jersey. International Journal of the Commons 4, 367–389.

NCPoliticalNewscom, 2013. Fisheries Managers Seek Public Input on Issues Impacting Commercial Summer Flounder Fishery. NCPoliticalNews.com, http://www.ncpoliticalnews.com/?p=1937.

Ostrom, E., 2009. A general framework for analyzing sustainability of social-ecological systems. Science 325, 419–422.

Palumbi, S.R., Barshis, D.J., Traylor-Knowles, N., Bay, R.A., 2014. Mechanisms of reef coral resistance to future climate change. Science 344, 895–898.

Pershing, A.J., Alexander, M.A., Hernandez, C.M., Kerr, L.A., Bris, A.L., Mills, K.E., Nye, J.A., Record, N.R., Scannell, H.A., Scott, J.D., Sherwood, G.D., Thomas, A.C., 2015. Slow adaptation in the face of rapid warming leads to collapse of the Gulf of Maine cod fishery. Science 350, 809–812.

Pinsky, M.L., Fogarty, M., 2012. Lagged social-ecological responses to climate and range shifts in fisheries. Climatic Change 115, 883–891.

Pinsky, M.L., Mantua, N.J., 2014. Emerging adaption approaches for climate ready fisheries management. Oceanography 27, 146–159.

Pinsky, M.L., Worm, B., Fogarty, M.J., Sarmiento, J.L., Levin, S.A., 2013. Marine taxa track local climate velocities. Science 341, 1239–1242.

Poloczanska, E.S., Brown, C.J., Sydeman, W.J., Kiessling, W., Schoeman, D.S., Moore, P.J., Brander, K., Bruno, J.F., Buckley, L.B., Burrows, M.T., Duarte, C.M., Halpern, B.S., Holding, J., Kappel, C.V., O'Connor, M.I., Pandolfi, J.M., Parmesan, C., Schwing, F., Thompson, S.A., Richardson, A.J., 2013. Global imprint of climate change on marine life. Nature Climate Change 3, 919–925.

Pörtner, H.O., Knust, R., 2007. Climate change affects marine fishes through the oxygen limitation of thermal tolerance. Science 315, 95–97.

Punt, A.E., A'mar, T., Bond, N.A., Butterworth, D.S., Moor, C.L.d., Oliveira, J.A.A.D., Haltuch, M.A., Hollowed, A.B., Szuwalski, C., 2014. Fisheries management under climate and environmental uncertainty: control rules and performance simulation. ICES Journal of Marine Science 71, 2208–2220.

Rhein, M., Rintoul, S.R., Aoki, S., Campos, E., Chambers, D., Feely, R.A., Gulev, S., Johnson, G.C., Josey, S.A., Kostianoy, A., Mauritzen, C., Roemmich, D., Talley, L.D., Wang, F., 2013. Observations: ocean. In: Stocker, T.F., Qin, D., Plattner, G.-K., Tignor, M., Allen, S.K., Boschung, J., Nauels, A., Xia, Y., Bex, V., Midgley, P.M. (Eds.), Climate Change 2013: The Physical Science Basis. Contribution of Working Group I to the Fifth Assessment Report of the Intergovernmental Panel on Climate Change. Cambridge University Press, Cambridge, pp. 255–316.

Robinson, L.M., Elith, J., Hobday, A.J., Pearson, R.G., Kendall, B.E., Possingham, H.P., Richardson, A.J., 2011. Pushing the limits in marine species distribution modelling: lessons from the land present challenges and opportunities. Global Ecology and Biogeography 20, 789–802.

Schindler, D.E., Hilborn, R.W., 2015. Prediction, precaution, and policy under global change. Science 347, 953–954.

Seebacher, F., White, C.R., Franklin, C.E., 2015. Physiological plasticity increases resilience of ectothermic animals to climate change. Nature Climate Change 5, 61–66.

Shackell, N.L., Frank, K.T., Nye, J.A., den Heyer, C.E., 2016. A transboundary dilemma: dichotomous designations of Atlantic halibut status in the Northwest Atlantic. ICES Journal of Marine Science: Journal du Conseil 73, 1798–1805.

Sunday, J.M., Pecl, G.T., Frusher, S., Hobday, A.J., Hill, N., Holbrook, N.J., Edgar, G.J., Stuart-Smith, R., Barrett, N., Wernberg, T., Watson, R.A., Smale, D.A., Fulton, E.A., Slawinski, D., Feng, M., Radford, B.T., Thompson, P.A., Bates, A.E., 2015. Species traits and climate velocity explain geographic range shifts in an ocean-warming hotspot. Ecology Letters 18, 944–953.

The Economist, September 4, 2010. Mackerel Wars: Overfished and Over There. The Economist. Economist Group, London.

Thomas, C.D., Gillingham, P.K., Bradbury, R.B., Roy, D.B., Anderson, B.J., Baxter, J.M., Bourn, N.a.D., Crick, H.Q.P., Findon, R.A., Fox, R., Hodgson, J.A., Holt, A.R., Morecroft, M.D., O'Hanlon, N.J., Oliver, T.H., Pearce-Higgins, J.W., Procter, D.A., Thomas, J.A., Walker, K.J., Walmsley, C.A., Wilson, R.J., Hill, J.K., 2012. Protected areas facilitate species' range expansions. Proceedings of the National Academy of Sciences of the United States of America 109, 14063–14068.

Turner, W.R., Bradley, B.A., Estes, L.D., Hole, D.G., Oppenheimer, M., Wilcove, D.S., 2010. Climate change: helping nature survive the human response. Conservation Letters 3, 304–312.

UNU-IHDP, UNEP, 2014. Inclusive Wealth Report 2014: Measuring Progress Toward Sustainability. Cambridge University Press, Cambridge.

van Keeken, O.A., van Hoppe, M., Grift, R.E., Rijnsdorp, A.D., 2007. Changes in the spatial distribution of North Sea plaice (*Pleuronectes platessa*) and implications for fisheries management. Journal of Sea Research 57, 187–197.

van Oppen, M.J.H., Oliver, J.K., Putnam, H.M., Gates, R.D., 2015. Building coral reef resilience through assisted evolution. Proceedings of the National Academy of Sciences of the United States of America 112, 2307–2313.

Chapter 3

The Future Species of Anthropocene Seas

Nicholas K. Dulvy[1], Holly K. Kindsvater[2]
[1]Simon Fraser University, Burnaby, BC, Canada; [2]Rutgers University, New Brunswick, NJ, United States

INTRODUCTION

We know to the nearest minute when the last Passenger Pigeon died (1 p.m. on September 1, 1914). Although not all terrestrial extinctions can be so precisely timed, it is clear that identifying extinction is vastly more difficult in the oceans than on land. The demise of a marine species cannot be seen. This does not mean that marine extinctions do not occur, but rather that their detection is exceedingly difficult.

The global marine extinctions that have been detected mainly have been air-breathing mammals and birds, such as the Caribbean Monk seal (*Neomonachus tropicalis*; McClenachan and Cooper, 2008). As far as we know there has been only one global extinction of a fully marine fish, that of the Galapagos Damselfish (*Azurina eupalama*; Dulvy et al., 2009). Yet leading indicators of extinction risk caution that large numbers of other marine populations and species may disappear. For example, few people realize that two species of sawfishes were once found in US waters—the last Largetooth Sawfish (*Pristis pristis*) sighting was in Texas in 1961, and it is 99% certain that this species is extinct from US waters (Fernandez-Carvalho et al., 2014). Only a fragmented population of Smalltooth Sawfish (*Pristis pectinata*) in Florida and the Bahamas remains, occupying less than 5% of its historic range (National Marine Fisheries Service, 2009).

A survey of local and regional marine extinctions showed that 133 populations of a wide array of taxa ranging from algae to mammals have disappeared, including 28 populations of sawfishes, skates, and angel sharks (Dulvy et al., 2003; Dulvy and Forrest, 2010). Such local extinctions reflect the loss of behaviorally, morphologically, and ecologically distinct segments of biological population diversity (Dulvy et al., 2003). This among-population within-species biocomplexity and response diversity underpins species resilience and ecosystem services (Hilborn et al., 2003; Anderson et al., 2013, 2015).

Conservation for the Anthropocene Ocean. http://dx.doi.org/10.1016/B978-0-12-805375-1.00003-9
39

A countervailing view is that these local disappearances represent natural metapopulation patch dynamics—the winking out of edge-of-range populations that will eventually be rescued as abundance increases and hence range occupancy expands (Del Monte-Luna et al., 2007). Since the global survey of marine extinctions (Dulvy et al., 2003), each year that elapses without recolonization strengthens the case for the local extinction hypothesis. With very few exceptions, the volume of evidence confirming the former presence of species and their continued absence grows, with sawfishes being a case in point (Dulvy et al., 2016).

Marine species face multiple threats, but the overwhelming causes of extinction risk are overexploitation, habitat loss and degradation, and climate change (McClenachan et al., 2012). This problem is acute for intrinsically sensitive species with large geographic ranges, such as large-bodied predators, and high value species, for which intense fishing is driven by globalized trade demand (McClenachan et al., 2016). It is now obvious to many that oceans are not inexhaustible and some marine species can be driven to collapse. Therefore our challenge is to predict and prevent marine species' extinctions before the opportunity passes us by.

Our understanding of the status of our oceans and their inhabitants is deeply intertwined with our values and perceptions, which can differ based on education, upbringing, and experience (Mace et al., 2014; Mace and Hudson, 1999). Our epistemology—our way of knowing or understanding the world—shapes our view of conservation solutions and goals. Diverse perspectives on ocean conservation thus span fisheries-focused and conservation-focused worldviews (Salomon et al., 2011). A survey of shark and ray biologists found that people with measurable fisheries expertise viewed sustainable fisheries management as a viable goal (Simpfendorfer and Dulvy, 2017); by contrast those with no prior fisheries experience eschewed fisheries solutions and viewed a complete ban on elasmobranch fishing as the ultimate conservation goal (Shiffman and Hammerschlag, 2016). At a larger scale, international policy demands that we confront trade-offs on the fisheries-focused versus conservation-focused axis (Veitch et al., 2012). These trade-offs directly affect whether governments agree to policies affording species protection.

The signatory Parties of the Convention on Biological Diversity (CBD) committed to meeting the 2020 Aichi targets, including Target 6, specifying that all fish and invertebrate stocks and aquatic plants are managed and harvested legally and sustainably, and Target 11, mandating the prevention of species' extinctions and the sustained improvement of threatened species, and the related Sustainable Development Goal 14 to "Conserve and sustainably use the oceans, seas and marine resources for sustainable development" (Brooks et al., 2015). Accomplishing these targets requires consensus on what qualifies as sustainable, or conversely, threatened. How do we reach consensus regarding relative extinction threat? The International Union for the Conservation of Nature (IUCN) Red List assessments place species into one of three threatened categories (Critically Endangered, CR; Endangered, EN; or Vulnerable, VU), or classify them as Near Threatened, NT; Least Concern, LC; or Data Deficient, DD. In the worst case, assessors must determine if the species is shiftenterExtinct, EX, or Extinct in the Wild, EW (IUCN, 2014). This global standard has been widely accepted as the definitive index of extinction risk. To date, 7563 marine fishes (Actinopterygii, Chondrichthyes, and Sarcopterygii) have been assessed using IUCN Red List Categories and Criteria. Of these, 20% (1511) are Data Deficient (http://www.iucnredlist.org/search/link/5808c733-ca6b5fe0). By comparison, only around 200 species have been fisheries assessed for commercial and recreational management (Ricard et al., 2012). Furthermore, there are few species-specific measures of fisheries catch, for example, only around one-third of the global catch of chondrichthyans is identified to species level (Davidson et al., 2016). A major impediment to developing international conservation policy is meeting the shortfall in knowledge and monitoring of our seas.

Marine conservation in the Anthropocene needs to (1) avoid extinctions, (2) recover threatened species, and (3) sustain abundance of species that play functional roles or provide ecosystem services. Our focus here is on identifying, predicting, and preventing species extinctions. We show when and why marine extinctions have been unbelievable, unseen, and unmanaged. Finally, we summarize the scientific and policy tools needed to prevent further declines.

WHY IS UNDERSTANDING MINDSETS IMPORTANT TO UNDERSTAND THE STATE OF THE SEAS?

Our ability to identify and predict potential marine extinctions will depend on our mindset. An *evidentiary mindset* has dominated the scientific discourse and policy surrounding the diagnosis of marine extinctions. The bar for accepting a hypothesized extinction is high, and false alarms, where a marine species is incorrectly declared extinct, are rare (Peterman and M'Gonigle, 1992; Dayton, 1998). There is an analogy to the type I error rate in statistical hypothesis testing. Tolerance of type I error is commonly set to an α of 0.05, meaning that the risk of accepting an alternative hypothesis—a false alarm—is 1 in 20. By contrast, a *precautionary mindset* requires tolerating a greater risk of type I error to minimize the risk of missing extinction, a true emergency, or type II error. Failing to diagnose a marine extinction even when it has occurred is easily done, because our power (in statistical terms, $1-\beta$) is limited by our ability to detect and measure population trajectories in the marine realm. We will show that the evidentiary mindset that has prevailed in marine management has led to type II errors (Dayton, 1998). To avoid further extinctions, and protect the future species of Anthropocene seas, tolerance of higher type I error rates is required to minimize the risk of missing true emergencies.

Are Marine Extinctions Unbelievable?

Local extinctions have happened, but our mindset and capacity to detect them are limited. Thus marine extinctions can be overlooked, and in hindsight it is clear we failed to take a sufficiently precautionary approach to their prevention. There are two reasons why a species might be absent at a location within its expected geographic range: either it is now extinct or it is undetected by the census method or sampling gear. Very often historical records show what was caught where, on what date. Until the discovery of shifting baseline syndrome (Pauly, 1995), there has been little consideration of what was *not* caught.

We saw a shifting baseline unfold in Fiji in 2002 while searching in vain for the Bumphead Parrotfish (*Bolbometopon muricatum*). Dulvy and Polunin (2004) asked islanders if they ever catch *kalia* (the indigenous name for this large parrotfish), to which they always answered affirmatively. Eventually we thought to ask, "when did you last catch *kalia*?" This question sparked discussion leading to the villagers' self-realization that this species had disappeared unnoticed and had not been caught for decades (Dulvy and Polunin, 2004). Historians, archaeologists, paleontologists, and now ecologists, use expedition reports, cookbook recipes, and other nontraditional sources to demonstrate the role of shifting baselines in masking species extinctions, mainly at local and regional scales (Jackson, 1997; Wolff, 2000; Levin and Dufault, 2010; Thurstan et al., 2015).

This historical ecological information was always present—why have we been blind to it? Marine science and especially fisheries science has traditionally had a highly evidentiary mindset where the absence of data could not be considered as evidence of absence (Diamond, 1987; del Monte-Luna et al., 2009). Following lessons from terrestrial conservation, those with a conservation-focused mindset have shifted toward the precautionary approach in risk assessments (IUCN, 2014; p. 20). However, those with the fisheries-focused mindset can still demand an evidentiary approach to identifying extinctions at local, regional, and global scales. The higher evidentiary bar required to enact conservation measures for exploited marine species is well documented (Cooke, 2011). A review of terrestrial species listed under the Convention on International Trade in Endangered Species (CITES) revealed highly precautionary judgments: many terrestrial species were listed without qualifying under the criteria for extinction risk or trade (Cooke, 2011). Until recently, few exploited marine fishes were listed despite abundant evidence showing the criteria were met—the result of the unrealistically high bar for evidence required to list marine fishes (Cooke, 2011).

Marine Extinctions Are Unseen

The evidentiary mindset has led to the false assumption that marine fishes are safe from extinction. An increasing number of local and regional extinctions have proved this assumption wrong. Therefore we next discuss the problem of identifying marine extinctions when taxonomic uncertainty and observation error lead to false negatives and positives.

The Challenges of Counting Marine Species Extinctions

Marine extinctions have been underestimated because they are discovered long after the fact. A review of the status of 29 terrestrial and marine lineages reveals the proportion of threatened species tends to increase with assessment effort. In the best-studied lineages, the percentage of threatened species converges at around 20%–25% in both terrestrial and marine realms (Webb and Mindel, 2015). This is likely because the median lag between the local or regional extinctions of 133 marine populations and the reporting date was 55 years (Dulvy et al., 2003). Thus scientific knowledge and capacity are critical to understanding the state of the oceans and extinction risk (McClenachan et al., 2012; Miloslavich et al., 2016). Identifying extinctions requires accounting for uncertainty due to taxonomic uncertainty, observation error, and process error, all of which can generate false positives and false negatives.

Taxonomic Uncertainty, False Positives and Negatives

False positives (type I error) in extinction estimates can arise from updated taxonomy. For example, taxonomic reconsideration means that the "extinct" Green Wrasse (*Anampses viridus*) in Mauritius never was a species, and hence there is one fewer global species extinction on the tally than reported in 2009 (Dulvy et al., 2009). The "extinction" of this "species" was first identified in an early summary of marine extinction risk (Roberts and Hawkins, 1999). The authors stated, "The wrasse *Anampses viridis* was described from Mauritius in 1839 (Randall, 1972) but has not been seen in recent years despite intensive sampling. It may now be extinct, possibly a victim of sedimentation and nutrient pollution that has been degrading the reefs of Mauritius since the 19th century" (Hawkins et al., 2000). This paper and subsequent propagation of this reported extinction (Dulvy et al. 2003, 2009) were based on a continual review of the evidence. Russell and Craig (2013) resolved this 180-year-old case of mistaken identity by showing that the Green Wrasse is actually the adult male terminal phase color form (and junior synonym) of the common species *Anampses caeruleopunctatus* Rüppell, 1829.

Clearly, if the Green Wrasse was not a valid species, then it should never have been declared extinct. Thus precautionary warning of the scale of the biodiversity crisis must be balanced against the cost of declaring a species extinct. Falsely categorizing a species as "extinct" undermines the credibility of scientists in the public eye (Del Monte-Luna et al., 2007). The conservation status of each species must be reviewed and revised continually to account for retrospective changes in taxonomy and underscores the vital role of taxonomy in understanding the Anthropocene (Keith and Burgman, 2004; Butchart et al., 2007).

A taxonomic false negative (type II error) arises when one extinct or near extinct species turns out to be a species complex—instead of one extinction, the tally increases by two or more. Although not a global extinction, the Common Skate complex (*Dipturus batis* spp.) provides a notable example. These large skates disappeared from the Northeast Atlantic shelf seas after decades of retained secondary take (bycatch; Rogers and Ellis, 2000). Their depletion went unnoticed as their catch biomass was stabilized by a portfolio effect due to the serial depletion of smaller, more productive members of the skate assemblage (Dulvy et al., 2000). Prescient and credible warnings of the disappearance of the largest "species" (Brander, 1981) went largely unheeded by managers (Holden, 1992). By the early 1990s, Common Skate had all but disappeared from the North and Irish Seas (Walker and Hislop, 1998). However, the expansion of the French deepwater trawl fleet to the West of Scotland led to new catches of Common Skate. This allowed a savvy taxonomist, Samuel Igésias, to compare skate specimens side by side, revealing taxonomic identification issues. It turned out that this skate is in fact two species: a smaller species that reaches maturity at 120 cm, and another maturing at 200 cm. When "common skate" landings from 2005 were reassigned to the correct species, less than 2% were of the larger species (Iglésias et al., 2010). Overlooking a new large vertebrate species is understandable in less well-studied areas of the world, but it was shocking when a new skate species was described on the doorstep of the United Kingdom, given the nation's long and proud natural history tradition (Dulvy and Reynolds, 2009). This underscores the difficulty of "seeing" marine extinctions. Furthermore, emerging taxonomic science suggests sibling species and complexes in marine fishes are more common than previously thought (Bickford et al., 2007).

Observation Error and Lazarus Species

A false positive (type II error) in extinction risk can also occur if insufficient effort has been expended to find the species presumed extinct (Diamond, 1987). In the oceans, the broad scale and depth range of species' ranges, which may encompass several political jurisdictions, make this a persistent concern. Hence, the classification of extinctions requires a balance of two risks: (1) that a species is extinct and has gone undetected and unreported, and (2) that a Lazarus species is categorized as extinct at some scale when it is still present and, embarrassingly, is sighted at a later date (Keith and Burgman, 2004). For example, the Barndoor Skate (*Dipturus laevis*) was declared near extinct based on its absence in Atlantic shelf trawl surveys (Casey and Myers, 1998), yet significant numbers were subsequently discovered on the continental slope, preventing its listing under the US Endangered Species Act (Kulka et al., 2002).

In the tropics, observation and monitoring is a persistent challenge, even in nearshore waters. One paper suggested sawfishes may be extinct in Mexico (del Monte-Luna et al., 2009). This seemed plausible, as the last Largetooth Sawfish was landed in 1997 at Mujeres Island, Quintana Roo, Mexico.

However, in 2016 the scientific world was stunned and relieved when a Largetooth Sawfish was reported in Veracruz, rewarding recent efforts to raise awareness, and providing hope that all is not lost for this species in the Gulf of Mexico (R. Bonfil, Personal Communication).

In South Africa, by contrast, extensive long-term sampling bolsters our confidence that sawfishes are regionally extinct. South Africa has long time series of elasmobranch abundance from netting programs designed to protect bathers from sharks. These data show that sawfishes were formerly common in KwaZulu-Natal, but that numbers declined (likely due to incidental mortality in trawl fisheries and degradation of juvenile habitat) (Everett et al., 2015). The last reported observation of any sawfish species in South Africa was in 1999 (Fig. 3.1). Although this fish was released alive, no sawfish has been recorded since, despite the presence of survey gear through 2012. Both sawfish case studies illustrate that our confidence in species' disappearance depends on "observation error," a rather prosaic term that encompasses awareness, search effort, and continued monitoring using appropriate methods and gears.

False Alarms Depend on Process Error and Risk Tolerance

The likelihood of detecting meaningful declines—the precursor to raising the alarm on an impending extinction—depends on both observation error and

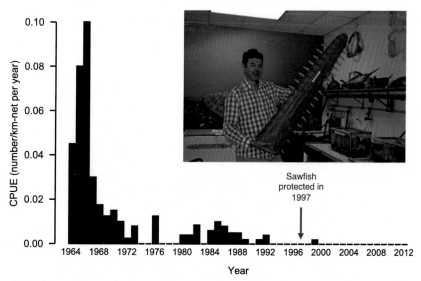

FIGURE 3.1 Sawfish catch-per-unit-effort from South African bather nets from 1964 to 2012. *(Redrawn from Everett, B.I., Cliff, G., Dudley, S.F.J., Wintner, S.P., van der Elst, R.P., 2015. Do sawfish Pristis spp. represent South Africa's first local extirpation of marine elasmobranchs in the modern era? African Journal of Marine Science 37, 275–284.)* Inset is the rostrum, probably of the Largetooth Sawfish *Pristis pristis*, of the last known captive South African sawfish that was housed in uShaka marine World, Durban. *(Photo credit Colin A. Simpfendorfer.)*

process error. Observation error stems from our ability to measure population abundance or range, whereas process error is the inherent "noise" in population dynamics that comes from natural environmental variation. Both present challenges for detecting population trends: process error can increase the chance a population randomly winks out, whereas observation error limits our power to determine the true population trend. There is no way to entirely eliminate either risk and again, risk tolerance depends on the mindset of the audience. Fisheries-focused scientists might try to minimize false alarms (false positives), in case they lead to unnecessary fisheries closures. Conservation scientists fear false negatives because, at best, the species' chances of recovery are diminished and costly; at worst extinctions are irreversible (Mace and Hudson, 1999; Matsuda et al., 1997; Reynolds and Mace, 1999). In reality, neither risk can be eliminated entirely without elevating the risk of the other error (Punt, 2000). The choice of balancing risks may not just be a scientific one (Dayton, 1998; Peterman and M'Gonigle, 1992). However, science can contribute by quantifying when a precautionary approach (minimizing false negatives) is warranted, based on life history traits and irreversibility of consequences.

Species' life history traits and population ecology affect the risk of false positives and negatives. Species with more variable dynamics were more likely to be misclassified in a study where both risks were estimated for the terrestrial species represented in the Global Population Dynamics database (Connors et al., 2014). This study determined the characteristics of the population time series associated with two outcomes: (1) incorrectly detecting a decline (type I error) and (2) failing to detect a true decline (type II error) (Connors et al., 2014). Shorter time series (<10 years) and shallow decline thresholds (<30%) lead to a moderate frequency of false alarms (45%) and true emergencies (60%) for populations with variable population dynamics. For populations with more predictable dynamics due to stronger density dependence, such as long-lived birds and exploited long-lived fishes, the frequency of false alarms and true emergencies was much lower (15% and 55%, respectively for 10-year-long time series) (Connors et al., 2014; Keith et al., 2015).

Only recently have we been able to quantify the risks and trade-offs between false negatives and positives (Fig. 3.2; Porszt et al., 2012; d'Eon-Eggertson et al., 2015). If a strong fisheries-focused ethic prevailed—adverse to a false positive—one might set the triggering threshold (at which one might declare a species to be threatened) to a 90% decline (gray diamond; Fig. 3.2A). This would guarantee zero false positives, but would result in a species being falsely classified as nonthreatened at least 20% of the time (end of downward gray arrow; Fig. 3.2B). If a strong conservation-focused ethic prevailed, then one might lower the triggering threshold to 40% to eliminate the false-negative risk of overlooking a threatened species (gray dot; Fig. 3.2B). This would mean false positives in at least half of the status assessments (end of upward gray arrow; Fig. 3.2A). Historically, the tendency has been to call for raised thresholds, indicating a fisheries-focused mindset. For example, in 1999 the American

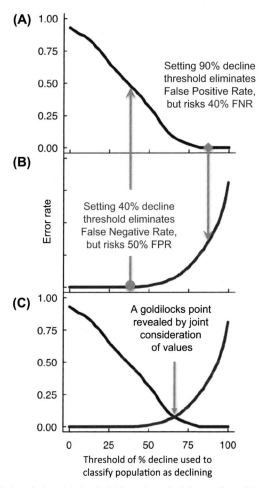

FIGURE 3.2 Choice of triggering threshold depends on the balance of two risks. (A) False Positive Rate (FPR) is the risk that a species is listed as threatened when it is sustainably fished. (B) False Negative Rate (FNR) is the risk that a species is classified as Least Concern when it is overfished and headed toward extinction. (C) A goldilocks point where both risks are equal, but the choice of threshold may be asymmetric depending on the relative costs and benefits of each risk. Extinction is forever; hence it could be argued that the 50% threshold (zero FNR, minimal FPR) should be used. *Redrawn from Figure S5 d'Eon-Eggertson, F., Dulvy, N.K., Peterman, R.M., 2015. Reliable identification of declining populations in an uncertain world. Conservation Letters 8, 86–96.*

Fisheries Society (AFS) proposed raising the threatened threshold from 70% to 99% decline to eliminate false positives. The empirically measured False-Negative Rate (the rate of failing to detect true emergencies) of this decision was 62% when AFS criteria were applied to EU fish stocks (Dulvy et al., 2005).

There are profound conservation costs to the fisheries-focused mindset. The lack of recognition of the trade-off in these risks has led to "too little, too late" conservation measures, as well as extreme management. *Too little, too late* occurred in South Africa, where sawfishes were protected only 2 years before the last sawfish capture in 1999 (Fig. 3.1) (Everett et al., 2015). By contrast, extreme management measures occurred in EU fisheries, but only after decades of alarms raised by fisheries scientists were ignored, including the disappearance of the Common Skate (Brander, 1981) and the very steep declines of Spiny Dogfish (called the Spurdog in Europe; *Squalus acanthias*) (Hammond and Ellis, 2005; Holden, 1974). Instead of gradually reducing take of these species using the quota management system, a zero Allowable Catch was set prohibiting take (Clarke, 2009). Hence, these species went from no management to prohibition almost overnight, a huge management challenge for a bycatch species! This type of management is *too much, too late*. Although such extreme measures could be appropriate in some cases, successful marine conservation requires policy makers, fisheries scientists, and conservation biologists to proactively navigate a middle road.

Predicting Species' Risk of Extinction

Until now we have focused on the challenges and trade-offs that come with identifying extinctions in the ocean. How then can one identify, or even better, predict species' risk of extinction before it causes a management crisis? In other words, what actions can be taken to prevent both *too little, too late,* and heavy-handed *too much, too late* measures?

The risk of a population or species extinction is a function of *intrinsic sensitivity* (biology) and *exposure* to an extrinsic threatening process. This risk can be offset by a species' adaptive capacity (Turner et al., 2003; Allison et al., 2009), whereby it can mitigate its sensitivity or exposure. This provides a conceptual framework that has great utility for framing species' risk of decline and extinction:

$$Vulnerability \propto \frac{Sensitivity + Exposure}{Adaptive\ capacity}$$

For our purposes, adaptive capacity might be an evolutionary response that rescues species by allowing adaptation to climate change (Bell and Gonzalez, 2009) or to new habitats. Although such responses are possible for organisms with faster generation times, evolutionary rescue will be too slow for large-bodied species currently at risk of disappearing within one or two generations (Vander Wal et al., 2013). Therefore we do not consider evolutionary adaptive capacity further, although it remains a pertinent issue.

Vulnerability is the combination of intrinsic sensitivity and exposure to an extrinsic threatening process. A large body size or a slow life history per se will not mean that a species is necessarily at greater risk, unless the species is exposed to a threat. Many large-bodied marine fishes are at risk because they are heavily fished; many small-bodied freshwater species are at risk from habitat degradation and loss because they have small geographic ranges (Arthington et al., 2016). Small-bodied freshwater fishes, however, are not necessarily at risk from overfishing (Reynolds et al., 2005). In birds, the largest species are at risk from overhunting and the smaller species are threatened by habitat degradation (Bennett and Owens, 1997). A species' response to one threat does not indicate its response or cotolerance to other threatening processes (Isaac and Cowlishaw, 2004; Vinebrooke et al., 2004; Graham et al., 2011); in addition, where more than one threatening process is operating, cumulative impacts are likely (Selkoe et al., 2015).

Trait-Based Predictions of Extinction Risk

By comparison with habitat loss, our understanding of the importance of hunting and fishing mortality is hindered by a lack of data on population-level mortality (Reynolds, 2003; Cowlishaw et al., 2009). Opportunities to understand mortality in marine species come from stock assessments, which in some cases estimate natural mortality and fishing mortality rates (F), and high-value species for which exposure to trade-driven extinction risk can be indexed by their market value (McClenachan et al., 2016). These species provide the best evidence for the interaction of exposure and sensitivity.

Traits Related to Exposure Fish behavior, particularly aggregation, can increase exposure by increasing catchability. Reef fishes that form spawning aggregations, salmon that return to natal rivers to spawn, and migratory fishes that follow their food sources are predictably concentrated in a small area. Indeed, many of the world's most commonly fished species (by weight) such as cod, pollock, mackerel, and herring migrate or aggregate to spawn, increasing their catchability (FAO, 2016). In a survey of exploited marine fishes, Sadovy de Mitcheson (2016) showed that global IUCN Red List status depends on (1) if the spawning season is short or long (indicating how predictably concentrated they are in time) and (2) if they aggregate to spawn (Fig. 3.3). The role of behavior in increasing exposure to threats like fishing is even more obvious when comparing two closely related, large-bodied Caribbean groupers: Nassau Grouper, *Epinephelus striatus*, and Red Grouper, *Epinephelus morio*. Historically, Nassau Grouper formed brief, large, and predictable aggregations (many of which have now vanished), whereas populations of Red Grouper remain viable in the Caribbean despite an ongoing fishery (Sadovy de Mitcheson, 2016). There is little doubt that this difference is due to the increased exposure of Nassau Grouper to fishing during their spawning aggregations. Thus behavior plays a strong role in determining whether fishes are at risk of overexploitation and extinction.

Traits Related to Sensitivity We now turn to the traits that predict intrinsic sensitivity, temporarily setting aside the issue of exposure. When fishing mortality is controlled for statistically, large-bodied species are the most likely

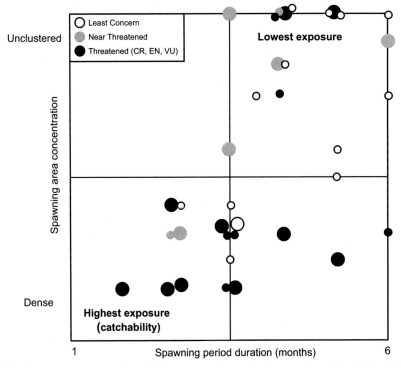

FIGURE 3.3 Thirty-six species of fished aggregating and nonaggregating fishes. Shade indicates International Union for the Conservation of Nature Red List Status. Size of point indicates body size: small points correspond to fish of less than 100 cm TL and large points are greater than 100 cm TL. *CR*, Critically Endangered; *EN*, Endangered; *VU*, Vulnerable. *Redrawn from Sadovy de Mitcheson, Y., 2016. Mainstreaming fish spawning aggregations into fishery management calls for a precautionary approach. BioScience 66, 295–306.*

to have declined steeply in temperate and coral-reef fish assemblages (Dulvy et al., 2000, 2004; Jennings et al., 1998, 1999a,b). However, when analyzing the response of 21 tuna populations to fishing, body size was slightly less important than time-related or "speed-of-life" traits such as growth rate or age at maturity (Juan-Jordá et al., 2013). Furthermore, environmental temperature sets the speed of life such that species with faster generation times are found in warmer habitats (Munch and Salinas, 2009). The relationship between temperature and speed of life suggests that species in cooler habitats and higher latitudes are intrinsically more prone to decline for a given level of mortality. This hypothesis is borne out in tunas: species with slower life histories such as the cold-water temperate bluefin tunas and deeper, tropical Bigeye Tuna (*Thunnus obesus*) are largely overfished (Collette et al., 2011; Juan-Jordá et al., 2011, 2015), whereas the tropical Yellowfin Tuna (*Thunnus albacares*) are more likely to be sustainable, despite their large body size. These patterns reveal an opportunity to understand the geographic patterning of intrinsic sensitivity. The connection between temperature and time-related traits suggests that biogeography provides the template for life history evolution (Southwood, 1977; Juan-Jordá et al., 2013). A challenge is that time-related traits—growth and maturation rates—can be more difficult to measure than morphological traits such as body size.So far we have discussed life histories in simplistic phenomenological terms of size- and time-related traits. We need a deeper understanding of life history sensitivity and how it relates to population regulation if we are to evaluate and justify these simple "rule-of-thumb" approaches. Variation in birth and death rates, which depend on life history, influences both the growth rate and the *compensatory capacity* of a population (its ability to compensate for additional mortality such as fishing; Kindsvater et al., 2016). Naturally, persistence of any population or species depends on the processes that regulate its population dynamics. Regulation arises from a combination of top-down processes, such as predation, and bottom-up processes, such as resources. Regulatory processes that depend on density can buffer populations against disturbance (the underlying principle enabling sustainable fisheries take). However, scientists have long understood that aquatic species have evolved multiple strategies for coping with their environment (Winemiller, 2005), which affects their density-dependent regulation.

To connect these insights to compensatory capacity, in Fig. 3.4 we introduce a conceptual framework to categorize species based on our knowledge of life

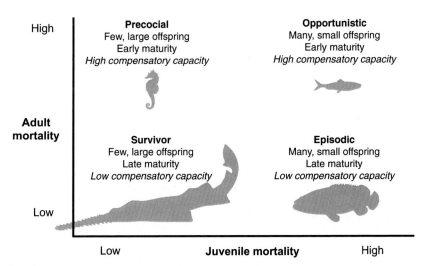

FIGURE 3.4 The Precocial, Opportunistic, Survivor, and Episodic (POSE) framework. Relative adult and juvenile mortality risk select for differences in age at maturation, offspring size, and number. We compared each species' ability to compensate for the same level of fishing mortality with simulation models of population dynamics, parameterized with demographic data from a representative species in each category (*gray silhouettes*). These analyses revealed that compensatory capacity increases with adult mortality. *Redrawn from Kindsvater, H.K., Mangel, M., Reynolds, J.D., Dulvy, N.K., 2016. Ten principles from evolutionary ecology essential for effective marine conservation. Ecology and Evolution 6, 2125–2138.*

history evolution. We categorize species as Precocial, Opportunistic, Survivor, and Episodic (which we refer to as the POSE framework; Kindsvater et al., 2016). Species such as forage fish mature early and capitalize on favorable conditions, attempting to reproduce before the environment changes (we call this an Opportunistic strategy). Alternatively, Episodic species such as cod grow slowly, mature late, and live a long time, allowing for a bet-hedging reproductive strategy. These species reproduce for many years, waiting for favorable environmental conditions that will allow their progeny to survive. Both Opportunistic and Episodic species have relatively high fecundity, producing thousands if not millions of progeny over their lifetime. Of the two, the slow-growing late-maturing Episodic species have lower compensatory capacity (Kindsvater et al., 2016), and they are much more likely to be overfished (Dulvy et al., 2014; Juan-Jordá et al., 2015). The relationship between fecundity and intrinsic sensitivity to extinction is weak (Dulvy et al., 2003; Hutchings et al., 2012), because populations of highly fecund species (Episodic and Opportunistic) are more likely to have strong density-independent juvenile mortality (reviewed in Kindsvater et al., 2016).

At the other end of the spectrum are species that have evolved under strong density-dependent regulation. Density-dependent competition among juveniles selects for large relative offspring size when there is a size advantage among competitors. For example, elasmobranchs such as skates are Survivors; they mature late and grow large. Seahorses are Precocial, meaning they have extreme parental investment in offspring, which allows them to mature early. These clades have similar fecundities, despite differences in body size. Historically, it has been unclear whether skates or seahorses are most vulnerable to overfishing. Using simulations that factored in the different dimensions of each of these life histories, in Kindsvater et al. (2016) we found that for the same level of fishing mortality, seahorses have a much greater intrinsic capacity to compensate than skates (setting aside the fact that seahorses may have elevated exposure to fishing mortality due to their habitat). In fact, large relative offspring size, which enables early maturation in Precocial species, confers the strongest compensatory capacity of any POSE category. Yet policy and management do not necessarily reflect this difference in sensitivity. Because declines in heavily traded charismatic seahorse species are more visible, their protection has received global support at CITES and they were listed under Appendix II before any other marine fish (Vincent et al., 2014). By stark contrast, measures to protect elasmobranchs have been absent or implemented only after their local extinction, as explained earlier. We should not protect seahorses less, but we should protect skates.

These deeper insights connecting life history and sensitivity are essential for conserving marine species where little is known about their population biology. For example, we know very little about the population sizes, movement, and behavior of most sea turtle species. Much of the initial efforts to conserve sea turtles focused on improving survival of hatchlings, as they are the life history stage that humans can see. Yet from Fig. 3.4 we can infer that sea turtles are Episodic species, as they mature late, yet produce hundreds or thousands of eggs in their lifetime. They have evolved under conditions with extremely low juvenile survival, whereas adult survival must be relatively high. Reducing incidental take of adult turtles was recognized to have more profound effect on their conservation in the long term (Crouse et al., 1993), and today there has been a large effort to reduce adult bycatch. This example underscores that the contributions of different life stages to population dynamics is a key component of optimizing conservation and management efforts.

How does the POSE framework relate to conservation status? In Fig. 3.5 we categorize fished species into POSE categories according to their age at maturation and fecundity. Age at maturation is inversely related to adult mortality rates, as delayed maturation is an indicator of low natural mortality (Kindsvater et al., 2016). Fecundity is related to juvenile survival; species with lower fecundity tend to have greater survival. We can then add the IUCN Red List Status, where available. The analysis in Fig. 3.5 highlights

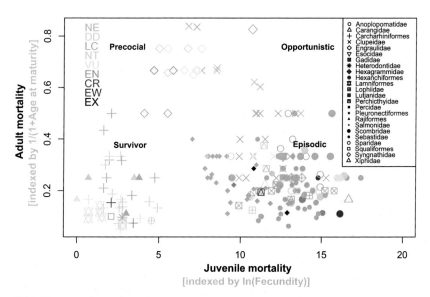

FIGURE 3.5 The relationship between life history and Precocial, Opportunistic, Survivor, and Episodic (POSE) category for 24 clades of marine fishes representing 204 species. The natural log of fecundity (x-axis) is assumed to correlate with juvenile mortality. The inverse of the age at maturity, standardized between 0 and 1, reflects adult mortality risk, as increased adult mortality leads to earlier maturation. Each point represents a species, color indicates most recent International Union for the Conservation of Nature (IUCN) Red List status, and character correspond to clade. Teleost life history data are species' means from FishBase (Froese and Pauly, 2016) and were collated using rfishbase (Boettiger et al., 2012). *CR*, Critically Endangered; *DD*, Data Deficient; *EW*, Extinct in the Wild; *EX*, Extinct; *EN*, Endangered; *LC*, Least Concern; *NE*, Not Evaluated; *NT*, Near Threatened; *VU*, Vulnerable. *Chondrichthyan life history data and all Red List status data are from the IUCN Red List website (*www.iucnredlist.org*).*

the disparity between life-history-driven sensitivity and conservation evaluation and status. Many, but not all, Episodic species are Least Concern, despite the fact that they are both intrinsically sensitive and fished (meaning they are exposed), yet others are Endangered. The Extinct and Extinct in the Wild species in this analysis are both European whitefish endemic to estuaries with small natural ranges. Finally, many heavily fished species are Not Evaluated (NE), despite the availability of data from fisheries. We can infer that exposure is the missing link that can explain much of the variation in Red List Status.

With this framework in mind, we can return to the question of determining exposure. Accessibility to humans, proximity to centers of human population density, contributes to exposure to anthropogenic threats (Jennings and Polunin, 1995). Marine species with the highest exposure are highly catchable species in shallow, nearshore habitat, such as the Bumphead Parrotfish in Fiji. Large-bodied species in these habitats are the strongest candidates for elevated risk (Dulvy and Polunin, 2004). Species in inaccessible habitats, such as the deep ocean, far from port, will be protected regardless of their life history (Dulvy et al., 2014).

Quantitative Predictions of Extinction Risk

It is one thing to explain declines post hoc; it is another to predict extinction risk a priori. Although many paper titles claim to "predict" risk, in reality they are fitting trait models to "explain" risk. True prediction is an entirely different beast that again involves balancing the risks of true and false positives (Fig. 3.2). There are two kinds of prediction: within and beyond sample prediction. The former is a form of cross-validation that simply measures accuracy or the degree to which a model fit to a subset of data can explain the remaining data (Anderson et al., 2011). In the quest to assess the IUCN Red List Status of the world's animals, plants, and fungi, beyond-sample prediction is most useful. This is because the primary obstacle to the completion of this quest is the large number of Data Deficient (DD) species for which there are insufficient data to assess whether they are threatened or not, much less the specific IUCN Red List category.

The simplest approach to this challenge is to model the binomial probability that a species is safe (Least Concern = 0) versus threatened (CR, EN, or VU = 1) in a mixed-effects modeling framework (Dulvy et al., 2014; Field et al., 2009). Using linear models incorporating maximum body size and geographic distribution traits (representing exposure to fishing mortality), Dulvy et al. (2014) estimated that 68 of 396 DD chondrichthyan species are potentially threatened. The prediction accuracy can be calculated as the Area Under the Curve (AUC) of the relationship between false-positive rates (α or p-value) and true-positive rates (β). The estimated AUCs were in the range of 77% with moderate explanatory power $R^2 = 0.3$, which is a good start but better statistical tools are available (Dulvy et al., 2014).

Two more-sophisticated approaches hold promise for predicting the IUCN status of Data Deficient species but require considerably more data. The "simplest" requires a phylogeny and distribution maps of all species, including the DD species. The premise of the approach is that sensitivity-related traits are phylogenetically clustered, whereas exposure is likely to be geographically clustered. By this reasoning, extinction risk depends on the combination of phylogenetic and geographic proximity. Data Deficient species that are related to and geographically near Critically Endangered species are also likely to be in the same IUCN category. Using this approach, scientists found that 331 of the 483 Data Deficient mammals might be in one of the IUCN threat categories; AUC was 0.9 and $R^2 = 0.4$ (Jetz and Freckleton, 2015).

In the second approach, new hierarchical statistical tools offer the opportunity to spread information from data-rich to Data Deficient taxa according to their shared characteristics, such as evolutionary lineage. This is most useful when quantifying population trajectories (i.e., fitting time-series models) in the case where *some* members of a given clade are data rich, meaning they are assessed regularly. These statistical methods can then fit model parameters from both direct information (e.g., Bayesian priors based on life history traits) and indirect information (e.g., abundance indices that are underpinned by interacting demographic processes; New et al., 2012; Matthiopoulos et al., 2014). Rather than fitting models to each species separately, data from multiple species may be used simultaneously (hierarchically), accounting for similarity among species based on their phylogeny, habitat, or geography (Maunder et al., 2015). This approach can capture dimensions of sensitivity and exposure that are not obvious to the naked eye, potentially predicting the risk of extinction of marine species that have so far escaped assessment. By leveraging multiple types of evolutionary, biogeographic, and socioeconomic information to predict species' extinction risk, this approach offers the chance to predict the looming threat of extinction in a more comprehensive way than has ever been accomplished.

Marine Extinctions Are Unmanaged

Much of the attention on the state of the world's fishes are focused on large-scale industrial fisheries and their effects on species that live in the waters of wealthy, developed nations. Indeed, there is increasing evidence of fishery sustainability in developed nations (Branch et al., 2011; Worm et al., 2009). Yet small-scale fisheries, an ambiguous category that can also include aquaculture, are far more important resources for the food security and livelihoods of developing nations (FAO, 2016). Half of the world's fish and 98 of every 100 fishers are part of small-scale fisheries that are far removed from national or global fisheries governance (FAO, 2016). Surveys have documented widespread unsustainability of fisheries of this type (Costello et al., 2012; Davidson et al., 2016).

Preventing Species' Extinctions

The primary tool we have to prevent extinction is a focus on saving species—this may sound circular but the reality is that most marine conservation efforts do not have the explicit aim of saving species! With few exceptions, many current conservation actions improve habitat quality or protect locations, but there has been little policy action to ensure that marine species do not go extinct (Redford et al., 2013). For example, the rapid rise in supersized marine protected areas (MPAs) is viewed as a conservation win (Lubchenco and Grorud-Colvert, 2015), but it is far from clear what the specific species conservation objectives are or which species will benefit (Edgar, 2011; Wilson, 2016). A focus on ecosystem function and services is important, but it does not save species. Ecosystem services are driven by numerical abundance and biomass, and hence the most abundant species, including invasive species, provide the bulk of function and services (Solan et al., 2004). Although monetizing ecosystems and biodiversity will unlock greater awareness of their value to governments, this does little to directly prevent extinction. To reiterate, to avoid extinctions we actually need to focus on saving species.

A focus on meeting the area coverage target of MPAs is important, but this focus alone currently does not save species. We are currently protecting the areas left over after industries have been allocated fishing and other extraction rights, which distracts from protected areas that are most valuable to biodiversity (Barnes, 2015; Devillers et al., 2015). The phenomenon of protecting areas, not saving species, has been described as the appearance that the naked emperor has clothes (Pressey, 2013). This is nowhere more prevalent than in the creation of "shark sanctuaries" that have borne few measurable benefits for sharks (Davidson, 2012). Almost one-third of the world's MPAs were designated for sharks by 2015, yet they do little to save those species most at risk of extinction—only 10 imperiled sharks and rays had more than 10% of their geographic range within a protected area (Davidson and Dulvy, 2017). The solution is to focus MPA expansion toward the outcome of avoiding extinctions (Aichi target 11), which would involve protecting those places that harbor the most endangered species, especially their most sensitive life stages (Devitt et al., 2015). Ideally the next wave of MPAs and associated CBD targets out to 2030 will capture a significant fraction of the remaining range of the most endangered marine species and set appropriate goals for their recovery (Venter et al., 2014).

A focus on fisheries sustainability is important for food security and ecosystem services (but it does not necessarily save species). We have seen local and regional extinction can happen both as a result of directed fishing (as in the case of Nassau Grouper) and due to incidental take (as in the case of Common Skate or South African sawfishes). A primary challenge is to minimize the mortality of threatened species occasionally taken alongside more productive target species. Minimizing incidental take for endangered species through improved bycatch management is one of the most effective ways for policy changes to prevent extinction. For example, the once-controversial 1994 law banning gillnets in Florida proved to be crucial in preventing the local extinction of the United States' remaining sawfish species, the Smalltooth Sawfish (Adams, 2000). Similar legislation in 1990 in California protected the Pacific Angelshark (*Squatina californica*), now one of the few angelshark populations that is not Threatened according to IUCN criteria (its Atlantic counterpart, *Squatina squatina*, was once found throughout the North Sea and eastern Atlantic but has been reduced to an isolated population in the Canary Islands) (Ferretti et al., 2015).

Serious issues in bycatch management continue to threaten vulnerable marine species. For example, Yellowfin Tuna managed by the Western Central Pacific Fisheries Council are MSC certified, yet the Oceanic Whitetip Shark (*Carcharhinus longimanus*), retained as valuable secondary catch in the fishery, is unmanaged and declining at a rate of 5% per year (Clarke et al., 2013). Often, insufficient data (and an evidentiary mindset) impede effective regulation of bycatch. Again, we need an alternative method of assessment and a precautionary approach to preventing species extinctions. As a last resort, CITES listings have been used to force trade regulations of bycaught species. However, a diagnosis or listing does not mean action will be taken to recover populations to sustainable levels. Instead of waiting until a species qualifies for CITES listing, we need preventative action.

CONCLUSION

The global future of marine species depends on our ability to pick and choose what species we eat, instead of indiscriminately scooping up whatever is available. Early maturing, fast-growing species hold the greatest promise for productive fisheries. Our first challenge in protecting threatened species is the identification and assessment of those at greatest risk. The next step is cooperation between parties with conservation-focused and fisheries-focused perspectives (Fig. 3.2C). Some progress in this direction has been made: fisheries agencies are no longer the sole custodians of ocean management, and Departments of Environment are expanding beyond terrestrial issues to confront marine conservation issues. The mandates of Multilateral Environmental Agreements, such as CITES and Convention of Migratory Species, and the need to deliver on the Convention of Biological Diversity's Aichi targets are measureable progress. Their broader remit and societal engagement has the power to drive fisheries improvements that can also help secure ecosystem services, alleviate poverty, and promote climate change adaptation, while also ensuring that species recover, rather than go extinct.

ACKNOWLEDGMENTS

We thank A.C. Stier and P.S. Levin for insightful comments that improved the quality of this chapter. Nicholas K. Dulvy was supported by Discovery and Accelerator grants from Natural Science and Engineering Research Council and a Canada Research Chair. Holly K. Kindsvater was supported by the US National Science Foundation (DEB-1556779).

REFERENCES

Adams, C., 2000. Since the Net Ban, Changes in Commercial Fishing in Florida. Florida Sea Grant College Program FLSGP-G-00–001.

Allison, E.H., Perry, A.L., Badjeck, M.-C., Neil Adger, W., Brown, K., Conway, D., Halls, A.S., Pilling, G.M., Reynolds, J.D., Andrew, N.L., Dulvy, N.K., 2009. Vulnerability of national economies to the impacts of climate change on fisheries. Fish and Fisheries 10, 173–196.

Anderson, S.C., Farmer, R.G., Ferretti, F., Houde, A.L.S., Hutchings, J.A., 2011. Correlates of vertebrate extinction risk in Canada. Bioscience 61, 538–549.

Anderson, S.C., Moore, J.W., McClure, M.M., Dulvy, N.K., Cooper, A.B., 2015. Portfolio conservation of metapopulations under climate change. Ecological Applications 25, 559–572.

Anderson, S.C., Cooper, A.B., Dulvy, N.K., 2013. Ecological prophets: quantifying metapopulation portfolio effects. Methods in Ecology and Evolution 4, 971–981.

Arthington, A.H., Dulvy, N.K., Gladstone, W., Winfield, I.J., 2016. Fish conservation in freshwater and marine realms: status, threats and management. Aquatic Conservation: Marine and Freshwater Ecosystems 26, 838–857.

Barnes, M., 2015. Aichi targets: protect biodiversity, not just area. Nature 526, 195.

Bell, G., Gonzalez, A., 2009. Evolutionary rescue can prevent extinction following environmental change. Ecology Letters 12, 942–948.

Bennett, P.M., Owens, I.P.F., 1997. Variation in extinction risk among birds; chance or evolutionary predisposition? Proceedings of the Royal Society B Biological Sciences 264, 401–408.

Bickford, D., Lohman, D.J., Sodhi, N.S., Ng, P.K.L., Meier, R., Winker, K., Ingram, K.K., Das, I., 2007. Cryptic species as a window on diversity and conservation. Trends in Ecology and Evolution 22, 148–155.

Boettiger, C., Lang, D.T., Wainwright, P.C., 2012. rfishbase: exploring, manipulating and visualizing FishBase data from R. Journal of Fish Biology 81, 2030–2039.

Branch, T.A., Jensen, O.P., Ricard, D., Ye, Y., Hilborn, R., 2011. Contrasting global trends in marine fishery status obtained from catches and from stock assessments. Conservation Biology 25, 777–786.

Brander, K., 1981. Disappearance of common skate *Raia batis* from irish sea. Nature 290, 48–49.

Butchart, S.H., Akçakaya, H.R., Chanson, J., Baillie, J.E., Collen, B., Quader, S., Turner, W.R., Amin, R., Stuart, S.N., Hilton-Taylor, C., 2007. Improvements to the red list index. PLoS One 2, e140.

Brooks, T.M., Butchart, S.H.M., Cox, N.A., Heath, M., Hilton-Taylor, C., Hoffmann, M., Kingston, N., Rodríguez, J.P., Stuart, S.N., Smart, J., 2015. Harnessing biodiversity and conservation knowledge products to track the Aichi targets and sustainable development goals. Biodiversity 16, 157–174.

Casey, J., Myers, R.A., 1998. Near extinction of a large, widely distributed fish. Science 281, 690–692.

Clarke, M.W., 2009. Sharks, skates and rays in the northeast Atlantic: population status, advice and management. Journal of Applied Ichthyology 25, 3–8.

Clarke, S.C., Harley, S.J., Hoyle, S.D., Rice, J.S., 2013. Population trends in Pacific Oceanic sharks and the utility of regulations on shark finning. Conservation Biology 27, 197–209.

Collette, B.B., Carpenter, K.E., Polidoro, B.A., Juan Jorda, M.J., Boustany, A., Die, D.J., Elfes, C., Fox, W., Graves, J., Harrison, L., et al., 2011. High value and long-lived – double jeopardy for tunas and billfishes. Science 333, 291–292.

Connors, B.M., Cooper, A.B., Peterman, R.M., Dulvy, N.K., 2014. The false classification of extinction risk in noisy environments. Proceedings of the Royal Society B Biological Sciences 281, 20132935.

Cooke, J.G., 2011. Application of CITES Listing Criteria to Commercially Exploited Marine Species. CITES, Geneva. Twenty-fifth meeting of the Animals Committee, 18–22 July 2011, AC25 Inf. 10 http://www.cites.org/common/com/AC/25/E25i-10.pdf.

Costello, C., Ovando, D., Hilborn, R., Gaines, S.D., Deschenes, O., Lester, S.E., 2012. Status and solutions for the world's unassessed fisheries. Science 338, 517–520.

Cowlishaw, G., Pettifor, R.A., Isaac, N.J.B., 2009. High variability in patterns of population decline: the importance of local processes in species extinctions. Proceedings of the Royal Society B Biological Sciences 276, 63–69.

Crouse, D.T., Crowder, L.B., Caswell, H., 1993. A stage-based population model for loggerhead sea turtles and implications for conservation. Ecology 68, 1412–1423.

Davidson, L.N.K., 2012. Shark sanctuaries: substance or spin? Science 338, 1538–1539.

Davidson, L.N.K., Krawchuk, M.A., Dulvy, N.K., 2016. Why have global shark and ray landings declined: improved management or overfishing? Fish and Fisheries 17, 438–458.

Davidson, L.N.K., Dulvy, N.K., 2017. Global marine protected areas to prevent extinctions. Nature Ecology & Evolution 1, 0040.

Dayton, P., 1998. Reversal of the burden of proof in fisheries management. Science 279, 821–822.

d'Eon-Eggertson, F., Dulvy, N.K., Peterman, R.M., 2015. Reliable identification of declining populations in an uncertain world. Conservation Letters 8, 86–96.

del Monte-Luna, P., Castro-Aguirre, J.L., Brooke, B.W., de la Cruz-Aguero, J., Cruz-Escalona, V.H., 2009. Putative extinction of two sawfish species in Mexico and the United States. Neotropical Ichthyology 7, 509–512.

del Monte-Luna, P., Lluch-Belda, D., Serviere-Zaragoza, E., Carmona, R., Reyes-Bonilla, H., Aurioles-Gamboa, D., Castro-Aguirre, J.L., Próo, S.A.G.D., Trujillo-Millán, O., Brook, B.W., 2007. Marine extinctions revisited. Fish and Fisheries 8, 107–122.

Devillers, R., Pressey, R.L., Grech, A., Kittinger, J.N., Edgar, G.J., Ward, T., Watson, R., 2015. Reinventing residual reserves in the sea: are we favouring ease of establishment over need for protection? Aquatic Conservation: Marine and Freshwater Ecosystems 25, 480–504.

Devitt, K.R., Adams, V.M., Kyne, P.M., 2015. Australia's protected area network fails to adequately protect the world's most threatened marine fishes. Global Ecology and Conservation 3, 401–411.

Diamond, J.M., 1987. Extant unless proven extinct? or, extinct unless proven extant. Conservation Biology 1, 77–79.

Dulvy, N.K., Davidson, L.N.K., Kyne, P.M., Simpfendorfer, C.A., Harrison, L.R., Carlson, J.K., Fordham, S.V., 2016. Ghosts of the coast: global extinction risk and conservation of sawfishes. Aquatic Conservation: Marine and Freshwater Ecosystems 26, 134–153.

Dulvy, N.K., Forrest, R.E., 2010. Life histories, population dynamics, and extinction risks in chondrichthyans. In: Carrier, J.C., Musick, J.A., Heithaus, M.R. (Eds.), Sharks and Their Relatives II: Biodiversity, Adaptive Physiology, and Conservation. CRC Press, Boca Raton, pp. 635–676.

Dulvy, N.K., Fowler, S.L., Musick, J.A., Cavanagh, R.D., Kyne, P.M., Harrison, L.R., Carlson, J.K., Davidson, L.N.K., Fordham, S., Francis, M.P., Pollock, C.M., Simpfendorfer, C.A., Burgess, G.H., Carpenter, K.E., Compagno, L.V.J., Ebert, D.A., Gibson, C., Heupel, M.R., Livingstone, S.R., Sanciangco, J.C., Stevens, J.D., Valenti, S., White, W.T., 2014. Extinction risk and conservation of the world's sharks and rays. eLIFE 3, e00590.

Dulvy, N.K., Jennings, S.J., Goodwin, N.B., Grant, A., Reynolds, J.D., 2005. Comparison of threat and exploitation status in Northeast Atlantic marine populations. Journal of Applied Ecology 42, 883–891.

Dulvy, N.K., Metcalfe, J.D., Glanville, J., Pawson, M.G., Reynolds, J.D., 2000. Fishery stability, local extinctions and shifts in community structure in skates. Conservation Biology 14, 283–293.

Dulvy, N.K., Pinnegar, J.K., Reynolds, J.D., 2009. Holocene extinctions in the sea. In: Turvey, S.T. (Ed.), Holocene Extinctions. Oxford University Press, Oxford, pp. 129–150.

Dulvy, N.K., Polunin, N.V.C., 2004. Using informal knowledge to infer human-induced rarity of a conspicuous reef fish. Animal Conservation 7, 365–374.

Dulvy, N.K., Polunin, N.V.C., Mill, A.C., Graham, N.A.J., 2004. Size structural change in lightly exploited coral reef fish communities: evidence for weak indirect effects. Canadian Journal of Fisheries and Aquatic Sciences 61, 466–475.

Dulvy, N.K., Reynolds, J.D., 2009. Biodiversity: skates on thin ice. Nature 462, 417.

Dulvy, N.K., Sadovy, Y., Reynolds, J.D., 2003. Extinction vulnerability in marine populations. Fish and Fisheries 4, 25–64.

Edgar, G.J., 2011. Does the global network of marine protected areas provide an adequate safety net for marine biodiversity? Aquatic Conservation-Marine and Freshwater Ecosystems 21, 313–316.

Everett, B.I., Cliff, G., Dudley, S.F.J., Wintner, S.P., van der Elst, R.P., 2015. Do sawfish *Pristis* spp. represent South Africa's first local extirpation of marine elasmobranchs in the modern era? African Journal of Marine Science 37, 275–284.

FAO, 2016. Small-scale Fisheries – Web Site. From Catch to Consumer. FI Institutional Websites. FAO Fisheries and Aquaculture Department, Rome. http://www.fao.org/fishery/topic/16610/en.

Ferretti, F., Morey, G., Serena, F., Mancusi, C., Fowler, S.L., Dipper, F., Ellis, J.E., 2015. Squatina squatina. The IUCN Red List of Threatened Species 2015: e.T39332A48933059.

Fernandez-Carvalho, J., Imhoff, J.L., Faria, V.V., Carlson, J.K., Burgess, G.H., 2014. Status and the potential for extinction of the largetooth sawfish *Pristis pristis* in the Atlantic Ocean. Aquatic Conservation: Marine and Freshwater Ecosystems 24, 478–497.

Field, I.C., Meekan, M.G., Buckworth, R.C., Bradshaw, C.J.A., 2009. Susceptibility of sharks, rays and chimaeras to global extinction. Advances in Marine Biology 56, 275–363.

Froese, R., Pauly, D., 2016. FishBase. World Wide Web electronic publication. www.fishbase.org.

Graham, N.A., Chabanet, P., Evans, R.D., Jennings, S., Letourneur, Y., Aaron Macneil, M., McClanahan, T.R., Ohman, M.C., Polunin, N.V., Wilson, S.K., 2011. Extinction vulnerability of coral reef fishes. Ecology Letters 14, 341–348.

Hammond, T.R., Ellis, J.R., 2005. Bayesian assessment of Northeast Atlantic spurdog using a stock production model, with prior for intrinsic population growth rate set by demographic methods. Journal of the Northwest Atlantic Fisheries Science 35, 299–308.

Hawkins, J.P., Roberts, C.M., Clark, V., 2000. The threatened status of restricted-range coral reef fish species. Animal Conservation 3, 81–88.

Hilborn, R., Quinn, T.P., Schindler, D.E., Rogers, D.E., 2003. Biocomplexity and fisheries sustainability. Proceedings of the National Academy of Sciences of the United States of America 100, 6564–6568.

Holden, M.J., 1974. Problems in the rational exploitation of elasmobranch populations and some suggested solutions. In: Harden Jones, F.R. (Ed.), Sea Fisheries Research. ELEK Science, London, pp. 117–137.

Holden, M.J., 1992. The Common Fisheries Policy: Origin, Evaluation and Future. Fishing News Books, London, p. 274.

Hutchings, J.A., Butchart, S.H.M., Collen, B., Schwartz, M.K., Waples, R.S., 2012. Red flags: correlates of impaired species recovery. Trends in Ecology and Evolution 27, 542–546. http://dx.doi.org/10.1016/j.tree.2012.06.005.

Iglésias, S.P., Toulhout, L., Sellos, D.P., 2010. Taxonomic confusion and market mislabelling of threatened skates: important consequences for their conservation status. Aquatic Conservation—Marine and Freshwater Ecosystems 20, 319–333.

Isaac, N.J., Cowlishaw, G., 2004. How species respond to multiple extinction threats. Proceedings of the Royal Society B Biological Sciences 271, 1135–1141.

IUCN, 2014. Guidelines for Using the IUCN Red List Categories and Criteria. IUCN Species Survival Commission, Gland, Switzerland, p. 87.

Jackson, J.B.C., 1997. Reefs since columbus. Coral Reefs 16, S23–S32.

Jennings, S., Polunin, N.V.C., 1995. Relationships between catch and effort in Fijian multispecies reef fisheries subject to different levels of exploitation. Fisheries Management and Ecology 2, 89–101.

Jennings, S., Greenstreet, S.P.R., Reynolds, J.D., 1999a. Structural change in an exploited fish community: a consequence of differential fishing effects on species with contrasting life histories. Journal of Animal Ecology 68, 617–627.

Jennings, S., Reynolds, J.D., Mills, S.C., 1998. Life history correlates of responses to fisheries exploitation. Proceedings of the Royal Society B Biological Sciences 265, 333–339.

Jennings, S., Reynolds, J.D., Polunin, N.V.C., 1999b. Predicting the vulnerability of tropical reef fishes to exploitation using phylogenies and life histories. Conservation Biology 13, 1466–1475.

Jetz, W., Freckleton, R.P., 2015. Towards a general framework for predicting threat status of data-deficient species from phylogenetic, spatial and environmental information. Philosophical Transaction of the Royal Society B Biological Sciences 370, 20140016.

Juan-Jordá, M.J., Mosqueira, I., Cooper, A.B., Freire, J., Dulvy, N.K., 2011. Global population trajectories of tunas and their relatives. Proceedings of the National Academy of Sciences of the USA 108, 20650–20655.

Juan-Jordá, M.J., Mosqueira, I., Freire, J., Dulvy, N.K., 2013. Life in 3-D: life history strategies of tunas, bonitos and mackerels. Reviews in Fish Biology and Fisheries 23, 135–155.

Juan-Jordá, M.J., Mosqueira, I., Freire, J., Dulvy, N.K., 2015. Population declines of tuna and relatives depend on their speed of life. Proceedings of the Royal Society B Biological Sciences 282, 20150322.

Keith, D.A., Burgman, M.A., 2004. The Lazarus effect: can the dynamics of extinct species lists tell us anything about the status of biodiversity? Biological Conservation 117, 41–48.

Keith, D., Akçakaya, H.R., Butchart, S.H.M., Collen, B., Dulvy, N.K., Holmes, E.E., Hutchings, J.A., Keinath, D., Schwartz, M.K., Shelton, A.O., et al., 2015. Temporal correlations in population trends: conservation implications from time-series analysis of diverse animal taxa. Biological Conservation 192, 247–257.

Kindsvater, H.K., Mangel, M., Reynolds, J.D., Dulvy, N.K., 2016. Ten principles from evolutionary ecology essential for effective marine conservation. Ecology and Evolution 6, 2125–2138.

Kulka, D.W., Frank, K.T., Simon, J.E., 2002. Barndoor Skate in the Northwest Atlantic off Canada: Distribution in Relation to Temperature and Depth Based on Commercial Fisheries Data. Department of Fisheries and Oceans, Canadian Science Advisory Secretariat Research Document. 2002/073, 17 pp.

Levin, P.S., Dufault, A., 2010. Eating up the food web. Fish and Fisheries 11, 307–312.

Lubchenco, J., Grorud-Colvert, K., 2015. Making waves: the science and politics of ocean protection. Science 350, 382–383.

Mace, G.M., Hudson, E.J., 1999. Attitudes towards sustainability and extinction. Conservation Biology 13, 242–246.

Mace, P.M., O'Criodain, C., Rice, J.C., Sant, G.J., 2014. Conservation and risk of extinction of marine species. In: Garcia, S.M., Rice, J.C., Charles, A.T.J.W.S. (Eds.), Governance of Marine Fisheries and Biodiversity Conservation: Interaction and Co-evolution, first ed.. John Wiley & Sons Ltd., London, pp. 181–194.

Matsuda, H., Yahara, T., Uozumi, Y., 1997. Is tuna critically endangered? Extinction risk of a large and overexploited population. Ecological Research 12, 345–356.

Matthiopoulos, J., Cordes, L., Mackey, B., Duck, C., Thompson, D., Thompson, P., 2014. State–space modelling reveals proximate causes of harbour seal population declines. Oecologia 174, 151–162.

Maunder, M.M., Deriso, R.B., Hanson, C.H., 2015. Use of state-space population dynamics models in hypothesis testing: advantages over simple log-linear regressions for modeling survival, illustrated with application to longfin smelt (*Spirinchus thaleichthys*). Fisheries Research 164, 102–111.

McClenachan, L., Cooper, A.B., 2008. Extinction rate, historical population structure and ecological role of the Caribbean monk seal. Proceedings of the Royal Society B: Biological Sciences 275, 1351–1358.

McClenachan, L., Cooper, A.B., Carpenter, K.E., Dulvy, N.K., 2012. Extinction risk and bottlenecks in the conservation of charismatic marine species. Conservation Letters 5, 73–80.

McClenachan, L., Cooper, A.B., Dulvy, N.K., 2016. Rethinking trade-driven extinction risk in marine and terrestrial megafauna. Current Biology 26, 1640–1646.

Miloslavich, P., Webb, T.J., Snelgrove, P., Vanden Berghe, E., Kaschner, E., Halpin, P.N., Reeves, R.R., Lascelles, B., Tarzia, M., Wallace, B.P., et al., 2016. Chapter 35 Extent of assessment of marine biological diversity. In: Inniss, L., Simcock, A. (Eds.), The First Global Integrated Marine Assessment: World Ocean Assessment I. United Nations, p. 58.

Munch, S.B., Salinas, S., 2009. Latitudinal variation in lifespan within species is explained by the metabolic theory of ecology. Proceedings of the National Academy of Sciences of the United States of America 106, 13860–13864.

National Marine Fisheries Service, 2009. Recovery Plan for Smalltooth Sawfish (*Pristis pectinata*). Silver Spring, Maryland, p. 648. Smalltooth Sawfish Recovery Team for the National Marine Fisheries Service.

New, L.F., Buckland, S.T., Redpath, S., Matthiopoulos, J., 2012. Modelling the impact of hen harrier management measures on a red grouse population. Oikos 121, 1061–1072.

Pauly, D., 1995. Anecdotes and the shifting baseline syndrome of fisheries. Trends In Ecology & Evolution 10, 430.

Peterman, R.M., M'Gonigle, M., 1992. Statistical power analysis and the precautionary principle. Marine Pollution Bulletin 24, 231–234.

Punt, A.E., 2000. Extinction of marine renewable resources: a demographic analysis. Population Ecology 42, 19–27.

Pressey, R.L., 2013. Australia's New Marine Protected Areas: Why They Won't Work. Available at: https://theconversation.com/australias-new-marine-protected-areas-why-they-wont-work-11469.

Randall, J.E., 1972. A revision of the labrid fish genus *Anampses*. Micronesica 8, 151–195.

Redford, K.H., Padoch, C., Sunderland, T., 2013. Fads, funding, and forgetting in three decades of conservation. Conservation Biology 27, 437–438.

Reynolds, J.D., 2003. Life histories and extinction risk. In: Gaston, K.J., Blackburn, T.J. (Eds.), Macroecology. Blackwell Publishing, Oxford, pp. 195–217.

Reynolds, J.D., Mace, G.M., 1999. Risk assessments of threatened species. Trends in Ecology and Evolution 14, 215–217.

Reynolds, J.D., Webb, T.J., Hawkins, L.A., 2005. Life history and ecological correlates of extinction risk in European freshwater fishes. Canadian Journal of Fisheries and Aquatic Sciences 62, 854–862.

Ricard, D., Minto, C., Jensen, O.P., Baum, J.K., 2012. Examining the knowledge base and status of commercially exploited marine species with the RAM Legacy Stock Assessment Database. Fish and Fisheries 13, 380–398.

Roberts, C.M., Hawkins, J.P., 1999. Extinction risk in the sea. Trends in Ecology and Evolution 14, 241–246.

Rogers, S.I., Ellis, J.R., 2000. Changes in the demersal fish assemblages of British coastal waters during the 20th century. International Council for Exploration of the Seas, Journal of Marine Science 57, 866–881.

Russell, B.C., Craig, M.T., 2013. *Anampses viridis* Valenciennes 1840 (Pisces: Labridae)—a case of taxonomic confusion and mistaken extinction. Zootaxa 3722, 83–91.

Sadovy de Mitcheson, Y., 2016. Mainstreaming fish spawning aggregations into fishery management calls for a precautionary approach. BioScience 66, 295–306.

Salomon, A.K., Gaichas, S.K., Jensen, O.P., Agostini, V.N., Sloan, N.A., Rice, J., McClanahan, T.R., Ruckelshaus, M.H., Levin, P.S., Dulvy, N.K., Babcock, E.A., 2011. Bridging the divide between fisheries and marine conservation science. Bulletin of Marine Science 87, 251–274.

Selkoe, K.A., Blenckner, T., Caldwell, M.R., Crowder, L.B., Erickson, A.L., et al., 2015. Principles for managing marine ecosystems prone to tipping points. Ecosystem Health and Sustainability 1, 1–18.

Shiffman, D.S., Hammerschlag, N., 2016. Shark conservation and management policy: a review and primer for non-specialists. Animal Conservation 19, 401–412.

Simpfendorfer, C.A., Dulvy, N.K., 2017. Bright spots of sustainable shark fishing. Current Biology 27, R97–R98.

Solan, M., Cardinale, B.J., Downing, A.L., Engelhardt, K.A.M., Ruesink, J.L., Srivastava, D.S., 2004. Extinction and ecosystem function in the marine benthos. Science 306, 1177.

Southwood, T.R.E., 1977. Habitat template for ecological strategies. Journal of Animal Ecology 46, 337–365.

Thurstan, R.H., McClenachan, L., Crowder, L.B., Drew, J.A., Kittinger, J.N., Levin, P.S., Roberts, C.M., Pandolfi, J.M., 2015. Filling historical data gaps to foster solutions in marine conservation. Ocean & Coastal Management 115, 31–40.

Turner 2nd, B.L., Kasperson, R.E., Matson, P.A., McCarthy, J.J., Corell, R.W., Christensen, L., Eckley, N., Kasperson, J.X., Luers, A., Martello, M.L., et al., 2003. A framework for vulnerability analysis in sustainability science. Proceedings of the National Academy of Science United States of America 100, 8074–8079.

Vander Wal, E., Garant, D., Festa-Bianchet, M., Pelletier, F., 2013. Evolutionary rescue in vertebrates: evidence, applications and uncertainty. Philosophical Transactions of the Royal Society B: Biological Sciences 368, 20120090.

Veitch, L., Dulvy, N.K., Koldewey, H., Lieberman, S., Pauly, D., Roberts, C.M., Rogers, A.D., Baillie, J.E.M., 2012. Avoiding empty ocean commitments at Rio +20. Science 336, 1383–1385.

Venter, O., Fuller, R.A., Segan, D.B., Carwardine, J., Brooks, T., Butchart, S.H.M., Di Marco, M., Iwamura, T., Joseph, L., O'Grady, D., et al., 2014. Targeting global protected area expansion for imperiled biodiversity. PLoS Biology 12, e1001891.

Vincent, A.C.J., Sadovy de Mitcheson, Y.J., Fowler, S.L., Lieberman, S., 2014. The role of CITES in the conservation of marine fishes subject to international trade. Fish and Fisheries 15, 563–592.

Vinebrooke, R.D., Cottingham, K.L., Norberg, J., Scheffer, M., Dodson, S.I., Maberly, S.C., Sommer, U., 2004. Impacts of multiple stressors on biodiversity and ecosystem functioning: the role of species co-tolerance. Oikos 104, 451–457.

Walker, P., Hislop, J., 1998. Sensitive skates or resilient rays? Spatial and temporal shifts in ray species composition in the central and north-western North Sea between 1930 and the present day. ICES Journal of Marine Science: Journal du Conseil 55, 392–402.

Webb, T.J., Mindel, B.L., 2015. Global patterns of extinction risk in marine and non-marine systems. Current Biology 25, 506–511.

Wilson, B., 2016. Might marine protected areas for mobile megafauna suit their proponents more than the animals? Aquatic Conservation-Marine and Freshwater Ecosystems 26, 3–8.

Winemiller, K.O., 2005. Life history strategies, population regulation, and implications for fisheries management. Canadian Journal of Fisheries and Aquatic Sciences 62, 872–885.

Wolff, W.J., 2000. Causes of extirpations in the Wadden Sea, an estuarine area in the Netherlands. Conservation Biology 14, 876–885.

Worm, B., Hilborn, R., Baum, J.K., Branch, T.A., Collie, J.S., Costello, C., Fogarty, M.J., Fulton, E.A., Hutchings, J.A., Jennings, S., et al., 2009. Rebuilding global fisheries. Science 325, 578–585.

Chapter 4

How Can the Oceans Help Feed 9 Billion People?

John Z. Koehn[1], Edward H. Allison[1], Nicole Franz[2], Esther S. Wiegers[2]

[1]University of Washington, Seattle, WA, United States; [2]Food and Agriculture Organization of the United Nations, Rome, Italy

INTRODUCTION

The ocean, like the land, is being transformed by large-scale biophysical, climatological, and anthropogenic drivers. Humans, clearly, no longer fill the hunter/fisher-gatherer role we occupied for 90% of our history; we have transformed so much of the Earth that some posit we have shifted global epochs from the Holocene to the Anthropocene (Steffen et al., 2011). While proactive management shows a pathway forward for fisheries, 63% of fish populations have been reduced to levels below those likely to generate sustainable harvest rates; many of these fisheries overlap with human communities most vulnerable to food security concerns (Costello et al., 2012). While global fish yields are likely near their limit for sustained exploitation, fish consumption continues to increase, supplemented by aquaculture. Furthermore, the hyper-connectivity induced by global trade ensures that decisions made by economic powers in the developed world will impact many of their suppliers and price-takers in the developing world, many of whom are sensitive to changes in the food system (Steffen et al., 2011). Managing fisheries and aquaculture production for the sustainable harvest of all stocks is of utmost importance, and progress is being made in doing so (Costello et al., 2016). However, beneath the ceiling set by maximum sustainable production, there should also be a foundation in social responsibility to ensure that resource management decisions do not disadvantage the poorest and most vulnerable populations. This echoes calls that global environmental governance should consider a "safe and just operating space for humanity" bounded by a ceiling set by "planetary boundaries" on natural systems and a floor set by considerations of social equity and fundamental human rights, including the Right to Food (Dearing et al., 2014).

Conservation for the Anthropocene Ocean. http://dx.doi.org/10.1016/B978-0-12-805375-1.00004-0

It is estimated that by 2050, production of animal-source protein will need to double to meet anticipated demand from a more numerous, wealthier and more urbanized global population (Steinfeld et al., 2006). The fisheries and aquaculture[1] sectors will continue to play a crucial role in food security, nutrition, and health. They contribute directly by providing micronutrient-rich fish, shellfish, and other aquatic foods[2] to billions of people. They contribute indirectly by providing a source of full-time, part-time, or supplementary income for over 100 million people in fish capture and farming operations, as well as seafood processing, trading, retailing, and ancillary industries (HLPE, 2014). Income and trade revenues, in turn, contribute to growth and prosperity at local to national levels.

Despite the importance of its contributions, seafood, paradoxically, is not commonly integrated into national and regional food policies. From local to global scales, decision-makers across the fisheries, economic and social development, and public health sectors face the following questions: *How can the fisheries sector (i) better contribute to increasing intake of nutritious and safe food or to improving livelihoods, and (ii) effectively contribute to food production in the midst of potentially declining resources, population growth, environmental impacts, and competition over land and water from other users?* To answer these questions, the chapter develops a four-step process to identify relevant policies to address food security and fisheries governance challenges and opportunities at local, national, and regional levels:

1. Describe how the fisheries sector integrates with the food system and contributes to nutrition and health.
2. Identify fishery policy options and their implications for food security.
3. Develop effective food security policies in the fishery sector.
4. Evaluate decision-making processes and the capacity to facilitate change.

The four-step approach outlined here can be simultaneously applied at local, country, or regional levels, and developed to identify whether nutrition-sensitive fisheries sector policy could help feed a growing population.

---

1. The term "fisheries" generally refers to capture of fish from the wild, while "aquaculture" refers to fish that are raised under farm conditions, either in the sea, in freshwater, or in land-based systems, such as enclosed tanks. There is a continuum from totally wild to intensively farmed, with some species raised in hatcheries but released to the wild, others species with juveniles caught in the wild to be reared on farms, and numerous other variations. The term "the fisheries sector" when used generically in this report, refers to fisheries and aquaculture. While secondary sectors are normally not governed under the same management regimes as the fisheries themselves, for the purposes of this paper they are also included in this broad definition. In instances where the specific content addresses processors or processing, those particular terms will be used.

2. In the rest of this chapter, "fish" is used to refer to finfish and shellfish (mollusks and crustaceans). Other aquatic animals (e.g., echinoderms, amphibians, reptiles, marine mammals) are also included where relevant. Aquatic plants (mostly seaweeds) are not included unless specified.

Fisheries-food security linkages

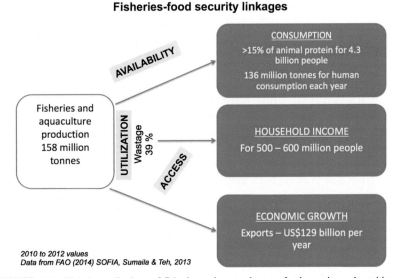

FIGURE 4.1   Global contributions of fisheries and aquaculture to food security and nutrition—main direct and indirect pathways.

Step 1: Describe How the Fisheries Sector Integrates With the Food System and Contributes to Nutrition and Health

Understanding the relative importance of the different pathways through which fisheries and aquaculture can contribute to food security, nutrition, and health (Fig. 4.1) is the first step in the process of identifying the most effective policies to support and enhance the sector's contributions.

Challenges to the efficacy of these pathways include weak linkages with global markets, food waste and food safety, poor management of capture fisheries, and negative impacts of global environmental change.

Pathways That Link the Fisheries Sector and Food Security

Fish is a primary source of animal protein in human diets. According to FAO estimates, fish accounted for 17% of the global population's intake of animal protein in 2013 (FAO, 2016). In addition to protein, certain micronutrients like iron and vitamin A—especially from small fish—are present in higher concentrations and in more bioavailable forms than they are in vegetables, fortified staples, and food supplements (Bogard et al., 2015; Thilsted et al., 2016). Continued sustainable supply of seafood is thus critical to food security: 1.39 billion people worldwide who are vulnerable to micronutrient deficiencies—defined as the intake levels for nutrients needed to meet the needs of a population[3]—currently secure nutrition via

---

3. In Golden et al.'s paper, a population was deemed nutritionally vulnerable if specific nutrient intake was less than double the estimated average requirement (EAR) (Golden et al., 2016). The EAR is defined as "the average daily nutrient intake level estimated to meet the requirements of half of the healthy individuals in a group" (Food and Nutrition Board, 2011).

their reliance on seafood. Thirty-three percent of the global population is Vitamin A deficient, 17% of the global population is zinc deficient, and 20% of pregnant women have iron-deficiency anemia. An additional 845 million people (11% of current global population) will become deficient in at least one of these nutrients by 2050 under projected impacts of climate change on fishery production potential, business-as-usual fishery management, and extrapolated trends for aquaculture growth (Golden et al., 2016). Box 4.1 summarizes recent evidence of the health benefits of fish consumption.

The growth of farmed-fish production has contributed to increased consumption of fish by lowering global fish prices, thereby increasing economic access for all but the poorest consumers (EU Seventh Framework Program, 2014). Farmed-fish production has been growing faster than all other food sectors and is expected to play an increasingly important role in the provision of essential nutrients from fish and fish-related products. However, differences in the nutritional value of farmed and wild fish, due in part to a switch to plant-based feeds as the cost of fish meal rises, suggest that one may not be a direct health substitute for the other (Beveridge et al., 2013).

Box 4.1 The Benefits of Fish Consumption to Human Health: Some Recent Evidence

Fish intake is associated with a 36% reduced mortality risk from heart disease, while consumption of 60 g fish/day is associated with a 12% reduction in mortality from all causes. Diets low in omega-3 fatty acids (which are largely seafood-derived) accounted for 1.4 million deaths in 2010 and are responsible for roughly 1% of the world's total burden of disease-related disability-adjusted life years (DALYs) (Lim et al., 2012). In addition, fish consumption in the USA is significantly associated with long-term weight loss. The benefits of fish are associated in part with high concentrations of minerals and vitamins that are easily digestible, as well as with essential fatty acids and animal protein (Thilsted et al., 2016).

Fish are rich in vitamin B_{12}, which is only found in animal-source foods and "is essential for multiple functions, including growth, brain function and nervous system maintenance" (Thilsted et al., 2016). Small fish are also rich in calcium, zinc, and iron: access to these minerals is often limited in less developed coastal regions. Fish are also a unique source of long chain omega-3 fatty acids which are associated with positive pre- and postnatal outcomes for children (Thilsted et al., 2016). High consumption of these fatty acids reduces the risk of pregnancy complications whereas low consumption during pregnancy increases the risk of the child's suboptimal neurodevelopmental outcomes, including cognition and fine motor skills (FAO and WHO, 2010; Thilsted et al., 2016).

For sources of this information, see: FAO, WHO, 2010. Report of the Joint FAO/WHO Expert Consultation on the Risks and Benefits of Fish Consumption. FAO Fisheries and Aquaculture Report No. 978. Food and Agriculture Organization of the United Nations, Rome. Available at: www.fao.org/docrep/014/ba0136e/ba0136e00.pdf; Thilsted, S.H., Thorne-Lyman, A., Webb, P., Bogard, J.R., Subasinghe, R., Phillips, M.J., Allison, E.H., 2016. Sustaining healthy diets: the role of capture fisheries and aquaculture for improving nutrition in the post-2015 era. Food Policy 61, 126–131. http://dx.doi.org/10.1016/j.foodpol.2016.02.005.

Fish and fish-related products also provide a primary source of income and livelihood for many communities worldwide: an estimated 45 million people are directly engaged in the production and harvesting sector. Altogether, it is estimated that fisheries and aquaculture support the livelihoods of some 10–12% of the world's population (FAO, 2014). Small-scale fisheries are especially important for women's livelihood, who constituted 50% of the small-scale fisheries workforce, compared with only 15% of all people engaged in the fisheries sector overall (FAO, 2015).

Small-scale fisheries are a key source for employment, income, and seafood. While large-scale operations land more fish, only 80% of this seafood is destined for direct human consumption compared to almost 100% in small-scale fisheries. In absolute global production terms, small- and large-scale fisheries contribute approximately the same amount to human consumption (HLPE, 2014). In addition, small-scale fisheries (including small-scale processing and marketing) employ 90% of the world's fisherfolk, the vast majority of whom are in developing countries (The World Bank, 2012). Globally, over 90% of those depending on commercial fisheries value chains operate in the small-scale fisheries sector (The World Bank, 2012), despite currently limited investment. For each million dollars that is invested, between 3 and 30 jobs are generated in large-scale fisheries as compared to between 200 and 10,000 in small-scale fisheries (HLPE, 2014). In addition, secondary sectors such as handling and processing provide employment for millions more.

Local and international fish trade, licenses to access fisheries' resources, and fees to lease sea areas for aquaculture also provide important contributions to national trade balances and government revenue. Further indirect contributions derive from profits gained in fisheries businesses ranging in size from household economies to multinational, vertically integrated seafood companies with diversified profits deriving from the capture, farming, processing, trading, and retail of fish.

Drivers Influencing the Potential of Seafood to Contribute to Food Security

Pathways linking the fisheries sector to nutrition are moderated by factors both external to and within the fishery system. Seafood production and supply to local and global markets are affected by four major drivers: Fishery management regimes and their ability to regulate harvest from target and nontarget species in seas and inland waters to sustainable; increasing contributions from the growing aquaculture sector, which is more closely linked to demand than is wild-fish production; global environmental change associated with anthropogenic greenhouse gas emissions, affecting distribution of wild capture fisheries and aquaculture production; and changes in supply chain management that affect access and postharvest losses, from investing in small-scale product marketing to products from currently discarded wastes.

Key drivers of changes to market demand include population or economic growth, urbanization, globalization of food markets and associated changes in consumption habits, awareness of health benefits, and perceptions of health risks of fish consumption. Pacific Island nations, for example, have some of the world's highest rates of obesity, due largely to a change in diet and lifestyles, driven by rapid rates of population growth and urbanization, scarcity of arable land, shifts in land and fishery resource tenure, and the growth of the global food trade. Seafood—rich in protein, essential fatty acids, vitamins, and minerals—and traditional root crops are being replaced in Pacific diets by cheap, energy-dense, and nutritionally poor imported foods such as white bread, sugar, fried chicken, and "mutton flaps" (Bell et al., 2015).

Fish is one of the most traded food commodities across the world, contributing not only to food security at the local level, but also at regional and international levels. Fish production and trade contribute significantly to global agricultural output. Fish production in 2012 exceeded 158 million metric tons, while the value of international fish trade amounted to USD 129 billion (HLPE, 2014). An increasing share of fish entering global markets derives from aquaculture, while much of the fish produced and traded within low-income countries still derive from capture fisheries. These production systems have complementary roles in meeting rising demand for fish and other products (such as fish feed and oil), and enhancing incomes and nutrition among smallholder producers, small-scale fishers and poor consumers. However, fisheries policies are increasingly orientated around value creation through export to urban and international markets. Capture fisheries governance reforms aim to prevent overfishing and capitalize on economies of scale to maximize economic output, largely by excluding access. This may bring benefits to resource conservation and trade, but risks decreasing the quantity of fish available on local markets and leaving many small-scale fishers excluded from livelihood opportunities (Béné et al., 2010). Aquaculture policies, similarly, tend to maximize productivity and economic efficiency (Hishamunda et al., 2009). If implemented too indiscriminately, these policies leave little room for promoting diversity of systems and alternative species, or accessibility of fish to poorer producers and consumers.

Availability of fish to consumers is affected by two other key issues: food waste and food safety. Loss and waste amount to 39% of fish landed globally (HLPE, 2014). Postharvest loss in low-income countries occurs mainly due to poor infrastructure, processing, refrigeration, and storage facilities; while in high-income countries waste is mainly seen at retail and consumer levels. In industrialized marine capture fisheries, a portion of catch from marine capture fisheries is discarded to the sea before landing; but, improved catch technologies and emerging markets for so-called "trash fish" have reduced seafood waste[4]. Policies such as the European Union's recent discard ban promise to further reduce waste.

---

4. This includes markets for novel species, surimi products, aquaculture, poultry, and pet feeds. Processing waste from factories may also enter these markets.

Food safety issues in fisheries result from postharvest spoilage, harvest from polluted waters, and from occurrence of environmental toxins—for example, ciguatera and paralytic and amnesiac shellfish poisoning, all of which originate from fish and shellfish that have ingested toxic microalgal species. Although some fish (e.g., longer lived, fatty-acid rich species swordfish) accumulate heavy metals like mercury and synthetic chemical pollutants risks to consumers from these pollutants usually arise only from high levels of consumption during vulnerable life stages including pregnant and lactating women (Dovydaitis, 2008). In aquaculture, safety issues can arise from the integration of livestock (particularly poultry and pigs) into fish farming systems, risking zoonotic disease transmission (Lima dos Santos and Howgate, 2011).

The potential for the fisheries sector to meet food security and livelihood needs of vulnerable and marginalized fisheries-dependent people are hampered by changes to the environments or habitats in which they are produced. Examples of these threats include pollution, the destruction of mangrove habitat critical for fish at early life stages, degradation of coral reefs and other sensitive areas, and pressures of development in ecologically important coastal zones (Halpern et al., 2008). For example, pressures on fisheries are expected to increase along with increasing human population densities inland and along coastlines in the tropics, and will alter the local environment if not mitigated by improved management (Sale et al., 2014). While aquaculture is growing rapidly, its role in meeting food security also faces similar environmental threats on land and water, including loss of biodiversity, consumer safety, and competition over resource access and use (Troell et al., 2014).

Gender-specific marginalization also presents a challenge, despite considerable reliance on the fisheries sector in certain regions. Research on the various roles women have from fishing to bookkeeping to processing is lacking (Sze Choo et al., 2008). Gender-specific understandings of the fisheries sector are important in a food security and nutrition context, as gender roles and norms determine access to resources and influence how revenues derived from fishing and aquaculture activities are distributed (Weeratunge et al., 2010). Outside of harvesting, recent gender research revealed that gender differentiation in the marketing and trading of fish limited access of women to higher value fish and markets, as well as social and economic resources (Fröcklin et al., 2013), indicating that women's livelihoods, and the benefits conferred to households continue to be negatively moderated by gender inequality.

Orientating policy in the fisheries and aquaculture sector may help to maintain and enhance its contributions to food security. Explicit nutrition-sensitive strategies in fisheries sector policy is currently lacking. To redress this neglect, policy makers in the fisheries sector must strike a balance between:

1. short-term gains and longer-term sustainability needs,
2. domestic priorities and international agreements signed by the country, and
3. diverse interest groups involved at local, regional, national, and international levels.

A broad spectrum of national policy instruments may be adopted to develop and regulate the fisheries sector, each with its own specific objectives, which may be competing or contradictory and could have different implications for food security and nutrition. Nutritionally sensitive fishery sector policymaking has the potential to substantially contribute to satisfying the nutrition demands facing a growing global population. Examples of promising policies will be examined in the Step 2. Identifying the challenges to develop food security fisheries pathways also pinpoints where policy changes could be most effective.

Step 2: Identify Fishery Policy Options and Their Implications for Food Security

Commonly Applied Fishery Policies

In most countries, the policy framework for managing capture fisheries is based on a simple model to match fishery production through harvest rates to the natural productivity of the system. Fishery management seeks to exploit fish stocks until they reach maximum sustained catch (e.g., total allowable catch programs) or maximum profit over the long term (e.g., catch share programs). Without regulations on entry, it has been shown that fishing effort tends to increase to levels beyond ecological sustainability, only stopping when fishing costs outweigh the benefits (Beddington et al., 2007). Fisheries policy sets out the priorities for the national fishery sector—for example, maximizing profitability of the sector, fish production, or maintaining as many fishing jobs as possible—and fishery assessments determine what level of fishing can be sustained at each of those target reference points for management. These fisheries policies have, in the past, been highly sectoral and poorly integrated with wider development initiatives (Thorpe et al., 2006), though efforts to make national policies[5] compatible with wider economic, social, and conservation goals have increased in recent years.

In aquaculture, most national policies set an aspirational production target and, if effective, they provide enabling policy and legislation for private sector (including small-holder) investment to contribute toward meeting that target (Ha and Bush, 2010). State and donor-supported aquaculture support programs aimed at the poorest have not been particularly successful, on the whole, but small- and medium-scale private enterprises have driven the rapid increase in aquaculture seen over the last 30 years, particularly in Asia (Belton and Little, 2011).

Different countries may pursue vastly different policy objectives, according to the size of their resource, its potential for generating macroeconomic

---

5. For example, the U.S. Magnuson Stevens Act's amendments to consider economic and social benefits and impacts when setting biological catch targets.

benefits, its importance to formal and informal employment, and the role fish plays in a nation's diet. Below, we outline examples from Africa and South Asia.

Export-based fisheries: Namibia and Angola have significant marine fisheries resources in the Benguela Current system (including sardines and hake), but do not have large populations with cultural preferences for marine fish in their diet. Neither do their industrial fisheries employ a large segment of the population, nor do they have significant small-scale marine fisheries. In these circumstances, leasing the right to fish to industrial fleets (domestic and foreign) and aiming to maximize the revenues (e.g., from license and access fees or tax payments to the government) relative to costs is often the preferred policy choice to achieve maximum economic yield. In Namibia, benefits derived from these macroeconomic fisheries development strategies then reach the poor through redistributive economic and social policy measures such as increased purchasing power (to secure alternative healthy foods) and welfare programs (Paterson et al., 2013). However, realized benefits appear to be lacking compared with these theoretical economic and social gains (Paterson et al., 2013).

Mixed export and artisanal fisheries: In Ghana and Senegal, there are off-shore fisheries caught by industrial vessels who pay the governments to legally access the territorial waters of these countries and nearshore fisheries caught by artisanal or small-scale fisheries, which have a higher likelihood to contribute to food security or alleviate poverty. Here, policies and management objectives vary among subsectors of the fishery, with an overall policy aim to minimize trade-offs among the conflicting objectives of maximizing revenues, catches, and employment. Aiming to maximize economic yield by granting or selling fishing rights only to larger-scale, export-oriented fisheries would likely lead to loss of employment and disruption of domestic marketing of fish, and may be politically unfeasible with reduce food security and positive health outcomes (Allison, 2011). Thus, policies in these countries are a trade-off between maximizing revenues, increasing fish availability for consumption, and increasing fish-related jobs.

Aquaculture-based regional production: Egypt is the largest aquaculture producer in Africa. The country farms tilapia and mullet to supply middle-class consumers in domestic and regional urban and rural markets, and has plans to continue to expand domestic supply (Hebisha and Fathi, 2014). Increasing production and profitability are Egypt's main policy objectives.

Aquaculture-based export and local production: In Bangladesh, aquaculture policy supports pond-based culture of fish to improve small-farm productivity and nutrition among the rural poor, small- and medium-enterprise aquaculture to supply domestic and regional urban markets, and the development of an export-oriented shrimp farming industry supplying developed-country markets (Allison, 2011). The challenge lies in balancing investments and support to these subsectors in ways that do not place them in competition with each other.

Linkages of Fishery Policy With Other Sectors and Scales

Fisheries policy and laws are usually set at national level. Although with decentralization prevalent in many countries, fisheries management has frequently been devolved to local or provincial levels, and fishing communities can be key partners in the management of the resource they depend upon. Fisheries policy does not occur in a vacuum, however, and it interacts with a range of other legal and policy instruments at multiple levels.

Fishing takes place within the broader frame of ocean governance under the UN Convention on the Law of the Sea, under which states have claimed Exclusive Economic Rights (including to fisheries) in the zone encompassing 200 nautical miles from the coast. In addition to national laws and policies and the UN Law of the Sea, fisheries and aquaculture are governed through a set of policy standards and guidelines such as the FAO Code of Conduct for Responsible Fisheries and the Voluntary Guidelines for Securing Sustainable Small-Scale Fisheries in the Context of Food Security and Poverty Eradication (Fig. 4.2). These instruments are endorsed by the member states of FAO through the Committee on Fisheries (COFI), which meets biennially. The codes and standards are meant to guide national governments, as well as other fisheries and aquaculture stakeholders.

Global, regional, and national laws and policies from other sectors (e.g., pollution, trade, food safety, environmental conservation, and development) also influence fisheries and aquaculture policies. There are also influential regional policy forums, such as the Inter-American Tropical Tuna Commission. Recently, human rights instruments have become influential in governing fisheries. Displacement and marginalization of the world's small-scale "fisherfolk" and the resulting loss of access is an increasing concern (Bennett et al., 2015). Recent media reports of slavery within shrimp value chains (Mason et al., 2015) caused public concern and are encouraging a shift in thinking and process by value chain players. Civil society organizations, such as the International Collective in Support of Fishworkers, have drawn on labor rights and various other provisions of international human rights law to campaign for justice for fish workers, and to oppose policy that seeks to privatize resources that are held in common by communities.

Challenges and Solutions in Contemporary Fishery Sector Policy

The complexity of the policy landscape risks incoherence but also presents multiple entry points for improvement of fisheries sector performance with respect to economic performance, maintenance of cultural practices, sustainability, food security, nutrition, and human health. A key part of mapping the sector's policies is thus to understand which of these other policy domains is influential in a particular country or for a particular issue.

A wave of reform has swept through fisheries policy over the last 20 years. Broadly, the "governance revolution" in fisheries is an ongoing

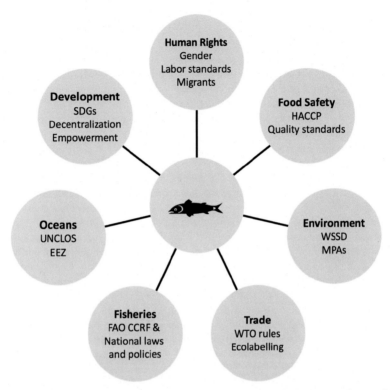

FIGURE 4.2 Examples of governance domains relevant to fisheries and aquaculture and its linkages to food security. *EEZ*, exclusive economic zone; *FAO CCRF*, Code of Conduct for Responsible Fisheries; *HACCP*, hazard at critical control point; *MPA, marine protected area; SDGs*, sustainable development goals; *UNCLOS,* UN Conference on the Law of the Sea; *VGSSF*, Voluntary Guidelines for Securing Sustainable Small-Scale Fisheries in the Context of Food Security and Poverty Eradication; *WSSD*, World Summit on Sustainable Development; *WTO*, World Trade Organization.

attempt to replace a system of largely unsuccessful efforts to manage conditions of access to a state-owned resource through licensing and technical measures, with a different system based on a combination of participatory local-level management, a variety of market-based instruments, and a set of global principles and codes of conduct. The overall goal of these reforms is to end the "race for fish," where each individual fishing enterprise has a limited quota share of the total catch allowed for the fishing season (e.g., a total allowable catch set by managers). This incentivizes participating fishermen to maximize their individual share of the resources in the state-managed or high-seas commons by racing other fishers to catch the last fish, placing lives of the fishermen at risk and causing economic inefficiency by oversupplying markets.

Box 4.2 Fish, Food Security and Nutrition Linkages – Knowledge and Data Gaps

Lack of data and monitoring: Data on the importance of fishing as a livelihood to food and nutrition security, its role in consumption, and its value to trade, is scarce. There is also an absence of a comprehensive database on nutrient composition of important species consumed and the number of people dependent on the fisheries and aquaculture sector for income and livelihood that are food insecure and malnourished. Furthermore, fish production may be underreported, as some of the sector's contribution to food security is informal (e.g., subsistence fisheries that are shared), and not communicated to government ministries.

Lack of research: There is a lack of ex-ante and ex-post evaluations of fisheries policies and investment programs, making it difficult to rigorously examine the impacts of key policy measures on household incomes, food security, nutrition, and health. One review of the state of our empirical understanding of linkages between fisheries and aquaculture, and food security and nutrition, concluded that a large and consistent literature demonstrates that fish contribute to nutrition and health; however, much of it is focused on Asia (specifically Cambodia and Bangladesh) and information from other low-income food-deficit countries (especially those in sub-Saharan Africa) is generally lacking (Béné et al., 2016). There is an additional lack of focus on vulnerable groups (e.g., infants, expectant mothers, the elderly) with respect to both the risks and benefits of fish consumption.

Complexity challenges research methodologies: The fish–food security–nutrition linkages are complex. Many interrelated factors determine the impact of fish on food security and nutrition: the price of fish, wages, and social security; intra-household allocation of food; competition for resources; insecurity of tenure or access to fishery resources; trade subsidies; environmental concerns; food safety concerns; and climate change impacts. Importantly, existing research does not allow the resolution of some key policy debates in the sector, such as whether the rise of aquaculture is resulting in a reduction in potential health benefits from fish consumption, and whether the global fish trade improves or undermines domestic food and nutrition security.

Adapted from Gillespie et al., 2015. Developing a Cross-Sectoral Policy Framework for Food and Nutrition Security: A Guide. Draft. FAO, Rome.

Step 3: Develop Effective Food Security Policies in the Fishery Sector

Existing fisheries policy is oriented toward commercial interests, concentrating mainly on balancing long-term exploitation of fish with its profitable exploitation. The importance of the sector for food security and nutrition is undervalued. With a few exceptions, specialist fisheries discussions tend to be mostly geared toward tackling issues of sustainable resource management and economic efficiency, while neglecting issues related to impact on food security, nutrition, and people's livelihoods (HLPE, 2014). Due to the overall lack of knowledge about and recognition of fish, food security, and nutrition linkages, as well as poor coordination across the respective policy domains, food security and nutrition concerns are rarely included in fisheries policy and research (Box 4.2).

Applying food security policies to the fishery sector requires consideration of supply- and demand-side innovation. Despite environmental strain and an increasing demand for seafood globally, increasing sustainable fish production is possible by considering more effective utilization of waste and discards. Alternative market arrangements might support fishermen and processor livelihoods variably, by potentially improving access to markets and increasing the value received for their efforts. Fisheries do not act in isolation when satisfying food security concerns. Cross-sectoral strategies can better include fisheries in nutrition-sensitive policymaking.

Sustainable Fish Production Amidst Increasing Demand and Environmental Stress

While advocating for greater inclusion of fish in healthy diets—and therefore greater coherence between food, health and fisheries sector policy—there is a need to be aware of the constraints the sector faces. In evaluating whether the fisheries sector can maintain or expand its contribution to food security, nutrition, and health, these constraints need to be identified and, where possible, assessed. These constraints range from rising sea level to high real-estate prices displacing seafood-related industry to competition with environmental or energy sector interests.

In capture fisheries, the task of meeting rising demand can be streamlined by making more efficient use of existing productive potential. This means looking for opportunities to reduce losses at sea (incidental killing of fish not targeted or kept, or fish rendered unfit for human use by destructive fishing gear), or spoilage of fish postharvest; and to improve utilization of fish by processing by-products that are currently considered "waste" (Box 4.3).

Nutrition-Sensitive Policies That Could Be Adopted in Fishery and Aquaculture Management and Policy

Adapting the idea of nutrition-sensitive agriculture to the fisheries sector provides a policy pathway to improving intake of fish by the malnourished—which includes nearly one billion people whose cognitive and physical development and health status is affected by micronutrient deficiencies and the half a billion people who are obese across more and less developed countries (Belton and Thilsted, 2014). Promoting the inclusion of fish in food-based strategies can address the "hidden hunger" of micronutrient deficiency as well as health concerns related to diets that are too high in refined carbohydrates and saturated fats. The potential role and importance of fish in nutrition strategies that build on people's existing food production and culinary systems needs to be promoted if it contributes to "food-based strategies" for addressing micronutrient deficiencies.

Fisheries can also contribute a vital line of nutrition to vulnerable populations or those in protracted crises or emergencies (Committee on World Food Security, 2015). There are a number of interlinked opportunities for enabling

Box 4.3 Examples of Initiatives to Reduce Wastage and Improve Utilization in Fisheries and Aquaculture

Reducing discards at sea: significant losses of fish at the harvesting stage partially result from using methods and gears that are destructive and/or not selective. This leads to capture of unsellable, unwanted and inedible products, which are subsequently discarded dead or debilitated. The volume of fish discards varies greatly between and within fisheries – making estimation of global discard volumes challenging – but the latest report published by FAO in 2005 on the issue estimates an 8 percent global discard rate of in the world's capture fisheries overall, with a lower rate of 3.7 percent for the small-scale fisheries sub-sector (Kelleher, 2005).

Improved utilization of tuna: an example linking food loss and waste reduction and food and nutrition security: The tuna canning and processing industries create large quantities of byproducts converted into for diversified products that either directly support food security or livelihoods. While uses vary by region, by-products of the Thai canning process include tuna meal, tuna oil, and tuna soluble concentrate. Similarly, the canning industry in the Philippines converts by-products to tuna meal, but also exports less desirable darker meats less desirable to alternative markets outside those generating demand like the raw fish sashimi sector. Tuna by-products from the fresh or frozen sector are utilized locally; heads and fins are used for soup, visceral organs are used for fish sauce production, and leftover meat and trimming are also consumed (Gamarro et al., 2013). Fish powders derived from tuna frames – the bones and tissue remaining after processing tuna – and other seafood processing byproducts rich in protein collagens, calcium, and phosphorous have also been studied for their capacity to fortify dishes for schoolchildren (Glover-Amengor et al., 2012).

vulnerable populations access, including: *improving the quality and quantity of fish supply*, which can improve diets of producers and consumers; *facilitating women's empowerment*, which improves maternal and child health and can ensure improved intra-household distribution of food, including fish; *promoting equitable trade and enhanced markets*, which can improve access by lowering price or offering higher quality seafood; and *including fish in nutrition programs targeting those at risk from malnutrition*, which includes bringing fish into school meal programs.

Policy analysis to identify possible food security, nutrition, and fishery sector synergies requires searching for existing successful and unsuccessful policy and programmatic interventions. Acknowledging the need for public health policymakers to actively engage with agricultural subsectors, the Second International Conference on Nutrition (ICN2) stated that "fisheries and aquaculture need to be addressed comprehensively through coordinated public policies" (FAO and WHO, 2014). The call for improved policy coordination, environmental protection, enhanced fish production, and reduced loss and waste represents a major opportunity to promote capture fisheries and aquaculture as a key nutrition-sensitive agricultural subsector, especially in their contribution of bioavailable

---

Box 4.4 Policy and Programmatic Interventions Aimed at Increasing Access and Utilization of Fish as Part of Diverse, Healthy Diets, and Contributing to Children's Physical and Cognitive Development

South Africa has had a primary school meal program since 1994 that continues to develop innovative approaches to improve child health and cognitive performance. Following a trial that showed the addition of omega-3 to children's diets improved their verbal learning and memory, a bread spread was developed from fish waste products in the major hake fishery to provide a rich source of long chain omega-3 polyunsaturated fatty acids. The idea was to give fatty acids to the schoolchildren in a form that would be more acceptable. An evaluation of the spread indicated that children who consumed it at school performed better in verbal learning and memory tests, including spelling and reading (Dalton et al., 2009).

Zambia's First 1000 Most Critical Days Program addresses malnutrition challenges facing women and children in the first 1000 days of the child's life. Limited intra-household surveys showed that women were far more likely to have eaten fish than other animal-source foods. The surveys also showed fish-derived nutrient contributions to children, though that contribution varied by age and by region.

---

minerals and vitamins, essential fatty acids, and animal protein as discussed in Box 4.1.

Fish is included in research on dietary needs for the "first 1000 days" of human life—from conception to second birthday—in Bangladesh, Cambodia, and India (Bogard et al., 2015). Trials and evaluations on child health and learning efficacy are underway in a number of countries (Box 4.4). These trials would benefit from policy coordination among fisheries, education and health ministries to scale out for highest impact among food-insecure and malnourished populations.

Nutrition education and improved access to fish have been promoted in populations where changing diets and/or lifestyles have led to a rise in noncommunicable diseases (e.g., heart disease or skeletal-muscular disorders associated with obesity). These initiatives could exclusively be focused on diet advice, or could be linked to community-led or community-private partnerships, enabling increased seafood supply to schools, hospitals, and religious institutions. For national "eat fish for health" campaigns, the popularity of TV cooking shows in semi-urbanized rural areas of developing countries offers novel opportunities, such as enlisting popular "celebrity chefs" who are influential opinion leaders that shape dietary habitats in developed and transitional countries.

Alternative Market Arrangements for Enabling Fisheries and Food Security Pathways

Improving access of small-scale fishers, aquaculturists, and processors to markets can help secure livelihoods for vulnerable communities faced with limited resource access by providing high-value buyers. For niche and high-value products with markets in developed countries (e.g., ornamental reef fish from Southeast Asia to EU and North America), programs primarily led by NGOs and multi-stakeholder groups help ensure that small-scale producers use resources sustainably, meet market standards, and are given fair prices. These approaches currently have limited application to worldwide markets to meet global demand, and have had mixed effectiveness (Sampson et al., 2015). One example of market-based reform that has made much progress is the limitation of shark fins to East Asia (Duggan, 2014), which shifted consumer purchasing away from shark fins by an appeal to ethical choice regarding whether or not to support an ecologically damaging and wasteful practice (often the rest of the shark is discarded). Clearly missing from current attempts to use market linkages for poverty reduction and sustainability is investment in fostering sustainable and ethical consumerism in developing and transitional countries, including China and Brazil. Shortening value chains from rural areas to regional urban centers has investment potential, especially for upgrading market value and promoting traceability in support of a move toward sustainable and ethical supply chains.

Finally, strengthening the connections between local supply and global markets can provide small-scale fisheries with additional livelihood options. Fisheries improvement programs, for example, aim to give producers access to sustainably minded high-value markets in return for a planned program of transition to sustainability. Unfortunately, such transitions may be slow and fisheries have gained access to these markets without delivering the promised "improvements" (Sampson et al., 2015).

Cross-Sectoral Strategies and Inclusion of the Fisheries Sector in Policy Reform

Policy forums considering these issues are increasingly engaging multiple economic sectors and stakeholders in planning the use of coastal and aquatic spaces (Le Cornu et al., 2014). These may include authorities for integrated coastal zone management, lake catchment, river basin and floodplain management, and marine spatial planning. Unfortunately, the fisheries sector is not always well represented in these forums: water resource and land-use planning tend to dominate lake and river catchment forums, and urban, industrial, tourism, and offshore oil and gas sectors tend to dominate coastal and marine spatial planning. In coral reef–dominated coastal areas of the tropics, marine biodiversity conservation interests increasingly drive the policy process. There is scope for integrating habitat and ecosystem conservation with improved fisheries governance, as functional ecosystems are required for healthy fisheries.

Organizations, government and nongovernmental entities creating nutrition-sensitive policies and interventions thus face a triple challenge: the fisheries sector needs to engage with the larger food security and nutrition sectors to create a joint agenda. Critical to national and regional policymaking, they must ensure nutrition issues remain on the agenda when considering all other stakeholder interests including trans-boundary negotiations over resource use and allocation.[6] A range of technical options may address some of these challenges and ensure the sector continues to supply nutrient-rich food at affordable prices. Many opportunities rest with the aquaculture sector, where there remains great potential for expanding efficient production and better-incorporating nutrition concerns. Competition between production and food security can be remediated in coastal and inland surface waters by investing in closed or recirculation systems like those used for shrimp aquaculture in Singapore and for salmon in Canada. Advances in plant- and insect-based feed technology can take pressure off nutritionally important wild-capture fish stocks (e.g., anchoveta) for fish-meal production. Multi-trophic aquaculture produces different species together in less intensive systems, such as growing algae (e.g., marketable seaweed) alongside herbivorous abalone who then fertilize the seaweed, to increase energy-use efficiency and reduce waste-treatment costs (Nobre et al., 2010). Developing nutrition-sensitive strategies still requires integration with wider policy objectives, and also requires successful understanding of political economy. The final section outlines how more effective nutrition-sensitive strategies could be successfully implemented.

Step 4: Evaluate Decision-Making Processes and the Capacity to Facilitate Change

This final step in assessing the scope for policy interventions to support integration of food security and nutrition in fisheries and aquaculture policy is to consider the overall policy process at relevant (usually national or regional) scales. Many policy prescriptions emanate from global think tanks and international organizations, or are the product of "lessons learned" from other countries, but their transferability should not be assumed. Changing policy is a political process and understanding what shapes and changes it is a necessary part of any reform. However, policy reform can have unintended consequences, and issues of distributive justice should be considered prior to implementation.

Identify and Engage Stakeholders in Policy Reform

An analysis of the potential for policy change to support transition to a more nutrition-focused fisheries sector begins with a cluster of questions to identify stakeholders, their interests, and their power to support or block policy change and policy implementation. Any policy reform should identify

---

6. One example of relevant regional forums where fisheries are engaged in such debates include the Mekong River Commission, which advises on the management of the water resources in the basin shared by Cambodia, China, PDR Lao, Thailand, and Vietnam. A particular concern on the Mekong River Delta is how hydropower developments upstream and their fisheries and food security consequences downstream.

policy-change champions and obstructers; local, national regional, and international influencers; and winners and losers under potential reform scenarios. Additionally, developing a clear understanding of how formal and informal policy processes are organized and how policy debates may include or exclude voices is essential to decision-making that reflects all dimensions of the policy reform space.

Fisheries and aquaculture policy is shaped by state and nonstate actors operating at multiple levels. Global standards, codes of conduct, and policy initiatives from FAO, UNEP, and other UN organizations have an important agenda-shaping function. These standards and codes are increasingly developed through extensive stakeholder consultations, including the recent FAO Voluntary Guidelines for Securing Sustainable Small-Scale Fisheries in the Context of Food Security and Poverty Reduction. The last one represents a global consensus on small-scale fisheries governance and development, and is the result of a long consultation process directly engaging over 4000 stakeholders from over 100 countries. In fisheries, the key process of ratifying these initiatives is the FAO's biennial Committee on Fisheries, comprising representatives from all 193 FAO member states. Policy initiatives are debated and, if approved, provide a mandate for their implementation through national and regional policies. Processes such as these work toward the goal of including food security and nutrition in fisheries sector policy reforms (Ratner and Allison, 2012).

Funding for fisheries policy reform and implementation has historically been provided by governments and bilateral and multilateral donors and investors such as the World Bank. In recent decades, philanthropic organizations and the private sector have rapidly grown into the space. From 2004 to 2009, philanthropic spending in marine and ocean conservation grew from $60 million to over $200 million in research, policy development, and pilot testing of new initiatives (Blue Earth Consultants, 2010). These fisheries and aquaculture initiatives have been most prominent in marine environmental conservation, investments in global market linkages, and promotion of sustainable fishing.

Civil society organizations in the social and environmental realms, influential market actors (e.g., large seafood buyers), and research organizations influence the content and implementation of fisheries policy. Organizations representing fish producers, such as the International Fishworkers and Fisherfolk Coalition, work on behalf of fishers and fishing communities to ensure their voices are heard in high-level negotiations. Numerous regional-, national-, and local-level producer and trade organizations also exist to support members at national and regional forums. As fisheries and aquaculture crossover into food security, nutrition, and health, other global and regional players become influential; much current work on fish and nutrition is taking place through international organizations like USAID's "Feed the Future" and "First 1000 Days" programs.

Implement Food-Security and Nutrition-Focused Fisheries Policies

A simple analytical approach, building on the identification of stakeholders, can help identify necessary steps toward more integrated policy at regional, national, and subnational levels. This approach aids in the process of identifying major influences in policymaking and in assessing the feasibility of reform.

First, analyze the dominant style of policymaking in the sector to identify approaches with higher likelihood of influencing that policy. If fisheries and aquaculture are identified as nutritionally significant but insufficiently represented in current fisheries policies, then it will be desirable to reform policy to better reflect the sector's nutrition and food security potential. Reform also requires an understanding of how policy is made in each country. For example, if the fisheries policy is implemented largely through traditional authority like Fijian inshore fisheries, then reforms to move policy toward greater nutrition and health orientations need to be co-produced with traditional leadership. If the national NGO sector is influential, as in Bangladesh, then major NGOs will need to be important partners in policy dialogue and implementation. In large, decentralized states, such as Indonesia, local and provincial governments may be the key agents of policy reform. In all cases, key ministries related to health, food, and the environment must be engaged in these policy analyses.

Second, assess the feasibility and cost of reforms in the context of the national political economy. In some countries, policy messages highlighting the nutrition contribution of fish may have strong receptivity. For example, Bangladesh has a large nutrition-sensitive population with direct or local-market access to fish. In countries where the state makes a large proportion of its revenue from the sale of fish licenses to domestic and foreign industrial fleets, policies to divert fish to low-income consumers have little traction because they challenge the interests of the state and influential private-sector actors. In states like Ghana and Senegal where fish yield for human consumption would increase if harvesting pressure could be reduced, the overcapacity is perpetuated by fuel subsidies that are politically difficult to withdraw. In this last example, political stability is maintained at the cost of improved nutrition security and optimized fishery sector financial performance and reform to nutrition-sensitive policy will likely face political obstacles.

Livelihood diversification out of fishing is often promoted as a means to reduce pressure on fisheries, allowing stocks to recover to the benefit of food security. These programs have mixed success, with few alternatives more lucrative than fishing—there is money to be made even in depleted fisheries—and strong occupational attachments to fishing as a way of life (Pollnac et al., 2015), both limit the appeal of livelihood diversification. Although idealized returns from transition to optimal fishery management globally can be calculated (Melnychuk et al., 2012) and management outcomes forecasted (Costello et al., 2016), to date such calculations have not accounted for the variable and potentially substantial transaction costs of policy reform. Political economy analysis offers a means by which the costs and benefits of reforms can be evaluated at national level.

Where there is sufficient evidence that fisheries and aquaculture could provide greater benefits to nutrition and human health via policy reform, states and their development partners could use the above analytical steps to identify policy objectives, processes of cross-sectoral engagement harmonization, and institutional and financial mechanisms for policy implementation.

Distributive Justice Issues and Unintended Consequences From Policy Reforms

Identifying those who stand to gain and lose from change is a key task in policy reform. When taken together with an analysis of relative stakeholder power and legitimacy, such policy analysis highlights where opposition and support is likely to come from and informs the design of any mitigation required to address those whose interests are damaged.

Policy analysis must also be cautious to unintended consequences of change. For example, research has demonstrated the high nutrition value of small indigenous fish species in Bangladesh, including the iron and vitamin A-rich *"mola."* This fish was once consumed only by the rural poor and its culture in pond systems has been advocated as part of food security and nutrition initiatives. In the meantime, it has developed a market among the urban middle class and is now more expensive, by weight, than farmed carp and tilapia (Fiedler et al., 2016). This might be termed the "superfood" effect—whereby an item important in the diet of the poor is identified as beneficial to health, and is marketed to wealthy health conscious consumers, raising the possibility that it becomes too expensive for the poor to eat.[7]

Fisheries sector productivity gains are insufficient, if we do not also reduce inequities created by focusing exclusively on environmental or economic yields. Moving into an uncertain future regarding access to sustainable and healthy foods, we must consider how the current seafood system moves fishery resources away from the vulnerable or health-deprived regions and communities with the greatest needs. Fishery sector policy must emphasize the amelioration of these inequities when designing and implementing nutrition-sensitive interventions—to transition into a more "safe and just" contribution toward sustaining humanity in the Anthropocene (Dearing et al., 2014). This does not mean abandoning ecological and economic goals, but adding to them a broader societal goal—to ensure that the distributional inequities of seafood supply are addressed, and that vulnerable and nutrition-poor communities are among the beneficiaries of these renewable natural resources.

---

7. The Andean grain quinoa (*Chenopodium quinoa*) is perhaps the best-known example of this phenomenon.

SYNTHESIS

Beyond maximum sustainable and economic yield, can fisheries and aquaculture systems also produce *maximum nutritious yield* to feed a growing global population? Answers rest in the ability to identify needs and craft policy that meets them. Fisheries and aquaculture policies have not historically focused on food security and nutrition. While claims are commonly made that fisheries conservation and aquaculture production benefits food security, nutrition and health, the details have been lacking. The sector is diverse; and its linkages to non-fisheries economic, environmental, and social dynamics, creates additional governance challenges. In this chapter we have sought to outline a process for analysis of policies, management actions, and drivers of change external to the fishery system. By thinking through fisheries and aquaculture in this way, we hope that analysts in the sector can provide guidance on the range of issues that must be taken into account when attempting to harmonize fisheries policies with food security and nutrition concerns–including the identification of effective implementation pathways and policy strategies. It is imperative that environmental justice concerns are not only voiced but accounted for. There are more pressing concerns than whether enough is available to supply luxury demand. Seafood's maximum nutrition benefit is contingent upon its access and affordability by those who need it most.

ACKNOWLEDGMENTS

JZK and EHA thank their colleagues Marisa Nixon and Hannah Russell, from MARINA lab (www.marinalab.org), Dr. Shakuntala Thilsted (WorldFish) and Denis Hellebrandt de Silva (University of East Anglia) for their assistance with work on previous projects—funding from Rockefeller Foundation, UN Food and Agricultural Organization, WorldFish, Wellcome Trust, the National Socio-environmental Synthesis Center (SESYNC), and support from the National Science Foundation's IGERT Program on Ocean Change under award no. 1068839.

REFERENCES

Allison, E.H., 2011. Aquaculture, Fisheries, Poverty and Food Security. Working Paper 2011–65.

Beddington, J.R., Agnew, D.J., Clark, C.W., 2007. Current problems in the management of marine fisheries. Science 316, 1713–1716. http://dx.doi.org/10.1126/science.1137362.

Bell, J.D., Allain, V., Allison, E.H., Andrefouet, S., Andrew, N.L., Batty, M.J., Blanc, M., Dambacher, J.M., Hampton, J., Hanich, Q., Harley, S., Lorrain, A., McCoy, M., McTurk, N., Nicol, S., Pilling, G., Point, D., Sharp, M.K., Vivili, P., Williams, P., 2015. Diversifying the use of tuna to improve food security and public health in Pacific Island countries and territories. Marine Policy 51, 584–591. http://dx.doi.org/10.1016/j.marpol.2014.10.005.

Belton, B., Little, D.C., 2011. Immanent and interventionist inland Asian aquaculture development and its outcomes. Development Policy Review 29, 459–484. http://dx.doi.org/10.1111/j.1467-7679.2011.00542.x.

Belton, B., Thilsted, S.H., 2014. Fisheries in transition: food and nutrition security implications for the global South. Global Food Security 3, 59–66. http://dx.doi.org/10.1016/j.gfs.2013.10.001.

Béné, C., Hersoug, B., Allison, E., 2010. Not by rent alone: analysing the pro-poor functions of small-scale fisheries in developing countries. Development Policy Review.

Béné, C., Arthur, R., Norbury, H., Allison, E.H., Beveridge, M., Bush, S., Campling, L., Leschen, W., Little, D., Squires, D., Thilsted, S.H., Troell, M., Williams, M., 2016. Contribution of fisheries and aquaculture to food security and poverty reduction: assessing the current evidence. World Development 79, 177–196. http://dx.doi.org/10.1016/j.worlddev.2015.11.007.

Bennett, N.J., Govan, H., Satterfield, T., 2015. Ocean grabbing. Marine Policy 57, 61–68. http://dx.doi.org/10.1016/j.marpol.2015.03.026.

Beveridge, M.C.M., Thilsted, S.H., Phillips, M.J., Metian, M., Troell, M., Hall, S.J., 2013. Meeting the food and nutrition needs of the poor: the role of fish and the opportunities and challenges emerging from the rise of aquaculturea. Journal of Fish Biology 83, 1067–1084. http://dx.doi.org/10.1111/jfb.12187.

Blue Earth Consultants, 2010. Ocean Conservation Strategic Funding Initiatives: A Study of Successes and Lessons Learned.

Bogard, J.R., Hother, A.L., Saha, M., Bose, S., Kabir, H., Marks, G.C., Thilsted, S.H., 2015. Inclusion of small indigenous fish improves nutritional quality during the first 1000 days. Food and Nutrition Bulletin 36 (3), 276–289. http://dx.doi.org/10.1177/0379572115598885.

Committee on World Food Security, 2015. Framework for Action for Food Security and Nutrition in Protracted Crises (CFS-FFA).

Costello, C., Ovando, D., Clavelle, T., Strauss, C.K., Hilborn, R., Melnychuk, M.C., Branch, T.A., Gaines, S.D., Szuwalski, C.S., Cabral, R.B., Rader, D.N., Leland, A., 2016. Global fishery prospects under contrasting management regimes. Proceedings of the National Academy of Sciences of the United States of America 113, 5125–5129. http://dx.doi.org/10.1073/pnas.1520420113.

Costello, C., Ovando, D., Hilborn, R., Gaines, S.D., Deschenes, O., Lester, S.E., 2012. Status and solutions for the world's unassessed fisheries. Science 338 (80), 517–520. http://dx.doi.org/10.1126/science.1223389.

Dalton, A., Wolmarans, P., Witthuhn, R.C., van Stuijvenberg, M.E., Swanevelder, S.A., Smuts, C.M., 2009. A randomised control trial in schoolchildren showed improvement in cognitive function after consuming a bread spread, containing fish flour from a marine source. Prostaglandins, Leukotrienes, and Essential Fatty Acids 80, 143–149. http://dx.doi.org/10.1016/j.plefa.2008.12.006.

Dearing, J.A., Wang, R., Zhang, K., Dyke, J.G., Haberl, H., Hossain, M.S., Langdon, P.G., Lenton, T.M., Raworth, K., Brown, S., Carstensen, J., Cole, M.J., Cornell, S.E., Dawson, T.P., Doncaster, C.P., Eigenbrod, F., Flörke, M., Jeffers, E., Mackay, A.W., Nykvist, B., Poppy, G.M., 2014. Safe and just operating spaces for regional social-ecological systems. Global Environmental Change 28, 227–238. http://dx.doi.org/10.1016/j.gloenvcha.2014.06.012.

Dovydaitis, T., 2008. Fish consumption during pregnancy: an overview of the risks and benefits. Journal of Midwifery and Women's Health 53, 325–330. http://dx.doi.org/10.1016/j.jmwh.2008.02.014.

Duggan, J., 2014. Sales of Shark Fin in China Drop by up to 70%. (Guard).

EU Seventh Framework Programme, 2014. Aquaculture for Food Security Poverty Alleviation and Nutrition Final Technical Report.

FAO, WHO, 2010. Report of the Joint FAO/WHO Expert Consultation on the Risks and Benefits of Fish Consumption.

FAO, 2015. Voluntary Guidelines for Securing Sustainable Small-scale Fisheries in the Context of Food Security and Poverty Eradication. Rome.

FAO, 2014. The State of World Fisheries and Aquaculture 2014. Food and Agriculture Oraganization of the United Nations. doi:92-5-105177-1.

FAO, WHO, 2014. Conference outcome Document: Rome Declaration on nutrition. In: Second International Conference on Nutrition, p. 6 Rome.

FAO, 2016. The State of World Fisheries and Aquaculture 2016. Contributing to food security and nutrition for all. Rome. 200 pp.

Fiedler, J.L., Lividini, K., Drummond, E., Thilsted, S.H., 2016. Strengthening the contribution of aquaculture to food and nutrition security: the potential of a vitamin A-rich, small fish in Bangladesh. Aquaculture 452, 291–303. http://dx.doi.org/10.1016/j.aquaculture.2015.11.004.

Food and Nutrition Board, 2011. Dietary Reference Intakes (DRIs): Recommended Dietary Allowances and Adequate Intakes, Vitamins Food and Nutrition Board. Institute of Medicine, National Academies. http://dx.doi.org/10.1111/j.1753-4887.2004.tb00011.x.

Fröcklin, S., de la Torre-Castro, M., Lindström, L., Jiddawi, N.S., 2013. Fish traders as key actors in fisheries: gender and adaptive management. Ambio 42, 951–962. http://dx.doi.org/10.1007/s13280-013-0451-1.

Gamarro, E.G., Orawattanamateekul, W., Sentina, J., Gopal, T.K.S., 2013. By-products of Tuna Processing. Rome, Italy.

Glover-Amengor, M., Ottah Atikpo, M.A., Abbey, L.D., Hagan, L., Ayin, J., Toppe, J., 2012. Proximate composition and consumer acceptability of three underutilised fish species and tuna frames. World Rural Observations 4, 71–108.

Golden, C.D., Allison, E.H., Cheung, W.W.L., Dey, M.M., Halpern, B.S., McCauley, D.J., Smith, M., Vaitla, B., Zeller, D., Myers, S.S., 2016. Fall in fish catch threatens human health. Nature News 534, 317. http://dx.doi.org/10.1038/534317a.

Ha, T.T.T., Bush, S.R., 2010. Transformations of Vietnamese shrimp aquaculture policy: empirical evidence from the Mekong Delta. Environment and Planning C: Government and Policy 28, 1101–1119. http://dx.doi.org/10.1068/c09194.

Halpern, B.S., Walbridge, S., Selkoe, K.A., Kappel, C.V., Micheli, F., D'Agrosa, C., Bruno, J.F., Casey, K.S., Ebert, C., Fox, H.E., Fujita, R., Heinemann, D., Lenihan, H.S., Madin, E.M.P., Perry, M.T., Selig, E.R., Spalding, M., Steneck, R., Watson, R., 2008. A global map of human impact on marine ecosystems. Science (80), 319.

Hebisha, H., Fathi, M., 2014. Small and Medium Scale Aquaculture Value Chain Development in Egypt: Situation Analysis and Trends.

Hishamunda, N., Ridler, N.B., Bueno, P., Yap, W.G., 2009. Commercial aquaculture in Southeast Asia: some policy lessons. Food Policy 34, 102–107. http://dx.doi.org/10.1016/j.foodpol.2008.06.006.

HLPE, 2014. Sustainable Fisheries and Aquaculture for Food Security and Nutrition, High Level Panel of Experts on World Food Security.

Kelleher, K., 2005. Discards in the World's Marine Fisheries: An Update. Food and Agriculture Organization of the United Nations, Rome.

Le Cornu, E., Kittinger, J.N., Koehn, J.Z., Finkbeiner, E.M., Crowder, L.B., 2014. Current practice and future prospects for social data in coastal and ocean planning. Conservation Biology 28, 902–911. http://dx.doi.org/10.1111/cobi.12310.

Lim, S.S., Vos, T., Flaxman, A.D., Danaei, G., Shibuya, K., et al., 2012. A comparative risk assessment of burden of disease and injury attributable to 67 risk factors and risk factor clusters in 21 regions, 1990–2010: a systematic analysis for the Global Burden of Disease Study 2010. Lancet 380, 2224–2260. http://dx.doi.org/10.1016/S0140-6736(12)61766-8.A.

Lima dos Santos, C.A.M., Howgate, P., 2011. Fishborne zoonotic parasites and aquaculture: a review. Aquaculture 318, 253–261. http://dx.doi.org/10.1016/j.aquaculture.2011.05.046.

Mason, M., McDowell, R., Mendoza, M., Htusan, E., 2015. Enslaved in Shrimp Sheds. Assoc. Press.

Melnychuk, M.C., Essington, T.E., Branch, T.A., Heppell, S.S., Jensen, O.P., Link, J.S., Martell, S.J.D., Parma, A.M., Pope, J.G., Smith, A.D.M., 2012. Can catch share fisheries better track management targets? Fish and Fisheries 13, 267–290. http://dx.doi.org/10.1111/j.1467-2979.2011.00429.x.

Nobre, A.M., Robertson-Andersson, D., Neori, A., Sankar, K., 2010. Ecological–economic assessment of aquaculture options: comparison between abalone monoculture and integrated multi-trophic aquaculture of abalone and seaweeds. Aquaculture 306, 116–126. http://dx.doi.org/10.1016/j.aquaculture.2010.06.002.

Paterson, B., Kirchner, C., Ommer, R.E., 2013. A short history of the Namibian hake fishery—a social-ecological analysis. Ecology and Society 18. http://dx.doi.org/10.5751/ES-05919-180466.

Pollnac, R.B., Seara, T., Colburn, L.L., 2015. Aspects of fishery management, job satisfaction, and well-being among commercial fishermen in the Northeast region of the United States. Society and Natural Resources 28, 75–92. http://dx.doi.org/10.1080/08941920.2014.933924.

Ratner, B.D., Allison, E.H., 2012. Wealth, rights, and resilience: an agenda for governance reform in small-scale fisheries. Development Policy Review 30, 371–398. http://dx.doi.org/10.1111/j.1467-7679.2012.00581.x.

Sale, P.F., Agardy, T., Ainsworth, C.H., Feist, B.E., Bell, J.D., Christie, P., Hoegh-Guldberg, O., Mumby, P.J., Feary, D.A., Saunders, M.I., Daw, T.M., Foale, S.J., Levin, P.S., Lindeman, K.C., Lorenzen, K., Pomeroy, R.S., Allison, E.H., Bradbury, R.H., Corrin, J., Edwards, A.J., Obura, D.O., Sadovy de Mitcheson, Y.J., Samoilys, M.A., Sheppard, C.R.C., 2014. Transforming management of tropical coastal seas to cope with challenges of the 21st century. Marine Pollution Bulletin 85, 8–23. http://dx.doi.org/10.1016/j.marpolbul.2014.06.005.

Sampson, G.S., Sanchirico, J.N., Roheim, C.A., Bush, S.R., Taylor, J.E., Allison, E.H., Anderson, J.L., Ban, N.C., Fujita, R., Jupiter, S., Wilson, J.R., MSC, Pérez-Ramírez, M., MSC, Jacquet, J., Bush, S.R., Mills, M., Micheli, F., Bush, S., Oosterveer, P., Deighan, L.K., Jenkins, L.D., Christian, C., Goyert, W., Pérez-Ramírez, M., Ponce-Díaz, G., Lluch-Cota, S., Ruddle, K., Hickey, F., Crona, B., Asche, F., Borit, M., Olsen, P., Helyar, S.J., Trumble, R.J., 2015. Secure sustainable seafood from developing countries. Science 348, 504–506. http://dx.doi.org/10.1126/science.aaa4639.

Steffen, W., Persson, Å., Deutsch, L., Zalasiewicz, J., Williams, M., Richardson, K., Crumley, C., Crutzen, P., Folke, C., Gordon, L., Molina, M., Ramanathan, V., Rockström, J., Scheffer, M., Schellnhuber, H.J., Svedin, U., 2011. The anthropocene: from global change to planetary stewardship. Ambio 40, 739–761. http://dx.doi.org/10.1007/s13280-011-0185-x.

Steinfeld, H., Gerber, P., Wassenaar, T.D., Castel, V., de Haan, C., 2006. Livestock's long shadow: environmental issues and options. Food and Agricultural Organization of the United Nations.

Sze Choo, P., Nowak, B.S., Kusakabe, K., Williams, M.J., 2008. Guest editorial: gender and fisheries. Development 51, 176–179. http://dx.doi.org/10.1057/dev.2008.1.

The World Bank, 2012. Hidden Harvest: The Global Contribution of Capture Fisheries. World Bank. Econ. Sect. Work 92.

Thilsted, S.H., Thorne-Lyman, A., Webb, P., Bogard, J.R., Subasinghe, R., Phillips, M.J., Allison, E.H., 2016. Sustaining healthy diets: the role of capture fisheries and aquaculture for improving nutrition in the post-2015 era. Food Policy 61, 126–131. http://dx.doi.org/10.1016/j.foodpol.2016.02.005.

Thorpe, A., Reid, C., Van Anrooy, R., Brugere, C., Becker, D., 2006. Poverty reduction strategy papers and the fisheries sector: an opportunity forgone? Journal of International Development 18, 489–517. http://dx.doi.org/10.1002/jid.1245.

Troell, M., Naylor, R.L., Metian, M., Beveridge, M., Tyedmers, P.H., Folke, C., Arrow, K.J., Barrett, S., Crépin, A.-S., Ehrlich, P.R., Gren, Å., Kautsky, N., Levin, S.A., Nyborg, K., Österblom, H., Polasky, S., Scheffer, M., Walker, B.H., Xepapadeas, T., de Zeeuw, A., 2014. Does aquaculture add resilience to the global food system? Proceedings of the National Academy of Sciences of United States of America 111, 13257–13263. http://dx.doi.org/10.1073/pnas.1404067111.

Weeratunge, N., Snyder, K.A., Sze, C.P., 2010. Gleaner, fisher, trader, processor: understanding gendered employment in fisheries and aquaculture. Fish and Fisheries 11, 405–420. http://dx.doi.org/10.1111/j.1467-2979.2010.00368.x.

Chapter 5

Social Resilience in the Anthropocene Ocean

Elena M. Finkbeiner[1], Kirsten L.L. Oleson[2], John N. Kittinger[3]

[1]Stanford University, Monterey, CA, United States; [2]University of Hawai'i at Mānoa, Honolulu, HI, United States; [3]Conservation International, Honolulu, HI, United States

INTRODUCTION

The oceans make up Earth's primary life support system and cover 70% of our planet's surface. Oceans make the planet livable and allow people and societies to prosper. Yet we have not been kind to the sea. And this has consequences for global humanity.

As our global society transitions into the Anthropocene era, we have made profound changes to the global ecosystems that support humanity (Rockstrom et al., 2009). More than a third of the world's fisheries are overexploited or depleted, threatening the ability of fish populations to replenish themselves, and placing at risk food security and livelihoods for 3 billion people (FAO, 2016). One-third of coastal ecosystems have been destroyed, reducing the coastal protection, food production, and carbon storage benefits they provide, making coastal communities more vulnerable to storms and sea level rise (Barnett and Adger, 2003). The ocean harbors incredible biodiversity, but we are in the midst of a mass extinction event, endangering these unique species and habitats (Estes et al., 2011; McCaulay et al., 2015). And the ocean has become the primary sink for our carbon dioxide problem, causing changes in our global climate that destabilize economies and threaten national security.

Our oceans are peopled seascapes that nourish communities, support household economies, protect coastal populations from storms, and give us lifesaving medicines. The intersecting threats of overharvesting, climate change, pollution, and habitat destruction continue to threaten the ocean's ability to support the well-being of global humanity. As we diminish the ocean, we diminish ourselves, and place at risk the most vulnerable people on earth.

This chapter explores what it means for people, communities, and societies to be resilient to a changing sea. We first provide a brief overview of the concept of resilience and its application in the social sciences and the emerging literature on coupled social-ecological systems. We then turn our attention

Conservation for the Anthropocene Ocean. http://dx.doi.org/10.1016/B978-0-12-805375-1.00005-2
89

to the real-world implications of social resilience, with a focus on three of the biggest threats that humanity faces in the Anthropocene ocean: overfishing and its impacts on food security and livelihoods; rising seas and coastal storms; and land-based pollution and habitat destruction. We then discuss potential solutions for enhancing social resilience in the context of these threats facing our Anthropocene ocean.

A PRIMER ON RESILIENCE THINKING AND THE SOCIAL SCIENCES

There is a rich history of inquiry into the concept of resilience from a wide array of fields and disciplines. This area of research spans from the natural sciences and the social, behavioral, economic, and medical sciences, to engineering and related design fields (Bahadur et al., 2013; Folke, 2006; Gunderson, 2000; Holling, 1961; Martin-Breen and Anderies, 2011; Scheffer, 2009; Walker and Salt, 2006). The many researchers who engage resilience as a concept—from different disciplines—often understand, investigate, and apply resilience thinking differently (Brown, 2014; Downes et al., 2013; Gallopín, 2006). However, one constant across this breadth of definitions and applications is a focus on system response to external change, and in particular how systems are able to recover from undesirable changes (e.g., environmental degradation). The common focus on response to change makes resilience a meaningful construct for a broad array of diverse fields, both theoretical and applied. A full review of the engagement of resilience and similar concepts by social science researchers and practitioners is beyond this book chapter, but we provide a brief overview in the following discussion.

The resilience perspective first emerged in the ecology literature in the 1960s, notably in Holling's (1961) research on predator/prey interactions characterized by tipping points, regime shifts, and alternate stable states. A general definition of ecological resilience is the capacity of an ecosystem to absorb disturbance and reorganize while retaining essentially the same functions, structure, identity, and feedbacks (Gunderson, 2000). But within the social and behavioral sciences resilience has an equally impressive longevity, including within economics (Perrings, 2006; Schumpeter, 1942), anthropology (Vayda and McCay, 1975), psychology (Masten, 2001; Masten and Obradovic, 2008), political ecology (Peterson, 2000), development (Brown, 2016, 2014), and more. Social resilience is the ability of society (individuals, groups, communities) to cope with change, including shocks to their social infrastructure (Adger, 2000). Resilience has been taken up in the interdisciplinary scholarship around coupled human–environmental systems, particularly with regards to natural resource-dependent communities where human and environmental linkages are prevalent (Adger, 2000). This body of literature defines resilience as the capacity for social-ecological systems to absorb disturbance and reorganize (Walker et al., 2004).

Social resilience differs from its ecological counterpart in a number of ways. First, adaptive capacity describes the ability of humans to *influence* their resilience (Nelson et al., 2007; Walker et al., 2004; Berkes and Ross, 2013), and sometimes leverage change and uncertainty to create opportunities for positive transformation (Klein et al., 2004). Although common metrics of ecological resilience include characteristics such as diversity, human systems are more complex, with resilience being influenced by features like human agency (the capacity of humans to act independently and make their own choices), power dynamics, and collective action (Davidson, 2010). Put more simply, humans have the ability to be forward looking, anticipate change, and transform their environment as well as their relationship with it.

Second, understanding social resilience therefore requires explicit attention to power relations and the unequal distribution of agency (Berkes and Ross, 2013; Brown, 2016; Davidson, 2010). For example, policy changes that are intended to foster ecological resilience (i.e., stricter conservation regulations) may undermine social resilience by excluding the most marginalized groups from accessing environmental resources (Mansfield, 2007; Marshall and Marshall, 2007). A related issue centers on economic inequality, which can undermine collective action, and thus societal adaptive behavior (Adger et al., 2002). Importantly, while researching or managing for resilience, we may be reinforcing or even exacerbating existing power, agency, and economic inequalities, creating winners and losers (Brown, 2016). Thus an important question for social scientists is resilience for whom (Brown, 2016)?

Third, whether or not individuals or communities have high resilience and can adapt to change depends on differences in perceived risk to different types of change and uncertainty. Psychosocial factors known to influence risk perception, such as culture and human relationships (Brown and Westaway, 2011), can determine how people cope and respond in the face of change, both long term and sudden. Thus disaster is experienced and processed differently across individuals, families, and communities, and social resilience must incorporate perceptions of risk and how they influence coping thresholds (Marshall and Marshall, 2007). In sum, integrating processes of politics, power, and agency into social resilience, or social-ecological resilience, can help to address issues of social justice, equity, and human well-being in research and practice (Brown, 2016).

Despite a burgeoning body of scholarship on social-ecological resilience, the link between ecological and social resilience is not entirely clear—do resilient ecosystems support resilient communities? And do resilient communities foster resilient marine environments? The nature of this implied theoretical relationship between social and ecological resilience needs deeper and more substantive empirical exploration (Adger, 2000; Kittinger et al., 2012). Resilient ecosystems surely have the potential to support resilient communities, but this relationship is often mediated by communities' resources, rights, and access to the environment (Finkbeiner, 2015; Leach et al., 1999). In addition, ecosystem outcomes are dependent on social factors such as tenure and management regimes, markets, technological factors, and more (Cinner et al., 2016). These efforts examining the complex link between social and ecological resilience compel more in-depth consideration of social institutions (i.e., the formal and informal rules governing human behavior) (Schlager and Ostrom, 1992), as well as the roles of power and politics in enabling individuals, households, and communities to benefit from ecological resilience (Brown, 2016; Leach et al., 1999).

One of the similarities in the way the resilience concept has been addressed in both the biophysical and social sciences regards the issue of scale. This is important because processes that influence resilience can manifest differently across scales (Adger, 2006; Adger et al., 2004; Vincent, 2007), and scales of resilience are not independent (Smit and Wandel, 2006). In the social sciences, resilience has been applied from the individual, to the social group, to the community level, and up to the societal level (see examples in Table 5.1). At the societal level, our increasingly globalized economy is becoming more resilient to shocks and disturbance, perhaps at the expense of resilience in local or regional economies (Perry et al., 2011). At the individual or family unit level, the factors that convey resilience are often different. For example, the capacity of a household to adapt is constrained by conditions at the community, country, or other scales (Cinner and Bodin, 2010; Smit and Wandel, 2006), depending on the afforded resources, rights, and opportunities, and the ability to mobilize these based on levels of human agency and autonomy (Leach et al., 1999).

SOCIAL RESILIENCE IN A CHANGING OCEAN

Even with the increased focus on resilience among scholars, the gap between knowledge and its application in the real world can lead to serious consequences for human communities. This "implementation gap" is the focus of significant attention by practitioners in research, governmental, nonprofit, and philanthropic communities. Put more simply, if our normative objective is to increase the resilience of people to a changing sea, what key factors are relevant? And more importantly, how do we design, implement, and amplify interventions that build social resilience? These kinds of questions are at the root of conservation and stewardship initiatives across the globe, from local to global scales. As authors, we wish we had the answers to these questions. But as a global community of practice we are still very much in a learning phase about what makes individuals, groups, and societies more able to adapt to change.

In this section, we focus on three primary threats to social resilience in the Anthropocene ocean, including overfishing and implications for food security and livelihoods, rising seas and coastal storms, and land-based pollution and associated habitat destruction. We approach these threats from the viewpoint of practitioners, focusing on their impact on human communities, as well as the social responses that people have taken to adapt to a changing ocean. In

TABLE 5.1 Examples of Definitions of Social Resilience at Different Scales of Social Organization

Individual/Family Scale

Social coherence: The way people view their life has a positive influence on their health: Three elements, comprehensibility (cognitive), manageability (instrumental/behavioral), and meaningfulness (motivational), formed the concept of sense of coherence (Antonovsky, 1987, 1979)

"The capacity of individuals, families, … to anticipate, withstand and/or judiciously engage with catastrophic events and/or experiences, actively making meaning out of adversity, with the goal of maintaining 'normal' function without fundamentally losing identity" (Almedom and Tumwine, 2008)

"The way in which individuals … adapt, transform, and potentially become stronger when faced with environmental, social, economic or political challenges" (Maclean et al., 2014)

Community Scale

The ability of groups or communities to cope with external stresses and disturbances manifested as social, political, and environmental change (Adger, 2000)

Community resilience, as defined herein, is the existence, development, and engagement of community resources by community members to thrive in an environment characterized by change, uncertainty, unpredictability, and surprise (Magis, 2010)

Aspects of social resilience include flexibility in livelihood strategies and the formal institutions governing marine resources, involvement in community organizations and decision making, assets, trust, formal levels of education and the opportunity to learn, and the capacity to organize (Cinner et al., 2009a,b)

Societal Scale

The capacity of a social system to absorb stress or disturbance without losing fundamental functions and structure (*functional resilience*);

A social system's ability to self-organize (*self-organization*); and

A social system's ability to build capacity for learning and adaptation (*adaptive capacity*; Berkes, 2003)

addition, we provide an overview of solutions, both old and new, and the evidence (both for and against) of these approaches to enhance or degrade social resilience to a changing ocean.

Overfishing, Food Security, and Livelihoods

About a third of the world's fisheries are overexploited (FAO, 2016), resulting in depletion of both commercially targeted and nontargeted marine species down and through the food chain (Essington et al., 2006; Pauly et al., 1998), with major ecological and evolutionary implications (Jackson et al., 2001). Some stocks have remained level over time or even increased as fisheries are rebuilt (Hilborn, 2007; Worm et al., 2009). For overexploited fisheries, overfishing threatens social resilience from the global to local scale, and particularly for people who directly rely on the ocean for livelihoods, food security, and well-being.

Drivers of overfishing are complex and difficult to manage, often resulting in severe consequences for the most marginalized and vulnerable human populations. Much of the global seafood production is driven by demand from developed countries (Swartz et al., 2010), and a small number of transnational seafood corporations disproportionately control the seafood supply chain (Österblom et al., 2015). The advent of distant water fleets, some equipped with onboard processing capacity, has provided many of these countries, who are net importers of seafood, access to high seas fisheries, as well as other countries' Exclusive Economic Zones (Berkes et al., 2006; Brashares et al., 2004). In this context, distant water fleets are often directly competing with local small-scale or subsistence fisheries by targeting the same stocks, resulting in overcapacity. Local small-scale fisheries are often unable to compete with highly mechanized fleets, and thus are less likely to retain the benefits of their local fish stocks, undermining food security and livelihoods (Alder and Sumaila, 2004; Bennett et al., 2015; Kaczynski and Fluharty, 2002). In particular, issues of access, distribution, and power are deeply embedded in overfishing (Finkbeiner et al., n.d.). Likewise, increasingly liberalized trade channels have facilitated export of seafood products from developing countries to the developed countries, driving demand (FAO, 2012). When coastal communities are deprived of access to their local fisheries as sources of income or protein, human agency and well-being are compromised, and resilience to change and disturbance becomes less likely.

Moreover, prescribed policies intended to address overfishing can further undermine social resilience, particularly for already vulnerable human populations, and may even exacerbate overfishing (Finkbeiner et al., n.d.). As overfishing is often perceived as a problem of "too many fishers chasing too few fish" (Pauly, 1990), policies often support a reduction in the number of fishers, or reduced access to the marine environment through spatial or temporal closures. Catch shares, or other forms of rights-based management, are often successful at achieving their primary goal (increasing economic efficiency in a fishery) (Costello et al., 2008), but have the potential to reallocate fishery rights and access to a small proportion of the original fleet, excluding the most marginalized and vulnerable fishers (Bromley, 2009; Pinkerton and Edwards, 2009). This can not only exacerbate socioeconomic disparity, but also force preexisting fishing activity to the illegal sector with implications for ecological resilience (Raemaekers et al., 2011). Likewise, spatial or temporal closures can lock out fishers from traditional fishing grounds, undermining livelihoods and food security, and may also displace fishing effort in space or time, instead of facilitating a net reduction (Bennett et al., 2015; Lele et al., 2010). In sum, top-down policies depriving access to the marine environment, and subsequently limiting the number of biological or socioeconomic options for human response in the face of change, can threaten social resilience and even exacerbate overfishing.

How can we facilitate adaptive capacity and social resilience for coastal communities contributing to or affected by overfishing, particularly in an era of global change? Instead of focusing management efforts solely on quantitative stock assessments, technological fixes, or specific tools, managing for resilient fisheries will also require the creation of flexible, local-level rules that are consistent with local realities (McClanahan et al., 2008), allowing fishers to adapt in the face of change (Chapin et al., 2010; Finkbeiner and Basurto, 2015; Jentoft, 1989). Approaches such as adaptive comanagement, emphasizing social learning and collaborative partnerships between fishers, government, and other stakeholders, are consistent with local and relevant rule making (Armitage et al., 2009). Including fishers in policy processes can also increase legitimacy and fairness of resulting rules and laws, thus increasing compliance and social equity (Hauck, 2008; Jentoft, 2000; Jentoft et al., 1998; Nielsen, 2003; Singleton, 2000). Importantly, policies such as marine reserves or catch shares have the potential to address overfishing while maintaining social resilience if fishers are involved in the process of crafting such policies (McCay et al., 2014; Micheli et al., 2012).

Maintaining a diversity of biological and socioeconomic options for human response is also critical in managing resilient fisheries (Chapin et al., 2010). Fishers with access to a diversity of species and fisheries are better able to cope and adapt during difficult years, especially in highly dynamic systems (Aguilera et al., 2015; Finkbeiner, 2015). Likewise, diversification across livelihoods provides an important mechanism for fishers to maintain an income when conditions in the fishery are unfavorable (Marschke and Berkes, 2006), and helps avoid overfishing poverty traps (Cinner et al., 2009a,b). In sum, addressing overfishing while maintaining social resilience will require increased local autonomy in decision making and greater and more flexible access to the marine environment or other livelihoods for local fishing communities, facilitating human agency and the ability to adapt to change (Leach et al., 1999).

Rising Seas and Coastal Storms

Global sea level has risen 10–25 cm over the last century and is expected to rise a half meter by 2100, representing a two- to fivefold acceleration (Nicholls and Mimura, 1998). It poses severe, even existential, threats to coastal communities and ecosystems worldwide. Coasts are already experiencing adverse consequences, such as coastal inundation, erosion, ecosystem loss, salinization, increased vulnerability to extreme storm events, and transmission of infectious diseases (Adger et al., 2005; Nicholls et al., 2007; Nicholls and Mimura, 1998). Over the coming decades, risks related to sea level rise and synergistic effects driven by climate change, such as warming seas, ocean acidification, and deoxygenation, are anticipated to increase (Nicholls et al., 2007). An anticipated 50% of the global population will live within 100 km of the coast by 2030, further increasing human vulnerability to coastal storms, flooding, and other disturbances (Adger et al., 2005; Nicholls et al., 2007). And these hazards are further likely to become disasters through erosion of social resilience (Adger et al., 2005).

Rising seas and coastal storms produce varied socioeconomic impacts that are unevenly distributed within and among nations, regions, communities, and individuals because of different exposures and vulnerabilities (Dolan and Walker, 2003). Differential access and distribution of resources, technology, information, wealth, risk perceptions, social capital, community structure, and institutions addressing climate change hazards interact with various exposure types, intensities, frequencies, and durations. Ultimately, these interactions produce a suite of different outcomes (Dolan and Walker, 2003). For example, many small island nations facing complete inundation will require forced displacement (Nicholls et al., 2011), and heavily populated coastal areas and vital agricultural land in Africa and Southeast Asia will need to be abandoned (Nicholls and Cazenave, 2010). By contrast, in Europe, although sea level rise will lead to huge negative economic effects, the resilience of the economy and capacity to adapt may ensure that the overall effect on the gross domestic product will be small (Bosello et al., 2012).

Ultimately, however, it is the poorest coastal regions that will bear the harshest consequences of climate change (Oliver-Smith, 2009). For example, sea level rise, flooding, and coastal storms are threatening the very existence of small atoll nations such as Kiribati, Tuvalu, and Maldives by undermining food security, habitability, and human health and safety (Barnett and Adger, 2003). Leaders from these vulnerable atoll nations are already scenario planning for relocation and reestablishment in new geographies, balancing the harsh reality of the stress that such actions will place on their people, while also developing alternatives to ensure the continuation of cultural ways of life and sovereignty.

While atoll nations search for ways to survive, coastal megacities (>10 million inhabitants) are also not exempt from sea level rise impacts (Klein et al., 2004), as many are located in coastal areas or in areas where saltwater intrusion will threaten water and food supplies. This is particularly dire considering some of the most marginalized and vulnerable communities within megacities may be disproportionately impacted from sea level rise due to proximity to coastlines and poor infrastructure incapable of buffering the effects of sudden inundation from extreme storm events (Klein et al., 2004). A major challenge for international climate change policy is to generate international norms of justice, sovereignty, and human and national security (Barnett and Adger, 2003).

Due to differential responsibility for, and impacts from, sea level rise across populations and geographies, adaptive responses and mitigation measures will have to occur across scales of organization. At the international and national scales, leaders have an ethical duty to address drivers of and impacts from climate change. The 2015 Paris Agreement represents important progress toward international consensus, responsibility, and accountability. The Agreement catalyzed development funding to aid climate adaptation efforts of the most vulnerable, but these funds will need to increase by orders of magnitude in the coming years. Although the role of national governments is critical, it is at the local and regional scale where vulnerability can be reduced and resilience increased through local action (Adger et al., 2005; Dolan and Walker, 2003). Local action increasing resilience to climate change and sea level rise may include stewardship of ecosystem function and diversity through sustainable use (i.e., mangrove restoration for storm buffering), diverse livelihood portfolios (i.e., economic alternatives outside of fishing), maintenance of local memory and social learning processes for responding to change (i.e., strong civil society organizations), and legitimate and inclusive local governance institutions (i.e., comanagement or other participatory processes) (Adger et al., 2005). Local action is contingent on the ability to access and mobilize resources and rights, and thus can be undermined or fostered by action at the regional, national, or international scale. These solutions represent more than just a technical fix for post hoc response, but rather focus on participatory and proactive adaptation through information development, awareness raising, planning, design, implementation, and monitoring (Klein et al., 2004).

Land-Based Pollution and Habitat Destruction

Estuarine and coastal ecosystems are some of the most threatened, and valuable, natural systems globally (Barbier et al., 2011; Costanza et al., 2014; Halpern et al., 2008; Lotze et al., 2006; Worm et al., 2006). Coastal habitats are being lost at an unprecedented rate, and pollution is degrading coastal areas worldwide (Burke et al., 2011; De'ath et al., 2012; Halpern et al., 2008). Over 35% of the world's mangroves have been lost (Valiela et al., 2001), 29% of the worldwide aerial extent of seagrass beds are gone (Waycott et al., 2009), and over 30% of coral reefs globally are degraded, whereas 60% are threatened (Burke et al., 2011). In Europe, there has been a near total loss of European wetlands, seagrass meadows, shellfish beds, reefs, and other coastal habitats (Airoldi and Beck, 2007).

Habitat loss and degradation are primarily caused by human activity. Changes in land cover, land use, and channelization have profoundly increased the coastal discharge rate of nutrients, sediment, toxins, and other pollutants (Chen and Hong, 2012; Foley et al., 2005; Howarth, 2012; Hupp et al., 2009; Jones et al., 2001). A diverse array of land-based source pollutants are discharged into coastal environments, ranging from sediment to pharmaceuticals to plastics (Fabricius, 2011; Gorman et al., 2009; Islam and Tanaka, 2004), which cause many deleterious effects on ecological function and human health (Islam and Tanaka, 2004). Estuarine and coastal habitats have been destroyed by coastal development, land reclamation, overfishing, invasive species, pollution, and climate change (Lotze et al., 2006).

Coastal systems support millions of people and generate more than 60% of global gross national product (UNEP, 2006), yet the increasing destruction and degradation of coastal habitats disrupt critical ecological functions underpinning delivery of ecosystem goods and services (Barbier et al., 2011; Burke et al., 2011; Orth et al., 2006; Pandolfi et al., 2011; Worm and Branch, 2012). Some of the most impacted ecosystem goods and services provided by wetlands, seagrass beds, oyster reefs, coral reefs, and other habitats are food fisheries (Edinger et al., 1998; Worm et al., 2006), nutrient and carbon cycling and filtration, regulation of waves (Orth et al., 2006; Worm et al., 2006), and tourism/recreation (Moberg and Ronnback, 2003; UNEP, 2006).

Addressing habitat destruction and degradation requires a systems perspective that spans the land–sea interface (Beger et al., 2010; Klcin et al., 2012). Increasingly coastal management is shifting from a single-species approach to a more holistic one that encompasses the entire ecosystem, including humans (Hughes et al., 2010; Leslie and Mcleod, 2007). The scale of ecosystem-based management reflects the geographic perimeters of the ecosystem (spatial), considers long-term historical and future trends (temporal), and encompasses all of the relevant ecological and socioeconomic processes underpinning system dynamics (functional).

However, the socioeconomic drivers of human activity can be complex, ranging from local population growth resulting in coastal development to meet housing needs and increased effluent discharges, to national or global pressures such as political unrest or loss of soil fertility that trigger coastal migration or deleterious upland watershed land use practices (Kittinger et al., 2012). Defining these drivers, as well as the scales of ecological and social processes, is challenging, revealing a distinct need for coordinated social and ecological research and monitoring of management effectiveness on commensurate scales, such that the management can be adapted in real time (Christie, 2011; Nobre, 2011; Wilkinson and Brodie, 2011).

The integration of participatory and collaborative processes with ecosystem-based management and habitat restoration can further increase the effectiveness of such approaches. For example, habitat restoration projects that foster collaboration between international nonprofit organizations and local community institutions can provide additional livelihood opportunities, increase the capacity of community organizations to address other threats to their coastal zone, and ultimately heighten community awareness and a broader sense of stewardship in the area, creating enabling conditions for collective community action (Kittinger et al., 2016). Finally, participatory and collaborative approaches within ecosystem-based management can increase agency and self-determination of local communities, thus improving their access to environment resources and decision-making processes, both important components for social resilience.

Overfishing, sea level rise, and habitat destruction are all anthropogenic processes that undermine the health of our oceans and the resilience of the communities that depend on them. Across all three examples, power relations are at play, creating winners and losers on a global scale. For example, the economies that are primary culprits in driving overfishing, global climate change, and pollution in our oceans are often disparate from the communities that absorb the brunt of the impact. Thus the study of, and management for, resilience needs to address these issues of power and politics as to not exacerbate existing social inequities and erode social resilience where it is needed most. Here we can elucidate several important lessons for addressing the role of power and politics in enabling or disabling social resilience. First, human agency can be increased through political inclusion and empowerment, where affected stakeholders are an integral or central component of decision-making processes. Second, adaptive capacity can be increased through greater and more flexible access to resources and rights surrounding the marine environment. And third, we need to increase accountability on behalf of all global citizens for disproportionate contributions to these degrading processes on our oceans. In other words, why are we asking the most vulnerable citizens to be "resilient" in the first place? Ultimately we hope to inform a research and governance agenda that empowers individuals and communities to become active agents of positive change, and also to mitigate the need for the most vulnerable societies to be the most resilient.

CONCLUSION

In conclusion, we suggest the importance of managing the Anthropocene ocean, not just for ecological sustainability but also for social resilience. Individuals, communities, and societies must have the resources necessary to anticipate and adapt to change to engage in healthy, stewardship behavior with our oceans. Whether or not the billions of people who rely on the ocean for basic means of survival can access the resources they need in a changing world depends on whether we collectively invest in building their adaptive capacity. This will undoubtedly require coordination of policy and implementation across international, national, and local actions, supported by the different disciplinary perspectives required of a strong scientific basis for action, and delivered through adequate funding that ensures the threats of climate change are met by an equally impressive set of resources. Enabling social resilience is not about the advent of technical fixes but about harnessing human ingenuity through participatory, integrative, transparent, and just processes. This will require a level of investment on par with the other massive international efforts, but ultimately it will require the empowerment of individuals to be important active agents of positive change for ocean and human resilience.

REFERENCES

Adger, W.N., 2000. Social and ecological resilience: are they related? Progress in Human Geography 24, 347–364. http://dx.doi.org/10.1191/030913200701540465.

Adger, W.N., 2006. Vulnerability. Global Environmental Change 16, 268–281. http://dx.doi.org/10.1016/j.gloenvcha.2006.02.006.

Adger, W.N., Brooks, N., Bentham, G., Agnew, M., Eriksen, S., 2004. New Indicators of Vulnerability and Adaptive Capacity. Technical Report 7. Norwich, UK.

Adger, W.N., Hughes, T.P., Folke, C., Carpenter, S.R., Rockström, J., 2005. Social-ecological resilience to coastal disasters. Science 309, 1036–1039. http://dx.doi.org/10.1126/science.1112122.

Adger, W.N., Kelly, P., Winkels, A., 2002. Migration, remittances, livelihood trajectories, and social resilience. AMBIO A Journal 31, 358–366.

Aguilera, S.E., Cole, J., Finkbeiner, E.M., Le Cornu, E., Ban, N.C., Carr, M.H., Cinner, J.E., Crowder, L.B., Gelcich, S., Hicks, C.C., Kittinger, J.N., Martone, R., Malone, D., Pomeroy, C., Starr, R.M., Seram, S., Zuercher, R., Broad, K., 2015. Managing small-scale commercial fisheries for adaptive capacity: insights from dynamic social-ecological drivers of change in Monterey bay. PLoS One 10, 1–22. http://dx.doi.org/10.1371/journal.pone.0118992.

Airoldi, L., Beck, M.W., 2007. Loss, status and trends for coastal marine habitats of Europe. Oceanography and Marine Biology: An Annual Review 45, 345–405.

Alder, J., Sumaila, U.R., 2004. Western Africa: a fish basket of Europe past and present. Journal of Environmental and Development 13, 156–178. http://dx.doi.org/10.1177/1070496504266092.

Almedom, A.M., Tumwine, J.K., 2008. Resilience to disasters: a paradigm shift from vulnerability to strength. African Health Sciences 8, 1–4.

Antonovsky, A., 1979. Health, Stress and Coping. Jossey-Bass Publishers, San Francisco, CA.

Antonovsky, A., 1987. Unraveling the Mystery of Health: How Do People Manage Stress and Stay Well. Jossey-Bass Publishers, San Francisco, CA.

Armitage, D.R., Plummer, R., Berkes, F., Arthur, R.I., Charles, A.T., Davidson-Hunt, I.J., Diduck, A.P., Doubleday, N.C., Johnson, D.S., Marschke, M., McConney, P., Pinkerton, E.W., Wollenberg, E.K., 2009. Adaptive co-management for social–ecological complexity. Frontiers in Ecology and the Environment 7, 95–102. http://dx.doi.org/10.1890/070089.

Bahadur, A.V., Ibrahim, M., Tanner, T., 2013. Characterizing resilience: unpacking the concept for tackling climate change and development. Climate and Development 5, 55–65.

Barbier, E.B., Hacker, S.D., Kennedy, C., Koch, E.W., Stier, A.C., Silliman, B.R., 2011. The value of Estuarine and coastal ecosystem services. Ecological Monographs 81, 169–193. http://dx.doi.org/10.1890/10-1510.1.

Barnett, J., Adger, W.N., 2003. Climate dangers and atoll countries. Climate Change 61, 321–337. http://dx.doi.org/10.1023/B.

Beger, M., Grantham, H.S., Pressey, R.L., Wilson, K.A., Peterson, E.L., Dorfman, D., Mumby, P.J., Lourival, R., Brumbaugh, D.R., Possingham, H.P., 2010. Conservation planning for connectivity across marine, freshwater, and terrestrial realms. Biological Conservation 143, 565–575. http://dx.doi.org/10.1016/j.biocon.2009.11.006.

Bennett, N.J., Govan, H., Satterfield, T., 2015. Ocean grabbing. Marine Policy 57, 61–68. http://dx.doi.org/10.1016/j.marpol.2015.03.026.

Berkes, F., 2003. Alternatives to conventional management: lessons from small-scale fisheries. Environments 31.

Berkes, F., Hughes, T.P., Steneck, R.S., Wilson, J.A., Bellwood, D.R., Crona, B., Folke, C., Gunderson, L.H., Leslie, H.M., Norberg, J., Nystrom, M., Olsson, P., Osterblom, H., Scheffer, M., Worm, B., 2006. Globalization, roving bandits, and marine resources. Science 311 (80), 1557–1558.

Berkes, F., Ross, H., 2013. Community resilience: toward an integrated approach. Society and Natural Resources 26, 5–20. http://dx.doi.org/10.1080/08941920.2012.736605.

Bosello, F., Nicholls, R.J., Richards, J., Roson, R., Tol, R.S., 2012. Economic impacts of climate change in Europe: sea-level rise. Climate Change 112, 63–81.

Brashares, J.S., Arcese, P., Sam, M.K., Coppolillo, P.B., Sinclair, A.R.E., Balmford, A., 2004. Bushmeat Hunting, wildlife declines, and fish supply in west Africa. Science 306 (80), 1180–1183.

Bromley, D.W., 2009. Abdicating responsibility: the deceits of fisheries policy. Fisheries 34, 280–290. http://dx.doi.org/10.1577/1548-8446-34.6.280.

Brown, K., 2014. Global environmental change I: A social turn for resilience? Progress in Human Geography 38, 107–117.

Brown, K., 2016. Resilience, Development and Global Change. Routledge.

Brown, K., Westaway, E., 2011. Agency, capacity, and resilience to environmental change: lessons from human development, well-being, and disasters. Annual Review of Environment and Resources 36, 321.

Burke, L., Reytar, K., Spalding, M., Perry, A., 2011. Reefs at Risk Revisited.

Chapin, F.S., Carpenter, S.R., Kofinas, G.P., Folke, C., Abel, N., Clark, W.C., Olsson, P., Smith, D.M.S., Walker, B., Young, O.R., Berkes, F., Biggs, R., Grove, J.M., Naylor, R.L., Pinkerton, E., Steffen, W., Swanson, F.J., 2010. Ecosystem stewardship: sustainability strategies for a rapidly changing planet. Trends in Ecology and Evolution 25, 241–249. http://dx.doi.org/10.1016/j.tree.2009.10.008.

Chen, N., Hong, H., 2012. Integrated management of nutrients from the watershed to coast in the subtropical region. Current Opinion in Environmental Sustainability 4, 233–242. http://dx.doi.org/10.1016/j.cosust.2012.03.007.

Christie, P., 2011. Creating space for interdisciplinary marine and coastal research: five dilemmas and suggested resolutions. Environmental Conservation 38, 172–186. http://dx.doi.org/10.1017/S0376892911000129.

Cinner, J.E., Bodin, O., 2010. Livelihood diversification in tropical coastal communities: a network-based approach to analyzing "livelihood landscapes". PLoS One 5, e11999. http://dx.doi.org/10.1371/journal.pone.0011999.

Cinner, J.E., Daw, T., McClanahan, T.R., 2009a. Socioeconomic factors that affect artisanal fishers' readiness to exit a declining fishery. Conservation Biology 23, 124–130. http://dx.doi.org/10.1111/j.1523-1739.2008.01041.x.

Cinner, J.E., Huchery, C., MacNeil, M.A., Graham, N.A.J., McClanahan, T.R., Maina, J., Maire, E., Kittinger, J.N., Hicks, C.C., Mora, C., Allison, E.H., D'agata, S., Hoey, A.S., Feary, D.A., Crowder, L.B., Williams, I., Kulbicki, M., Vigliola, L., Wantiez, L., Edgar, G., Stuart-Smith, R., Sandin, S., Green, A., Hardt, M.J., Beger, M., Friedlander, A.M., Campbell, S.J., Holmes, K.E., Wilson, S.K., Brokovich, E., Brooks, A.J., Cruz-Motta, J.J., Booth, D.J., Chabanet, P., Gough, C., Tupper, M., Ferse, S.C.A., Sumaila, U.R., Mouillot, D., 2016. Bright spots among the world's coral reefs. Nature 535, 416–419.

Cinner, J.E., McClanahan, T.R., Daw, T.M., Graham, N.A., Maina, J., Wilson, S.K., Hughes, T.P., 2009b. Linking social and ecological systems to sustain coral reef fisheries. Current Biology 19, 206–212.

Costanza, R., De Groot, R., Sutton, P., Van Der Ploeg, S., Anderson, S.J., Kubiszewski, I., Farber, S., Turner, R.K., 2014. Changes in the global value of ecosystem services. Global Environmental Change 26, 152–158. http://dx.doi.org/10.1016/j.gloenvcha.2014.04.002.

Costello, C., Gaines, S.D., Lynham, J., 2008. Can catch shares prevent fisheries collapse. Science 321 (80), 1678–1681. http://dx.doi.org/10.1126/science.1159478.

Davidson, D.J., 2010. The applicability of the concept of resilience to social systems: some sources of optimism and nagging doubts. Society and Natural Resources 23, 1135–1149. http://dx.doi.org/10.1080/08941921003652940.

De'ath, G., Fabricius, K.E., Sweatman, H., Puotinen, M., 2012. The 27-year decline of coral cover on the Great Barrier Reef and its causes. Proceedings of the National Academy of Sciences of the United States of America 109, 17995–17999. http://dx.doi.org/10.1073/pnas.1208909109.

Dolan, A.H., Walker, I.J., 2003. Understanding vulnerability of coastal communities to climate change related risks. Journal of Coastal Research (SI 39).

Downes, B.J., Miller, F., Barnett, J., Glaister, A., Ellemor, H., 2013. How do we know about resilience? An analysis of empirical research on resilience, and implications for interdisciplinary praxis. Environmental Research Letters 8, 14041.

Edinger, E.N., Jompa, J., Limmon, G.V., Widjatmoko, W., Risk, M.J., 1998. Reef degradation and coral biodiversity in Indonesia: effects of land-based pollution, destructive fishing practices and changes over time. Marine Pollution Bulletin 36, 617–630.

Essington, T.E., Beaudreau, A.H., Wiedenmann, J., 2006. Fishing through marine food webs. Proceedings of the National Academy of Sciences of the United States of America 103, 3171–3175. http://dx.doi.org/10.1073/pnas.0510964103.

Estes, J.A., Terborgh, J., Brashares, J.S., Power, M.E., Berger, J., Bond, W.J., Carpenter, S.R., Essington, T.E., Holt, R.D., Jackson, J.B.C., Marquis, R.J., Oksanen, L., Oksanen, T., 2011. Trophic downgrading of planet earth. Science 333 (80), 301–306. http://dx.doi.org/10.1126/science.1205106.

Fabricius, K.E., 2011. Factors determining the resilience of coral reefs to eutrophication: a review and conceptual model. In: Coral Reefs: An Ecosystem in Transition. Springer, Netherlands, pp. 493–505.

FAO, 2012. State of World Fisheries and Aquaculture. Rome.

FAO, 2016. The State of World Fisheries and Aquaculture 2016: Contributing to Food Security and Nutrition for All. Rome.

Finkbeiner, E., Bennett, N., Brooks, C., Frawley, T., Mason, J., Ng, C., Ourens, R., Seto, K., Swanson, S., Urteaga, J., Briscoe, D., Crowder, L., n.d. Reconstructing overfishing: moving beyond Malthus for comprehensive and equitable solutions. Fish Fish.

Finkbeiner, E.M., 2015. The role of diversification in dynamic small-scale fisheries: lessons from Baja California Sur, Mexico. Global Environmental Change 32, 139–152.

Finkbeiner, E.M., Basurto, X., 2015. Re-defining co-management to facilitate small-scale fisheries reform: an illustration from northwest Mexico. Marine Policy 51, 433–441. http://dx.doi.org/10.1016/j.marpol.2014.10.010.

Foley, J.A., DeFries, R., Asner, G.P., Barford, C., Bonan, G., Carpenter, S.R., Chapin, F.S., Coe, M.T., Daily, G.C., Gibbs, H.K., Helkowski, J.H., 2005. Global consequences of land use. Science 309 (80), 570–574.

Folke, C., 2006. Resilience: the emergence of a perspective for social–ecological systems analyses. Global Environmental Change 16, 253–267. http://dx.doi.org/10.1016/j.gloenvcha.2006.04.002.

Gallopín, G.C., 2006. Linkages between vulnerability, resilience, and adaptive capacity. Global Environmental Change 16, 293–303. http://dx.doi.org/10.1016/j.gloenvcha.2006.02.004.

Gorman, D., Russell, B.D., Connell, S.D., 2009. Land-to-sea connectivity: linking human-derived terrestrial subsidies to subtidal habitat change on open rocky coasts. Ecological Applications 19, 1114–1126.

Gunderson, L.H., 2000. Ecological Resilience—in theory and application. Annual Review of Ecology and Systematics 425–439.

Halpern, B.S., Walbridge, S., Selkoe, K.A., Kappel, C.V., Micheli, F., D'Agrosa, C., Bruno, J.F., Casey, K.S., Ebert, C., Fox, H.E., Fujita, R., 2008. A global map of human impact on marine ecosystems. Science 319 (80), 1149345. http://dx.doi.org/10.1126/science.1149345.

Hauck, M., 2008. Rethinking small-scale fisheries compliance. Marine Policy 32, 635–642. http://dx.doi.org/10.1016/j.marpol.2007.11.004.

Hilborn, R., 2007. Reinterpreting the state of fisheries and their management. Ecosystems 10, 1362–1369. http://dx.doi.org/10.1007/s10021-007-9100-5.

Holling, C., 1961. Principles of insect predation. Annual Review of Entomology 11, 163–182.

Howarth, R., 2012. Nitrogen Cycling in the North Atlantic Ocean and Its Watersheds: Report of the International Scope Nitrogen Project. Springer Science & Business Media.

Hughes, T.P., Graham, N.A.J., Jackson, J.B.C., Mumby, P.J., Steneck, R.S., 2010. Rising to the challenge of sustaining coral reef resilience. Trends in Ecology and Evolution 25, 633–642. http://dx.doi.org/10.1016/j.tree.2010.07.011.

Hupp, C.R., Pierce, A.R., Noe, G.B., 2009. Floodplain geomorphic processes and environmental impacts of human alteration along coastal plain rivers, USA. Wetlands 29, 413–429.

Islam, M.S., Tanaka, M., 2004. Impacts of pollution on coastal and marine ecosystems including coastal and marine fisheries and approach for management: a review and synthesis. Marine Pollution Bulletin 48, 624–649.

Jackson, J.B., Kirby, M.X., Berger, W.H., Bjorndal, K.a, Botsford, L.W., Bourque, B.J., Bradbury, R.H., Cooke, R., Erlandson, J., Estes, J.A., Hughes, T.P., Kidwell, S., Lange, C.B., Lenihan, H.S., Pandolfi, J.M., Peterson, C.H., Steneck, R.S., Tegner, M.J., Warner, R.R., 2001. Historical overfishing and the recent collapse of coastal ecosystems. Science 293 (80), 629–637. http://dx.doi.org/10.1126/science.1059199.

Jentoft, S., 1989. Fisheries co-management: delegating government responsibility to fishermen's organizations. Marine Policy 13, 137–154.

Jentoft, S., 2000. Legitimacy and disappointment in fisheries management. Marine Policy 24, 141–148.

Jentoft, S., McCay, B.J., Wilson, D.C., 1998. Social theory and fisheries co-management. Marine Policy 22, 423–436. http://dx.doi.org/10.1016/S0308-597X(97)00040-7.

Jones, K.B., Neale, A.C., Nash, M.S., Van Remortel, R.D., James, D., Riitters, K.H., Neill, R.V.O., 2001. Predicting nutrient and sediment loadings to streams from landscape metrics: a multiple watershed study from the United States Mid-Atlantic Region. Landscape Ecology 16, 301–312.

Kaczynski, V.M., Fluharty, D.L., 2002. European policies in West Africa: who benefits from fisheries agreements? Marine Policy 26, 75–93. http://dx.doi.org/10.1016/S0308-597X(01)00039-2.

Kittinger, J.N., Bambico, T.M., Minton, D., Miller, A., Mejia, M., Kalei, N., Wong, B., Glazier, E.W., 2016. Restoring ecosystems, restoring community: socioeconomic and cultural dimensions of a community-based coral reef restoration project. Regional Environmental Change 16, 301–313.

Kittinger, J.N., Finkbeiner, E.M., Glazier, E.W., Crowder, L.B., 2012. Human dimensions of coral reef social-ecological systems. Ecology and Society 17, 17.

Klein, C.J., Jupiter, S.D., Selig, E.R., Watts, M.E., Halpern, B.S., Kamal, M., Roelfsema, C., Possingham, H.P., 2012. Forest conservation delivers highly variable coral reef conservation outcomes. Ecological Applications 22, 1246–1256.

Klein, R.J.T., Nicholls, R.J., Thomalla, F., 2004. Resilience to Natural Hazards: How Useful Is This Concept? (No. 9). EVA Working Paper, Potsdam, Germany.

Leach, M., Mearns, R., Scoones, I., 1999. Environmental entitlements: dynamics and institutions in community-based natural resource management. World Development 27, 225–247. http://dx.doi.org/10.1016/S0305-750X(98)00141-7.

Lele, S., Wilshusen, P., Brockington, D., Seidler, R., Bawa, K., 2010. Beyond exclusion: alternative approaches to biodiversity conservation in the developing tropics. Current Opinion in Environmental Sustainability 2, 94–100. http://dx.doi.org/10.1016/j.cosust.2010.03.006.

Leslie, H.M., Mcleod, K.L., 2007. Confronting the challenges of implementing marine ecosystem-based management. Frontiers in Ecology and the Environment 5, 540–548.

Lotze, H.K., Lenihan, H.S., Bourque, B.J., Bradbury, R.H., Cooke, R.G., Kay, M.C., Kidwell, S.M., Kirby, M.X., Peterson, C.H., Jackson, J., 2006. Depletion, degradation, and recovery potential of estuaries and coastal seas. Science 312 (80), 1806–1809.

Maclean, K., Cuthill, M., Ross, H., 2014. Six attributes of social resilience. Journal of Environmental Planning and Management 57, 144–156. http://dx.doi.org/10.1080/09640568.2013.763774.

Magis, K., 2010. Community resilience: an indicator of social sustainability. Society and Natural Resources 23, 401–416. http://dx.doi.org/10.1080/08941920903305674.

Mansfield, B., 2007. Property, markets, and dispossession: the western Alaska community development quota as neoliberalism, social justice, both, and neither. Antipode 39, 479–499.

Marschke, M., Berkes, F., 2006. Exploring strategies that build livelihood resilience: a case from Cambodia. In: Survival of the Commons: Mounting Challenges and New Realities, the Eleventh Conference of the International Association for the Study of Common Property Bali, Indonesia.

Marshall, N., Marshall, P., 2007. Conceptualizing and operationalizing social resilience within commercial fisheries in northern Australia. Ecology and Society 12.

Martin-Breen, P., Anderies, J.M., 2011. Resilience: A Literature Review.

Masten, A.S., 2001. Ordinary magic: resilience processes in development. American Psychologist 56, 227.

Masten, A.S., Obradovic, J., 2008. Disaster preparation and recovery: lessons from research on resilience in human development. Ecology and Society 13, 9.

McCaulay, D., Pinsky, M., Palumbi, S., Estes, J., Joyce, F., Warner, R., 2015. Marine defaunation: animal loss in the global ocean. Science 347 (80), 1255641.

McCay, B.J., Micheli, F., Ponce-Díaz, G., Murray, G., Shester, G., Ramirez-Sanchez, S., Weisman, W., 2014. Cooperatives, concessions, and co-management on the Pacific coast of Mexico. Marine Policy 44, 49–59. http://dx.doi.org/10.1016/j.marpol.2013.08.001.

McClanahan, T.R., Cinner, J.E., Maina, J., Graham, N.A.J., Daw, T.M., Stead, S.M., Wamukota, A., Brown, K., Ateweberhan, M., Venus, V., Polunin, N.V.C., 2008. Conservation action in a changing climate. Conservation Letters 1, 53–59. http://dx.doi.org/10.1111/j.1755-263X.2008.00008.x.

Micheli, F., Saenz-Arroyo, A., Greenley, A., Vazquez, L., Espinoza Montes, J.A., Rossetto, M., De Leo, G., 2012. Evidence that marine reserves enhance resilience to climatic impacts. PLoS One 7, e40832. http://dx.doi.org/10.1371/journal.pone.0040832.

Moberg, F., Ronnback, P., 2003. Ecosystem services of the tropical seascape: interactions, substitutions and restoration. Ocean and Coastal Management 46, 27–46.

Nelson, D.R., Adger, W.N., Brown, K., 2007. Adaptation to environmental change: contributions of a resilience framework. Annual Review of Environment and Resources 32, 395–419. http://dx.doi.org/10.1146/annurev.energy.32.051807.090348.

Nicholls, R., Cazenave, A., 2010. Sea-level rise and its impact on coastal zones. Science 328 (80), 1517–1520.

Nicholls, R.J., Marinova, N., Lowe, J.A., Brown, S., Vellinga, P., De Gusmao, D., Hinkel, J., Tol, R.S., 2011. Sea-level rise and its possible impacts given a "beyond 4 C world" in the twenty-first century. Philosophical Transactions of the Royal Society A: Mathematical, Physical and Engineering Sciences 369, 161–181.

Nicholls, R.J., Mimura, N., 1998. Regional issues raised by sea-level rise and their policy implications. Climate Research 11, 5–18.

Nicholls, R.J., Wong, P.P., Burkett, V.R., Codignotto, J.O., Hay, J.E., McLean, R.F., Ragoonaden, S., Woodroffe, C.D., 2007. Coastal systems and low-lying areas. In: Parry, M.L., Canziani, O.F., Palutikof, J.P., van der Linden, P.J., Hanson, C.E. (Eds.), Climate Change 2007: Impacts, Adaptation and Vulnerability. Contribution of Working Group II to the Fourth Assessment Report of the Intergovernmental Panel on Climate Change. Cambridge University Press, Cambridge, UK, pp. 315–356.

Nielsen, J.R., 2003. An analytical framework for studying: compliance and legitimacy in fisheries management. Marine Policy 27, 425–432. http://dx.doi.org/10.1016/S0308-597X(03)00022-8.

Nobre, A.M., 2011. Scientific approaches to address challenges in coastal management. Marine Ecology Press Series 434, 279–289. http://dx.doi.org/10.3354/meps09250.

Oliver-Smith, A., 2009. Sea Level Rise and the Vulnerability of Coastal Peoples: Responding to the Local Challenges of Global Climate Change in the 21st Century (No. 7). Interdisciplinary Security Connections Publication Series of UNU-EHS, Bonn, Germany.

Orth, R.J., Carruthers, T.J., Dennison, W.C., Duarte, C.M., Fourqurean, J.W., Heck, K.L., Hughes, A.R., Kendrick, G.A., Kenworthy, W.J., Olyarnik, S., Short, F.T., 2006. A global crisis for seagrass ecosystems. Bioscience 56, 987–996.

Österblom, H., Jouffray, J.-B., Folke, C., Crona, B., Troell, M., Merrie, A., Rockström, J., 2015. Transnational corporations as "keystone actors" in marine ecosystems. PLoS One 10, e0127533. http://dx.doi.org/10.1371/journal.pone.0127533.

Pandolfi, J.M., Connolly, S.R., Marshall, D.J., Cohen, A.L., 2011. Projecting coral reef futures under global warming and ocean acidification. Science 333 (80), 1204794. http://dx.doi.org/10.1126/science.1204794.

Pauly, D., 1990. On Malthusian overfishing. Naga, ICLARM Quarterly 13, 3–4.

Pauly, D., Christensen, V., Dalsgaard, J., Froese, R., Torres Jr., F., 1998. Fishing down the food webs. Science. http://dx.doi.org/10.1126/science.279.5352.860.

Perrings, C., 2006. Resilience and sustainable development. Environment and Development Economic 11, 417–427.

Perry, R.I., Ommer, R.E., Barange, M., Jentoft, S., Neis, B., Sumaila, U.R., 2011. Marine social-ecological responses to environmental change and the impacts of globalization. Fish Fish 12, 427–450. http://dx.doi.org/10.1111/j.1467-2979.2010.00402.x.

Peterson, G., 2000. Political ecology and ecological resilience: an integration of human and ecological dynamics. Ecological Economics 35, 323–336. http://dx.doi.org/10.1016/S0921-8009(00)00217-2.

Pinkerton, E., Edwards, D.N., 2009. The elephant in the room: the hidden costs of leasing individual transferable fishing quotas. Marine Policy 33, 707–713. http://dx.doi.org/10.1016/j.marpol.2009.02.004.

Raemaekers, S., Hauck, M., Bürgener, M., Mackenzie, A., Maharaj, G., Plagányi, É.E., Britz, P.J., 2011. Review of the causes of the rise of the illegal South African abalone fishery and consequent closure of the rights-based fishery. Ocean and Coastal Management 54, 433–445. http://dx.doi.org/10.1016/j.ocecoaman.2011.02.001.

Rockstrom, J., Steffen, W., Noone, K., Persson, A., Chapin, F.S., Lambin, E.F., Lenton, T.M., Scheffer, M., Folke, C., Schellnhuber, H.J., Nykvist, B., de Wit, C.A., Hughes, T., van der Leeuw, S., Rodhe, H., Sorlin, S., Snyder, P.K., Costanza, R., Svedin, U., Falkenmark, M., Karlberg, L., Corell, R.W., Fabry, V.J., Hansen, J., Walker, B., Liverman, D., Richardson, K., Crutzen, P., Foley, J.A., 2009. A safe operating space for humanity. Nature 461, 472–475.

Scheffer, M., 2009. Critical Transitions in Nature and Society. Princeton University Press.

Schlager, E., Ostrom, E., 1992. Property-rights regimes and natural resources: a conceptual analysis. Land Economy 68, 249–262.

Schumpeter, J., 1942. Creative destruction. In: Captialism, Socialism and Democracy. Harper, New York, pp. 82–85.

Singleton, S., 2000. Co-operation or capture? The paradox of co-management and community participation in natural resource management and environmental policy-making. Environmental Politics 9, 1–21.

Smit, B., Wandel, J., 2006. Adaptation, adaptive capacity and vulnerability. Global Environmental Change 16, 282–292. http://dx.doi.org/10.1016/j.gloenvcha.2006.03.008.

Swartz, W., Rashid Sumaila, U., Watson, R., Pauly, D., 2010. Sourcing seafood for the three major markets: the EU, Japan and the USA. Marine Policy 34, 1366–1373. http://dx.doi.org/10.1016/j.marpol.2010.06.011.

UNEP, 2006. Marine and Coastal Ecosystems and Human Well-Being: A Synthesis Report Based on the Findings of the Millennium Ecosystem Assessment. Nairobi, Kenya.

Valiela, I., Bowen, J.L., York, J.K., 2001. Mangrove forests: one of the World's threatened major tropical environments. Bioscience 51.

Vayda, A., McCay, B., 1975. New directions in ecology and ecological anthropology. Annual Review of Anthropology 4, 293–306.

Vincent, K., 2007. Uncertainty in adaptive capacity and the importance of scale. Global Environmental Change 17, 12–24. http://dx.doi.org/10.1016/j.gloenvcha.2006.11.009.

Walker, B., Holling, C., Carpenter, S., Kinzig, A., 2004. Resilience, adaptability and transformability in social–ecological systems. Ecology and Society 9.

Walker, B.H., Salt, D., 2006. Resilience Thinking: Sustaining People and Ecosystems in a Changing World. Island Press, Washington, DC.

Waycott, M., Duarte, C.M., Carruthers, T.J.B., Orth, R.J., Dennison, W.C., Olyarnik, S., Calladine, A., Fourqurean, J.W., Heck, K.L., Hughes, A.R., Kendrick, G.A., 2009. Accelerating loss of seagrasses across the globe threatens coastal ecosystems. Proceedings of the National Academy of Sciences of the United States of America 106, 12377–12381.

Wilkinson, C., Brodie, J., 2011. Catchment Management and Coral Reef Conservation: A Practical Guide for Coastal Resource Managers to Reduce Damage from Catchment Areas Based on Best Practice Case Studies.

Worm, B., Barbier, E.B., Beaumont, N., Duffy, J.E., Folke, C., Halpern, B.S., Jackson, J.B.C., Lotze, H.K., Micheli, F., Palumbi, S.R., Sala, E., Selkoe, K.A., Stachowicz, J.J., Watson, R., 2006. Impacts of biodiversity loss on ocean ecosystem services. Science 314 (80), 787–790. http://dx.doi.org/10.1126/science.1132294.

Worm, B., Branch, T.A., 2012. The future of fish. Trends in Ecology and Evolution 27, 594–599. http://dx.doi.org/10.1016/j.tree.2012.07.005.

Worm, B., Hilborn, R., Baum, J.K., Branch, T.A., Collie, J.S., Costello, C., Fogarty, M.J., Fulton, E.A., Hutchings, J.A., Jennings, S., Jensen, O.P., Lotze, H.K., Mace, P.M., McClanahan, T.R., Minto, C., Palumbi, S.R., Parma, A.M., Ricard, D., Rosenberg, A.A., Watson, R., Zeller, D., 2009. Rebuilding global fisheries. Science 325 (80), 578–585. http://dx.doi.org/10.1126/science.1173146.

Section II

Principles for Conservation in the Anthropocene

Chapter 6

Principles for Interdisciplinary Conservation

Heather M. Leslie
University of Maine, Walpole, ME, United States

I have been intrigued by people's connections to nature for as long as I can remember. I grew up in Plymouth, Massachusetts, just 15 minutes from the town beach. This narrow barrier beach protects the harbor, and what remains of Plymouth Rock, where the Pilgrims are said to have landed in 1620. At its 3-mile tip, deep bare swales and grass-covered dunes shelter nesting gulls and terns in the summertime. Striped bass, bluefish, and other predatory fish swim over the banks just offshore, as do silversides and other small fish that terns feed their young.

People are drawn to this beach as well. More than a dozen houses dot its spine and in the summertime, beachgoers crowd its shores. People are allowed to drive on the beach itself, not just on the road that bisects the narrow peninsula. When I was a girl, my mom would drive my brother, sister, and me more than two miles out the beach's spine. In between the long, hot, jostling rides in the back of my dad's jeep, we would spend the day covered in sand and salt, peering in tide pools, and splashing in the cold water.

As a college student, I left my car behind and walked this same beach, counting terns, plovers, and other migratory beach birds for a statewide conservation organization. My favorite days on the job were those when I was out at the tip of the peninsula before the sun rose. I enjoyed the quiet and deliberate search for the young plover chicks, who were much more likely to be active at dawn than during the bustle and heat of midday. But many days, I walked the beach later, when the sun was overhead and the high tide line was covered by a row of cars and trucks like my dad's jeep.

While far from meditative, these hotter workdays were a challenge I enjoyed. My job was still to count birds, but I now had an opportunity to show them to others. Most people had come for the view, the water, and the chance to let their kids run free. Many had never seen young plovers. Some did not care to or were downright hostile (I saw more than one bumper sticker proclaiming "Piping Plover Tastes Like Chicken"), but most were curious about the diminutive birds and pleased to catch a look at them through my spotting scope.

Conservation for the Anthropocene Ocean. http://dx.doi.org/10.1016/B978-0-12-805375-1.00006-4
109

The people I played next to on Plymouth Beach as a little girl, and later spoke with as a young biologist, were from many different walks of life. They had different ways of knowing the place that I loved, and different reasons why it was important to them. I sought to make sense of their passions and to more deeply understand their origins and what they meant for the future of the beach and the birds. I did not know it then, but in my first foray as a conservation scientist, I was seeking an interdisciplinary approach to the connections I observed among people and nature on Plymouth Beach.

WHAT MAKES CONSERVATION INTERDISCIPLINARY?

When conservation science emerged in the 1980s, the field focused on biology and allied disciplines, in order to understand how and why species and ecosystems were changing (Kareiva and Marvier, 2012). Today, the scope of the field has broadened. Conservation scientists are interested in a wide array of connections between people and nature, i.e., socio-ecological systems (see Fig. 6.1, from Mace, 2014).

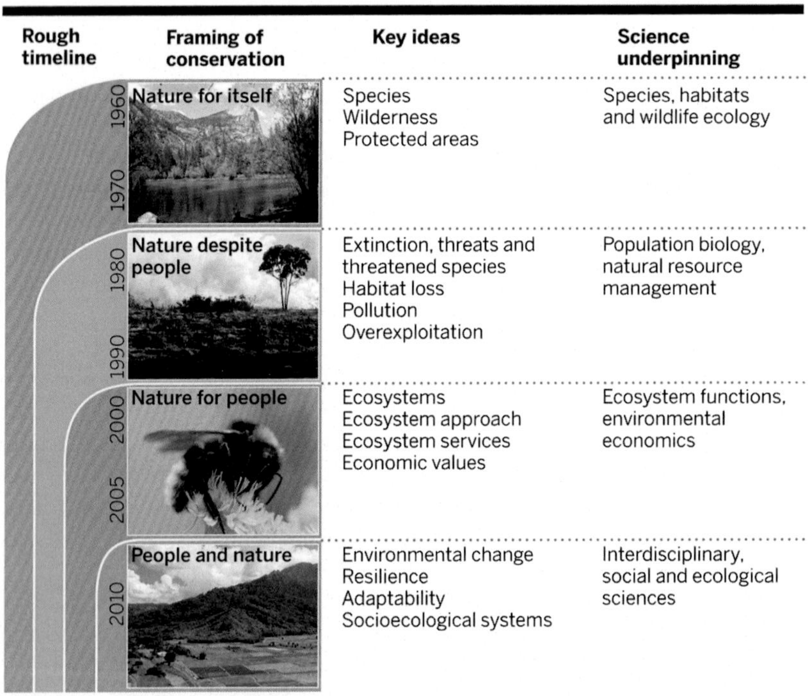

Rough timeline	Framing of conservation	Key ideas	Science underpinning
1960–1970	Nature for itself	Species, Wilderness, Protected areas	Species, habitats and wildlife ecology
1980–1990	Nature despite people	Extinction, threats and threatened species, Habitat loss, Pollution, Overexploitation	Population biology, natural resource management
2000–2005	Nature for people	Ecosystems, Ecosystem approach, Ecosystem services, Economic values	Ecosystem functions, environmental economics
2010	People and nature	Environmental change, Resilience, Adaptability, Socioecological systems	Interdisciplinary, social and ecological sciences

FIGURE 6.1   The framing of conservation—and the science that supports it—has evolved in the last 50 years. Note that none of the framings have been discarded, and thus contemporary conservation includes multiple framings. *Reprinted with permission from Mace, G., 2014. Whose conservation? Science 345, 1558–1560.*

TABLE 6.1 Principles of Engagement for Conservation Science and Practice

1. Attention and openness when defining the coupled natural and human systems of interest. That is, a willingness to do interdisciplinary scholarship;
2. Commitment to useful, solutions-oriented scholarship;
3. Mindful engagement with collaborators, including community members; and
4. Humility.

As the chapters in this section aptly illustrate, the science and practice of conservation leverages knowledge and approaches from both the biophysical sciences and social sciences. Some of these chapters were motivated primarily by approaches from ecology and evolution, whereas others were written by individuals with depth in anthropology, human geography, and other social science disciplines. Together they create a rich understanding of social–ecological systems (which are also referred to as coupled human–natural systems or socioenvironmental systems or human–environment systems). This understanding is deeper than one discipline could possibly provide.

A program officer at a foundation that has been deeply engaged in conservation science once made a comment to me that really stuck. We were discussing how best to craft an investigation of ecosystem-based management in practice, a topic I will explore further later in this chapter. He cautioned me (and I am paraphrasing): We are only interested in interdisciplinary work, if that is what is needed to address the challenge before us. Interdisciplinarity for its own sake is not the goal.

This is an important point. We can learn so much about the world through both disciplinary and interdisciplinary scholarship. The question before us is—at least those of us who seek to translate our science into action—what approaches are needed to help craft durable solutions to the marine conservation challenges we face? That brings me to Principle #1: I pay explicit attention when defining the coupled natural and human systems of interest, with the intention that the research will be both intellectually exciting and societally relevant (see Table 6.1). Thus the scales of both ecological and social dynamics—and particularly how people make decisions about the use and stewardship of ocean resources—guide much of my work. I find that more often than not that this approach to defining my study system compels me to do interdisciplinary scholarship.

LINKING KNOWLEDGE TO ACTION

Like many conservation scientists trained in the 1980s and 1990s, my training is in biology. After post baccalaureate field biology and conservation experiences in Maine, Massachusetts, and Mexico, I pursued a PhD in ecology and conservation biology at Oregon State University. I went on to do a postdoc at

Princeton University. Over the last 20 years, I have applied my ecological training to contribute to conservation in practice in multiple settings, from the design of marine reserves to the implementation of ecosystem-based management.

Now as faculty and director of University of Maine's marine laboratory, the Darling Marine Center, I still love to study barnacles, birds, and other organisms living in coastal ecosystems. But at this stage in my career, I have a hard time crafting a research project that does not involve human dynamics. That means that the majority of my research is interdisciplinary, leveraging both social and natural sciences approaches.

I seek out research questions and collaborators—including other conservation scientists, conservation practitioners, and community members—that enable me to be as responsive and relevant to conservation in practice as possible. Sometimes that means working directly on a problem identified by decision makers working at the local or regional level, whereas at other times that means working on a challenge I have identified, with the expectation that the connection to "the real world" will come later, and not necessarily through my direct actions. This spectrum of science is specific to the circumstances within which I work, as an academic scientist at a Land and Sea Grant institution. I recognize that others have different opportunities to pursue fundamental versus applied projects, depending on their institutional context and interests.

This spectrum of research is in keeping with Donald Stokes' articulation of the three types of research and the need to have a wide range of approaches, including the use-inspired basic research (Stokes, 1997; Clark, 2007) (see Fig. 6.2). This attention to both knowledge and solutions is not the sole purview of conservation science. But, for me, it is a vital one that guides my engagement in both conservation science and practice. Principle #2 is: A commitment to solutions-oriented scholarship (see Table 6.1).

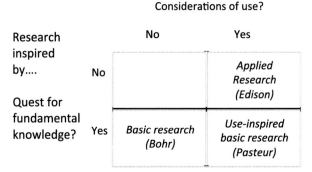

FIGURE 6.2 Stokes (1997), as cited by Clark (2007) articulates three distinct and complementary motivations for conducting research. Conservation science fits squarely in the use-inspired basic research quadrant, but also can include activities in the other two quadrats.

Sustainability science has a similar orientation. In fact, these two fields have influenced one another over the last two decades in particular. Conservation science sprung from the field of conservation biology (Soule, 1985), while sustainability science began with geographers (Kates et al., 2001; Clark, 2007). Scientists in both of these areas have a strong systems orientation. In other words, they recognize that the pieces of the puzzle that they are most interested in—whether it is an endangered whale or the fisherman who is incidentally catching those whales, or both—are linked, and that in order to understand the impacts of these connections on both people and nature, one needs to have an understanding of the whole (Levin, 1999).

LEARNING BY DOING, WITH OTHERS

My efforts to translate knowledge into action have definitely been a process of trial and error. I have not always "hit my mark," in terms of actually observing an impact of my science on conservation policy or action. However, I have learned by doing and also by sharing stories with colleagues—many of whom have contributed to this volume—about effective ways to engage in conservation practice. This brings me to Principle #3: Mindful engagement with collaborators. I want to dive a bit more deeply into how this has played out in my own work, as I believe that collaboration is essential for both the science and practice of conservation (see Table 6.1). This belief comes from both personal experience, as well as my reading of the literature, including works by Layzer (2008) and Wondolleck and Yaffee (2000).

Most recently, my work has focused on ecosystem-based management. People have been talking about ecosystem management on land for decades, but it was not until the late 1990s that the concept of ecosystem-based management took root in the marine domain. The US "oceans agency," NOAA (National Oceanic and Atmospheric Administration), began to focus on a more integrated approach to fisheries management, one that connected efforts to sustain one stock with rules related to other stocks, as well as with protected species like whales in the late 1990s (NOAA, 1999). That work has led to exciting scientific and practical advances, both in fisheries management (Levin et al., 2009) as well as in ocean protection and stewardship more broadly (Lubchenco and Grorud-Colvert, 2015).

Interest in marine ecosystem-based management was growing in many parts of the world as I was wrapping up my PhD in 2004. Together with Karen McLeod, I brought together a group of more than 40 scientists and practitioners working on different dimensions of ecosystem-based management in coastal and ocean areas. We gathered to write the first book on the science and practice of marine ecosystem-based management, *Ecosystem-Based Management for the Oceans* (McLeod and Leslie, 2009). In defining ecosystem-based management, we followed the definition affirmed by more than 200 scientists in 2005: "Ecosystem-based management is an integrated approach to management that considers the entire ecosystem, including humans. The goal is to maintain an ecosystem in a healthy, productive and resilient condition so that it can provide the services humans want and need" (McLeod et al., 2005).

While connecting the dots among our diverse experiences and opinions about what the concept means, what science is needed, and where the concept has been tried and tested in the water, we helped to create a community of scholars. Some of these individuals had been working on the topic for a long time, while others were fairly new to the theme. What I am most proud of about the five-year project that produced our book was the network of scientists and practitioners that it helped to catalyze. Through the book project and efforts that followed, we were able to catalog and better learn from the impressive contributions of many of the scientists engaged in ecosystem-based management from the local to international scale. While Karen and I cannot take credit for the many impacts of our colleagues' work, I believe that our co-creation of the book and the conversations that it sparked contributed in some small way to where we are now with ecosystem-based management, both in the United States and other parts of the world.

One of the most concrete advances of the last decade was the creation of the first-ever US Ocean Policy, through Executive Order 13547 (Obama, 2010). The National Ocean Policy was established "to ensure the protection, maintenance, and restoration of the health of ocean, coastal, and Great Lakes ecosystems and resources, enhance the sustainability of ocean and coastal economies, preserve our maritime heritage, support sustainable uses and access, provide for adaptive management to enhance our understanding of and capacity to respond to climate change and ocean acidification, and coordinate with our national security and foreign policy interests." In addition, it mandated the development of coastal and marine spatial plans designed to "enable a more integrated, comprehensive, ecosystem-based, flexible, and proactive approach to planning and managing sustainable multiple uses across sectors and [to] improve the conservation of the ocean, our coasts, and the Great Lakes."

In 2010, when the policy was released, New England—where I live—was in a particularly strong position to respond to the new mandate. Decision makers from both the state and federal governments, as well as from universities and environmental nonprofits, had been working in a voluntary forum since 2005 to integrate science related to ecosystem-based management and translate it into action through the Northeast Regional Ocean Council (NROC). For example, NROC developed the Northeast Ocean Data Portal, which was launched in June 2011 to integrate the varied data available on ocean uses and the environment (http://www.NortheastOceanData.org). In addition to the portal, NROC's current areas of focus include ocean and coastal ecosystem health, coastal hazards resilience, and ocean planning.

Building on the foundational work of NROC, the Northeast region was the first of the nation's nine regions to establish the Regional Planning Body (RPB) mandated by the 2010 ocean policy. The RPB was established in

November 2012, and includes representatives from the six New England states, 10 federally recognized tribes, 10 federal agencies, and the New England Fishery Management Council. In December 2016, the Northeastern Ocean Plan was certified by the federal government. The group has no authority to create new regulations. Its mandate is to create a plan and oversee its implementation, with many opportunities for public participation. The completion of the Northeast Ocean Plan is a major accomplishment, both for the region and for the nation. We are among the first to translate the national mandate into in-the-water action and from this experiment, we will no doubt learn a great deal.

I have been observing the work of the RPB with great interest. I want to better understand how best to translate knowledge into action and whether the knowledge my colleagues and I have generated about EBM in practice actually contributes to policy and practice. I have found that learning by doing, in collaboration with those who implement policy and conservation measures like those enabled by the national ocean policy, is one of the most satisfying albeit challenging dimensions of my work as a conservation scientist.

INTERDISCIPLINARY SCIENCE: A MESSY AND MINDFUL PROCESS

The plans emerging in the Northeast and parallel efforts in other parts of the country and world are not without precedent. People have been working to translate the concept of ecosystem-based management into practice in varied social–ecological contexts for a number of years. Research that I conducted with human geographer Lisa Campbell and environmental anthropologist Leila Sievanen showed that the discourse in these efforts has tended to be dominated by biological scientists (Sievanen et al., 2012), similar to what was observed in the early development of the field of conservation science (Kareiva and Marvier, 2012). I observed this predilection in my own early research on human–environment connections; it was easier to reach out to people who had ways of looking at the world similar to my own. During my first years in the field, I did not yet have the knowledge or experience to identify which of the many branches of the social sciences—anthropology, economics, and geography, among others—and fields within them—specifically political ecology, environmental economics, and institutional design—could help me better articulate and answer the questions that emerged for me in Massachusetts and Mexico. Here we are fortunate to have contributions from scholars with deep knowledge in many of these areas (see Chapters 7–9 and 12, in particular).

While a postdoc at Princeton University, I was fortunate to be working in a group with incredibly varied disciplinary backgrounds. Economists, ecologists, mathematicians, and physicists sat around the same table. Our common language was mathematics, and we sought to describe how ecosystems worked and what the implications of those dynamics were, for both nature and people.

Together with Maja Schlüter, Richard Cudney-Bueno, and Simon Levin, I investigated how different patterns of connectivity between people and ecosystems influenced the sustainability of those connections. Specifically, we were interested in whether fishermen with different interests and fishing strategies would respond differently to climate variability, and whether that variation translated into variation in catch and economic returns. We found that they did; based on linked biological and economic models inspired by small-scale fisheries in the northern part of Mexico's Gulf of California (Leslie et al., 2009).

I maintained this focus on linked ecological and economic systems for a number of years, for several reasons. For one, ecologists and economists use many of the same approaches to gather and analyze data. While ecologists recognize the importance of natural history and economists need to validate their models with real-world observations, both disciplines emphasize quantitative, or numbers-based, analyses. Second, since I aspire to connect my science to conservation action, I try to work on a scale that is relevant to managing people's interactions with coastal and marine environments. That often means leveraging government statistics and other secondary data sources that encompass larger spatial and temporal scales than I could gather myself. It was easier to locate secondary economic and ecological data sources, than sources on some other aspects of coastal marine social–ecological systems. Finally, having been trained as an ecologist, ecologists dominate my professional network. As time went on, I sought to deliberately engage with a more diverse group of collaborators with expertise in both the social and natural sciences.

While some of these connections have been forged through professional meetings and seminars, I have found that time together in the field has been incredibly important to build both scientific understanding and trust. One paper I recently published, operationalizing Ostrom's social–ecological systems framework to assess sustainability of small-scale fisheries in Baja California Sur, Mexico (Leslie et al., 2015), originated from time in the field. This paper emerged from several years of research and engagement with in-the-water conservation and management.

One of the turning points in this collaboration occurred during our "all team" meeting in La Paz, Mexico in mid-2012. After meeting over PowerPoint and coffee for a couple of days, part of the team broke away from the city and traveled by boat to a remote part of Espíritu Santo National Park (known in Spanish as Parque Nacional Archipélago de Espíritu Santo), off the coast of La Paz. Camping on the beach, in view of a seasonal fishing camp and breaching whales, we sketched out what we thought we had learned thus far about the social and ecological systems supporting the region's fisheries. We had lots of details, from more than a decade of ecological surveys of reef fish communities, new interview and survey data from fishing communities, and reams of government fisheries and census statistics. We drew from a range of experiences and disciplinary perspectives, too. Some of us had lived and worked in the region for years; and others had only just arrived for the first time. Some of us had been working on these questions for decades; whereas for others, this was a first foray into interdisciplinary conservation science.

We cooked together. We swam together, dodging beautiful but stinging jellyfish, in the warm salty water just offshore from our tents. When we needed relief from the beach, we picked our way up a narrow trail among the cacti. We were rewarded with a view of the entire park and a sheltered mangrove in the next cove. And we talked late into the night, about science and life, marveling at the sky lit only by stars. In the course of these conversations, we not only pieced together what we knew about the coupled social–ecological systems that we were studying and trying to help sustain; we also realized what we did not know. Among the hikes and swims, we were able to set a course for how to tackle those unknowns, as a team.

BRIDGING DISCIPLINARY DIVIDES

The camping trip at Espíritu Santo highlighted the importance of deepening knowledge of the social system related to Baja's small-scale fisheries. Thanks to more than a decade of intensive fieldwork by Octavio Aburto and other marine scientists working in the region, the reef fish assemblages and dynamics of the natural system were fairly well known. Surprisingly, however, the human dimensions of the fisheries were much less well studied.

While there are real challenges to integrating social science into conservation science and practice, I have found that linking the perspectives of those who rely on quantitative data (whether they concern ecological, economic, or social systems) and qualitative approaches (e.g., interviews, focus groups, ethnography, discourse, and document analysis) can be just as if not more challenging. It was not until another field trip, perhaps a year after the camping trip, that I gained a full appreciation for the complementarity of qualitative and quantitative approaches. Sustainability scientist (and author of Chapter 15) Xavier Basurto and others with expertise in institutional and social dynamics had fortuitously begun work in some of the same areas where my collaborators and I were working.

In preparation for conducting surveys of the region's fishermen, I traveled with Xavier's graduate student Mateja Nenadovic and my postdoc Leila Sievanen to Cabo Pulmo, in the region known as "East Cape." Cabo Pulmo is a jarring two and a half hours south from La Paz. Like Espíritu Santo, the area includes a national park. We spent our first afternoon visiting with fishermen and other community members whom Mateja and Leila had met in the past. These old friends introduced us to new friends. Within a couple of days, we had met with enough people and the right people to convene a focus group in the municipal hall.

In the early evening, while the day was still hot, we arrived at the municipal hall with a bag of still-warm tacos, in plenty of time to make coffee and set up the room. Fishermen began to arrive as the heat of the day began to recede, after dinner. They greeted each other quietly and settled around the meeting table with their coffee and tacos, waiting. We went around the table, introducing ourselves. Mateja explained that we were interested in learning more about their perspectives on the fisheries and the park. Questions and observations came quickly after that. *Yes, the creation of the park has changed how we fish*, the fishermen affirmed. Individuals shared their thoughts on how the rules of the park had changed through time, and where and how they and others fished. *What do you know about climate change?* several people asked. They talked about how their fishing organizations and the permitting process had changed through the years, and their observations of seasonal change in the environment and the species they fished.

This focus group was a critical precursor to a formal survey, from which we ultimately collected information from 160 of the estimated 250 fishermen working in the vicinity of the park (Leslie et al., 2015). We developed similar surveys for communities in the vicinity of La Paz and Loreto, to the north. In all, Mateja, Leila, and the enumerators trained as part of this project, surveyed more than 430 fishermen over 18 months. Results from this substantial research effort contributed to a recent collaborative paper (Leslie et al., 2015) and are reported on in greater depth in other papers (Sievanen, 2014; Nenadovic and Epstein, 2016). From these data, along with complementary interviews, observations of activities in the communities and on the water, and analyses of peer-reviewed and popular literature, we were able to assemble a reasonable picture of the social and institutional variation of the region's small-scale fisheries.

This baseline knowledge then enabled us to ask a bigger, theoretically grounded question: *How does the capacity for sustaining small-scale fisheries differ throughout the state of Baja California Sur?* We used the social–ecological systems framework developed by Nobel laureate Elinor Ostrom and colleagues in order to answer this question (Ostrom, 2009; Basurto et al., 2013b). We found that the potential for sustainability varies geographically (Leslie et al., 2015). This in itself was not a surprise; social–ecological variation among different places in Baja had been documented previously (Basurto et al., 2013a; Reddy et al., 2013; McCay et al., 2014). However, given that the region's fisheries are currently managed on the scale of the state of Baja California Sur, our results provide additional evidence for the importance of tailoring management to a finer spatial scale. We also found that different parts of the state—what we term social–ecological system regions—vary in nature of their capacity for sustainability, based on Ostrom's four dimensions (see Fig. 6.3). This dissonance among the social and ecological dimensions of sustainability within and across Baja California Sur's social–ecological system regions suggests how fisheries management may be tailored to local conditions in order to foster both sustainable coastal ecosystems and human communities.

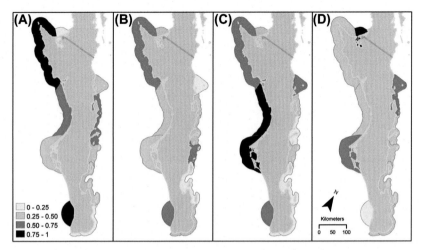

FIGURE 6.3   The potential for sustaining fisheries varies geographically among the 12 social–ecological system regions of the Mexican state of Baja California Sur. These maps provide a visualization of the four dimensions associated with sustainability. The dimensions are: (A) governance system, (B) actors, (C) resource units, and (D) resource system. Values range from 0 to 1, where a larger value is associated with a greater probability that fisheries will be sustainably managed. Details regarding the quantification of these dimensions and underlying data and theory can be found in the supporting information associated with this publication. *Reprinted with permission from Leslie, H.M., Basurto, X., Nenadovic, M., Sievanen, L., Cavanaugh, K.C., Cota-Nieto, J.J., Erisman, B.E., Finkbeiner, E., Hinojosa-Arango, G., Moreno-Báez, M., Nagavarapu, S., Reddy, S.M., Sánchez-Rodríguez, A., Siegel, K., Ulibarria-Valenzuela, J.J., Weaver, A.H., Aburto-Oropeza, O., 2015. Operationalizing the social–ecological systems framework to assess sustainability. Proceedings of the National Academy of Sciences of United States of America 112, 5979–5984.*

LOOKING AHEAD

In collaboration with many others, I am still expanding the edges of social–ecological systems knowledge in Mexico, and seeking ways to translate knowledge into action. Through partnerships with members of the fishing communities where we work as well as with partners in government and regional environmental and community development organizations, I am exploring how to make the knowledge we create most relevant. These exchanges highlight Principle #4: humility. Just because we have published papers on the environment and fishing communities of Baja does not give us a special place in crafting solutions to the challenges facing this region and the wider world (see Table 6.1). Our work represents but some of many needed pieces. Remembering that and demonstrating humility and respect for the value of different knowledge domains is critical to our success as conservation scientists.

Each member of our team is pursuing an interdisciplinary agenda to generate new knowledge and contribute in some way to conservation and management in Baja (see, for example, data stories from Octavio and his team at http://datamares.ucsd.edu/eng/). Through community meetings and one-on-one conversation on the beach, feedback from practitioners and community members are helping to shape the next stages of the research. Working together, we are able to answer questions that we could never tackle alone. Our individual research agendas are also enriched by this collaboration; as we seek answers to questions within our specific disciplines, our approaches are informed by a broader and deeper understanding of the social–ecological system than we could possibly have otherwise.

I believe that the principles of engaging in interdisciplinary conservation science and practice that I have highlighted here are transferable to many contexts. Attention and openness to defining the boundaries of the systems of interest; commitment to useful, solutions-oriented scholarship; mindful engagement with collaborators; and humility are what I have taken away from my field experiences in the northeastern United States and Mexico. I hope that your journey is equally enriching and that you will make the time and space to share it with others in the conservation science community. Together, not only by publishing our science, but also by reflecting on how we engage others in both doing the work and understanding how that informs solutions, we have great opportunities to shape the rapidly changing and incredibly important field of conservation science.

When I first outlined this chapter, I ended with the question *Will we still need the term "interdisciplinary conservation" in 20 years?* I think not. In 20 years, I suspect that we will be hard pressed to find examples of robust conservation science that are *not* interdisciplinary. The National Science Foundation's Advisory Committee for Environmental Research & Education articulated a similarly bold view in their 10-year outlook for the future (AC-ERE, 2015). In 10 years, we will have trained a generation of scholars and practitioners who expect, if not demand, interdisciplinarity in the journals they read and publish in, and from their students, colleagues, and professional societies (see, for example, the prescient March 2016 special issue of *Oceanography* on graduate education). I anticipate that there will still be a spectrum of engagement in policy and management; some research projects will be more closely linked with practice than others. There will always be a need to push the bounds of fundamental knowledge as well as to link knowledge to action (Clark et al., 2016). With conservation scientists and practitioners active in all three of the quadrats described by Stokes (1997), the field of conservation science will continue to develop productively and ultimately enable what we all seek: meaningful contributions to the science and practice of sustaining both nature and people.

ACKNOWLEDGMENTS

I am grateful to many collaborators, neighbors, friends, and family members, who have helped to shape my thinking and actions related to conservation science and practice, as well as to the David and Lucile Packard Foundation and US National Science Foundation for vital support. I also thank the editors and Karen McLeod, Leila Sievanen, and two anonymous reviewers for their constructive comments on earlier versions of this chapter.

REFERENCES

AC-ERE, 2015. America's Future: Environmental Research and Education for a Thriving Century. A Report by the NSF Advisory Committee for Environmental Research and Education (AC-ERE). http://www.nsf.gov/geo/ere/ereweb/advisory.jsp.

Basurto, X., Bennett, A., Weaver, A.H., Rodriguez-Van Dyck, S., Aceves-Bueno, J.-S., 2013a. Cooperative and noncooperative strategies for small-scale fisheries' self-governance in the globalization era: implications for conservation. Ecology Society 18, 38.

Basurto, X., Gelcich, S., Ostrom, E., 2013b. The social–ecological system framework as a knowledge classificatory system for benthic small-scale fisheries. Global Environmental Change 23, 1366–1380.

Clark, W.C., 2007. Sustainability Science: a room of its own. Proceedings of the National Academy of Sciences of United States of America 104, 1737–1738.

Clark, W.C., van Kerkhoff, L., Lebel, L., Gallopin, G.C., 2016. Crafting usable knowledge for sustainable development. Proceedings of the National Academy of Sciences of United States of America 113, 4570–4578.

Kareiva, P., Marvier, M., 2012. What is conservation science? BioScience 62, 962–969.

Kates, R.W., Clark, W.C., Corell, R., Hall, J.M., Jaeger, C.C., Lowe, I., McCarthy, J.J., Schellnhuber, H.J., Bolin, B., Dickson, N.M., Faucheux, S., Gallopin, G.C., Grubler, A., Huntley, B., Jager, J., Jodha, N.S., Kasperson, R.E., Mabogunje, A., Matson, P., Mooney, H., Moore III, B., O'Riordan, T., Svedlin, U., 2001. Sustainability science. Science 292, 641–642.

Layzer, J.A., 2008. Natural Experiments: Ecosystem-Based Management and the Environment. The MIT Press, Cambridge, MA.

Leslie, H., Schlüter, M., Cudney-Bueno, R., Levin, S., 2009. Modeling responses of coupled social–ecological systems of the Gulf of California to anthropogenic and natural perturbations. Ecological Research 24, 505–519.

Leslie, H.M., Basurto, X., Nenadovic, M., Sievanen, L., Cavanaugh, K.C., Cota-Nieto, J.J., Erisman, B.E., Finkbeiner, E., Hinojosa-Arango, G., Moreno-Báez, M., Nagavarapu, S., Reddy, S.M.W., Sánchez-Rodríguez, A., Siegel, K., Ulibarria-Valenzuela, J.J., Weaver, A.H., Aburto-Oropeza, O., 2015. Operationalizing the social-ecological systems framework to assess sustainability. Proceedings of the National Academy of Sciences of United States of America 112, 5979–5984.

Levin, P.S., Fogarty, M.J., Murawski, S.A., Fluharty, D., 2009. Integrated ecosystem assessments: developing the scientific basis for ecosystem-based management of the ocean. PLoS Biology 7, e1000014. http://dx.doi.org/10.1371/journal.pbio.1000014.

Levin, S., 1999. Fragile Dominion. Perseus Books, Reading, MA.

Lubchenco, J., Grorud-Colvert, K., 2015. Making waves: the science and politics of ocean protection. Science. http://dx.doi.org/10.1126/science.aad5443.

Mace, G., 2014. Whose conservation? Science 345, 1558–1560.

McCay, B.J., Micheli, F., Ponce-Díaz, G., Murray, G., Shester, G., Ramirez-Sanchez, S., Weisman, W., 2014. Cooperatives, concessions, and co-management on the Pacific coast of Mexico. Marine Policy 44, 49–59.

McLeod, K.L., Lubchenco, J., Palumbi, S.R., Rosenberg, A.A., 2005. Scientific Consensus Statement on Marine Ecosystem-Based Management. Signed by 221 academic scientists and policy experts with relevant expertise and published by the Communication Partnership for Science and the Sea at http://www.compassonline.org/pdf_files/EBM_Consensus_Statement_v12.pdf.

McLeod, K.L., Leslie, H.M. (Eds.), 2009. Ecosystem-Based Management for the Oceans. Island Press, Washington, DC.

Nenadovic, M., Epstein, G., 2016. The relationship of social capital and fishers' participation in multi-level governance arrangements. Environmental Science & Policy 61, 77–86.

NOAA, 1999. Turning to the Sea: America's Ocean Future. Department of Commerce, Washington, DC, USA.

Obama, B., July 19, 2010. In: House, T.W. (Ed.), Executive Order 13547–Stewardship of the Ocean, Our Coasts, and the Great Lakes.

Ostrom, E., 2009. A general framework for analyzing sustainability of social-ecological systems. Science 325, 419–422.

Reddy, S.M.W., Wentz, A., Aburto-Oropeza, O., Maxey, M., Nagavarapu, S., Leslie, H.M., 2013. Evidence of market-driven size-selective fishing and the mediating effects of biological and institutional factors. Ecological Applications 23, 726–741.

Sievanen, L., 2014. How do small-scale fishers adapt to environmental variability? Lessons from Baja CaliforniaSur,Mexico.MaritimeStudies13,9.http://dx.doi.org/10.1186/s40152-40014-40009-40152.

Sievanen, L., Campbell, L.M., Leslie, H.M., 2012. Challenges to interdisciplinary research in ecosystem-based management. Conservation Biology 26, 315–323.

Soule, M.E., 1985. What is conservation biology? BioScience 35, 727–734.

Stokes, D.E., 1997. Pasteur's Quadrant: Basic Science and Technological Innovation. Brookings Institution Press.

Wondolleck, J.M., Yaffee, S.L., 2000. Making collaboration work: lessons from innovation. In: Natural Resource Management. Island Press.

Chapter 7

Creating Space for Community in Marine Conservation and Management: Mapping "Communities-at-Sea"

Kevin St. Martin[1, 2], Julia Olson[3]
[1]Rutgers University, New Brunswick, NJ, United States; [2]University of Tromsø, Tromsø, Norway; [3]NOAA Fisheries, Woods Hole, MA, United States

INTRODUCTION

The challenge of the Anthropocene to marine resource management, the challenge of a marine environment thoroughly modified by human practices, will require more than a technology for maximizing the harvest of single species, which itself appears not only less possible than once thought but also less desirable as a path toward human and environmental well-being (Fogarty, 2014). As elsewhere (Steffen et al., 2011), the challenge of the Anthropocene within the marine context will require a more holistic approach and a foregrounding of biophysical interactions and relations over space; indeed, it will require an ecosystem approach cognizant of not only the role of humans in modifying past ecosystems but also our responsibility to create felicitous conditions for future ecosystems (Pitcher, 2001).

As social scientists we are heartened and emboldened by the "turn toward ecosystems" (USCOP, 2004) insofar as it emphasizes process, interrelationality, context, and diversity, long-held key concepts, starting points, and hallmarks of critical social science (Hicks et al., 2016). Furthermore, we are pleased to find that what constitutes an ecosystem approach remains open, even contested, because, this suggests there remains room for experimentation, new integrations, and possibility. Indeed, "ecosystem" enables scientists, managers, and other members of human communities to imagine alternative and inventive scientific methods, solutions to degradation, and forms of production or engagement beyond those provided by the confines of single species management (Sievanen et al., 2011). In this sense, the shift in marine and fisheries science and management toward an ecosystems approach is a

Conservation for the Anthropocene Ocean. http://dx.doi.org/10.1016/B978-0-12-805375-1.00007-6
123

hopeful move that makes visible and analytically incorporates interrelationalities, multiscalar processes, assemblages and aggregations of species, mutual dependencies, and a shared ecosystem health or well-being as a foundation for sustainability.

Such ecosystem-based work is emerging in a number of contexts relevant to marine science and management and is increasingly prioritized in policy development and planning practices, despite deeply institutionalized modes of knowledge production and management focused on single species (Pitcher et al., 2009). In practice, an ecosystem approach will require, and we clearly see many signs of, new streams of data, new synergistic metrics, and spatial methods and technologies that not only reveal but also allow for the assessment and modeling of ecosystem relationships, dynamics, and trajectories across space from particular estuaries to continental-scale ecosystems (Berkes, 2012; Koehn et al., 2013).

A parallel shift in marine social science will also be needed, one that shifts beyond analyses that presume and privilege a singularly driven individual human actor, the fisherman, harvesting from a single stock, to an understanding of human dimensions that, like ecosystem-based approaches, foregrounds interactions and relations, multiscalar processes, assemblages and aggregations, mutual dependencies, and a shared health or well-being as the basis for sustainability. The concept from social science that best captures these sensibilities is that of "community" (Jentoft, 2000); whether physical or epistemological (Creed, 2006), critical notions of community stress relationships and processes, and the connections within and between places. Furthermore, not unlike "ecosystem" or "ecosystem-based approach," "community" is notoriously ill-defined yet wonderfully productive: a wellspring for shared action and innovation (Cameron and Gibson, 2005) and the locus of adaptation and sustainable futures (Gibson-Graham and Roelvink, 2010).

In this chapter we propose a novel method to map and measure "communities-at-sea." Our goal is to create an approach that defines actual spaces at-sea where we can document the presence of community as it relates to fisheries (e.g., shared ecological knowledge, history and culture, common fishing grounds and practices, and coproduced adaptations and innovations). "Communities-at-sea" differ from social science approaches that have until now focused on community as a shore-side phenomenon largely divorced from the actions and practices of fishing offshore and, importantly, from the ecological systems within which such actions and practices occur. On the contrary, communities-at-sea represent the clustering of practices and processes that necessarily link onshore and offshore precisely because of the way that shared practices, experiences, and mutual dependence impact particular habitats, ecosystem elements, and environments.

Communities-at-sea foreground a host of scalable variables that we might understand as expressions of community processes and practices corresponding to fishers' lived experiences. These include metrics and measures that let us examine where and how communities fish (e.g., how far they travel, where and which species they target, what gear they use, and how much labor is expended). It also lets us examine fishing pattern (e.g., which communities consistently fish together on the same fishing grounds, which are distributed widely, and which overlap with other communities and to what degree). And, importantly, with a time series of spatial data, we can characterize not only practice and pattern but change over time (e.g., which communities have changed fishing grounds, species, or gear, to what degree and when). These variables can be mapped, explored in graphs, or used as input into statistical procedures demonstrating trends across a region or by type of community.

In what follows we further specify the concept of communities-at-sea and its technical manifestation. We then demonstrate how it can be used to document change over time as it relates to the fate of fishing communities, to integrate knowledge of community processes and practices into fisheries and ecosystem analyses and modeling, and to expand community beyond its association with only local concerns to a region-wide foundation for analysis and action. We are hopeful that communities-at-sea might contribute to the interdisciplinary and creative work by scientists, managers, and communities themselves that will be needed to effectively address the challenges of the Anthropocene.

LOCATING COMMUNITY IN FISHERIES SCIENCE AND MANAGEMENT

Community as both a concept in the social sciences, and as a participant in policy processes, long predates making the maps of communities-at-sea proposed here, a history that underscores why community maps matter in the advent of the more contemporary move to ecosystem-based management. Certainly the notion of community itself has long been seen as a "warmly persuasive" (Williams, 1976) but vague idea beset by multiple meanings (Hillery, 1955). The concept found an ally among scholars working in common property theory (e.g., Ostrom, 1990; McCay and Acheson, 1987), as a space between an inevitable tragedy of the commons and restrictive government intervention, but was also quickly critiqued within those same circles for a lack of attention of power and heterogeneity (e.g., Agrawal and Gibson, 1999). Indeed, the generic call to conceive of ecosystems as including humans leads to similar questions: but which humans? Which humans are linked to ecosystems, engaging with them, altering them, affecting them, and living with them?

Within US fisheries management more specifically, National Standard 8 (NS8) to the Magnuson–Stevens Act requires the assessment of impacts to fishing communities from regulatory measures. National Oceanic and Atmospheric Administration Guidelines for NS8 define such communities as "a social or economic group whose members reside in a specific location and share a common dependency" and further note that regulatory measures "may economically benefit some communities while adversely affecting others" (50 C.F.R. § 600.34).

In practice though, communities appear in most fisheries management efforts as "simply the places that get impacted" (Olson, 2005, p. 249), potential players in a zero-sum game, on land rather than linked to territories in the ocean. As a result, despite the relevance to ecosystem-based management of the many processes and practices that constitute community, which constrain the behaviors of its members or activate their potentials, they do not figure in the ecosystem analyses or management measures themselves despite their overlap and integration with marine environments, habitats, species, and so on. Indeed, if humans are part of ecosystems (Berkes, 2004), then where and how communities utilize and inhabit the marine environment must be considered essential information by which we can address the many challenges of the Anthropocene.

Therefore to move beyond community as shore-side container, we start from an understanding of community as a site of processes and practices that include knowledge exchange, reciprocal relation, mutual support, and shared well-being.[1] In this case, community draws attention to how fishers engage in similar fishing practices on shared fishing grounds, maintain and share local ecological knowledge (LEK) of marine species and habitats, develop a common understanding and topology of marine space, are bound by mutual dependencies at-sea and on shore, and experience a shared social, economic, and cultural well-being. The boundary of these groups with shared concerns is necessarily fuzzy and fluid even as the shared processes that link fishers into communities have concrete effects that range from livelihood maintenance to ecosystem health.

Community has to date been represented primarily by anthropologists and geographers via ethnographic methods and other forms of qualitative analysis such as histories and biographies (St. Martin, 2006). This work is mostly port based and, although participants might refer to at-sea experiences, such experiences and detailed knowledge of place, habitat, and territory are rarely mapped or specified in ways that coalesce with nonqualitative analyses. As a result, the many processes of society, culture, and local economy, despite being well documented by social scientists in ports, are rarely integrated into a fisheries science and management concerned with fish populations, habitats, and the predicted behaviors of fishers at-sea. Furthermore, the knowledge of fishers' lives and livelihoods produced by social scientists is most often in narrative form rather than represented in metrics or measures, making its integration into science or policy development (as opposed to impact analysis) challenging at best. As a result, community is made real and known in-depth across a range of experience

---

1. Community, as noted, has many senses in both popular and scholarly uses. Certainly one understanding of community is as an inward-looking and bounded entity, beset by reactionary and exclusionary impulses. What Massey calls a "progressive sense of place," however, asks us to focus on shared but diverse specificities that necessarily derive their force from multiscalar relations and processes. For further discussion of community as process and relation rather than timelessness and boundaries, see arguments in, for example, Massey (2005) and Cameron and Gibson (2005). Clay and Olson (2008) provide a review of the literature on fishing communities specifically.

and difference but it remains distant and vague as a measured phenomenon affecting fishing practices in particular locations or areas at-sea. Stakeholder engagement might bring voices from communities into impact analyses or decision making, but it does not put community-level processes and practices into the maps and metrics that inform science and policy.

A METHOD FOR MAPPING COMMUNITIES-AT-SEA

Mapping those areas upon which fishers and fishing communities rely grew out of research that sought to document and make usable to management the LEK of experienced fishers (e.g., Murray et al., 2008). Acknowledging the spatial contingency and limits of LEK given mobility and territorial constraints of fishers made it clear that mapping (of habitats and other ecological phenomena, seasonal harvesting rounds, fisher biographies, and fisher topographies) was an essential mode of documentation as well as an effective form of communication with fishers.

Research that draws on fishers' experiences at-sea is often interview based and engages fishers using standard nautical charts upon which fishers can indicate where and how they fish, detailed environmental histories, social topographies, and the locations of particular incidents or phenomena. Interviews might also solicit fishers' long-term experiences and knowledge of change over time: change in fishing locations, species compositions, environmental parameters, and human community use and socioeconomic context. Such extraordinarily rich and detailed information offers a distinctly geographic addition to fisher oral histories and community ethnographies.

Our work shares the goal of mapping human use, dependence, and experiences at-sea but, rather than interviews, it starts with vessel logbook data collected by the National Marine Fisheries Service. Vessel Trip Report (VTR) data are collected for all trips taken by commercially licensed fishing vessels fishing in federal waters (from 3 to 200 miles from the coast). VTRs have been collected since 1994 and contain information on date of sail and date of landing, catch and bycatch, number of crew, and, importantly, trip location. These self-reported data are often questioned for its accuracy, yet studies have shown location reporting approximates that estimated from observer records, especially for less mobile gear on shorter duration trips (DePiper, 2014). Moreover, when groups of vessels taking many trips over many years are used as the basis for mapping rather than any individual vessel, autocorrelated patterns and clusters at-sea become evident. Spatial outliers and misreported locations become minor concerns to analysis when the data are used only in aggregate. The question then is how to aggregate: how to group vessels in ways meaningful to social and scientific analysis, management, and fishers themselves.

Although the use of secondary data, such as VTR, is subject to various limitations including the "silences" about phenomena of interest to critical social science[2] as well as interpretational difficulties (St. Martin and Pavlovskaya, 2010), it can be productively reread using understandings gained in ethnographic fieldwork and, importantly, in cross-referenced readings with fishermen themselves. Our ethnographic and community-based fieldwork in the Northeast (St. Martin, 2001; Olson, 2011) made clear that fishing practices and locations are largely a function of community-level processes of communication, knowledge exchange, and mimicry. Fishers working from the same port, using similar gear, and sailing on vessels of similar length and design, tend to fish for the same species, on the same fishing grounds, and at the same time of year. Although these tendencies vary from fishery to fishery and/or port to port, they suggest community as a unit of analysis for understanding the role of shared socioeconomic and cultural processes relative to fishing practices and locations at-sea, a unit of analysis that we found resonates with fishers themselves. Peer groups of vessels with these shared characteristics—originating from a shared socioeconomic and cultural milieu—are our communities-at-sea, whose various spatial ranges turn fishing vessels into vehicles that link and bind communities to ecosystems.[3] Like ecosystems, communities-at-sea should be understood as starting points for analysis and not as bound entities; they are an ontological statement about the existence of processes and practices of community in places (e.g., sharing of environmental information) rather than a definition of membership or denotation of division.

To study communities-at-sea, we developed an algorithm by which we can aggregate vessels and vessel trips to best fit the peer groups of fishers. In our work such a "community" is a function of gear type and port. VTRs specify gear but knowing the port with which a vessel is associated is less straightforward. VTRs specify the port of landing ([PORTLAND]), whereas permit data specify two other port variables suggesting port association and which are declared by the vessel owner: home port ([HPORT]) and principal port ([PPORT]). Our interviews with fishers in the Northeast suggest that home port is most often understood as where a vessel is registered, whereas principal port is most often understood as where a vessel is most active. In addition to the port declared by the vessel owner (what we call [DECPORT]), a vessel could be seen to be associated with a port if it lands there often. Therefore we also specify port association based on a 50% or greater frequency of landings ([PERPORT]). In all cases, the port of community association ([COMMUNITYPORT]) must

---

2. Silences might include informal activities (such as bartering or sharing) that fall outside the purview of formal data collections, or more "epistemic" silences, which result from the inability to represent cultural understandings and categories that differ from mainstream ones (cf. Harley and Laxton, 2002).

3. Although the call to ecosystem-based approaches in marine science and management invariably includes a call to consider human communities as part of those systems (e.g., Leslie and McLeod, 2007), in practice it remains uncommon to find human communities as elements within ecosystem analysis (St. Martin and Hall-Arber, 2008a).

also be an actual fishing port ([OFFICIALPORT]), rather than inland or other municipality (e.g., based on a fisher's home address).

Therefore after joining home port and principal port (found in permit data) to the VTR records, we use the following rules to associate vessels and trips with particular port locations:

> *If [PPORT] = [OFFICIALPORT] then [DECPORT] = [PPORT]*
>   *Else If [HPORT] = [OFFICIALPORT] then [DECPORT] = [HPORT]*
>   *Else [DECPORT] = [PORTLAND].*
> *Calculate [PERPORT] as port with 50% or greater landings by a vessel.*
> *If [PORTLAND] for a given trip = ([DECPORT] or [PERPORT])*
>   *then [COMMUNITYPORT] = [PORTLAND].*

This algorithm allows us to group trips into particular peer groups of vessels that are most likely to engage in a host of community processes. Furthermore, the fishing locations of these communities can be mapped such that the pattern and practice of different communities-at-sea become available to visual and spatial statistical analyses. Mapping peer groups of vessels based on trip locations usefully delimits the range of a community; weighting locations by labor time, however, allows us to map degree of community presence and, therefore, dependency upon particular locations. We refer this key variable as [FISHERDAYS].

[FISHERDAYS] for a given trip = number of crew [CREW] x trip length [TRIPDAYS]

Fisherdays offers a measure of invested time on the part of a community in particular trips and trip locations at-sea. It is distinct from other common variables indicating fishing effort (e.g., catch or value of catch) insofar as it emphasizes labor inputs, rather than harvest outputs or profit, as a measure of community engagement and dependence upon particular fishing grounds. Although effort in terms of labor often correlates with effort in terms of harvest, there are fisheries and locations where they diverge. Divergence could, for example, help us to identify those fisheries or fishing grounds where fishers consistently make a living yet catch relatively fewer fish than other large-volume fisheries or overfished fishing grounds; the implication is that some communities may already be fishing "sustainably." Fisherdays allows us to explore both the socioeconomic and spatial-ecological implications of different community fishing practices and traditions, ecological knowledge, and harvesting priorities; it creates an opening for analyses that begin to measure sustainability in terms of the continued existence and health of both fish populations and local livelihoods.

Using the techniques defined previously, we can develop various communities-at-sea databases where individual fishing trips are assigned to particular communities and fisherdays are calculated for each trip. Once developed, such data can then be used to characterize communities and entire regions in terms of community characteristics and change over time. Furthermore, using the geographic locations of trips (weighted by fisherdays) we can examine the spatial patterns of communities-at-sea by deriving density surfaces ("heat maps") and percent volume contours that we interpret as depicting community presence at-sea (see also St. Martin, 2004). As we describe in the following sections, communities-at-sea data (in a variety of forms: tabular, graphic, and map-based) can then be integrated into social, economic, and ecological analyses.

UTILIZING COMMUNITIES-AT-SEA: DATA AND TOOLS FOR ANALYSIS

As with any research or policy development project, data development can be one of the most challenging for a variety of reasons. Access to VTR data, for example, is restricted by the National Marine Fisheries Service (NMFS) because of confidentiality. Although communities-at-sea data are derived from VTRs, it is necessarily aggregated information made up of trips from all vessels in a given peer group (see earlier discussion). When a "community" has too few vessels needed to mask individual practices (currently three as mandated by NMFS), then such peer groups of vessels are simply not recorded as a community-at-sea. Such data might be pooled into a new category (e.g., trips by independent vessels) but only if appropriately aggregated. The result is a database of community-level information accessible to analysis and policy development. Furthermore, our experience also suggests that these data are easily understood and trusted by fishers themselves.[4] Indeed, insofar as communities-at-sea data can be used to depict the relationship between community well-being and access to particular fishing grounds, fishers see the maps produced by communities-at-sea data as a way to legitimate their claims against, for example, conservation closures, energy development lease blocks, and other spatial management measures (St. Martin and Hall-Arber, 2008b).

Representing communities in terms of metrics, descriptive statistics, and maps allows us to visualize and integrate communities into forms of bioeconomic and ecological analysis previously closed to community concerns due to the requirements of quantification and discrete units of analysis. To be clear, we do not wish to undermine ethnographic and narrative forms; quite the opposite, we wish to use communities-at-sea as a space where ethnographies, histories, and biographies might be located and associated with other quantified and mapped processes, be they regional economic or ecosystem dynamics, habitat transformations, conservation initiatives, energy development, or climate change. In its

---

4. Communities-at-sea maps and other summarizations of community activity were presented in several workshops as part of the Atlas project (see Final Report, Northeast Consortium subcontract 06-028, *An Atlas-based Audit of Fishing Territories, Local Knowledge, and Potential for Community Participation in Fisheries Science and Management*, Available at: www.northeastconsortium.org) as well as the MARCO data portal project (see www.portal.midatlanticocean.org). Fishing community members responded positively to the map products from these projects. They were not concerned about confidentiality once it was clear that the maps did not directly display VTR data but only data aggregated into communities-at-sea.

most literal form, such a community-based database would allow geographic information system–based querying of locations at-sea that would be populated with community-specific information and result in community-level metrics.

Understanding community dynamics as they relate to fishing practices, dependencies upon particular resources, and histories (both social and environmental) in particular locations at-sea would seem to be fundamental to both policy development and impact analyses, yet this is rarely the case. Although communities-at-sea do not reveal all aspects of community, they nevertheless create a foundation for investigation, corroboration with other data sources (qualitative and quantitative), and community engagement; they present a geography and social seascape (see Fig. 7.1) that we can query and quantify to ask the following:

Given a communities-at-sea database we might query by location to ask:

- Which community or communities utilize this area?
- To what degree is community employment dependent upon this area?
- What species do they catch in this area?
- What is the history of community use of this area?

Alternatively we might query by attribute to ask:

- Do vessels from this community fish in the same locations and when?
- What are the primary, secondary, or tertiary fishing grounds that support this community in terms of employment, catch, or value?

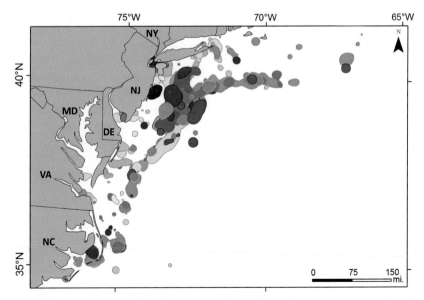

FIGURE 7.1   Select communities-at-sea. Outlines based on percent volume contours (75% of fisherdays).

- What is this community's spatial pattern of fishing (e.g., tightly clustered, distributed, or random) and has it changed over time?
- How does this community's fishing overlap with morphologic features, habitats, or water depth?

These basic queries can be linked to other qualitative (e.g., interview) information to ask:

- Why does your community fish in these locations and how did you come to know them?
- In what ways and when does your community depend upon these locations?
- What is the social and environmental history of these locations?
- What environmental changes have you seen here and how are they linked to socioeconomic processes and change?

These questions and many others suggest a usefulness of communities-at-sea to impact not just analysis but also policy development. For example, a region-wide analysis might use a communities-at-sea database in conjunction with other spatial decision-making methods that incorporate a wide range of environmental information to ask: Which areas can we close to fishing to meet our conservation goals and minimize negative impacts on employment in vulnerable communities (cf. St. Martin and Hall-Arber, 2008a)?

COMMUNITIES-AT-SEA IN PRACTICE

From the impacts of wind energy siting on local economy to the links between changing fish distributions and fishing practices, it is vital that we know in each case and in each location who maintains key LEK, which fishing practices and harvesting techniques are dominant, and whose livelihood and well-being are most directly relevant. Communities-at-sea give us the means to link systems previously disconnected and supplement analyses in ways previously devoid of community concerns, such that the uneven impacts of area management initiatives and, alternatively, the spatial effects of nonspatial regulations (e.g., constraining fishing of a particular species) become clear and available to decision makers and communities alike.

Communities-at-sea make available the concerns of communities to a wide range of analyses. For example, social scientists interested in the effects of a global transformation of marine space driven by policies promoting quota systems in fisheries and rights-based distributions of resources generally might use communities-at-sea data to examine socioeconomic outcomes at the level of community rather than economy as a whole (cf. St. Martin, 2007). A clear antecedent of our work is Olson's use of VTR data to group peer groups of vessels and thereby foreground the heterogeneity and social spaces of fishing in the Northeast and the potential for subsequent uneven (and unjust) distributional outcomes from neoliberal policies (Olson, 2010, 2011).

Using a community-level sensibility, Olson reveals fishers' dependencies on particular locations at-sea rather than catch or value as a measure of effort (ibid.).

The latter tend to obscure social relations, processes, and practices. For example, a map of fishing effort based on catch or value may show "hot spots" of activity, but these hot spots cannot always be equated to the areas that are important to many fishermen. A hot spot could represent the effort of one vessel or a thousand, so without further information its social and economic importance is unclear. Furthermore, the relation between specific fishing practices and ecological outcome in terms other than quantity caught is left unexamined.

The idea of resource dependency, on the other hand, was used by Olson to ascertain what percentage of a vessel's annual catch is landed in what areas, giving rise to a way of visualizing relative spatial dependence that showed the nearshore waters along the Northeastern coast were the most important for most fishing communities, despite not having the highest effort overall. Density maps comparing the spatial practices of those fishermen most dependent on a single statistical area with those fishermen who are more mobile clearly suggested distinct areas of importance and a seascape differentiated by a range of sociospatial practices (see Fig. 7.2).

Olson's work reminds us that there are many different reasons that underlie fishers' choice of where to fish. Some may be economic, such as lower fuel costs, but there are also sociocultural motivations, such as wishing to return home daily. These decisions of where to fish are socially embedded, and they provide insight into different socioeconomic processes and practices depending on the scale and unit of analysis.

Although a heterogeneous seascape of human communities and territories allows us to better address questions related to the uneven distribution of the costs of marine policies, it also makes accessible a space of intertwined human and environmental processes. As a result, research projects investigating complex coupled systems, but challenged by the call for interdisciplinarity, now have a common ground where both socioeconomic and ecological processes are not only manifest but also interrelated. In such cases,

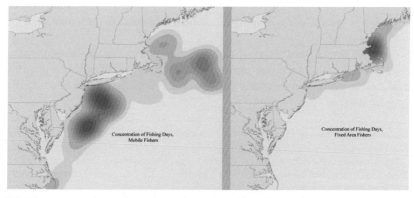

FIGURE 7.2   Kernel density maps comparing fishing grounds of mobile and nonmobile fishermen. Data are based on NMFS 2008 Vessel Trip Report data.

communities-at-sea act as an ontological foundation linking discipline-specific modes of investigation, understandings of process, and recommendations for action.

For example, we are engaged in an interdisciplinary project focused on climate-induced shifts in the distribution of commercial fish species (on the latter see Pinksy and Fogarty, 2012; Pershing et al., 2015).[5] The project seeks to understand both the ecological and socioeconomic dimensions of species' shifts, and it hopes to inform both policy development and community-level adaptation strategies. Beginning from the overlap of communities-at-sea with shifting biomass, we are performing a variety of analyses that range from spatial statistics to in-depth interviews with fishers from those communities most dependent upon apparently shifting species. Our goal is to assess both quantitatively and qualitatively the relationship between fishing practices and ecological change; in particular, we seek to better understand the effects of fishing on range shift as well as the nature of adaptation to range shift at the level of community.

Our final example demonstrates the immediate use of communities-at-sea to develop a best practices approach to wind energy siting in Virginia's coastal waters. The aim of the Collaborative Fisheries Planning for Virginia's Offshore Wind Energy Area[6] was to identify those communities most affected by wind energy development in a designated Virginia Wind Energy Area and to establish a protocol for community engagement and information exchange. In a series of meetings in key ports, commercial and recreational fishers were presented communities-at-sea maps depicting their use of proximate fishing grounds. They were asked about both the accuracy of the maps as well as what information, given the limits of the data, might be missing from the maps such as seasonal movements, patterns predating the mid-1990s, and so on. The maps clearly showed which communities depend on areas within the study area (in terms of fisherdays, see earlier discussion) as well as the degree of dependence (percent fisherdays) and how dependence has changed over time (Final Report, 2016). In this case, the maps worked well as a forum for communication; they resonated with fishing communities and they clearly showed the uneven impacts that would occur should the fishing grounds of the study area be no longer available.

The aforementioned examples of communities-at-sea analyses demonstrate how this approach can make processes and practices more tangible in place. As such, communities-at-sea can serve as data vital to scientific

---

5. *Adaptations of Fish and Fisheries to Rapid Climate Velocities* (National Science Foundation #1426891). See also Fenichel E.P. et al. (2016).

6. This is a project developed by Bureau of Ocean Energy Management, Department of Mines, Minerals and Energy, and Virginia Coastal Zone Management. The final report demonstrates a variety of impact analyses made available by a communities-at-sea approach and is available at: http://www.deq.virginia.gov/Portals/0/DEQ/CoastalZoneManagement/Virginia-Wind-Energy-Area-Collaborative-Fisheries%20Planning-Final-Report.pdf.

understanding, policy development, and community-level strategies and innovations. "Community" need not remain onshore embedded only in retrospective analyses; it can be integrated into modes of analysis seeking to understand key processes and dynamics that are transforming marine space.

FROM LOCAL CONCERNS TO REGIONAL ANALYSIS AND ACTION

Community experiences, knowledge, and concerns are often solicited, welcomed, and enthusiastically expressed at regional fisheries council meetings or similar marine decision-making fora. Yet they are also as often dismissed as anecdotal or as unobservable in data streams collected and analyzed at the level of the region—the level of marine management itself (St. Martin, 2001). To consider them more effectively relative to region-wide science and management initiatives, a unit of analysis closely corresponding to the experiences of fishers, yet expressed in a standardized and scalable form, is required. Although our work has used a communities-at-sea approach to facilitate the exploration of fishers' experiences, knowledge, and concerns in situ (e.g., via participatory mapping practices), it has done so cognizant of the need to "scale up" to make such information accessible and, indeed, vital to region-wide assessments. In practice, we can begin from the most general of fishers' concerns, the survival of communities themselves. For example, at the scale of the region, it is clear that the dramatic declines in catch over the last decade in the Northeast United States—whether due to overfishing and subsequent cutbacks in fishing effort or changes in ecological conditions that have changed fish habits—have negatively impacted many livelihoods and fishing communities across the region. As social scientists we can begin to document this decline, for example, in terms of the number of active vessels (see Fig. 7.3A), which might then be examined more closely by sector or in terms of trips taken. Here (see Fig. 7.3B) we have divided data on groundfishing by vessel size and can clearly see that the cost of reductions in fishing expressed in numbers of trips is borne primarily by the small vessel sector.

Understood as indicators of employment opportunity, production capacity, or relative distribution of wealth, these summary statistics usefully document and corroborate fishers' concerns for their own livelihoods as well as the survival of their communities. Yet how change is distributed and how it is experienced by fishers within particular communities remains unexamined in these common statistics. Using a communities-at-sea approach we can see that change not only affects some sectors more than others, but also apparently affects some communities more than others with implications not only for fishing effort but also for localized cultural, historical, and socioeconomic transformations (see Figs. 7.4 and 7.5).

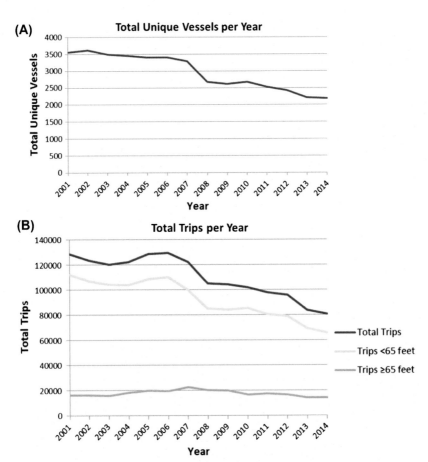

FIGURE 7.3 (A) Total number of unique vessels active in commercial fishing. (B) Trips taken by commercial vessels using gear types primarily associated with groundfishing.

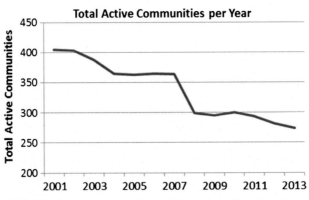

FIGURE 7.4 Active communities based on communities-at-sea data.

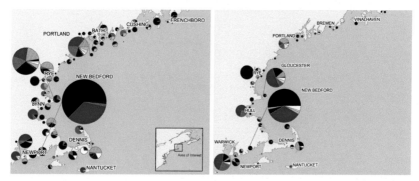

FIGURE 7.5   Port-based fishing activity in 2000 and 2013. Circle size is total fisherdays by port. Pie slices are major gear groups.

Communities-at-sea are, as discussed, more than just port-based experiences; they also represent a shared at-sea experience thereby linking cultural and socioeconomic lives onshore to offshore processes and practices of fishing (the raison d'être for communities-at-sea). Using the entire series of data we can do analyses at the regional level concerning changes in fishing effort and location and then examine what might be experienced locally relative to those broader changes. For example, we might perform a Median Trend analysis using rasterized data developed from communities-at-sea data and representing labor time (i.e., fisherdays) by location to show hot spots (and cold spots) of those locations where most change has occurred. Analyzing small-trawler groundfish vessels, whose patterns of fishing relative to larger vessels are often contentious and who have also long advocated for more science and management directed at near-shore fisheries, clearly shows the locations that have changed most in terms of fisherdays expended (Fig. 7.6). Fisherdays, as community investment in place, can also be mapped by community to better understand the relationship between regional trends and the experiences of particular communities. Indeed, over the time period examined many small vessel communities have not changed their fishing locations, whereas others have moved to new fishing grounds, perhaps following fish responding to changes in water temperature, expanding harvest in new areas due to new technologies or retracting from areas of conflict or overfishing.

To better understand how change is experienced by communities themselves, we can query the locations of most change by community to see who fishes in these locations and to what degree over time (Fig. 7.7). Analyzing community-at-sea variables lets us integrate the practices of individual communities (and the experiences and concerns of individual fishers) into typologies of communities across a region, which can then be seen in relation to region-wide phenomena (e.g., distributions of species, environmental and climate variability, or regulations effecting fishing locations, species, and quantities). Our intention in this "scaling up" is not to overwrite the diversity of experience across space but to be attentive to diverse practices, responses,

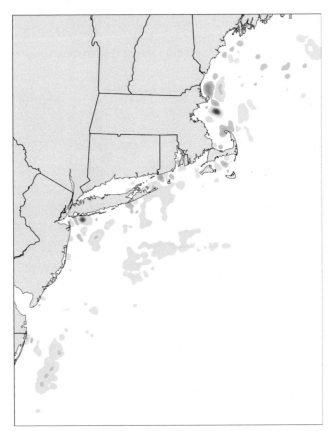

FIGURE 7.6 Median Trend analysis for Northeast region developed from annual density surfaces representing small-vessel trawling activity (i.e., fisherdays). Hot spots (*red*) represent areas of significant increase in fisherdays. Cold spots (*deep green*) represent areas of significant decrease in fisherdays.

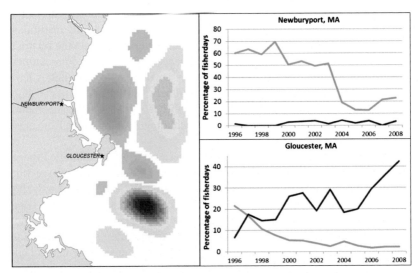

FIGURE 7.7 Hot (*red*) and cold (*green*) spots near Cape Ann, Massachusetts. Graphs show the degree to which particular communities expended effort (i.e., fisherdays) in these spots over time. Not all communities participated equally in the movement from the green to red spot despite their close proximity and use of these fishing grounds.

adaptations, and innovations as we engage in region-wide analyses and management decision making.

CONCLUSION

Communities-at-sea data and techniques are accessible, complementary, and potentially informative to both ethnographic and modeling (economic and ecological) forms of analysis, and it can work to link such analyses via a common ground, common space, and common community-level unit of analysis. Furthermore, communities-at-sea avoid many of the problems of spatial or otherwise local information solicited directly from fishers, which, despite its richness and analytical purchase, is challenging to use as input into current forms of analyses and policy development that demand standardization, quantification, and an ability to scale regionally. The problem of confidentiality is also avoided as communities-at-sea data are by definition aggregated and representative of only community-level patterns and spatial practice. The result is a novel form of data designed to foreground community attributes and spatial practice thereby making them available to a range of statistical and spatial analyses and depictions.

Furthermore, a communities-at-sea approach addresses the conceptual and in-practice gap that exists between onshore communities and offshore ecosystems, between the livelihoods of fishers and the management of fish, and between ethnographic and ecosystem knowledge production. This approach is not intended to supplant either ethnographic or other forms of critical social scientific knowledge, but rather to ensure that such knowledge can be more firmly anchored to the methods and models that predominate in fisheries management today. Our goal is to make the processes and practices that constitute communities (i.e., shared ecological knowledge, history and culture, common fishing grounds and practices, and coproduced adaptations and innovations) present within the space of fisheries science and management to help visualize, model, and study community as a site of shared well-being and potential relative to sustainable futures. We do not wish to reduce community to a cartographic and statistical object, but to insert community as a measurable and scalable set of concerns and practices within forms of analysis and management where it has been long absent.

Communities-at-sea are gaining momentum in interdisciplinary projects attempting to address the challenges of climate change, environmental degradation, industrial decline, and community crisis. Furthermore, it is proven to resonate strongly with communities themselves as a mode of self-reflection and as a device engendering innovation (Snyder and St. Martin, 2015). Finally, while the Anthropocene makes clear the need for assessing impacts and adaptations at the level of human communities, it also provides an impetus to develop actions and practices that will enhance and ensure community and environmental well-being. In this case, it is vital that we understand the processes and practices of communities insofar as it is from these sites that key adaptations and innovations will emerge.

REFERENCES

Agrawal, A., Gibson, C., 1999. Enchantment and disenchantment: the role of community in natural resource conservation. World Development 27 (4), 629–649.

Berkes, F., 2004. Rethinking community-based conservation. Conservation Biology 18 (3), 621–630.

Berkes, F., 2012. Implementing ecosystem-based management: evolution or revolution? Fish and Fisheries 13 (4), 465–476.

Cameron, J., Gibson, K., 2005. Alternative pathways to community and economic development: the Latrobe Valley community partnering project. Geographical Research 43 (3), 274–285.

Clay, P.M., Olson, J., 2008. Defining "fishing communities": vulnerability and the Magnuson-Stevens fishery conservation and management act. Human Ecology Review 15 (2), 143–160.

Creed, G.W., 2006. The seductions of community: reconsidering community. In: Creed, G.W. (Ed.), The Seductions of Community: Emancipations, Oppressions, Quandaries. SAR Press, Santa Fe, pp. 1–20.

DePiper, G.S., 2014. Statistically assessing the precision of self-reported VTR fishing locations. NOAA Tech Memo NMFS NE, vol. 229.

Fenichel, E.P., Levin, S.A., McCay, B., St Martin, K., Abbott, J.K., Pinsky, M.L., 2016. Wealth reallocation and sustainability under climate change. Nature Climate Change 6 (3), 237–244.

(Final Report) Virginia Coastal Zone Management Program, 2016. Collaborative Fisheries Planning for Virginia's Offshore Wind Energy Area (Final Report). US Dept. of the Interior, Bureau of Ocean Energy Management, Office of Renewable Energy Programs, Herndon. OCS Study BOEM 2016-040. 129 pp.

Fogarty, M.J., 2014. The art of ecosystem-based fishery management. Canadian Journal of Fisheries and Aquatic Sciences 71 (3), 479–490.

Gibson-Graham, J.K., Roelvink, G., 2010. An economic ethics for the Anthropocene. Antipode 41 (S1), 320–346.

Harley, J.B., Laxton, P., 2002. The New Nature of Maps: Essays in the History of Cartography (No. 2002). JHU Press.

Hicks, C.C., Levine, A., Agrawal, A., Basurto, X., Breslow, S.J., Carothers, C., Charnley, S., Coulthard, S., Dolsak, N., Donatuto, J., Garcia-Quijano, C., Mascia, M.B., Norman, K., Poe, M.R., Satterfield, T., St Martin, K., Levin, P.S., 2016. Engage key social concepts for sustainability. Science 352 (6281), 38–40.

Hillery, G., 1955. Definitions of community: areas of agreement. Rural Sociology 20, 111–123.

Jentoft, S., 2000. The community: a missing link of fisheries management. Marine Policy 24 (1), 53–59.

Koehn, J.Z., Reineman, D.R., Kittinger, J.N., 2013. Progress and promise in spatial human dimensions research for ecosystem-based ocean planning. Marine Policy 42, 31–38.

Leslie, H.M., McLeod, K.L., 2007. Confronting the challenges of implementing marine ecosystem-based management. Frontiers in Ecology and the Environment 5 (10), 540–548.

Massey, D., 2005. For Space. Sage, London.

McCay, B., Acheson, J. (Eds.), 1987. The Question of the Commons: The Culture and Ecology of Communal Resource. University of Arizona Press, Tucson.

Murray, G., Neis, B., Palmer, C.T., Schneider, D.C., 2008. Mapping cod: fisheries science, fish harvesters' ecological knowledge and cod migrations in the Northern Gulf of St. Lawrence. Human Ecology 36 (4), 581–598.

Olson, J., 2005. Re-placing the space of community: a story of cultural politics, policies, and fisheries management. Anthropological Quarterly 78 (1), 233–254.

Olson, J., 2010. Seeding nature, ceding culture: redefining the boundaries of the marine commons through spatial management and GIS. Geoforum 41 (2), 293–303.

Olson, J., 2011. Producing nature and enacting difference in ecosystem-based fisheries management: an example from the Northeastern US. Marine Policy 35 (4), 528–535.

Ostrom, E., 1990. Governing the Commons: The Evolution of Institutions for Collective Action. Cambridge University Press, Cambridge.

Pershing, A.J., Alexander, M.A., Hernandez, C.M., Kerr, L.A., Le Bris, A., Mills, K.E., Sherwood, G.D., 2015. Slow adaptation in the face of rapid warming leads to collapse of the Gulf of Maine cod fishery. Science 350 (6262), 809–812.

Pinksy, M.L., Fogarty, M., 2012. Lagged social-ecological responses to climate and range shifts in fisheries. Climatic Change 115 (3–4), 883–891.

Pitcher, T.J., 2001. Fisheries managed to rebuild ecosystems? Reconstructing the past to salvage the future. Ecological Applications 11 (2), 601–617.

Pitcher, T.J., Kalikoski, D., Short, K., Varkey, D., Pramod, G., 2009. An evaluation of progress in implementing ecosystem-based management of fisheries in 33 countries. Marine Policy 33 (2), 223–232.

Sievanen, L., Leslie, H.M., Wondolleck, J.M., Yaffee, S.L., McLeod, K.L., Campbell, L.M., 2011. Linking top-down and bottom-up processes through the new U.S. National Ocean Policy. Conservation Letters 4 (4), 298–303.

Snyder, R., St. Martin, K., 2015. A fishery for the future: the Midcoast Fishermen's Association and the work of economic being-in-common. In: Roelvink, G., St Martin, K., Gibson-Graham, J.K. (Eds.), Making Other Worlds Possible: Performing Diverse Economies. University of Minnesota Press, pp. 26–52.

St. Martin, K., 2001. Making space for community resource management in fisheries. Annals of the Association of American Geographers 91 (1), 122–142.

St. Martin, K., 2004. GIS in marine fisheries science and decision making. In: Fisher, W.L., Rahel, F.J. (Eds.), Geographic Information Systems in Fisheries. American Fisheries Society, pp. 237–258.

St. Martin, K., 2006. The impact of "community" on fisheries management in the U.S. Northeast. Geoforum 37 (2), 169–184.

St. Martin, K., 2007. The difference that class makes: neoliberalization and noncapitalism in the fishing industry of New England. Antipode 39 (3), 527–549.

St. Martin, K., Hall-Arber, M., 2008a. The missing layer: geo-technologies, communities, and implications for marine spatial planning. Marine Policy 32 (5), 779–786.

St. Martin, K., Hall-Arber, M., 2008b. Community in New England fisheries. Human Ecology Review 15 (2), 161–170.

St. Martin, K., Pavlovskaya, M., 2010. Secondary data. In: Gomez, B., Jones III, J.P. (Eds.), Research Methods in Geography: A Critical Introduction. Wiley-Blackwell, Malden, pp. 173–193.

Steffen, W., Persson, Å., Deutsch, L., Zalasiewicz, J., Williams, M., Richardson, K., Crumley, C., Crutzen, P., Folke, C., Gordon, L., Molina, M., Ramanathan, V., Rockström, J., Scheffer, M., Schellnhuber, H.J., Svedin, U., 2011. The anthropocene: from global change to planetary stewardship. Ambio 40 (7), 739–761.

U.S. Commission on Ocean Policy (USCOP), 2004. An Ocean Blueprint for the 21st Century, Final Report Washington, D.C..

Williams, R., 1976. Keywords: a Vocabulary of Culture and Society. Oxford University Press, New York.

Chapter 8

Conservation Actions at Global and Local Scales in Marine Social–Ecological Systems: Status, Gaps, and Ways Forward

Natalie C. Ban[1], Charlotte Whitney[1], Tammy E. Davies[2], Elena Buscher[1], Darienne Lancaster[1], Lauren Eckert[1], Chris Rhodes[1], Aerin L. Jacob[1]

[1]University of Victoria, Victoria, BC, Canada; [2]BirdLife International, Cambridge, United Kingdom

INTRODUCTION

Global drivers of change are affecting marine ecosystems and the people who depend on them at increasing rates and severities. In particular, climate change and associated increases in ocean temperature and acidification, changing ocean current and upwelling/downwelling patterns, and hypoxic areas are severely affecting some ocean ecosystems, and will soon affect many if not all. Climate change impacts on fisheries have the potential for disastrous effects on human communities dependent on marine fisheries (Golden et al., 2016). Climate impacts are also likely to exacerbate existing issues confronting fisheries management (Ogier et al., 2016), yet many marine conservation actions—interventions, strategies, and activities designed to reduce threats on marine ecosystems (Salafsky et al., 2008)—were developed before climate change was widely recognized as a major driver of change. Globalization, a global driver of change that is closely linked to climate change, also affects marine ecosystems, especially with the emergence of global markets that increase demands on, and use of, marine resources (Berkes et al., 2006).

Marine conservation actions at local scales [e.g., marine protected areas (MPAs), species-level protections and restoration] were developed to address mostly localized threats (overfishing, habitat destruction, pollution), and have been among the most effective approaches to marine conservation over the past decades. We define local scales as spatial areas that encompass single ecosystems (e.g., coral reefs, kelp forests, estuaries) and where the jurisdictional authority is at the subnational level (e.g., villages, municipalities, states). We consider conservation actions to be effective if they meet the objectives set

Conservation for the Anthropocene Ocean. http://dx.doi.org/10.1016/B978-0-12-805375-1.00008-8
143

out for them. At a minimum, this should mean slowing and ideally reversing declines of depleted species and ecosystems by reducing or eliminating threats. Yet conservation objectives will vary widely, be context-specific, and should also ideally consider the impacts and benefits to people affected by them. Most local actions were not designed to address global drivers of change, and hence to curtail the effects of global ecological changes, conservation actions must reach beyond local scales. We refer to global scales as those at the spatial scale of multiple countries (e.g., ocean basins, large marine ecosystems) or the world, encompassing multiple complex ecosystems. While local actions remain essential to address local threats, unfortunately global pressures have the potential to negate local conservation efforts, and hence attention is needed at multiple scales. Globalization exacerbates and confounds local conservation impacts and responses, through market forces across regional and national boundaries, labor movement across scales, and neoliberal governance policies that affect local social structures (McClanahan et al., 2015; Perry et al., 2011). This new era of the Anthropocene is one of previously unknown cross-scalar dynamics and interactions among human and ecological systems (Smith and Zeder, 2013), across jurisdictions and spatial scales (Österblom et al., 2016).

Marine conservation actions at local and global scales—and all others in-between—affect both ecosystems and people, and the interactions people have with marine systems. Thus it can be useful to think of conservation actions as interventions in complex social–ecological systems (Ostrom, 2009) (Fig. 8.1). Social–ecological systems comprise people (referred to as actors or stakeholders), ecosystems and species (also known as resources systems and resource units), and the institutions, rules, or management that govern how people behave (i.e., the governance systems). These core components of social–ecological systems interact to produce outcomes (e.g., ecosystem degradation, reduced human well-being) in what is known as the action situation. For example, if a local area is designated as a no-take marine protected area, the fishers' response can range from fishing elsewhere, fishing harder, switching livelihoods, to fishing illegally within the area (Cinner et al., 2011). In turn, such choices influence the effectiveness of the conservation action. A social–ecological system is also influenced by external biophysical and social factors (e.g., climate-driven changes; the political landscape); these external drivers of change can influence conservation outcomes. The conceptualization of social–ecological systems recognizes that social and ecological systems are intricately connected, and acknowledges that interventions within a social–ecological system have inherent linkages and trade-offs among the components of that system (Fig. 8.1).

The purpose of this chapter is to highlight the conservation approaches that build greater social–ecological resilience for both large and small systems in the Anthropocene. We focus on local and global scales to highlight the diversity of approaches at these extremes, while recognizing that actions apply at all scales in-between. We synthesize categories of marine conservation actions and their relevance at local and global scales, discuss linkages between scales, the gaps that exist, and provide recommendations for this epoch of anthropogenic global change.

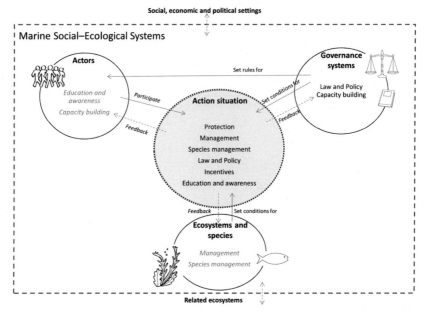

FIGURE 8.1 Conceptual model of a marine social–ecological system, detailing where categories of marine conservation actions occur within this framework. *Italic gray text* denotes predominant local-scale actions. *Black text* refers to local and global scale actions. The components of the social–ecological system are denoted by *circles* and labeled with *larger bold text*: actors are the people and stakeholder groups; governance systems refer to the organizations and rules that govern the system; ecosystems and species are the resource systems and units; and the action situation is where the components interact to produce outcomes. The *arrows* between these components indicate interactions. The *dashed outline border* conceptually denotes the extent of the social–ecological system, and the related ecosystems and external drivers such as social, economic, and political settings that influence the system.

CONSERVATION ACTIONS AT GLOBAL AND LOCAL SCALES

In this section, we use a proposed common lexicon and categorization of conservation actions (Salafsky et al., 2008) to outline the suite of marine conservation actions that exist at local and global scales. We divide the actions into those that directly address threats (direct actions: protection, species management, and habitat management), and those that are supporting or enabling actions that facilitate direct actions (supporting actions: law and policy; education and awareness; livelihood, economic and other incentives; and external capacity building). We review these categories as relevant to marine ecosystems, and provide a brief assessment of their relevance and effectiveness at global and local scales (see examples in Table 8.1) and the threats they directly or indirectly address (Table 8.2). We use examples to illustrate actions, and thus do not provide a comprehensive review of all possible marine conservation actions.

TABLE 8.1 Examples of Marine Conservation Actions at Global and Local Scales

	Conservation Actions	Global Action	Global Example	Local Action	Local Example
Direct actions	Protection	Networks of large MPAs. High seas protected areas agreements	CBD targets; OSPAR (Charlie-Gibbs MPA in international waters in the North Atlantic)	Community-based protected areas under customary marine tenure	Locally managed marine areas in the Indo-Pacific (e.g., LMMAs in Fiji) (Govan, 2009)
	Species-level management	RFMOs with ecosystem-based management systems	CCAMLR (Ruckelshaus et al., 2008)	Ecosystem-based local-level management	Barangay cooperation on Cebu Island, Philippines (Eisma-Osorio et al., 2009)
		Trade restrictions or treaties	CITES IWC (e.g., success of de-listing humpback whale from IUCN Red List in 2008)	Fisheries management Shark sanctuaries	Setting total allowable catch limits to zero (e.g., Spiny dogfish (*S. acanthias*) in European waters) (Nieto et al., 2015)
		Voluntary measures	MSC certification process Prohibitions on cargo (e.g., Delta, United & American Airlines for trophy hunt cargo; Qantas and Singapore Airlines for shark fins)		Banning species take (e.g., banning commercial and recreational shark landings in Bahamas) (Graham et al., 2016)
	Management	International treaties to prevent pollution and invasive species spread	IMO international port clean up treaties	Invasive species management	Attempted removal of the Indo-Pacific lionfish (*Pterois* sp.) from Caribbean reefs (Frazer et al., 2012).

Indirect actions	Law and policy	International legislation and biodiversity conventions, including agreements for marine areas beyond national jurisdiction	CBD CITES CMS UNCLOS ICAAT CCAMLR OSPAR	Indigenous law/tribal laws;	Customary law which forms the basis for community managed MPAs (e.g., Adat law in Indonesia, which was the first declaration for the Raja Ampat Marine Network)
				Policies and regulations	Sustainable harvest limits Zoning regulations within MPAs
				Voluntary standards and professional codes	MSC
	Education/awareness	Public, media-based outreach and creation of education materials	UNESCO's MAB UN's MOE The National Geographic Society The Ocean Conservancy	Scientific schooling through formal education	Primary or secondary schools
		Development of standardized education materials	NMEA	Local hands-on exposure to marine environments	Local ecotourism companies
	Incentives (livelihoods)	Economic incentives	Blue Carbon Initiative	Conservation payments	Seychelles' debt-for-nature-swap (Weymouth Ullman, 2015)
		Market forces such as eco-certifications	MSC	Linked enterprises	
				Livelihood alternatives	Ecotourism, whale-watching (Williams et al., 2002)

Continued

TABLE 8.1 Examples of Marine Conservation Actions at Global and Local Scales—cont'd

Conservation Actions	Global		Local	
	Action	Example	Action	Example
External capacity building	Internationally organized capacity-building initiatives	IMO developing nations capacity-building initiative The Conservation Leadership Program Russell E. Train Fellowship	Regional cooperation and involvement in international capacity-building initiatives	Joint IMO/South East Asia Cooperation with capacity-building initiatives (Zhu, 2006) NOAA's International Marine Protected Area Management Capacity Building Program Coral Triangle Initiative
	Private foundations Development organizations Bridging organizations	GEF Big Ocean		

CBD, Convention on Biological Diversity; *CCAMLR*, Convention on the Conservation of Antarctic Marine Living Resources; *CITES*, Convention on International Trade in Endangered Species of Wild Fauna and Flora; *CMS*, Convention on Migratory species; *GEF*, Global Environment Facility; *ICAAT*, International Commission for the Conservation of Atlantic Tunas; *IMO*, International Maritime Organization; *IUCN*, International Union for the Conservation of Nature; *IWC*, International Whale Commission; *MSC*, Marine Stewardship Council; *NMEA*, National Marine Educators Association; *OSPAR*, Convention for the Protection of the Marine Environment of the North-East Atlantic; *RFMOs*, Regional fisheries management organizations; *UNCLOS*, The United Nations Convention on the Law of the Sea; *UNESCO's MAB*, United Nations Educational, Scientific, and Cultural Organization's Man and Biosphere Program; *UN's MOE*, United Nations' Mainstreaming Ocean Education.

TABLE 8.2 Threats Addressed by the Categories of Marine Conservation Actions

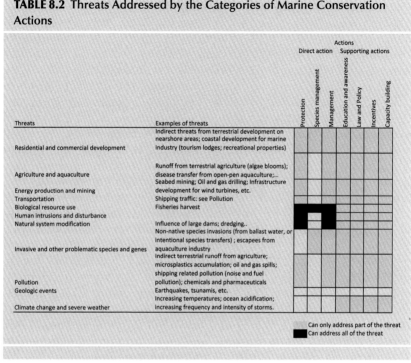

Threats	Examples of threats	Protection	Species management	Management	Education and awareness	Law and Policy	Incentives	Capacity building
		Direct action			Supporting actions			
Residential and commercial development	Indirect threats from terrestrial development on nearshore areas; coastal development for marine industry (tourism lodges; recreational properties)							
Agriculture and aquaculture	Runoff from terrestrial agriculture (algae blooms); disease transfer from open-pen aquaculture;...							
Energy production and mining	Seabed mining; Oil and gas drilling; Infrastructure development for wind turbines, etc.							
Transportation	Shipping traffic: see Pollution							
Biological resource use	Fisheries harvest							
Human intrusions and disturbance			■					
Natural system modification	Influence of large dams; dredging..							
Invasive and other problematic species and genes	Non-native species invasions (from ballast water, or intentional species transfers) ; escapees from aquaculture industry							
Pollution	Indirect terrestrial runoff from agriculture; microsplastics accumulation; oil and gas spills; shipping related pollution (noise and fuel pollution); chemicals and pharmaceuticals							
Geologic events	Earthquakes, tsunamis, etc.							
Climate change and severe weather	Increasing temperatures; ocean acidification; increasing frequency and intensity of storms.							

Can only address part of the threat
■ Can address all of the threat

DIRECT CONSERVATION ACTIONS

Protection

Land and water protection also encompasses site and area protection (i.e., national parks, town wildlife sanctuaries, private reserves, tribal-owned hunting grounds), and resource and habitat protection (i.e., easements, development rights, water rights, instream flow rights, wild and scenic river designation, securing resource rights) (Salafsky et al., 2008). Most relevant in the ocean are MPAs, "a clearly defined geographical space, recognized, dedicated and managed, through legal or other effective means, to achieve the long-term conservation of nature with associated ecosystem services and cultural values" (IUCN, 2008). MPAs can range from fully protected areas that prohibit all extractive activities to areas managed sustainably. MPAs, and zones within them, are commonly classified under the IUCN categories I–VI (Kelleher, 1999). MPAs thus work to prevent or reduce the direct threats to marine species and habitats by spatially restricting human activities. Spatial management has been used as a conservation tool for marine resources for centuries (Ballantine, 1994), and more recent global commitments have mandated nations to set aside at least 10% of the oceans for conservation (Aichi Target 11, Convention on Biological Diversity (https://www.cbd.int/)).

Most MPAs are relatively small in size (median size $3.3\,km^2$ (Boonzaier and Pauly, 2016)), although a number of very large MPAs—larger than $100,000\,km^2$ (Spalding et al., 2013)—have been designated in recent years (Boonzaier and Pauly, 2016; Maxwell et al., 2014; Spalding et al., 2013). Most common are locally implemented and managed small MPAs [e.g., within locally managed marine areas (LMMAs) in the Pacific (Léopold et al., 2009; Rocliffe et al., 2014)], implemented through customary marine tenure agreements (Ruddle et al., 1992), or by local jurisdictions (e.g., municipalities in the Philippines, Chaigneau and Daw, 2015). These locally managed MPAs are usually nearshore and coastal areas, managed by coastal communities, partner organizations, sea owners, and/ or through collaborative governance (Rocliffe et al., 2014). Some of these areas (e.g., LMMAs) are managed for the sustainable use of marine resources, rather than specifically for biodiversity conservation (Govan, 2009). More recently, very large MPAs, some larger than $1\,million\,km^2$, are being implemented (Boonzaier and Pauly, 2016). These large MPAs can encompass entire ecosystems and create synergistic links to adjacent ecosystems (Sheppard et al., 2012; Toonen et al., 2011), are more resilient to large-scale disturbances (Airamé et al., 2003; McLeod et al., 2009; Toonen et al., 2013), and are more likely to benefit highly ranging species (Maxwell and Morgan, 2013). Most of these areas have been designated within the exclusive economic zones (EEZs) of countries, although some designations exist in the high seas. In particular, the Convention for the Protection of the Marine Environment of the North-East Atlantic (OSPAR convention, named after the 1992 Oslo and Paris Commissions on Marine Environmental Management; www.ospar.org) resulted in six MPAs across the northeastern Atlantic Ocean, the world's first network of MPAs in the high seas. Under the legal framework of the United Nations Convention on the Law of the Sea, OSPAR is a key example of international cooperation to protect marine areas (O'Leary et al., 2012).

MPAs are effective marine conservation tools, with strong evidence that well-enforced no-take MPAs increase biomass, abundance, and diversity of targeted species compared to areas without protection (Edgar et al., 2014; Halpern et al., 2010; Lester et al., 2009; but see Botsford et al., 2009; Gaines et al., 2010). A recent review of MPA ecological effectiveness showed that areas that are no-take, adequately enforced, old, large, and isolated are most effective (Edgar et al., 2014). MPAs that offer only partial protection (i.e., IUCN categories III–VI) have been shown to be less successful for biodiversity values than no-take areas (Ban et al., 2014b). Well-managed MPAs are generally considered to be effective at protecting against fishing pressures and other extractive activities inside their boundaries by supporting population and species regeneration (Botsford et al., 2009). Climate change adds a layer of nuance to the sustainability goals of MPAs: there is some evidence that MPAs can build ecological resilience to large-scale stressors, including climate change and disease (Graham et al., 2008; Hansen et al., 2010; Mawdsley et al., 2009). From a social perspective, there is some evidence that partially protected MPAs can offer livelihood benefits to local communities (Fox et al., 2012; Roberts et al., 2001). Yet MPAs are not a failsafe for social–ecological systems. Illegal harvesting, poorly regulated or managed MPAs, weak regulations, poor siting, or inadequate size can lead to less effective or ineffective MPAs (Bergseth et al., 2013; Boonzaier and Pauly, 2016), and ultimately undermine local community support for conservation. There are also concerns that not enough is being done in an appropriate time frame to protect the marine environment. The 2010 biodiversity targets were missed, and the current goal to protect 10% of the global ocean by 2020 (Aichi Target 11) is unlikely to be achieved, with calculations by Boonzaier and Pauly (2016) predicting that this target will not be met until 2035.

Species Management

Species management actions are aimed at management or restoration of species, with particular focus on species of concern (Salafsky et al., 2008), and include harvest strategies, efforts aimed at species recovery, species reintroduction, and ex situ conservation initiatives. Global action on species management relies mostly on trade restrictions and the management of species' populations worldwide, such as through international treaties, including the Convention on International Trade in Endangered Species of Wild Fauna and Flora (CITES) and the International Whaling Commission (IWC). Global voluntary measures include embargos of endangered species, for example, by airlines refusing to transport cargo from trophy hunts or shark fins (Parry, 2013).

In marine systems, single or multispecies fisheries managements are common (Walters et al., 2005). It is currently estimated that approximately 58% of fish stocks are fully fished, and 31% are overfished (FAO, 2016), while other estimates suggest the overexploitation rate to be much higher (Pauly and Zeller, 2016). In some cases, failure of single species management can lead to the collapse and total closure of a fishery [e.g., Atlantic cod (*G. morhua*) fishery on the East coast of Canada (Hutchings and Myers, 1994)]. However, when fisheries management is effective—as is increasingly the case in developed countries—populations have been noted to increase (Costello et al., 2016). Fisheries management tools include stock assessments, input and output controls, and are increasingly diversifying to include measures such as territorial user rights for fisheries, individual or community (transferable) quotas, and bycatch reduction devices (Walters and Martell, 2004).

Single-species management is limited in its ability to address system-wide concerns, and thus some jurisdictions are moving toward ecosystem-based fisheries management both locally and globally, a necessary change for complex, global systems in an era of globalization and climate change (Ogier et al., 2016; Pikitch et al., 2004; Ruckelshaus et al., 2008). For example, while the majority of Regional Fisheries Management Organizations (RFMOs) use single-species fisheries management (Cullis-Suzuki and Pauly, 2010), some RFMOs are beginning to implement ecosystem-based fisheries management in order to manage for ecosystem health as a whole [e.g., Commission for the Conservation of Antarctic Marine Living Resources (CCAMLR); Link,

2010; Ruckelshaus et al., 2008]. A shift toward ecosystem-based management can also be seen at a local scale. For example, municipalities on Cebu Island, Philippines—responsible for independently managing local areas—began scaling up their fisheries management to an intermunicipal level to better account for larger ecosystem-level threats (Eisma-Osorio et al., 2009). These small-scale actions link well with the larger international coral triangle initiative (CTI), which seeks to create a comprehensive network of ecosystem-based management in the region (Rosen and Olsson, 2013). Fisheries comanagement (when fishers and managers work together to regulate a fishery) is also increasingly applied to solve the numerous problems of fisheries management. Successful co-managed fisheries occur when strong leadership exists with the community, along with a functioning individual or community-based fisheries quota system, good social cohesion, and some integration of protected areas supporting the fishery (Gutiérrez et al., 2011). Across these methods, fisheries' success or long-term sustainability depends on the ability of the fishery management tool of choice to adjust fishing pressure to sustainable levels, an inherently complex idea which needs further integration of social–ecological systems theory (Hilborn et al., 2015).

Threatened species management is aimed at the recovery of listed species (IUCN, 2015). For example, the IUCN Red List efforts are aimed at informing conservation planners on the status of threatened or endangered species, and can help to inform and set conservation targets (Nieto et al., 2015). Threatened species management actions can also be aimed at the recovery and/or avoidance of high-risk bycatch species by setting total allowable catch limits to zero [e.g., spiny dogfish (*Squalus acanthias*) in European waters (Nieto et al., 2015)]; or by entirely banning a species take (e.g., in 2011 the Bahamas banning commercial and recreational shark landings and thereby effectively creating a shark sanctuary around their EEZ (Graham et al., 2016)). However, commercially important marine fish have been slow to be included in globally recognized lists such as CITES (Vincent et al., 2014), and thus fisheries management remains predominantly a national responsibility.

Habitat Management

Management actions focus on the conservation and/or restoration of broad habitats and sites (Salafsky et al., 2008). These actions include MPAs, described in the "protection" section, as well as invasive species management and restoration. Marine management actions can directly eliminate or reduce biological resource use, human disturbances, and natural systems modifications (e.g., dredging) by prohibiting or regulating these activities, and partially address many other marine threats such as aquaculture, invasive species, and pollution by regulating them (Table 8.2).

Management of invasive species and restoration activities are considerably more challenging in marine than terrestrial systems, because much of the ocean remains relatively inaccessible to humans. Only diveable depths (~30 m) are accessible for people to take direct actions such as removing invasive species or restorating habitat; deeper areas require specialized submersibles and do not readily facilitate such activities. Consequently, marine restoration activities to date have been limited to nearshore habitats such as estuaries, mangroves, rocky and coral reefs (Elliott et al., 2007). While marine invasive species are a widespread problem for the global oceans (Molnar et al., 2008), they are also very difficult and costly to manage or attempt to control (Williams and Grosholz, 2008). Invasive species management actions have so far only occurred at local scales, such as the attempted (but so far unsuccessful) removal of the Indo-Pacific lionfish (*Pterois* sp.) from Caribbean reefs (Frazer et al., 2012). Invasive species management is likely to gain increasing importance in the coming decades as globalization continues to spread marine species through shipping and other industries.

SUPPORTING CONSERVATION ACTIONS

Law and Policy

Laws and policies are governance actions to protect biodiversity by developing, changing, influencing, and implementing formal legislation, regulations, and voluntary standards (Salafsky et al., 2008). Laws and policies can be applied at all scales from global to local, and include legislation (e.g., promoting conventions on biodiversity, including international wildlife trade conventions, such as CITES, and national laws such as the US Endangered Species Act); policies and regulations (e.g., zoning regulations, sustainable harvest limits); or voluntary standards and professional codes (e.g., Marine Stewardship Council, Conservation Measures Partnership). Laws and policies can provide the framework for a coordinated and comprehensive approach to marine conservation to address a range of threats, but their effectiveness at both the global and local scale depends on implementation, monitoring, and enforcement.

Since the 1960s, legislative action has been introduced worldwide, covering issues such as marine pollution, conservation of species and habitats, and protection of fish stocks (Bell et al., 2013). Over 500 global and regional multilateral environmental agreements are in force today (Scott, 2011). Much of the early marine legislation was introduced before current global threats were fully understood, and consequently do not explicitly consider anthropogenic climate change (Frost et al., 2016). However, there are national to local-level mechanisms that create flexibility in how legislation can be implemented (e.g., policies, review cycles, bylaws), and it is at this implementation level where emerging threats can most readily be taken into account. This flexibility also applies to the implementation of international conventions, e.g., the 1992 Convention for Biological Diversity (CBD) does not mention climate change or environmental variability. However, decisions made by the governing body of the convention (the Conference of Parties) can, and do, directly address climate change impacts on biodiversity, and it is these resulting decisions that allow climate change to be accounted for in programs of work (e.g., national-level biodiversity strategies (Frost et al., 2016). Likewise, this flexibility also allows nations to manage and exploit their natural resources appropriate to their own economic needs and priorities. Although international laws and policies provide no formal, centralized enforcement mechanism for appropriate implementation, they do serve as a means to build and demonstrate international consensus on environmental issues. Overall, the ability of law and policies to contribute effectively to marine conservation depends on their fit to both the ecological and social system, which is dependent on appropriate implementation at the national to local level.

Education and Awareness

Education and awareness actions are defined as those "directed at people to improve understanding and skills, and influence behavior" (Salafsky et al., 2008). Education can be formal, i.e., undertaken in schools or higher education, or informal, such as training programs outside of a standard school setting. International education and awareness programs are developed and implemented by organizations such as the United Nations' Mainstreaming Ocean Education program and UNESCO's Man and Biosphere Program. Both programs seek to improve the relationships of people to their environments and resources through the use of strategic scientific marine education (Kuijper, 2003; United Nations Educational Scientific and Cultural Organization, 2016). Locally, the formalized structure of universities, colleges, and primary and secondary schools provide some opportunities for marine education. Formal programs are overseen by bodies such as the US National Marine Educators Association, which works to standardize, improve, and forward marine content for youth. However, marine education and ocean awareness is likely lacking in many formal education structures. Finally, ecotourism companies (e.g., whale and dolphin watching, snorkeling and scuba diving, and eco-friendly boating) generate local citizen education opportunities for educational connections with ocean systems through interpretive programs and lived-experiences (Zeppel, 2008).

Education and awareness has been broadening in some places to include traditional and local ecological knowledge, which can be highly relevant for informing stewardship for marine conservation and social–ecological systems management (Thornton and Scheer, 2012). In particular, as the importance of incorporating indigenous worldviews and stewardship strategies is recognized, awareness programs for policy makers and other relevant professionals are needed.

Measuring the efficacy or benefit of educational initiatives is difficult (Alder, 1996). Studies attempting to quantify the effect of formal education on conservation are limited, but on the whole they agree that it has a beneficial effect (Alix-Garcia, 2007). There is also evidence for the importance of informal education, including marine citizenship, ecotourism, and media. However, increased knowledge of marine ecosystems and conservation does not necessarily translate into behavioral change, especially in places where people feel more disconnected from the ocean (Waylen et al., 2009).

Livelihood, Economic and Other Incentives

Livelihood, economic and other incentives are actions that aim to motivate or influence behavioral change (Salafsky et al., 2008). To encourage behaviors that benefit conservation, positive incentives have included the development of new businesses that depend on viable natural resources, the promotion of environmentally friendly products and services that substitute for environmentally damaging ones, and the application of market mechanisms and direct payment schemes for ecosystem protection. Negative incentives that seek to limit the harm to biodiversity or to prevent overuse of natural resources include catch limits (see Species Management section), fines, and taxes. In marine systems, however, the main economic incentive (subsidies) and increasing globalization (new markets for marine products) have promoted fisheries exploitation rather than conservation. Fisheries subsidies are estimated to exceed US $30 billion per year globally (Sumaila et al., 2010) mostly as fuel and capacity enhancements, which incentivize the overexploitation of marine resources by paying the fishing industry to fish when it would otherwise not be profitable (Hatcher and Robinson, 1999; Munro and Sumaila, 2012).

Incentives to promote positive marine conservation outcomes at local and global scales are varied and becoming increasingly prevalent. For example, the Blue Carbon Initiative is a global program coordinated by Conservation International, IUCN, and UNESCO's Intergovernmental Oceanographic Commission: they use economic incentives, among other measures, to promote the conservation, restoration, and sustainable use of coastal blue carbon ecosystems to help mitigate the threat of climate change (http://theblucarboninitiative. org). Conservation payments and livelihood incentives have been used on local-scale projects around the world to motivate pro-conservation outcomes. For example, in the Seychelles, a debt-for-nature-swap is underway whereby the island nation agreed to protect 30% of their ocean territory to enhance marine conservation and climate adaptation. In return, the foreign investor group The Paris Club, agreed to cancel US $31 million of the Seychelles' debt (Weymouth Ullman, 2015). However, such initiatives have been criticized as being unsustainable with adverse impacts on local communities, including erosion of local culture, values, and livelihoods (e.g., "ocean grabbing"; De Schutter, 2012). Linked enterprises and livelihood alternatives such as ecotourism provide incentives to develop businesses that directly depend on the maintenance of natural resources or provide substitute livelihoods as a means of changing behaviors and attitudes. This has been more successful with larger, charismatic species; for example, the International Whaling Commission, historically a fishing body, addresses the potential and encourages the further development of whale watching to become a "sustainable use of cetacean resources" (Williams et al., 2002). Cross-scalar social movements that have had some success in promoting positive consumer-based action on fisheries sustainability include market-based fisheries certification programs (e.g., Marine Stewardship Council, MSC (Jacquet et al., 2009); but see Hadjimichael and Hegland, 2016 for critique).

External Capacity Building

Successful scoping, planning, implementing and managing conservation actions require capacity, and technical and other expertise. External capacity building is aimed at building physical infrastructure and human and social capacities to carry out effective conservation, and it supports all the categories of direct conservation action. Capacity building includes creating or providing financial and non-financial support, forming and facilitating partnerships, alliances, and networks of organizations (Salafsky et al., 2008). Capacity building does not address any environmental threats directly, but rather increases financial, technical, and field capacity to improve planning and successful implementation of direct actions, and increase the alignment of those actions with regional conservation and development efforts. While most efforts to build capacity are context-specific and aimed at local scales, some global capacity-building efforts exist. The Conservation Leadership Program (http://www.conservationleadershipprogramme.org/) and Russell E. Train Fellowship (http://www.worldwildlife.org/initiatives/russell-e-train-education-for-nature), provide professional development for individual early career conservationists, sometimes with a "train the trainer" component to disseminate lessons learned in the recipient's home country. Programs like the Global Environment Facility support large-scale, national-level capacity building with priorities and mechanisms driven by recipient countries. Somewhere in-between, groups such as Flora & Fauna International (Bensted-Smith and Kirkman, 2010) and NOAA's International Marine Protected Area Management Capacity Building Program (http://sanctuaries.noaa.gov/international/welcome.html#international), use a stepwise approach to train groups of in-country partners over a number of years in different phases of training at successively broader spatial and temporal scales.

As far as we are aware, no comprehensive assessment of capacity building exists at global or local scales, but bridging organizations show promise in helping to build capacity, increase connections and peer learning (Alexander et al., 2016; Berdej and Armitage, 2016; Cohen et al., 2012). For example, the "Big Ocean" network is a peer learning organization that facilitates learning among managers of large-scale MPAs (http://bigoceanmanagers.org/). This group provides a forum for communication, networking, and social learning among practitioners and managers involved in all aspects of large MPAs. The International Maritime Organization (IMO) also works internationally on capacity-building efforts to assist developing countries to meet international pollution reduction targets (Zhu, 2006). Similarly, the CTI has helped to build capacities in the six member countries. The CTI is administered through a Regional Secretariat whose main activities include: organizational development, outreach and communication, regional coordination and mechanisms, technical and thematic working groups, and capacity development. At local scales, marine traditional ecological knowledge might be developed through collaborative management and bridging organizations that build trust and community involvement for improving marine conservation and adaptive actions (Thornton and Scheer, 2012).

LINKS BETWEEN GLOBAL AND LOCAL ACTIONS

In this section, we reflect on the scale and scope of conservation actions, focusing on existing linkages between global and local marine conservation actions. We discuss conservation actions (or categories of actions) that are already occurring and are relatively effective at reducing threats at global scales, and highlight situations where local actions are complementing global actions or where linkages between scales have been made. We also reflect on the threats that the categories most effectively address, and where the actions are intended to intervene in social–ecological systems.

All categories of conservation actions are already applied at a global scale, although many have focused on smaller scales. All of the direct actions—protection (e.g., MPAs), species management, and habitat management—target both scales. MPAs, for example, are commonly used as local, small conservation tools, although large MPAs, including some in the high seas, are starting to operate at global scales. Fisheries management is generally done by countries in their EEZs, but ocean-basin and wide-ranging species scale fisheries management also exists. Supporting actions also occur at local and global scales. Education and awareness, incentives, and capacity building have mostly been focused on local scales, although in all cases some examples of global efforts exist. Law and policy is applicable at local and global scales. Locally laws and policies provide the tools to implement direct conservation actions. Globally, laws and policies can assist with limiting global threats, and can encourage international cooperation to address ocean-basin threats. However, in comparison to the scale of global threats, to date none of the actions have been effective enough. Ongoing tensions between national jurisdiction over marine resources and global agreements for biodiversity and conservation targets (e.g., Convention on Biological Diversity goals) are certainly an issue. Indeed, many international agreements are voluntary, and even those that are binding have no or limited consequences for noncompliance (Jay et al., 2016; Warner, 2014). Furthermore, conservation in international waters is particularly challenging, as exemplified by the ongoing struggle to manage areas beyond national jurisdiction using area-based protection tools (Ban et al., 2014a).

Some conservation actions are explicitly being used to attempt to bridge scales. MPAs are a great example. Small-scale and locally managed MPAs have proliferated in the Indo-Pacific (Govan, 2009). Sites are usually chosen by local communities based on factors such as accessibility and perceived decline of resources, and therefore they may not be as well placed to achieve global biodiversity conservation objectives (Mills et al., 2012). In an effort to have individual MPAs comprise an ecologically meaningful network, the science of systematic conservation planning is being used to inform where additional MPAs can be placed to better meet ecological objectives (Game et al., 2011; Mills et al., 2010). For instance, the CTI was created to meet multiple objectives, including biodiversity conservation, at the scale of multiple countries. Similarly, this initiative is also helping to scale up local fisheries management actions to be more effective at an ecosystem scale (Eisma-Osorio et al., 2009; Rosen and Olsson, 2013). Most commonly, capacity-building efforts facilitate linking of scales, through funding and by supporting or creating bridging organizations, such as the CTI Secretariat. These organizations provide the capacity to link people and management actions.

Theoretically, marine conservation actions can intervene in multiple places in social–ecological systems to produce desirable outcomes, but in reality, interventions are quite focused (Fig. 8.1). The categories of marine conservation actions we reviewed predominantly serve to influence the way people interact with the ocean (i.e., they act on the action situation). This is to be expected, as it is in the action situation that people directly interact with the environment, for example through fishing and mining. Conservation actions that directly influence people's behavior are protection (by excluding or limiting people's actions from an area), species management (by limiting when/where/how people can use species), management (by also limiting when/where/how people do things that might harm the ocean), incentives and education (by influencing people's behavior), and law and policy (by creating the framework for the other actions).

Few conservation actions affect the characteristics of actor groups or ecosystems directly. Only two categories of actions have the potential to directly influence the attributes of actors in the social–ecological system (e.g., social capital, trust, norms, knowledge of the social–ecological system): education and awareness, and capacity building (Fig. 8.1). Education and awareness might influence people's worldviews, thereby changing their beliefs and influencing their actions. Capacity building can change the attributes of stakeholder groups, for example, increasing social capital. Similarly, only two categories of actions intervene in the governance system (e.g., network structure, operational rules, collective-choice rules): capacity building, and law and policy. Law and policy can create the institutions that then create the governance system. Capacity building can also create new institutions, such as bridging organizations. Only management directly influences the characteristics of the resource system (e.g., productivity of the system, mobility of species), through restoration activities.

GAPS AND WAYS FORWARD

In an era where global influences increasingly affect marine ecosystems, many gaps exist in conservation actions to effectively address all threats, especially climate change (Table 8.2). Threats related to climate change are global, and no marine conservation action is capable of directly addressing the sources of the problem. A concerted global effort is urgently needed to reduce greenhouse gas emissions. Marine conservation actions can do relatively little to address the stressors of climate change, such as ocean acidification, hypoxia, changing ocean currents, and increasing temperatures. Indeed, marine conservation actions that are otherwise effective at local scales may be for naught if rapid climate change continues. However, there is some evidence that healthier marine ecosystems are more resilient to climate change related stressors. Hence a common recommendation—one we endorse—is to have sufficient protection and effective ecosystem-based management to give ecosystems and species the best chance under a changing climate. More controversial geo-engineering options are also being considered, ranging from iron fertilization, to liming the oceans, to shading coral reefs, to altering water flows locally to reduce water temperatures. These options are risky, and without fully understanding marine social–ecological systems, have the potential to do more harm than good. Small-scale experiments in controlled settings are needed before any marine geo-engineering interventions should be pursued.

Another important gap lies in reducing land-based threats to marine systems (Table 8.2). Land-based stressors include input of nutrients, pesticides, and other potential harmful point and nonpoint sources of pollution. New and emerging threats from land-based sources include pharmaceutical products, nanoparticles, and microplastics (Rochman et al., 2016). While these threats are daunting, theoretically land-based influences on marine systems are readily addressed through integrated land–sea conservation (Álvarez-Romero et al., 2011; Pittman and Armitage, 2016). Yet in most countries, land and ocean jurisdictions are separate, and hence integrated land–sea conservation is challenging to implement effectively because of the many jurisdictions involved. Political will and capacity is needed to implement land–sea conservation.

Globally, the biggest spatial gap in marine conservation actions is in areas beyond national jurisdictions, which have little effective actions (Ban et al., 2014a). While the 18 RFMOs covering high-sea areas are intended to manage fish stocks sustainably, they are failing in two ways: (1) to comply with performance standards expected of RFMOs under international agreements; and (2) to prevent unsustainable fishing of the fish stocks under their management (Gjerde et al., 2013). Similarly, there is currently no straightforward mechanism for creating MPAs in the high seas (Ban et al., 2014a).

Gaps also exist in our review of marine conservation actions. We focused on categories outlined by Salafsky et al. (2008), and while their lexicon was intended to be comprehensive, there are likely gaps. Furthermore, in our overview of marine conservation actions, we relied on the peer-reviewed and searchable gray literature to identify examples. There are likely many more examples that are not reflected in the literature. We suspect this is particularly the case for education and awareness, capacity building, and alternative livelihoods.

Examining where in social–ecological systems the categories of marine conservation actions intervene illustrates additional gaps: few actions attempt to change the characteristics of actors, the ecosystem and species, or the governance system (Fig. 8.1). For the former two, interventions have been local, revealing further gaps. Opportunities thus exist to enhance effective interventions in these aspects of social–ecological systems. An expanded social–ecological approach has been suggested to make marine restoration and management more relevant in the Anthropocene (Abelson et al., 2015; Österblom et al., 2016).

The current suite of marine conservation actions is relatively effective at reducing localized direct threats (e.g., fishing, mining), but much more can and must be done. Many countries have existing laws and policies, and established actions such as MPAs and fisheries regulations, yet are either not fully making use of the tools or lack enforcement (Ainsworth et al., 2012). A key starting point to improving the outlook for marine ecosystems in an era of the Anthropocene is to effectively implement and manage the various laws and policies that already exist.

There are challenges and barriers to conservation action at all scales. Common barriers include limited financial means to implement marine conservation actions, opposition from stakeholders who perceive a potential negative impact on their livelihoods or interests, limited capacity to implement and enforce existing and proposed conservation actions, and lack of political will to follow through on commitments. Even in places where bridging of scales is being attempted, challenges arise. For example, scaling down international legislation is complicated by the different government departments and sectors across countries (e.g., fisheries, versus coastal, environment departments within governments). This has been acknowledged as a barrier to regional implementation in the CTI (Fidelman et al., 2012).

Technological innovations provide promises to improve enforcement and monitoring, and enhance communications, and potentially to take pressures off extraction of wild seafood. Telemetry tools and tracking networks (e.g., Ocean Tracking Network) that allow researchers and managers to track migrating species to understand patterns of habitat use or interference in migration (e.g., fisheries bycatch; entanglement) can help us to improve our understanding of complex systems and meet management objectives (Lennox et al., 2016). Satellite-based vessel monitoring systems and automatic identification systems can now monitor fishing activities and identify illegal or unregulated fishing by monitoring vessel presence and movement (e.g., SkyTruth). Social media tools are used to raise awareness of conservation actions and needs, to report poachers, and to shame violators of environmental norms or agreements

(Jacquet et al., 2011). Social media can also be valuable for rapid knowledge transfer and effective diffusion of innovations and conservation actions. Finally, emerging aquaculture technologies such as aquaponics and closed containment aquaculture techniques have the potential to support food security while diminishing the demand on, and threat to, wild stocks (Jennings et al., 2016). Overall, a shift toward a more integrated food system for aquatic food security will likely increase sustainability and support conservation goals.

However, emphasis on technological innovations, and indeed on conservation actions—as we have done in this chapter—miss the importance that good governance and process has in enabling conservation measures that are transparent and legitimate (Burt et al., 2014). Underpinning all effective marine conservation actions are processes that make those actions legitimate in the eyes of those affected by them. Such legitimacy in turn leads to better compliance, and hence improved conditions in the water and for people who rely on marine resources.

The future of the oceans, and people who rely on marine resources, depends upon a concerted global effort to slow climate change, while continuing and enhancing local and global actions to reduce other threats. Many effective marine conservation actions exist, and all can be implemented more broadly and at multiple scales. Linking scales—scaling up local actions, and making global actions locally relevant—will be increasingly required to effectively address drivers of change in the Anthropocene. Humans have shaped the current state of oceans, and we will continue to shape them into the future. If we prioritize marine conservation, we can ensure that our influence will be a positive one for all marine life and for generations to come.

REFERENCES

Abelson, A., Halpern, B.S., Reed, D.C., Orth, R.J., Kendrick, G.A., Beck, M.W., Belmaker, J., Krause, G., Edgar, G.J., Airoldi, L., Brokovich, E., France, R., Shashar, N., de Blaeij, A., Stambler, N., Salameh, P., Shechter, M., Nelson, P.A., 2015. Upgrading marine ecosystem restoration using ecological–social concepts. Bioscience. http://dx.doi.org/10.1093/biosci/biv171. XX, biv171.

Ainsworth, C.H., Morzaria-luna, H.N., Kaplan, I.C., Levin, P.S., Fulton, E.A., 2012. Full compliance with harvest regulations yields ecological benefits: Northern Gulf of California case study. Journal of Applied Ecology 49, 63–72. http://dx.doi.org/10.1111/j.1365-2664.2011.02064.x.

Airamé, S., Dugan, J.E., Lafferty, K.D., Leslie, H., McArdle, D.A., Warner, R.R., 2003. Applying ecological criteria to marine reserve design: a case study from the California channel islands. Ecological Applications 13, 170–184.

Alder, J., 1996. Costs and effectiveness of education and enforcement, cairns section of the great barrier reef marine park. Environmental Management 20, 541–551. http://dx.doi.org/10.1007/BF01474654.

Alexander, S.M., Andrachuk, M., Armitage, D., 2016. Navigating governance networks for community-based conservation. Frontiers in Ecology and Environment 14, 155–164. http://dx.doi.org/10.1002/fee.1251.

Alix-Garcia, J., 2007. A spatial analysis of common property deforestation. Journal of Environmental Economics and Management 53, 141–157. http://dx.doi.org/10.1016/j.jeem.2006.09.004.

Álvarez-Romero, J.G., Pressey, R.L., Ban, N.C., Vance-Borland, K., Willer, C., Klein, C.J., Gaines, S.D., 2011. Integrated land-sea conservation planning: the missing links. Annual Review in Ecology, Evolution and Systematics 42, 381–409. http://dx.doi.org/10.1146/annurev-ecolsys-102209-144702.

Ballantine, B., 1994. The practicality and benefits of a marine reserve network. In: Gimel, K.L. (Ed.), Limiting Access to Marine Fisheries: Keeping the Focus on Conservation. Center for Marine Conservation and World Wildlife Fund, US, Washington DC, pp. 205–223.

Ban, N.C., Bax, N.J., Gjerde, K.M., Devillers, R., Dunn, D.C., Dunstan, P.K., Hobday, A.J., Maxwell, S.M., Kaplan, D.M., Pressey, R.L., Ardron, J.A., Game, E.T., Halpin, P.N., 2014a. Systematic conservation planning: a better recipe for managing the high seas for biodiversity conservation and sustainable use. Conservervation Letters 7, 41–54. http://dx.doi.org/10.1111/conl.12010.

Ban, N.C., McDougall, C., Beck, M., Salomon, A.K., Cripps, K., 2014b. Applying empirical estimates of marine protected area effectiveness to assess conservation plans in British Columbia, Canada. Biological Conservervation 180, 134–148. http://dx.doi.org/10.1016/j.biocon.2014.09.037.

Bell, S., McGillivray, D., Pedersen, O., 2013. Environmental Law, eighth ed. Oxford University Press, Oxford.

Bensted-Smith, R., Kirkman, H., 2010. Comparison of Approaches to Management of Large Marine Areas, pp. 1–144 Fauna & Flaura International, Cambridge, UK and Conservation International, Washington DC.

Berdej, S.M., Armitage, D.R., 2016. Bridging organizations drive effective governance outcomes for conservation of Indonesia's marine systems. PLoS One 11, 1–26. http://dx.doi.org/10.1371/journal.pone.0147142.

Bergseth, B.J., Russ, G.R., Cinner, J.E., 2013. Measuring and monitoring compliance in no-take marine reserves. Fish and Fisheries 16, 240–258. http://dx.doi.org/10.1111/faf.12051.

Berkes, F., Hughes, T.P., Steneck, R.S., Wilson, J.A., Bellwood, D.R., Crona, B., Folke, C., Gunderson, L.H., Leslie, H.M., Norberg, J., Nystrom, M., Olsson, P., Osterblom, H., Scheffer, M., Worm, B., 2006. Globalization, roving bandits, and marine resources. Science 311, 1557–1558. http://dx.doi.org/10.1126/science.1122804.

Boonzaier, L., Pauly, D., 2016. Marine protection targets: an updated assessment of global progress. Oryx 50, 27–35. http://dx.doi.org/10.1017/S0030605315000848.

Botsford, L.W., Brumbaugh, D.R., Grimes, C., Kellner, J.B., Largier, J., O'Farrell, M.R., Ralston, S., Soulanille, E., Wespestad, V., 2009. Connectivity, sustainability, and yield: bridging the gap between conventional fisheries management and marine protected areas. Review in Fish Biology and Fisheries 19, 69–95. http://dx.doi.org/10.1007/s11160-008-9092-z.

Burt, J.M., Akins, P., Latham, E., Salomon, A.K., Ban, N.C., 2014. Marine Protected Area Design Features that Support Resilient Human–Ocean Systems: Applications for British Columbia, Canada. Simon Fraser University, Burnaby, BC, pp. 1–159.

Chaigneau, T., Daw, T.M., 2015. Individual and village-level effects on community support for Marine Protected Areas (MPAs) in the Philippines. Marine Policy 51, 499–506. http://dx.doi.org/10.1016/j.marpol.2014.08.007.

Cinner, J.E., Folke, C., Daw, T., Hicks, C.C., 2011. Responding to change: using scenarios to understand how socioeconomic factors may influence amplifying or dampening exploitation feedbacks among Tanzanian fishers. Global Environmental Change 21, 7–12. http://dx.doi.org/10.1016/j.gloenvcha.2010.09.001.

Cohen, P.J., Evans, L.S., Mills, M., 2012. Social networks supporting governance of coastal ecosystems in Solomon Islands. Conservation Letters 5, 376–386. http://dx.doi.org/10.1111/j.1755-263X.2012.00255.x.

Costello, C., Ovando, D., Clavelle, T., Strauss, C.K., Hilborn, R., Melnychuk, M.C., 2016. Global fishery prospects under contrasting management regimes. PNAS 113, 5125–5129. http://dx.doi.org/10.1073/pnas.1520420113.

Cullis-Suzuki, S., Pauly, D., 2010. Marine protected area costs as "beneficial" fisheries subsidies: a global evaluation. Coastal Management 38, 113–121. http://dx.doi.org/10.1080/08920751003633086.

De Schutter, O., 2012. The Right to Food. New York.

Edgar, G.J., Stuart-Smith, R.D., Willis, T.J., Kininmonth, S., Baker, S.C., Banks, S., Barrett, N.S., Becerro, M.a, Bernard, A.T.F., Berkhout, J., Buxton, C.D., Campbell, S.J., Cooper, A.T., Davey, M., Edgar, S.C., Försterra, G., Galván, D.E., Irigoyen, A.J., Kushner, D.J., Moura, R., Parnell, P.E., Shears, N.T., Soler, G., Strain, E.M.a, Thomson, R.J., 2014. Global conservation outcomes depend on marine protected areas with five key features. Nature 506, 216–220. http://dx.doi.org/10.1038/nature13022.

Eisma-Osorio, R.-L., Amolo, R.C., Maypa, A.P., White, A.T., Christie, P., 2009. Scaling up local government initiatives toward ecosystem-based fisheries management in Southeast Cebu Island, Philippines. Coastal Management 37, 291–307. http://dx.doi.org/10.1080/08920750902851237.

Elliott, M., Burdon, D., Hemingway, K.L., Apitz, S.E., 2007. Estuarine, coastal and marine ecosystem restoration: confusing management and science – a revision of concepts. Estuarine, Coastal and Shelf Science 74, 349–366. http://dx.doi.org/10.1016/j.ecss.2007.05.034.

FAO, 2016. The State of World Fisheries and Aquaculture. FAO, Rome, Italy. 92-5-105177-1.

Fidelman, P., Evans, L., Fabinyi, M., Foale, S., Cinner, J.E., Rosen, F., 2012. Governing large-scale marine commons: contextual challenges in the Coral Triangle. Marine Policy 36, 42–53. http://dx.doi.org/10.1016/j.marpol.2011.03.007.

Fox, H.E., Mascia, M.B., Basurto, X., Costa, A., Glew, L., Heinemann, D., Karrer, L.B., Lester, S.E., Lombana, A.V., Pomeroy, R.S., Recchia, C.A., Roberts, C.M., Sanchirico, J.N., Pet-Soede, L., White, A.T., 2012. Reexamining the science of marine protected areas: linking knowledge to action. Conservation Letters 5, 1–10. http://dx.doi.org/10.1111/j.1755-263X.2011.00207.x.

Frazer, T.K., Jacoby, C.a., Edwards, M.a., Barry, S.C., Manfrino, C.M., 2012. Coping with the lionfish invasion: can targeted removals yield beneficial effects? Reviews in Fisheries Science 20, 185–191. http://dx.doi.org/10.1080/10641262.2012.700655.

Frost, M., Bayliss-Brown, G., Buckley, P., Cox, M., Dye, S.R., Sanderson, W.G., Stoker, B., Withers Harvey, N., 2016. A review of climate change and the implementation of marine biodiversity legislation in the United Kingdom. Aquatic Conservation: Marine and Freshwater Ecosystems 595, 576–595. http://dx.doi.org/10.1002/aqc.2628.

Gaines, S.D., White, C., Carr, M.H., Palumbi, S.R., 2010. Designing marine reserve networks for both conservation and fisheries management. PNAS 107, 18286–18293. http://dx.doi.org/10.1073/pnas.0906473107.

Game, E.T., Lipsett-Moore, G., Hamilton, R., Peterson, N., Kereseka, J., Atu, W., Watts, M., Possingham, H., 2011. Informed opportunism for conservation planning in the Solomon Islands. Conservation Letters 4, 38–46. http://dx.doi.org/10.1111/j.1755-263X.2010.00140.x.

Gjerde, K.M., Currie, D., Wowk, K., Sack, K., 2013. Ocean in peril: reforming the management of global ocean living resources in areas beyond national jurisdiction. Marine Pollution Bulletin 74, 540–551. http://dx.doi.org/10.1016/j.marpolbul.2013.07.037.

Golden, C.D., Allison, E.H., Cheung, W.W.L., Dey, M.M., Halpern, B.S., McCauley, D.J., Smith, M., Vaitla, B., Zeller, D., Myers, S.S., 2016. Nutrition: fall in fish catch threatens human health. Nature 534, 317–320.

Govan, H., 2009. Status and Potential of Locally-managed Marine Areas in the Pacific Island Region: Meeting Nature Conservation and Sustainable Livelihood Targets Through Wide-spread Implementation of LMMAs. SPREP/WWF/WorldFish-Reefbase/CRISP. 95pp + 5 annexes.

Graham, N.A.J., McClanahan, T.R., MacNeil, M.A., Wilson, S.K., Polunin, N.V.C., Jennings, S., Chabanet, P., Clark, S., Spalding, M.D., Letourneur, Y., Bigot, L., Galzin, R., Ohman, M.C., Garpe, K.C., Edwards, A.J., Sheppard, C.R.C., 2008. Climate warming, marine protected areas and the ocean-scale integrity of coral reef ecosystems. PLoS One 3, e3039. http://dx.doi.org/10.1371/journal.pone.0003039.

Graham, F., Rynne, P., Estevanez, M., Luo, J., Ault, J.S., Hammerschlag, N., 2016. Use of marine protected areas and exclusive economic zones in the subtropical western North Atlantic Ocean by large highly mobile sharks. Diversity and Distribution 22, 534–546. http://dx.doi.org/10.1111/ddi.12425.

Gutiérrez, N.L., Hilborn, R., Defeo, O., 2011. Leadership, social capital and incentives promote successful fisheries. Nature 470, 386–389. http://dx.doi.org/10.1038/nature09689.

Hadjimichael, M., Hegland, T.J., 2016. Really sustainable? Inherent risks of eco-labeling in fisheries. Fisheries Research 174, 129–135. http://dx.doi.org/10.1016/j.fishres.2015.09.012.

Halpern, B.S., Lester, S.E., McLeod, K.L., 2010. Placing marine protected areas onto the ecosystem-based management seascape. PNAS 107, 18312–18317. http://dx.doi.org/10.1073/pnas.0908503107.

Hansen, L., Hoffman, J., Drews, C., Mielbrecht, E., 2010. Designing climate-smart conservation: guidance and case studies. Conservation Biology 24, 63–69. http://dx.doi.org/10.1111/j.1523-1739.2009.01404.x.

Hatcher, A., Robinson, K., 1999. Overcapacity, overcapitalization and subsidies in European fisheries. In: Proceedings of the First Workshop Held in Portsmouth, UK, October 28–30, 1998. CEMARE, University of Portsmouth, Portsmouth.

Hilborn, R., Fulton, E.A., Green, B.S., Hartmann, K., Tracey, S.R., Watson, R.A., 2015. When is a fishery sustainable? Canadian Journal of Fisheries and Aquaculture Science 1441, 1–46. http://dx.doi.org/10.1139/cjfas-2015-0062.

Hutchings, J., Myers, R., 1994. What can be learned from the collapse of a renewable resource? Atlantic cod, Gadus morhua, of Newfoundland and Labrador. Canadian Journal of Fisheries and Aquaculture Science 51, 2126–2146.

IUCN, 2008. Establishing Resilient Marine Protected Area Networks—Making It Happen. IUCN-WPCA, Washington DC.

IUCN, 2015. The IUCN Red List of Threatened Species. Version 2015–4.

Jacquet, J., Hocevar, J., Lai, S., Majluf, P., Pelletier, N., Pitcher, T., Sala, E., Sumaila, R., Pauly, D., 2009. Conserving wild fish in a sea of market-based efforts. Oryx 44, 45–56. http://dx.doi.org/10.1017/S0030605309990470.

Jacquet, J., Hauert, C., Traulsen, A., Milinski, M., 2011. Shame and honour drive cooperation. Biology Letters 7, 899–901. http://dx.doi.org/10.1098/rsbl.2011.0367.

Jay, S., Alves, F.L., O'Mahony, C., Gomez, M., Rooney, A., Almodovar, M., Gee, K., de Vivero, J.L.S., Gonçalves, J.M.S., da Luz Fernandes, M., Tello, O., Twomey, S., Prado, I., Fonseca, C., Bentes, L., Henriques, G., Campos, A., 2016. Transboundary dimensions of marine spatial planning: fostering inter-jurisdictional relations and governance. Marine Policy 65, 85–96. http://dx.doi.org/10.1016/j.marpol.2015.12.025.

Jennings, S., Stentiford, G.D., Leocadio, A.M., Jeffery, K.R., Metcalfe, J.D., Katsiadaki, I., Auchterlonie, N.A., Mangi, S.C., Pinnegar, J.K., Ellis, T., Peeler, E.J., Luisetti, T., Baker-Austin,

C., Brown, M., Catchpole, T.L., Clyne, F.J., Dye, S.R., Edmonds, N.J., Hyder, K., Lee, J., Lees, D.N., Morgan, O.C., O'Brien, C.M., Oidtmann, B., Posen, P.E., Santos, A.R., Taylor, N.G.H., Turner, A.D., Townhill, B.L., Verner-Jeffreys, D.W., 2016. Aquatic food security: insights into challenges and solutions from an analysis of interactions between fisheries, aquaculture, food safety, human health, fish and human welfare, economy and environment. Fish and Fisheries 17, 893–938.

Kelleher, G., 1999. Guidelines for Marine Protected Areas. World Commission on Protected Areas. http://dx.doi.org/10.1017/S0376892901290304. Best Practice Protected Area Guidelines Series.

Kuijper, M.W.M., 2003. Marine and coastal environmental awareness building within the context of UNESCO's activities in Asia and the Pacific. Marine Pollution Bulletin 47, 265–272. http://dx.doi.org/10.1016/S0025-326X(02)00469-1.

Lennox, R.J., Chapman, J.M., Souliere, C.M., Tudorache, C., Wikelski, M., Metcalfe, J.D., Cooke, S.J., 2016. Conservation physiology of animal migrations. Conservation Physiology 4, 1–15. http://dx.doi.org/10.1093/conphys/cov072.

Léopold, M., Cakacaka, a., Meo, S., Sikolia, J., Lecchini, D., 2009. Evaluation of the effectiveness of three underwater reef fish monitoring methods in Fiji. Biodiversity and Conservation 18, 3367–3382. http://dx.doi.org/10.1007/s10531-009-9646-y.

Lester, S., Halpern, B., Grorud-Colvert, K., Lubchenco, J., Ruttenberg, B., Gaines, S., Airamé, S., Warner, R., 2009. Biological effects within no-take marine reserves: a global synthesis. Marine Ecology Progress Series 384, 33–46. http://dx.doi.org/10.3354/meps08029.

Link, J., 2010. Ecosystem-Based Fisheries Management: Confronting Tradeoffs. Cambridge University Press.

Mawdsley, J.R., O'Malley, R., Ojima, D.S., 2009. A review of climate-change adaptation strategies for wildlife management and biodiversity conservation. Conservation Biology 23, 1080–1089. http://dx.doi.org/10.1111/j.1523-1739.2009.01264.x.

Maxwell, S.M., Morgan, L.E., 2013. Foraging of seabirds on pelagic fishes: implications for management of pelagic marine protected areas. Marine Ecology Progress Series 481, 289–303. http://dx.doi.org/10.3354/meps10255.

Maxwell, S., Ban, N., Morgan, L., 2014. Pragmatic approaches for effective management of pelagic marine protected areas. Endangered Species Research 26, 59–74. http://dx.doi.org/10.3354/esr00617.

McClanahan, T., Allison, E.H., Cinner, J.E., 2015. Managing fisheries for human and food security. Fish and Fisheries 16, 78–103. http://dx.doi.org/10.1111/faf.12045.

McLeod, E., Salm, R., Green, A., Almany, J., 2009. Designing marine protected area networks to address the impacts of climate change. Frontiers in Ecology and Environment 7, 362–370. http://dx.doi.org/10.1890/070211.

Mills, M., Pressey, R.L., Weeks, R., Foale, S., Ban, N.C., 2010. A mismatch of scales: challenges in planning for implementation of marine protected areas in the Coral Triangle. Conservation Letters 3, 291–303. http://dx.doi.org/10.1111/j.1755-263X.2010.00134.x.

Mills, M., Adams, V.M., Pressey, R.L., Ban, N.C., Jupiter, S.D., 2012. Where do national and local conservation actions meet? Simulating the expansion of ad hoc and systematic approaches to conservation into the future in Fiji. Conservation Letters 5, 387–398. http://dx.doi.org/10.1111/j.1755-263X.2012.00258.x.

Molnar, J.L., Gamboa, R.L., Revenga, C., Spalding, M.D., 2008. Assessing the global threat of invasive species to marine biodiversity. Frontiers in Ecology and Environment 6, 485–492. http://dx.doi.org/10.1890/070064.

Munro, G.R., Sumaila, R.U., 2012. Subsidies and the Sustainability of Offshore Fisheries: An Economist's Perspective. Forum: Dialogue on Fisheries Subsidies in Mexico, La Paz.

Nieto, A., Ralph, G.M., Comeros-Raynal, M.T., Kemp, J., García Criado, M., Allen, D.J., Dulvy, N.K., Walls, R.H.L., Afonso, P., 2015. European Red List of Marine Fishes. Luxembourg.

O'Leary, B.C., Brown, R.L., Johnson, D.E., Von Nordheim, H., Ardron, J., Packeiser, T., Roberts, C.M., 2012. The first network of marine protected areas (MPAs) in the high seas: the process, the challenges and where next. Marine Policy 36, 598–605. http://dx.doi.org/10.1016/j.marpol.2011.11.003.

Ogier, E.M., Davidson, J., Fidelman, P., Haward, M., Hobday, A.J., Holbrook, N.J., Hoshino, E., Pecl, G.T., 2016. Fisheries management approaches as platforms for climate change adaptation: comparing theory and practice in Australian fisheries. Marine Policy 71, 82–93. http://dx.doi.org/10.1016/j.marpol.2016.05.014.

Österblom, H., Crona, B.I., Folke, C., Nyström, M., Troell, M., 2016. Marine ecosystem science on an intertwined planet. Ecosystems 20, 54–61.

Ostrom, E., 2009. A general framework for analyzing sustainability of social-ecological systems. Science 325, 419–422. http://dx.doi.org/10.1126/science.1172133.

Parry, S., 2013. Qantas Puts Total Ban on All Shark Fin. South China Morning Post.

Pauly, D., Zeller, D., 2016. Catch reconstructions reveal that global marine fisheries catches are higher than reported and declining. Nature Communications 7, 10244. http://dx.doi.org/10.1038/ncomms10244.

Perry, R.I., Ommer, R.E., Barange, M., Jentoft, S., Neis, B., Sumaila, U.R., 2011. Marine social-ecological responses to environmental change and the impacts of globalization. Fish and Fisheries 12, 427–450. http://dx.doi.org/10.1111/j.1467-2979.2010.00402.x.

Pikitch, E.K., Santora, C., Babcock, E.a, Bakun, A., Bonfil, R., Conover, D.O., Dayton, P., Doukakis, P., Fluharty, D., Heneman, B., Houde, E.D., Link, J., 2004. Ecosystem-based fishery management. Science 300, 2003. http://dx.doi.org/10.1126/science.1106929.

Pittman, J., Armitage, D., 2016. Governance across the land-sea interface: a systematic review. Environmental Science and Policy 64, 9–17. http://dx.doi.org/10.1016/j.envsci.2016.05.022.

Roberts, C.M., Bohnsack, J.A., Gell, F., Hawkins, J.P., Goodridge, R., 2001. Effects of marine reserves on adjacent fisheries. Science 294, 1920–1923. http://dx.doi.org/10.1126/science.294.5548.1920.

Rochman, C.M., Brown, M.A., Underwood, A.J., van Franeker, J.A., Thompson, R.C., Amaral-Zettler, L.A., 2016. The ecological impacts of marine debris: unraveling the demonstrated evidence from what is perceived. Ecology 97, 302–312. http://dx.doi.org/10.1890/14-2070.

Rocliffe, S., Peabody, S., Samoilys, M., Hawkins, J.P., 2014. Towards a network of locally managed marine areas (LMMAs) in the Western Indian Ocean. PLoS One 9. http://dx.doi.org/10.1371/journal.pone.0103000.

Rosen, F., Olsson, P., 2013. Institutional entrepreneurs, global networks, and the emergence of international institutions for ecosystem-based management: the Coral Triangle Initiative. Marine Policy 38, 195–204. http://dx.doi.org/10.1016/j.marpol.2012.05.036.

Ruckelshaus, M., Klinger, T., Knowlton, N., Demaster, D.P., 2008. Marine ecosystem-based management in practice: scientific and governance challenges. Bioscience 58, 53. http://dx.doi.org/10.1641/B580110.

Ruddle, K., Hviding, E., Johannes, R.E., 1992. Marine resources management in the context of customary tenure. Marine Resource Economics 7, 249–273.

Salafsky, N., Salzer, D., Stattersfield, A.J., Hilton-Taylor, C., Neugarten, R., Butchart, S.H.M., Collen, B., Cox, N., Master, L.L., O'Connor, S., Wilkie, D., 2008. A standard lexicon for biodiversity conservation: unified classifications of threats and actions. Conservation Biology 22, 897–911. http://dx.doi.org/10.1111/j.1523-1739.2008.00937.x.

Scott, K.N., 2011. International environmental Governance: managing fragmentation through institutional connection. Melbourne Journal of International Law 12, 177–216. http://dx.doi.org/10.1525/sp.2007.54.1.23.

Sheppard, C.R.C., Ateweberhan, M., Bowen, B.W., Carr, P., Chen, C.A., Clubbe, C., Craig, M.T., Ebinghaus, R., Eble, J., Fitzsimmons, N., Gaither, M.R., Gan, C.H., Gollock, M., Guzman, N., Graham, N.A.J., Harris, A., Jones, R., Keshavmurthy, S., Koldewey, H., Lundin, C.G., Mortimer, J.A., Obura, D., Pfeiffer, M., Price, A.R.G., Purkis, S., Raines, P., Readman, J.W., Riegl, B., Rogers, A., Schleyer, M., Seaward, M.R.D., Sheppard, A.L.S., Tamelander, J., Turner, J.R., Visram, S., Vogler, C., Vogt, S., Wolschke, H., Yang, J.M.C., Yang, S.Y., Yesson, C., 2012. Reefs and islands of the Chagos Archipelago, Indian Ocean: why it is the world's largest no-take marine protected area. Aquatic Conservation: Marine and Freshwater Ecosystems 22, 232–261. http://dx.doi.org/10.1002/aqc.1248.

Smith, B.D., Zeder, M.A., 2013. The onset of the Anthropocene. Anthropocene 4, 8–13. http://dx.doi.org/10.1016/j.ancene.2013.05.001.

Spalding, M.D., Meliane, I., Milam, A., Fitzgerald, C., Hale, L.Z.., , 2010. Protecting marine spaces: global targets and changing approaches. Ocean Yearbook 27, 213–248.

Sumaila, U.R., Khan, A.S., Dyck, A.J., Watson, R., Munro, G., Tydemers, P., Pauly, D., 2010. A bottom-up re-estimation of global fisheries subsidies. Journal of Bioeconomics 12, 201–225. http://dx.doi.org/10.1007/s10818-010-9091-8.

Thornton, T.F., Scheer, A.M., 2012. Collaborative engagement of local and traditional knowledge and science in marine environments: a review. Ecology and Society 17. http://dx.doi.org/10.5751/ES-04714-170308.

Toonen, R.J., Andrews, K.R., Baums, I.B., Bird, C.E., Concepcion, G.T., Daly-Engel, T.S., Eble, J.A., Faucci, A., Gaither, M.R., Iacchei, M., Puritz, J.B., Schultz, J.K., Skillings, D.J., Timmers, M.A., Bowen, B.W., 2011. Defining boundaries for ecosystem-based management: a multispecies case study of marine connectivity across the Hawaiian Archipelago. Journal of Marine Biology 2011, 1–13. http://dx.doi.org/10.1155/2011/460173.

Toonen, R.J., Wilhelm, T.A., Maxwell, S.M., Wagner, D., Bowen, B.W., Sheppard, C.R.C., Taei, S.M., Teroroko, T., Moffitt, R., Gaymer, C.F., Morgan, L., Lewis, N., Sheppard, A.L.S., Parks, J., Friedlander, A.M., 2013. One size does not fit all: the emerging frontier in large-scale marine conservation. Marine Pollution Bulletin 77, 7–10. http://dx.doi.org/10.1016/j.marpolbul.2013.10.039.

United Nation's Educational Scientific and Cultural Organization, 2016. Man and Biosphere Programme. http://www.unesco.org/new/en/natural-sciences/environment/ecological-sciences/man-and-biosphere-programme/.

Vincent, A.C.J., Sadovy de Mitcheson, Y.J., Fowler, S.L., Lieberman, S., 2014. The role of CITES in the conservation of marine fishes subject to international trade. Fish and Fisheries 15, 563–592. http://dx.doi.org/10.1111/faf.12035.

Walters, C.J., Martell, S.J., 2004. Fisheries Ecology and Management. Princeton University Press.

Walters, C.J., Christensen, V., Martell, S.J., Kitchell, J.F., 2005. Possible ecosystem impacts of applying MSY policies from single-species assessment. ICES Journal of Marine Science 62, 558–568. http://dx.doi.org/10.1016/j.icesjms.2004.12.005.

Warner, R.M., 2014. Conserving marine biodiversity in areas beyond national jurisdiction: co-evolution and interaction with the law of the sea. Frontiers in Marine Science 1, 1–11. http://dx.doi.org/10.3389/fmars.2014.00006.

Waylen, K.A., McGowan, P.J.K., Milner-Gulland, E.J., 2009. Ecotourism positively affects awareness and attitudes but not conservation behaviours: a case study at Grande Riviere, Trinidad. Oryx 43, 343. http://dx.doi.org/10.1017/S0030605309000064.

Weymouth Ullman, K., 2015. Debt Swap to Finance Marine Conservation in the Seychelles. The Nature Conservancy. http://www.nature.org/newsfeatures/pressreleases/debt-swap-to-finance-marine-conservation-in-the-seychelles.xml.

Williams, S.L., Grosholz, E.D., 2008. The invasive species challenge in estuarine and coastal environments: marrying management and science. Estuaries and Coasts 31, 3–20. http://dx.doi.org/10.1007/s12237-007-9031-6.

Williams, R., Trites, A.W., Bain, D.E., 2002. Behavioural responses of killer whales (*Orcinus orca*) to whale-watching boats: opportunistic observations and experimental approaches. Journal of Zoology 256, 255–270. http://dx.doi.org/10.1017/S0952836902000298.

Zeppel, H., 2008. Education and conservation benefits of marine wildlife Tours: developing free-choice learning experiences. Journal of Environmental Education 39, 3–18. http://dx.doi.org/10.3200/JOEE.39.3.3-18.

Zhu, J., 2006. Asia and IMO technical cooperation. Ocean and Coastal Management 49, 627–636. http://dx.doi.org/10.1016/j.ocecoaman.2006.06.021.

Chapter 9

Ocean Cultures: Northwest Coast Ecosystems and Indigenous Management Systems

Darcy L. Mathews, Nancy J. Turner
University of Victoria, Victoria, BC, Canada

INTRODUCTION

In our efforts to sustain and restore our marine and coastal ecosystems on the Northwest Coast of North America and beyond, how can we learn from Indigenous People's traditional management systems? Marine species and habitats are critical elements of the world's biodiversity and are essential for human well-being. Yet, humans have caused major impacts on biodiversity, through overharvesting, habitat degradation, introducing invasive species, pollution and now, through induced climate change. Marine biodiversity is particularly vulnerable. However, there are also examples from many regions of people sustaining and even enhancing marine productivity while still using marine resources (Berkes, 2012; Butler and Campbell, 2004; Caldwell et al., 2012; Comberti et al., 2015; Lepofsky and Caldwell, 2013; Thornton and Deur, 2015; Turner and Clifton, 2006). The approaches used are obviously of great interest today, perhaps more than ever before, as our oceans and coastal ecosystems continue to be threatened.

In this chapter, we examine the diversity of strategies developed over millennia by Northwest Coast First Peoples to maintain and enhance marine and coastal species and habitats. These form a continuum with traditional terrestrial management systems, and tend to reflect the same overarching values and protocols applied and upheld by Indigenous People of the region "since time immemorial" (Brown et al., 2009; Turner, 2005). Our work is based on compilations and analyses of oral histories and ethnographic accounts from Indigenous environmental experts, along with reviews of published literature, journals and field notes of surveyors, colonial officials and others, as well as surveys and documentation by ourselves and colleagues of the physical, archaeological, and

Conservation for the Anthropocene Ocean. http://dx.doi.org/10.1016/B978-0-12-805375-1.00009-X
169

biological evidence of traditional management systems in various sites along the coast (Deur and Turner, 2005; Lepofsky et al., 2015; Lightfoot et al., 2013; Turner and Clifton, 2006; Turner et al., 2013b; Turner, 2014).

In the following sections we define the features, both tangible and intangible, of traditional land and resource management systems, including the geographical extent and the potential time depth of their development on the Northwest Coast. We consider how these various management practices have been integrated across different ecosystems and over seasonal and broader time scales, and how they have contributed to people's food security, cultural complexity, adaptation, and resilience. In addition to this information, the question of "learning"—both past and present—concerns not just knowledge transmission through time, but also restoring social relationships and developing new collaborations. To illustrate these points, we present three examples of marine resource management systems, with their physical, biological, ecological, and social attributes. Finally, we consider the implications of these traditional management systems for restoring productivity and well-being of coastal ecosystems, and for supporting Indigenous peoples' food security, food sovereignty, and cultural identity.

The Northwest Coast is a distinctive region but with high geographic, ecological, and cultural diversity. From the Gulf of Alaska southward to California, the Northwest Coast includes a narrow, convoluted length of rugged, mist-covered coastline. Alaska and British Columbia coastlines are a complex mosaic of submergence and emergence—a heterogeneous and dynamic environment resulting from the interplay between rugged topography, climate, and the sea. Interfacing terrestrial and marine ecosystems, the shoreline has always been in flux, due to eustatic and isostatic processes associated with deglaciation at the end of the Pleistocene, as well as ongoing tectonic events causing uplift and subsidence. Parallel to the coast are immense mountains, punctuated by valleys, lowlands, and the deltas and estuaries of the coast's great rivers—the Fraser, Skeena, Nass, and Stikine (Fig. 9.1). This often-irregular shoreline, characterized by Mackie et al. (2011) as Fjordland archipelago, incorporates numerous lagoons, tidal marshes, islands, islets, and nearshore reefs. The continental shelf is itself a drowned coastal plain between the mountains and the abyssal deep of the Pacific Ocean. Offshore islands, such as Haida Gwaii, are situated along the seaward edge of this shelf. The Northwest Coast climate is maritime, with cool summers, wet winters, and mild temperatures year-round. Much of the coast is dominated by vast temperate rainforest, mainly with conifers including western hemlock (*Tsuga heterophylla*), Sitka spruce (*Picea sitchensis*), Pacific silver fir (*Abies amabilis*), Douglas-fir (*Pseudotsuga menziesii*), and western redcedar (*Thuja plicata*). The combined effects of the cold, nutrient-rich northern Pacific waters meeting warm, offshore currents originating near Japan, result in abundant natural resources, both in the sea and in the adjacent deltas, estuaries, wetlands, and forests (Turner, 2014).

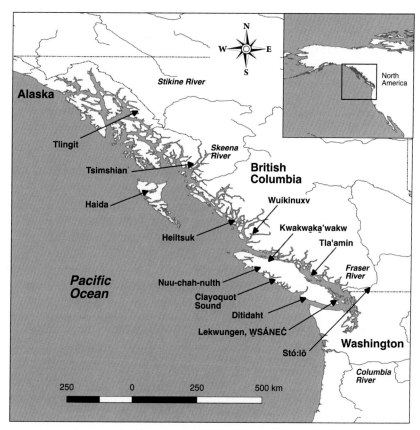

FIGURE 9.1   Geographic features and Indigenous peoples of the Northwest Coast.

Archeological and Indigenous understandings of the earliest periods of human habitation on the Northwest Coast are not necessarily mutually exclusive. Parallels are emerging between Indigenous oral histories and archaeologically derived theories concerning the arrival of the First People to this coast (e.g., Atleo, 2004; Kii7iljuus/Wilson and Harris, 2005). Stories of the beginning of the world, such as the Haida *Tl'guuhiga Gan Xaayda Gwaayaay*, or the creating of Haida Gwaii, describe a world of mostly water (Kii7iljuus/Wilson and Harris, 2005, p. 125). Archaeological, linguistic, and genetic evidence supports the Pacific Coast Migration Model, a theory proposing that First Peoples in the Americas originated in northeast Asia and entered the continents following the Pacific coastline along the then-exposed Beringian landmass extending between present-day Siberia and Alaska (Fladmark, 1979). These were groups of marine-oriented, hunter-gatherer-fishers who traveled in boats or along the shoreline, subsisting primarily on marine resources (Fladmark, 1979; Gruhn, 1994).

For example, they may have brought with them a highly developed system of salmon use, with associated ceremonial and ritual activities, evident in the archaeological record of Northeast Asia beginning by 16,000 years BP (Tabarev, 2011). The earliest record of people in the Americas comes from Monte Verde, Chile—the southern extent for people traveling to the Americas via Beringia. At this inland site, dated around 14,000 years ago, people were using seaweeds from beaches and estuarine environments for food and medicine (Dillehay et al., 2008). It may be that kelp "forests" and other North Pacific coastal ecosystems facilitated the peopling of the Americas, as widespread, productive kelp beds existed at that time along much of the Pacific Coast, from Japan to California (Erlandson et al., 2007). These coastal systems also supported other marine organisms, including sea otters, seals and other marine mammals, numerous fish and shellfish, seabirds, and marine algae. The earliest archaeological so far evidence of inhabitation on the Northwest Coast identified dates to about 12,700 years ago (Mackie et al., 2011). This heterogenous and dynamic coastal environment, however, makes finding early sites very difficult; many potential early sites would have been inundated by rising sea levels. Recently, sidescan sonar applied to the bottom of Hecate Strait—coastal plain exposed 13,700 years ago—have tentatively identified possible stone fish traps and house platforms, now under 120 m of water. More work, however, is required to explore these "drowned landscapes" of the Continental Shelf.

It is during this period when Haida oral history and geological records may align. As the coast warmed, starting around 17,000 years ago, and Cordilleran ice melted from the mountainous coastlines, seafaring people could have lived within and traversed the Pacific Rim entirely at sea level, without major obstructions, and within an ecosystem providing terrestrial and marine resources. According to Haida oral tradition, the beginning of the world was a "Time of Transformation" when Raven made the world the way it is today (Reid and Bringhurst, 1984). Raven caused an island to form in the primeval ocean, which grew and became populated by plants and animals and the First People, while the sun and moon were placed in the sky. Supernatural beings shaped landforms and, through time, the world slowly solidified into its present condition, while human cultures evolved through a series of trials and crises often including actions of heroic ancestors. Both the archaeological records and Haida narratives suggest a time characterized by profound, sometimes rapid geological and ecological change. In some places, like Haida Gwaii, sea level changes were rapid enough to have been perceptible; people returning to a seasonal camp might find it partially inundated by eustatically rising seas (Mackie et al., 2011). Ultimately, the rising ocean enveloped entire landscapes and low coastal plains, with fauna and flora experiencing multiple successions and extirpations. Other geological forces such as tectonism and resulting earthquakes, debris flows, and tsunamis have had profound effects on coastal people over this time, and, like the earliest inhabitation of the coast, are paralleled both in archaeological and Indigenous oral histories (e.g., McMillan and Hutchinson, 2002).

For First Peoples, such as the Haida, since the late Pleistocene/Time of Transformation, the intertidal—or *xhaaydla*—is the nexus between ocean and forest that people call home. Archaeological and ethnographic evidences indicate that people were inhabiting and fluently using many or all ecological niches, but communities were strongly maritime-adapted, developing technologies, techniques, knowledge, and rituals (e.g., Ames, 1998; Ames and Maschner, 1999; Moss, 2011) to enable not just use of marine and intertidal resources but also, over time, the active maintenance and management of these environments.

The land–shore interface is foundational to the worldview of Northwest Coast First Peoples. For some groups, such as the Ts'msyen (Tsimshian) and Haida, the intertidal is also a liminal space between the spirit guardians of the upper world—including the sky, the land, and the forests—and the underworld of rivers and the sea. Spirits intermingle daily with the ebb and flow of the tide. The intertidal is therefore a place of transformation between human and supernatural worlds, and either world is simply a few steps away into the woods or a few paddle strokes from the shore (Bringhurst, 2000, p. 155; MacDonald, 2006, pp. 17–18). For marine resources such as red laver seaweed (*Pyropia abbottiae*), salmon, clams, and marine mammals, there are numerous narratives and traditions that reflect the spiritual aspects of these resources (cf. Boas, 2002; Turner and Berkes, 2006; Turner, 2014).

TRADITIONAL LAND AND RESOURCE MANAGEMENT SYSTEMS ON THE NORTHWEST COAST

Early anthropological concepts of "hunter-fisher-gatherer" societies assumed natural species abundance and low human populations allowed people to simply harvest food and other resources as needed, requiring little intervention in natural ecosystem productivity. We now realize that there are many widespread practices with both social and ecological dimensions that have been developed across both nonagricultural and agricultural societies to regulate, expand, and increase abundance and quality of "wild" culturally important species at a range of spatial, temporal, and ecological scales (Anderson, 2005; Berkes, 2012; Ford, 1985; Smith, 2011). This recognition has extended firmly into the Northwest Coast culture area, where Indigenous Peoples' resource management in both terrestrial and aquatic ecosystems is increasingly recognized (e.g., Anderson, 2009; Caldwell et al., 2012; Cullis-Suzuki et al., 2015; Derr, 2014; Deur and Turner, 2005; Peter and Shebitz, 2006; Thornton, 2015; Thornton et al., 2015; Turner, 2014; Turner et al., 2013a,b). There is, in fact, a continuum of habitats that have been managed by humans on the Northwest Coast and these are interrelated in many ways at various scales and through various processes (Fig. 9.2).

Lertzman (2009, p. 339) defines management as "… a set of actions taken to guide a system toward achieving desired goals and objectives." He suggests reframing the concept from "management" to "management *system*," which

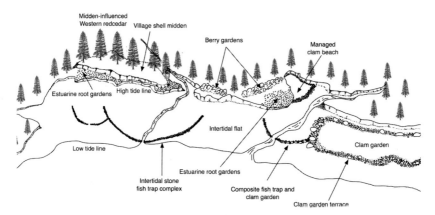

FIGURE 9.2   Diagram of coastal and adjacent habitats showing some features of human management.

is "the sum of these actions, goals and objectives, the process through which they are legitimized by social norms, and the institutions and actors involved in carrying them out." In other words, "management system" embodies both social and ecological dimensions and dynamics in the development of managed ecosystems.

Within the Northwest Coast region, a broad spectrum of activities, actions, and strategies are integrated within management systems. Furthermore, there is increasing evidence that at least some of these systems and practices extend far back into the past (Campbell and Butler, 2010; Lepofsky and Lertzman, 2008; Weiser and Lepofsky, 2009). Many of these are "mariculture systems," specifically functioning to sustain and enhance production of marine and associated habitats and species. They fit in with definitions of cultivation, and extend far beyond simple resource conservation or limiting use of certain resources (Thornton and Deur, 2015; Thornton et al., 2015). They reflect a web of material, social, technical, ecological, and spiritual approaches that result in enhanced quality, higher abundance, higher productivity, greater availability and accessibility, more equitable sharing, and greater diversity of marine and coastal resources (Smith, 1940; Suttles, 1987a,b) (Table 9.1). Caldwell et al. (2012), too, identified a remarkable array of modifications in marine ecosystems, especially in the intertidal zone, that expand numbers, diversity, and productivity of marine taxa. Many of the species featured in these management systems are dietary staples and cultural keystone species (see Garibaldi and Turner, 2004). By this, we mean plants and animals that have a fundamental role in the culture of a specific people. These are iconic species, so central to a people's diet, material culture, or health, that the societies they support would be radically different without them (Garibaldi and Turner, 2004). On the Northwest Coast, salmon are an example, figuring prominently not only in diet, but in language, ritual, narratives, and social structure.

TABLE 9.1 Marine Management: Associated Practices, Strategies and Approaches Used by Indigenous Peoples of British Columbia and Neighboring Areas

Practice, Strategy or Approach	References
I. Ecological Management Strategies	
Landscape burning: Burning of prairies to promote growth of camas, deer forage, etc., timed to work with ocean tides, which in turn determined the strength and direction of winds (upcoming tide, wind too strong to burn)	Dr. Luschiim Arvid Charlie (Hul'qumi'num), pers. comm. to NT 2016; Turner (1999)
Clearing, "cleaning": Manual removal of large rocks, driftwood, etc. from estuaries, clam beaches, canoe runs (e.g., Klahoose *wuxwuthin* "rocks piled up when digging clams, or rock corral used to store fish caught in fish traps"); shellfish, fish, and octopus harvested	Lepofsky et al. (2015), Deur et al. (2015a,b), Caldwell et al. (2012) and Ellis and Swan (1981)
Habitat creation, extension, or alteration: Creating new habitats through rock and log terracing, ditching, etc., in clam gardens and estuarine root gardens (see above)	Deur (2005), Lepofsky et al. (2015) and Thornton (2015)
Bounding of resource areas: Laying of plot boundaries or establishing borders, in estuarine root gardens, crabapple trees, edible red laver seaweed picking areas, clam gardens	Turner (2003), Turner et al. (2005) and Deur (2005)
Tilling soil (usually with digging stick): Aerates soil; enhances moisture penetration; helps recycle nutrients, etc., in estuarine root "gardens"	Deur (2002), Deur and Turner (2005) and Turner and Turner (2008)
Dissemination: Planting or scattering seeds, fruits, or other propagules	Turner and Peacock (2005), Joseph (2012) and Pukonen (2008)
Transplanting: Moving young fish, larvae, root fragments, etc. from one location to another, including transplanting salmon eggs, herring eggs, spawning herring and eulachon, clover rhizomes, riceroot bulbs, and possibly seaweed	Thornton (2015), Thornton et al. (2015), Turner (2014) and Turner et al. (2013b)
Selective, partial, rotational, or nondamaging harvesting: Taking only a portion of a plant, or only some individuals from a population, seaweed, kelp fronds, kelp stipes, shellfish	Deur et al. (2013), Turner and Peacock (2005) and Turner et al. (2013b)

Continued

TABLE 9.1 Marine Management: Associated Practices, Strategies and Approaches Used by Indigenous Peoples of British Columbia and Neighboring Areas—cont'd

Practice, Strategy or Approach	References
Fertilizing, mulching: Adding nutrients or moisture retaining materials to soil, as in estuarine root gardens	Deur (2005) and Turner and Peacock (2005)
Feeding: Providing food for growing fish; putting fishguts, bones, and dead salmon back into the river to nourish young fish, crabs, etc.	Heiltsuk Nation, ongoing interviews
II. Social Management Strategies	
Ownership/proprietorship: Individuals or cultural groups hold rights (usually inherited) to use particular resources or harvesting areas, as estuarine root gardens, clam gardens, halibut fishing grounds, fish traps, weirs and nets; tended and harvested; resource owner responsible for ensuring that management and harvesting is undertaken at the appropriate times and under the right conditions	Claxton and Elliott (1994), Deur et al. (2013, 2015a,b), George (2003) and Suttles (1974)
Monitoring: Groups or individuals have the responsibility to watch over certain resources and harvesting areas; as red laver seaweed, abalone, salmon; use of phenological indicators; fish traps used to monitor the size, gender, and health of fish, and to select which ones are harvested and which allowed to go upriver to spawn	Turner (2005), Turner et al. (1983, 2005) and Thornton et al. (2015)
Socially determined conservation: Ceremonial promotion or protection of particular places, species, and populations; edible seaweed, salmon	Claxton and Elliott (1994), Turner et al. (2005), Turner (2003), Turner and Clifton (2006) and George 2003
Teamwork and division of labor: Different task groups within a community specializing in different aspects of harvesting and processing plant resources, for all types of harvesting—seabird eggs, fish, shellfish, root vegetables, seaweed	Claxton and Elliott (1994), Turner and Clifton (2006) and George (2003)
Distributed seasonal access to resource areas: "Seasonal rounds"; spring harvesting of edible red laver seaweed, chitons, spring salmon, halibut, seabird eggs, seal; summer and fall salmon harvests; winter shellfish harvests	Widely practiced by BC First Nations; Turner (2003), Turner and Clifton (2006), George (2003) and Hunn et al. (2003)

TABLE 9.1 Marine Management: Associated Practices, Strategies and Approaches Used by Indigenous Peoples of British Columbia and Neighboring Areas—cont'd

Practice, Strategy or Approach	References
Trade and exchange: Kin-based trade networks; trading of surplus, for seaweed, clams, root vegetables, oulachen grease, salmon, dried halibut, black ducks	Widely practiced by NW Coast First Nations (Turner, 2014; Turner and Loewen, 1998; Turner et al., 1983)
Feasting and sharing: Feasting, sharing, with elites and leaders taking on primary roles; a way of distributing resources; eelgrass, halibut, salmon, seal and sea lion, root vegetables, etc.	Widely practiced by NW Coast First Nations (cf. Turner et al., 2012)
Knowledge transmission: Passing on knowledge and experiences relating to resource management and conservation through participatory and experiential learning, stories, ceremonies, art, discourse, and focused instruction	Widely practiced by NW Coast First Nations (e.g., Turner and Berkes, 2006; Turner and Turner, 2008)
III. Technological Management Strategies	
Increasing access: Finding more efficient ways to access particular resources (e.g., building trails, camp shelters, better canoes)	Widely practiced by NW Coast First Nations (Lepofsky and Lertzman, 2008)
Technical innovations: Improvements in tools and approaches for harvesting, processing, and storing food and other materials (e.g., improved digging sticks, baskets, mats, drying racks, smoking, pit-cooking)	Widely practiced by NW Coast First Nations (Lepofsky and Lertzman, 2008)
IV. Integrated Multiresource Management	
Combined management strategies: Effects and outcomes of two or more management strategies, applied to two or more species or entire habitats, over time and space. Estuarine root gardens, eelgrass beds, clam gardens, salmon fish traps etc.	Deur and Turner (2005) and Turner et al. (2013b)

Compiled from: Deur and Turner (2005), Lepofsky and Lertzman (2008), Thornton (2015), Thornton et al. (2015), Turner (2005, 2014); Turner and Hebda (2012), Turner and Peacock (2005) and Turner et al. (2005, 2013b); based on knowledge shared by Indigenous plant experts, especially Dr. Arvid Charlie (*Luschiim*), Clan Chief Adam Dick (*Kwaxsistalla*), Dr. Daisy Sewid-Smith (*Mayanilth*), John Thomas (*Tlishaal*).
Adapted from Turner, N.J., Deur, D., Lepofsky, D., 2013b. Plant management systems of British Columbia's first peoples. In: Turner, N.J., Lepofsky, D. (Eds.), Ethnobotany in British Columbia: Plants and People in a Changing World. Special Issue, BC Studies, vol. 179, pp. 107–133.

Table 9.2 lists some of the key Northwest Coast marine and coastal species whose harvesting, range, productivity, diversity, intensity of use, habitat, or other parameters of use are managed through traditional systems. The material correlates of these management systems include various types of fish traps and weirs (for selective harvesting, habitat enhancement, and monitoring), stone walls for clam gardens, and anchor stones associated with nets, which can be assessed by surveys, including, most recent aerial photographic surveys of intertidal features and shoreline features (Caldwell et al., 2012, p. 220).

The various management systems are culturally, temporally, spatially, and ecologically interconnected food webs. For example, harvesting eelgrass rhizomes

TABLE 9.2 Some Key Marine and Coastal Species Whose Harvesting, Range, Productivity, Diversity, Intensity of Use, Habitat or Other Parameters of Use Are Managed Through Traditional Systems Along the Northwest Coast

Habitat/Species	Material and Social Factors	References
Open Ocean		
Halibut	Ownership of prime halibut fishing grounds	Jones (1999), Turner et al. (1983, 2005) and Thornton et al. (2015)
Scoters/black ducks	Ownership of prime harvesting areas	Turner et al. (1983) and Turner and Hebda (2012)
Whales	Intensive training from infancy for whale hunters; restricted rights to hunt whales	Atleo (2004)
Seals	Managed as part of herring and eulachon fishery; associated with stone traps	Brown et al. (2009) and Haggan et al. (2006)
Sea lions	Managed as part of herring and eulachon fishery; associated with stone traps	Brown et al. (2009), Haggan et al. (2006) and Thornton (2015)
Crabs	Selection by size, sex; season; feeding with fish and sea mammal remains	Haggan et al. (2006); Emma Reid, interview with J. Bhattacharyya. May 2014
Kelp Beds and Reefs		
Rockfish	Ownership of prime harvesting areas; size selection; seasonal selection	McKechnie (2007)

TABLE 9.2 Some Key Marine and Coastal Species Whose Harvesting, Range, Productivity, Diversity, Intensity of Use, Habitat or Other Parameters of Use Are Managed Through Traditional Systems Along the Northwest Coast—cont'd

Habitat/Species	Material and Social Factors	References
Herring, herring roe	Ownership of prime harvesting areas; boughs placed in spawning areas to increase range and extent and success of spawning; harvesting giant kelp fronds with spawn; eggs rather than whole fish harvested; protected against excessive noise	Haggan et al. (2006), Thornton et al. (2010), Thornton et al. (2015), Turner (2014) and Turner et al. (2013b)
Salmon (reefnet fishery)	Ownership of prime harvesting areas; directing fish runs; size selection; seasonal selection; escape mechanisms	Claxton (2015) and Suttles (1974)
Bull kelp	Harvesting for fishing lines, storage vessels and other purposes	Turner et al. (1983)
Sea urchins	Ownership of prime harvesting areas; size selection; seasonal selection	Ellis and Swan (1981) and Ellis and Wilson (1981)
Intertidal Zone		
Eelgrass beds	Harvested in patchy clumps that promote regeneration	Cullis-Suzuki et al. (2015)
Crabs	Size selected; sex selected?	Haggan et al. (2006); Emma Reid, interview with J. Bhattacharyya. May 2014
Sea cucumbers	Habitat created in clam garden rock walls	Deur et al. (2015b)
Flounders, sculpins, salmon, flatfish, herring	Fish traps for selective harvesting	Caldwell et al. (2012)
Octopus	Ownership of prime harvesting areas; shelters build to attract and house octopi	Ellis and Wilson (1981); Barbara Wilson, pers. comm. 2015
Rocky Shoreline; Intertidal		

Continued

TABLE 9.2 Some Key Marine and Coastal Species Whose Harvesting, Range, Productivity, Diversity, Intensity of Use, Habitat or Other Parameters of Use Are Managed Through Traditional Systems Along the Northwest Coast—cont'd

Habitat/Species	Material and Social Factors	References
Northern abalone	Harvested only at lowest tide; only large size selected; seasonal harvesting; protected against excessive noise	Ellis and Swan (1981), Ellis and Wilson (1981) and Turner (2005)
Edible seaweed (e.g., *Pyropia abbottiae*)	Harvested by pulling off rocks; allows regeneration; patches cleaned off; control over the labor to extract resources	Turner and Clifton (2006) and Turner and Turner (2008)
Mussels (2 spp.)	Taboo against eating during seaweed harvesting; seasonal harvesting and tending: "semi-cultivation"	Turner and Clifton (2006), Ellis and Swan (1981) and Ellis and Wilson (1981)
Limpets, chitons	Selected by size; large size only; seasonal harvesting	Ellis and Swan (1981), Ellis and Wilson (1981) and Haggan et al. (2006)
Sandy/Silty Beaches		
Butter clams	Clam garden construction; selected by size; left alone during reproductive stage	Harper et al. (1995), Recalma-Clutesi, K. (2005), Williams (2006), Woods and Woods (2005) and Lepofsky et al. (2015)
Cockles	Clam gardens; control over the labor to extract resources	Harper et al. (1995), Recalma-Clutesi, K. (2005), Williams (2006), Woods and Woods (2005) and Lepofsky et al. (2015)
Narrows		
Ducks	Constrained areal access for monitoring and harvesting	Suttles (1974), Turner et al. (1983) and Turner and Hebda (2012)
Salmon	Control over the labor to extract resources	Suttles (1974)
Seals	Constrained areal access for monitoring and harvesting	Turner et al. (1983) and Turner and Hebda (2012)
Lagoons, Estuaries and Tidal Marshes and Ponds		
Ducks, geese, brants, swans	Ownership of prime harvesting areas; control over the labor to extract resources	Suttles (1974)

TABLE 9.2 Some Key Marine and Coastal Species Whose Harvesting, Range, Productivity, Diversity, Intensity of Use, Habitat or Other Parameters of Use Are Managed Through Traditional Systems Along the Northwest Coast—cont'd

Habitat/Species	Material and Social Factors	References
Root gardens	Ownership of prime harvesting areas; control over the labor to extract resources	Deur (2005) and Turner and Turner (2008)
Salmon	Control over the labor to extract resources; Fish weirs, traps and holes/pools; ceremonial control; selection of net size and location to capture specific taxa	Gunther (1926), Claxton and Elliott (1994), Claxton (2015), Haggan et al. (2006), Langdon (2006a,b, 2007) and Thornton and Deur (2015)
Deer hunting	Selection of net size and location; ownership of harvesting areas	Suttles (1974)
Crabapple thickets	Ownership of prime harvesting areas; pruning	Turner and Peacock (2005)
Rivers, River Shorelines, Waterfalls		
Eulachon	Ownership of prime harvesting areas	Haggan et al. (2006) and Thornton et al. (2015)
Salmon (diff species)	Ownership of prime harvesting areas; care of salmon streams; selective harvesting; management by size and sex	Jones (2002), George (2003) and Thornton et al. (2015)
Coastal Islands, Cliffs, Rocky Shoreline		
Gull eggs, oystercatcher eggs	Ownership of prime harvesting areas; selective harvesting rules; taboos and restrains	Elsie Claxton, pers. comm. 1999; Hunn et al. (2003)
Forests		
Cedar for canoes, boxes, baskets, etc.	Ownership of prime harvesting areas; selective harvesting for canoes, wood, bark	Stewart (1984) and Turner (2014)
Yew wood	Management to regulate overharvesting; selective harvesting of branches	Turner and Cocksedge (2001); Luschiim Arvid Charlie pers. comm. 2004

Continued

TABLE 9.2 Some Key Marine and Coastal Species Whose Harvesting, Range, Productivity, Diversity, Intensity of Use, Habitat or Other Parameters of Use Are Managed Through Traditional Systems Along the Northwest Coast—cont'd

Habitat/Species	Material and Social Factors	References
Spruce pitch	Harvested from designated pitch-trees over many years; used for waterproofing and gluing implements	Turner et al. (1983) and Turner (2014)
Cherry bark	Selective harvesting for wrapping around implements	Turner et al. (1983) and Turner (2014)
Berries, berry gardens	Ownership of prime harvesting areas	Turner and Berkes (2006) and Turner and Turner (2008)
Stinging nettles	Transplanting, tending	Turner (2014)
Sword fern	Selective harvesting; ceremonial control	Turner (2014)

can be expected to enhance the growth of young eelgrass shoots, which would, in turn, attract geese and ducks that feed on eelgrass. Estuarine root gardens also attract geese, swans, and ducks, which can then be hunted selectively from blinds, sometimes using the roots themselves as bait (Edwards, 1979). Restricted waterways, or narrows, can also provide constrained areal access, which present prime opportunities for monitoring and harvesting fish, seals, and other marine life, as well as ducks and other seabirds (Suttles, 1974; Turner et al., 1983; Turner and Hebda, 2012). Stones piled at the lowest tide line with a primary purpose to create clam habitat also support many other marine creatures, including octopus and sea cucumbers, and create a substrate for marine algae, which provide additional habitat for larval shellfish and other life (Caldwell et al., 2012). This principle also applies to fish weirs, which help to create sandbars where clams and oysters can thrive, and intertidal fish traps used to select a variety of different species, such as herring, sculpins, and flatfish (Anthony, 1976; Mobley and McCallum, 2001; White, 2006). Wolves and bears also benefit from the fish in human-built fish traps, and deer are also drawn to estuarine root gardens. All of these species are managed to some extent through the protocols and practices of their harvesting and use, such as imposed restrictions on size, numbers, or harvesting season (Turner, 2014; Turner et al., 2013b).

Terrestrial and marine ecosystems cannot be separated on the basis of management; both are required for their resources and are linked together culturally and ecologically. A direct example is the use of western hemlock (*Tsuga heterophylla*) boughs and young trees from the forest for capturing herring spawn, and of hemlock poles from the dense shaded understory to make eelgrass twisting poles. Knots from hemlock and some other trees provide the tough wood used for bentwood fishhooks (Turner et al., 1983). The remains of salmon and other fish from processing are either dug into the ground and help to fertilize berry bushes and increase berry production, or are replaced into the water, to feed crabs, young fish, and seabirds, in turn enhancing their growth and numbers. Western redcedar provides the wood for dugout canoes that allow the harvesting of many marine and coastal resources, as well as for boxes to transport, cook, and store seaweed and other marine foods (Stewart, 1984). Harvesting and processing salmon from fish traps requires fuel wood supplies from the forest, as well as the need for wood and pitch from forest species such as Sitka spruce, Douglas-fir, Pacific yew (*Taxus brevifolia*), and oceanspray ("ironwood"; *Holodiscus discolor*), for harpoon and spear shafts, fishhooks, and other implements. Fibers from stinging nettle (*Urtica dioica*), bitter cherry bark (*Prunus emarginata*), and tree roots and bark are used for nets, lines, and wrapping implements.

Sword fern (*Polystichum munitum*) fronds are harvested and used ritually, such as in the First Salmon Ceremony; they are widely considered as sacred and are harvested and replaced in the forest carefully following their use. The fronds were also used in a contest—removing as many leaflets one at a time with a single breath, saying "pala" each time—that trained young Ditidaht and Nuu-chah-nulth men to hold their breaths for a long time, so that they could dive and select the best bull kelp stipes to use as fishing line in the carefully managed halibut fishery (Turner et al., 1983).

As outlined by Berkes (2012), different elements of traditional management systems generally build up over time, starting with accumulation of basic knowledge about a newly encountered environment, including the usefulness of various species, through observation and experience. Knowledge of ecological relationships, life cycles, phenology, and habitats of species accrues and is passed on over generations. More complex social practices, including specialized roles, division of labor, task groups, and proprietary rights also develop, ensuring that access to resources is coupled with responsibility to share with family and community. Methods to increase resource productivity and quality through management of species and habitats are also developed over time, and this knowledge is passed on to others and to future generations, often through experiential processes. With time, as knowledge, social organization, and technologies continue to develop, the species and environments are entwined into complex belief systems and worldviews in which cultural practices and perspectives become encoded in language, stories, taboos, ceremonies, art, and ethics (Berkes, 2012; Turner and Berkes, 2006; Turner et al., 2013b). This may be the most significant outcome, leading to habits and ethics that allow for the development and long-term maintenance of sustainable cultural landscapes and seascapes.

MARINE RESOURCE MANAGEMENT CASE STUDIES

The scope and nature of marine management by First People on the Northwest Coast are exemplified by three well-studied examples: clam gardens; salmon production; and estuarine root gardens. The different aspects of these management systems are summarized in Table 9.3.

Clam Gardens

The accessibility and protein content of shellfish such as clams made them valuable resources to people of the Northwest Coast (Erlandson, 1988; Moss, 1993). From coastal Alaska to Washington State and possibly beyond, people of the Northwest Coast improved the productivity of clam habitat by both clearing stones from existing clam beds, as well as creating or enhancing clam habitat with intertidal rock wall terraces (Fig. 9.3) (Caldwell et al., 2012; Deur et al., 2015a,b; Groesbeck et al., 2014; Harper et al., 1995; Lepofsky and Caldwell, 2013; Lepofsky et al., 2015; Williams, 2006). These features, called "clam gardens," concentrated shellfish resources in accessible locations. Physical and ecological factors likely determined whether a wall was built, including elements such as size and slope of the beach and the location of the sea shelf (Caldwell et al., 2012). Social factors were also likely at play, with clam gardens often situated close to villages or other places within the purview of individual or family resource managers (Deur et al., 2015a,b). However, analysis of growth–stage profiles from clamshells within village sites in Heiltsuk Territory suggests that over the past 7000 years, clam harvesting was less intensive in the vicinity of residential sites, presumably to meet anticipated future needs (Cannon and Burchell, 2009). Other management practices included selective harvesting, and leaving behind small clams "to keep the populations productive" (Turner, 2005). It is possible that clam beds were periodically left fallow to allow populations to reach harvestable size (Cannon and Burchell, 2009).

The size, morphology, and locations of these clam garden features exist on a continuum, ranging from small sections of rocky intertidal beach cleared of stones, to elaborate and immense rock walls or terraces stretching hundreds of meters and comprising thousands of tons of stones. According to Wuikinuxv expert Johnny Johnson, of the central BC coast, the largest stones were rafted into place using canoes or logs on a falling tide (pers. comm. to DM, May 8, 2015). Efforts to build the largest of these features may have involved leaders capable of organizing labor, and a detailed knowledge of very localized and seasonal tidal cycles. Techniques employed to build these features ranged from clearing stones from beaches while gathering clams, to extensive construction of rock walls or terraces. Some of these features are truly monumental in scale, extending more than a kilometer in length with a meter or higher built walls. The intention was to enhance clam production while also providing habitat for a range of other marine taxa such as octopus and chitons along the seaward edge of the rock terrace. In fact, building a clam garden wall, or any intertidal

TABLE 9.3 Characteristics of Three Key Marine and Coastal Management Systems of Northwest Coast First Nations

Type of Management System	Clam Gardens	Salmon Production	Estuarine Root Gardens
Indigenous term(s)	"*lúxwxiwey*" (cf. *lúxwəkw* "rolled rocks to clear an area" (Kwak'wala) *t'iimiik* ("something being thrown"; "move aside rocks") Bouchard and Kennedy (1990, pp. 386) *wúxwuthin* (Tla'amin) "rocks piled on beach" "like a breakwater" Kennedy and Bouchard (1983, pp. 147–148)	Many terms in all NW Coast languages; e.g., *shíshitl'ech* "tidal weir-trap' and *tékwus*" "river weir-trap" (Tla'amin: Kennedy and Bouchard, 1974, pp. 22–23); *wúxwuthin* (Tla'amin: Kennedy and Bouchard, 1983, pp. 147–148)	*t'ekilakw* (Kwak'wala)
Habitat/location	Sandy beaches/ shallow bays	Reef channels; river estuaries	Estuarine flats/ upper tidal
Physical/ biological manifestations	Rocky wall constructed along lowest tide line; dense populations of butter clams, littleneck clams, cockles; often large amounts of barnacle shell	Seasonal presence of nets; evidence of stone walls or wooden weirs along estuary shore and channels; pools dug in river channel	Wooden or stone markers/ walls; presence of edible root species: Pacific silverweed, springbank clover, northern riceroot
Associated tools/ features	Yew wood digging sticks; cedar withe clam baskets; cooking boxes; clam steaming pits; clamshell middens	Cedar canoes; Spears of yew wood, Douglas-fir or other wood; nets of stinging nettle, willow bark or other fiber; fish drying racks; hearth features for drying salmon; smoke houses; salmon spreaders; baskets or other containers	Cedar canoes or boats for travel to estuary; yew wood digging sticks; baskets for washing and transporting roots; pitcooking features or steaming in bentwood boxes

Continued

TABLE 9.3 Characteristics of Three Key Marine and Coastal Management Systems of Northwest Coast First Nations—cont'd

Type of Management System	Clam Gardens	Salmon Production	Estuarine Root Gardens
Antiquity of use; evolution	Radiocarbon dating barnacles on underside of rolled rocks	Radiocarbon dating of wood in weirs	Pollen analyses and dating of associated organic matter in cores
Associated narratives, ceremonies, beliefs	*lúxwxiwey*'s mentioned in Kwak'wala story (Boas and Hunt, 1906, p. 93); and song (Woods and Woods, 2005)	Salmon estuarine and reefnet fishing widely included in narratives, songs, ceremonies, such as First Salmon Ceremony; returning salmon bones to river	Estuarine roots mentioned in many narratives; often associated with ducks, geese
Seasonality	Mostly accessed and used in winter months—November through March	Mostly accessed June through November for different runs of salmon	Cleaned and weeded in spring; harvested in fall through early spring
Basic principles	Creates and expands suitable area for clam/cockle production, as well as habitat for crabs, octopus, etc.; allows selective harvesting and aeration of substrate	Allows selective harvesting by size, sex and monitoring of spawning salmon runs	Creates and expands suitable area for edible root production; aeration of soil; removes competing species; allows selective harvesting and spreading of propagules for succeeding years
Results of management	Increases the numbers/quality and possibly size of clams; creates perpetual harvesting cycle	Maintains healthy runs of large salmon of various strains and species	Increases the numbers/quality of root vegetables; thins out "root-bound" populations and allows regrowth of tender shoots; creates perpetual harvesting cycles

TABLE 9.3 Characteristics of Three Key Marine and Coastal Management Systems of Northwest Coast First Nations—cont'd

Type of Management System	Clam Gardens	Salmon Production	Estuarine Root Gardens
Connections to other management systems/species	Eelgrass beds; seabirds; octopus; crabs; sea cucumbers; barnacles	Berry gardens (fish as fertilizer); CMTs; crab fishery (remains as crab food)	Crabapples; seal hunting; duck, goose and swan hunting; fishing
Potential genetic effects	Unknown; possible selection for particular desired species such as cockles	Particular strains of salmon promoted through selection; transplanting salmon	Transplanted populations of riceroot, springbank clover
Long-term social/economic implications	Trade/ownership; fallback food for times of salmon shortage	Trade/ownership; food security	Trade/ownership/nutrition
Future research; what we still need to learn	Climate change concerns; red tide; land rights; pollution (e.g., oil spills)	Impacts of commercial large-scale fishing; GM and Atlantic salmon impacts; fish farming impacts; fisheries regulations; pollution	Climate change concerns; invasive species; pollution
Potential/initiatives for future application/reinstatement	Reintroduction of clam gardens, such as in Salish Sea	W̱SÁNEĆ reefnet fishing; origin of salmon narratives; experimental restoration	Reestablishment of the root gardens at various estuaries
Key references	Augustine and Dearden (2014), Deur et al. (2015a,b), Harper et al. (1995) and Lepofsky et al. (2015)	Claxton (2015), Claxton and Elliott (1994), Thornton et al. (2015), Turner and Berkes (2006) and Turner and Hebda (2012)	Deur (2005), Deur and Turner (2005), Deur et al. (2013) and Turner et al. (2013b)

FIGURE 9.3 Two clam gardens on the Central Coast of British Columbia. Note the productive clam habitat of exposed silt/sand/shell hash created between the seaward edge of the built wall and the shoreline cleared of boulders.

or sub-tidal rock feature, likely created habitat for a variety of valued species, such as sea cucumbers, crabs, and small fish (Caldwell et al., 2012). These were places providing abundant and predictable sources of intertidal foods. For example, Groesbeck et al. (2014) has conducted controlled ecological experiments to test the connection between clam garden tidal height and slope modification, indicating that these features enhanced clam survivorship, growth rates, and densities compared to nonwalled beaches. Furthermore, clam gardens can be part of larger intertidal complexes of overlapping stone features. For example, Tla'amin Elder Charlie Bob noted that beaches cleared for clams could be further modified to produce a fish trap, with rocks moved during clamming piled up into walls and a gap left in the middle for fish to enter. On an incoming tide when the clam garden flooded, fish could pass through the gap, but once the opening was sealed on an outgoing tide, stranded fish could subsequently be collected (Caldwell et al., 2012).

When and how did the practice of building clam gardens come to pass? Archaeologists are excavating down to the interface between the original beach surface and the in-filled garden sediments, radiocarbon dating the clams that died in this position, as well as remnant barnacle scars on old beach surface rocks. So far, most clam gardens have dated to the past 1000 years. However, if these features parallel the historical trajectory of intertidal stone fish traps, we may find much older clam gardens, and observe that their numbers increase over the course of the last 3000 years (Lepofsky et al., 2015).

Clams harvested from managed beaches were brought ashore for processing and consumption, the shells then removed. The sheer amount of clam and other shell produced by First People on the coast has resulted in ubiquitous shell-midden sites along the entire length of the Northwest Coast. This was often not a simple act of discard, but was in some instances an intentional and structured practice. Shells were deposited on land to enlarge livable space, create high (and monumental) house platforms, in-fill wetlands adjacent to houses, make paths more visible at night, and improve drainage. Shell midden also provided a matrix for the burial of the dead, as midden burials became common on the coast after about 3400 BC (Ames and Maschner, 1999, p. 90). While shell middens are sources of information concerning the record of inhabitation, subsistence, ritual, and mortuary practice, the shells have also changed the soil chemistry of these nearshore places. Soils at habitation sites, for example, are higher in calcium—a limiting resource in coastal temperate rainforests. Consequently, western redcedars growing on these sites benefit from higher wood calcium, with resulting greater radial growth and decreased top die-back (Trant et al., 2016). These shells may also have potential to mitigate ocean acidification effects associated with carbonate chemistry in living shellfish (Waldbusser et al., 2013).

Salmon Production

Pacific salmon (*Oncorhynchus* spp.) are anadromous fish; born in fresh water, they migrate to the ocean and return to fresh water to reproduce and die. First Peoples of the Northwest Coast understood the cyclical and predictable nature of these salmon runs, knew where the salmon would be, and had technological, social, and ritual practices in place to intercept these runs. The predictability and prodigious numbers of salmon migrating and swimming up both major rivers and streams were in large part the basis for development of the coast's subsistence economy (Ames, 1994; Ames and Maschner, 1999). The technology of salmon storage and preservation, for example, facilitated the establishment of permanent or semi-sedentary winter villages. While some have argued that salmon storage economy came into existence around 3500 years ago (e.g., Matson, 1992), more recent DNA analysis of salmon suggests that on the Central Coast at least, salmon storage economy and permanent multi-season settlement was in place around 7000 years ago (Cannon and Yang, 2006). Facilitating this salmon economy involved methods to enhance productivity and limit overexploitation (e.g., Berkes, 2015; Claxton, 2015; Kennedy and Bouchard, 1983; Thornton et al., 2015). A review of approximately 7500 years of salmon use in the archaeological record from the central Northwest Coast and Interior Plateau east of the Cascade Crest, reveals stability in salmon use, which Campbell and Butler (2010) interpret as resilience in the management of not only salmon, but the food web it comprises.

Ethnographic evidence for sustaining or enhancing salmon populations includes people transplanting salmon eggs and creating spawning populations in areas where there was no prior spawning population, or to bolster declining stocks (e.g., Carpenter et al., 2000; Jones, 2002; Lepofsky and Caldwell, 2013; Thornton et al., 2010). The Tlingit removed beaver dams on rivers that blocked upriver sockeye (*Oncorhynchus nerka*) and coho (*Oncorhynchus kisutch*) salmon spawning, as well as rearranged rocks to improve stream flow and increase salmon spawning habitat (Langdon, 2006b as cited in Lepofsky and Caldwell, 2013, pp. 5–6). Traditional salmon harvesting restrictions were also effective, including the choice of net size and mesh to regulate the overall size of the catch, the species caught, and the size and age of harvested fish. Ethnographic accounts as well as waterlogged archaeological sites indicate a wide variety of net types and meshes. Mesh size was standardized by using net gauges, which have been recorded in ethnographic and archaeological contexts (Lepofsky and Caldwell, 2013, p. 4). People stopped fishing a run before the number of salmon captured might offend the salmon (Losey, 2010; Thornton et al., 2015). These reciprocal relationships between human and animals are inherent in the very constitution of subsistence practices—both human and salmon are moral beings with immortal spirits that require protocols for respectful engagement (Langdon, 2007).

Central to salmon management was tenure at productive fishing sites, as well as technological innovation, social organization, and ritual expertise. This is exemplified in the Salish Sea by Lekwungen and WSÁNEĆ (Fig. 9.1) reefnet fishing (Claxton, 2015; Claxton and Elliott, 1994; Suttles, 1974; Turner and Berkes, 2006). The Coast Salish people of southern Vancouver Island and the adjacent archipelago of islands had little access to riverine salmon runs, so they developed a means of intercepting salmon while they were still at sea. To do this required a keen awareness of the underwater geography, tidal currents and flow, and the habits of the salmon (Claxton, 2015). Paraphrasing Suttles (1974), the reefnet location was often nearshore on a kelp-covered reef in the path of migrating sockeye salmon. Each reefnet location had an owner who inherited it from his ancestors (Suttles, 1974). Taking advantage of tides, the net was anchored opposite a headland that caused a backward sweep of current, thus carrying the salmon toward the facing net. The fishing gear consisted of a net held between two parallel canoes that were anchored by four sets of stone anchors each. Extended between the two canoes was the large rectangular net. At its forward end, facing the current, the net was held down by two net weights at each corner, the rest billowing out with the current. Salmon followed a route along interconnected reefs and pathways cut through kelp. Much like a terrestrial drive lane used elsewhere in North America to funnel herd animals, this directed streams of salmon from the main migration toward the canoe-supported reefnets at the end of path. When the net was full, the canoes were swung together and the net was pulled into one canoe and the salmon dropped into the other. For this to work, water conditions had to be clear to see the salmon below, but also to see the placement of the anchors. These piles of stone anchors on the ocean floor are still archaeologically visible today (Easton, 1985; Moore and Mason, 2011).

Lekwungen and WSÁNEĆ worldview is an entangled whole that enmeshes belief and values with reefnet fishing. Proper social relations between people and salmon are underlain by the knowledge that salmon were like people and that they had come to feed the people with their own flesh. As such, people showed their respect for the salmon with the First Salmon ritual. According to Berkes (2015, p. 233), there were also management values in this ritual practice (see also Turner and Berkes, 2006). The First Salmon ritual allowed the ritual and fishery leaders to delay the opening of the fishery if the run was deemed weak, thus allowing the migration leaders to escape upstream before commencing fishing. Furthermore, the reefnet captains could make year-to-year adjustments to the methods and intensity of the fishery based on the strength of the salmon run. This allowed a sufficient number of salmon to escape upstream. A similar process occurred throughout the entire length of the salmon run, as the salmon leaders left the Salish Sea and swam upstream to the Interior Plateau and the St'at'imc people.

On the Northwest Coast, there were two basic kinds of intertidal fishing structures: latticework basketry traps and weirs in rivers to catch salmon swimming upstream; and intertidal traps constructed of stakes or rocks near the mouths of streams to catch returning salmon (Ballard, 1957; Kennedy and Bouchard, 1990, p. 444; Langdon, 2006a; Stewart, 1977). A weir is a linear obstruction or wall constructed to impede or direct the movement of salmon in some fashion. The weir assists in concentrating the salmon so that other devices, such as baskets, can be used to catch them. Stó:lō (Fig. 9.1) ancestors were taught how to construct fish weirs, or *siyak*, by Tamia, the wren (Hill-Tout and Maud, 1978, p. 56). Tamia instructed young cedar tree branches to twist themselves into withes, and short branches to sharpen one of their ends to a point. These were then firmly placed in the bed of the salmon-bearing stream in the form of a tripod, fastened at the top by the withes. Tamia then called upon other boughs to fasten themselves in the lower legs of these tripods, till the weir spanned the entire stream. The salmon soon congregated in great numbers and Tamia then instructed the people to make their salmon-weirs thereafter in this manner (Hill-Tout and Maud, 1978, p. 56). Langdon (2006a) has argued that this method of salmon harvesting was not damaging to salmon abundance and was likely designed to ensure adequate escapement to the spawning grounds. Langdon asserts that the Tlingit (Fig. 9.1) were selectively harvesting salmon stocks in a manner that ensured the survival of sufficient number of spawners to assure a continuing supply in the future.

A trap, in contrast, captures the salmon by drawing them into a structure on an outgoing tide from which they are unlikely to escape until the tide rises again (Fig. 9.4). Traps may be low curvilinear walls built with stones (Fig. 9.2), and are typically less than 30 m long. Most consist of fewer than three layers of stone piled up (Anthony, 1976; White, 2006). Traps were also made with wooden stakes and ranged from simple curvilinear features to complex chevron and heart-shaped wooden walls composed of hundreds of stakes (e.g., Greene et al., 2015). Radiocarbon dated wood-stake trap and weir sites were constructed between c.5500 cal. BP and the 20th century (Moss, 2013). These materials were sometimes combined—for example, in Tlingit territory, a stone trap with an associated wooden stake was radiocarbon dated to AD 1050 (Langdon, 2006a, p. 31). Salmon congregate at the mouth of creeks waiting for runoff or rains to raise water levels enough to ease their migration upstream. With each tidal rise, the waiting salmon drift shoreward, swimming over the top of the

FIGURE 9.4 Aerial photo of intertidal resource management features fronting the ancestral Heiltsuk village of Hauyat, on the Central Coast of British Columbia.

traps. With the ebbing tide, the salmon would become trapped behind the stone walls, where they could be selectively speared or gaffed. Stone traps continue to work whether they are attended to or not—the fish survive in the trap until the tide comes back in again and they can swim away. While initial labor to make them was substantial—occasionally requiring resources and organization at a monumental scale—these traps were easily accessible and people could visit multiple traps in a single tidal low tide, returning the catch from several traps to a central processing camp. This would have solved the problem of being in a territory with many small salmon runs at many small streams (Langdon, 2006a), in contrast to being on a river with major salmon runs, better suited to a weir.

Estuarine Root Gardens and Intertidal Plant Management

Fish traps and clam gardens are often part of an integrated network of other ethnoecological features. Commonly associated with these are estuarine root gardens (Fig. 9.2, Table 9.3). Clan Chief Adam Dick/Kwaxsistalla states that the Kwakwaka'wakw people specialized in estuarine agriculture—*t'ekilakw*, or "places of man-made soil." *T'ekilakw* is a system of perennial root gardens constructed along coastal estuaries using natural inputs from the sea. Estuarine salt marshes and flats near the mouths of rivers and streams were common sites for plots of Pacific silverweed (*Argentina egedii*) (Fig. 9.5), springbank clover (*Trifolium wormskioldii*), northern riceroot lily (*Fritillaria camschatcensis*), and Nootka lupine (*Lupinus nootkatensis*) (Deur, 2005). At least at Alberni Inlet, edible camas (*Camassia quamash*) was also produced in estuarine gardens (Turner and Bhattacharyya, 2016). The edible underground "roots" from these species were an important source of dietary carbohydrates, but were also

FIGURE 9.5 Archaeological testing of estuarine root garden near Bella Bella with Pacific silver-weed (*Argentina egedii*).

entangled with social, economic, and spiritual relationships including trade, feasting, social stratification, cosmology, and ritual (Boas, 1921; Turner and Kuhnlein, 1982). In particular, the carbohydrate-rich root vegetables were chiefly foods: the production of which was one way for elite families and individuals to meet ceremonial obligations. Important in feasts, potlatches, and as wedding gifts (Boas, 1910; Drucker, 1951; McIlwraith, 1948), these estuarine root vegetables could serve as a means of transforming material capital into social standing, in a stratified society founded on respect (Deur et al., 2013).

The soils, plants, and hydrology of estuarine gardens were modified to enhance the quality and productivity of these root foods (Derr, 2014; Deur, 2005; Deur et al., 2013; Turner et al., 2013b). Root gardens were places designed to capture debris brought by tidal waves, such as dead fish, wrack, and other nutrients. These inputs nurtured the perennial roots, capitalizing on natural processes like tides and salmon runs. Deur (2005, p. 304) developed the argument that the nature and scale of estuarine root management is best characterized as cultivation, entailing the seeding or transplanting of propagules, the intentional fertilization of soils, improvements in drainage or irrigation, mounding or terracing, and the clearing ("weeding") of competing plants. Furthermore, there were practices of tenure over these places, with estuarine root gardens often divided into numerous family-owned plots, which were sometimes delineated by low rock walls, posts, or other markers. Deur (2005) outlined the importance of this tenure, adding that "digging in a chief's estuarine garden or a family's sub-plot without their permission was a grave offense" (2005:305). Furthermore, this rigid system of tenure both underscores the value of this resource and perhaps also alludes to the possibility that at times there were shortages or overharvesting of these estuarine gardens (Deur, 2005, p. 305).

While additional archaeological work is required (Fig. 9.5), soil horizons believed to be associated with a root garden in Clayoquot Sound (Fig. 9.1) have been radiocarbon dated between AD 1479 and 1575 (Deur, 2005, p. 319). It is likely that these practices have a much greater antiquity. This method of the perpetual production of nutritionally, economically, and socially important root foods challenges the orthodoxy of cultivation. These are practices of resource enhancement and management that may have developed in parallel to, or were recursively related to, the establishment of large and increasingly sedentary winter village sites, social inequality, and the rise of elites (Ames, 1994, 1998). Root gardens, clam gardens, and fish traps often occur together (Fig. 9.4) and form a continuum of ecological management within these complex coastal food webs. Root gardens also tied households together within larger overlapping social and economic spheres, as these rhizomes, and the intensification and storage of salmon, clams, and other resources were integral to the power and prestige of corporate groups (Ames, 1994, 1998), a basis for reciprocal exchange and building of not just wealth and alliances but social solidarity (Mauss, 1990), an institution that supported resilience and sustainability (Trosper, 2011), and food systems as a gift economy to ameliorate risk and uncertainty in dynamic environments (LeCompte-Mastenbrook, 2015). In turn, this production contributes to the establishment of lineages and other long-term and historical connections to places.

DISCUSSION

These examples illustrate an entire range of features and characteristics of coastal Indigenous management reflecting long-standing social–ecological systems in place for at least several millennia. Not only have these systems enhanced people's food security and dietary diversity but they have provided products for trade and exchange (Turner, 2014), and have underlain complex ceremonial and economic systems that characterize Northwest Coast cultures. As indicated in Table 9.3, facets of management range from diverse social constraints to ecological and temporal constraints that influence resource distribution, productivity, and quality.

It is perhaps a reflection of cultural biases and, at the very least, colonial attitudes, that these management systems, their complexity, and their effectiveness have only recently been recognized by those outside these cultures. It was only around 1995 that geomorphologist John Harper, flying at low tide over the beach features that were later explained as clam gardens, began to wonder whether these could be human constructions, rather than glacial moraines or some other natural feature. Not even the archaeologists of the time recognized them as human-created (Recalma-Clutesi, 2005). Salmon weirs and traps were long regarded by Colonial and fisheries government officials as destructive and unsustainable, yet, as pointed out by Eugene Anderson (1996) and Butler and Campbell (2004), the First Nations had the capacity, with their weirs and traps situated across salmon streams and rivers all along the Northwest Coast, to completely annihilate the countless runs of salmon. While there are stories of depletion due to weir construction among Northwest groups, such as the loss of the small sockeye in several streams near Sitka after they became insulted due to their way being blocked by the weir and "left" (Thornton, 2008, pp. 173–4), these examples served to encode moral-ecological principles in particular places and fisheries. The sustainability of Indigenous fisheries is exemplified by the fact that when the Europeans arrived on the coast, salmon were present in immense numbers that darkened the waters during spawning times (Harris, 2001). It was only later when the commercial salmon fishery was established, with its canneries all along the coast, and when clear-cut logging resulted in habitat loss, changes in food web dynamics, damaged spawning rivers and creeks, that the salmon became depleted (Scientific Panel for Sustainable Forest Practices in Clayoquot Sound, 1995).

Today, even with modern salmon enhancement programs and fish hatcheries, salmon populations are much lower than in the past. Some populations are critically endangered or have gone extinct. There are also many threats to the clam gardens and estuarine root gardens, as well as to other management systems, due mainly to industrial activities (Turner et al., 2013a,b; Cullis-Suzuki et al., 2015). The losses of access to and destruction of their key food sources and managed habitats have been devastating for many First Nations communities, threatening not only their food security, but also their cultural identity and their ability to pass on important cultural and ecological knowledge to their children (Turner et al., 2008; Turner and Turner, 2008). The attitudes of caring and gratitude, and the kinship relationships that First Nations had, and continue to have, toward other species, including those they depend upon for food and other resources, have been little recognized or appreciated by those outside the cultures (Brown et al., 2009; Turner, 2005).

First People have been an integral part of coastal ecosystems since the late Pleistocene Late/Time of the Transformer. We have created a vicious cycle in which, because people's food systems and associated managed habitats have been disrupted, the knowledge and skills needed to harvest and process the foods and to maintain the management systems can no longer be effectively applied or passed on to succeeding generations. Without tending and harvesting, the resource systems themselves may deteriorate, and gradually become less productive. With the clam gardens, the beaches accumulate silt and potentially become anoxic when the clams are not dug and harvested selectively. Root gardens can become "root-bound" and less productive; the edible roots gradually replaced with tough-rhizomed sedges and grasses. Without the pools and careful tending of the salmon, the salmon will not thrive. All of the other species that depend on salmon—wolves, bears, eagles—are also affected. In this sense, the absence of Indigenous management becomes a form of ecological perturbation (Berkes et al., 1995).

What is needed is collaborative research and experimentation on these diverse management systems, to better understand how they have operated and to determine how they might be restored for the future. Of course, this is not a straightforward proposition. As Caldwell et al. (2012, p. 220), emphasize, "the terms [fish trap, clam garden] do not adequately describe the formal and functional variation often encompassed within these archaeological features," and this variation needs to be documented. Furthermore, much has changed over the past decades, from habitat destruction, pollution, siltation, overharvesting, introduced species impacts, and now ongoing climate change with predicted significant sea-level rise (Murray et al., 2015). Salmon farms along the coast, for example, with introduced Atlantic salmon, have impacted wild Pacific salmon, through sea lice infestations and other diseases (cf. Krkosek et al., 2006). Toxic algal blooms, even in the winter months, are preventing people from harvesting clams in places where they have done so for generations (Chris Picard, Science Director for the Gitga'at Nation, pers. comm. to NT, March 19, 2015; Thompson and Picard, 2016). Edible seaweed was far less abundant on the coast in the spring of 2016, according to many accounts, from Gitga'ata of Hartley Bay to Heiltsuk of Bella Bella. Concerns about spills from oil tankers, such as *Exxon Valdez*, are also widespread for coastal communities (Satterfield et al., 2011).

Nevertheless, there is tremendous interest in restoring and reengaging with traditional management systems as a means of enhancing food security on the coast, and for ethnoecological restoration and cultural renewal (Cuerrier et al., 2015). This is congruent with the potential fulfillment of settler commitments and responsibilities in the principles recognized by the United Nations Declaration on the Rights of Indigenous Peoples (2007) and the recent Truth and Reconciliation Commission of Canada (2015, p. 4), including: "Supporting Aboriginal peoples' cultural revitalization and integrating Indigenous knowledge systems, oral histories, laws, protocols, and connections to the land into the reconciliation process are essential" (guiding principle #8).

Although some of the knowledge about these traditional coastal management systems has been suppressed, there are still individuals such as Clan Chief Adam Dick (Kwaxsistalla) of the Kwakwaka'wakw Nation, Louis Claxton of the Tsawout W̱SÁNEĆ (Saanich) Nation, and other elders along the coast who participated in the maintenance and use of managed habitats and resources, including clam gardens and root gardens, and the traditional salmon reefnet fishery of the Straits Salish (Claxton, 2015; Claxton and Elliott, 1994; Deur et al., 2013; Recalma-Clutesi, 2005; Turner et al., 2013b). Their experiences and recollections are invaluable in efforts to revitalize these production systems. Furthermore, as noted previously, there are archaeological signatures for coastal management systems that can give us clues as to how they have functioned, and how they could function again in the future. Use of global positioning systems (GPS), light detection and ranging (LiDAR), remote controlled drones, and other new technologies for survey and photography, as well as application of standard archaeological, ecological, and ethnographic methods (interviews; literature reviews; survey plots; ecosystem mapping; repeat photography; field monitoring), has allowed a more complete understanding of these systems, now that they have been identified and recognized more widely (e.g., Caldwell et al., 2012; Lecompte-Mastenbrook, 2015).

As Taiaiake Alfred (2009, p. 43) states, colonialism "is best conceptualized as an irresistible outcome of a multi-generational and multi-faceted process of forced dispossession and attempted acculturation—a disconnection from land, culture, and community." In the case of British Columbia, this has meant, for example, the replacement of traditional economies for a cash economy (Lutz, 2008), the banning of reefnet fishing and other traditional and sustainable fisheries (Berkes, 2015), land and marine dispossession, residential schools removing children from their communities (Truth and Reconciliation Commission of Canada, 2015), banning of spiritual and cultural practices such as the potlatch (Cole and Chaikin, 1990), and a pervasive racism that underscores all of the above. The examples of Indigenous management we have outlined in this chapter are not limited to the past. Today, there is the experimental construction of clam gardens, building and reclaiming of estuarine root gardens, and re-implementation of the Straits Salish reefnet fishery, providing important and ongoing insights into the structure and function of these management systems (e.g., Augustine and Dearden, 2014; Claxton, 2015; Joseph, 2012; Lepofsky et al., 2015; Pukonen, 2008). In all of these cases, youth and other members of local communities have participated, with profound opportunities for revitalizing, learning, and celebrating and long-held cultural knowledge, which, through multiple and cumulative impacts relating to colonization and industrialization of lands and resources, have been seriously impacted over the past century and a half (Turner and Turner, 2008). These are part of the process of "everyday acts of resurgence" (Corntassel, 2012) which contribute to the decolonization and regeneration of Indigenous knowledge and ways of life. Hopefully, such restorative activities will continue, through collaborative research and adaptive comanagement. First Nations' Coastal Guardian Watchmen and related programs, as well as participation of Indigenous experts and students of all ages, will enrich and expand our relationships with coastal habitats and resources, foster appreciation, stewardship, and conservation of these precious elements, and will provide the opportunities necessary for maintaining these features and practices as well as enhancing food security and food sovereignty for Indigenous communities.

As restorative programs develop, communication about them is essential to their long-term success. Videos, public talks, on-the-ground tours, school field trips and science camps, and inclusion in language revitalization programs are all ways in which the knowledge, practice, and associated worldviews of respect and appreciation can be conveyed more widely (Anderson, 2014; Turner, 2014).

At the same time, it is imperative that our fragile and critically important coastal ecosystems be protected, in all their diversity. We also need to acknowledge and support the inextricable relationships that Indigenous People have had with them since the earliest times of their entry into the "New World", and to continue the process of respectful reconciliation with First Nations (Atleo, 2011; Deur et al., 2013).

CONCLUSION

We return to the question first posed in this chapter: How can we learn from Indigenous Peoples' traditional management systems in our efforts to sustain and restore our marine and coastal ecosystems on the Northwest Coast of North America and beyond?

Clam gardens, fish traps, estuarine root gardens, and other managed marine features and places continue to work, ebbing and flowing with each tidal cycle. Created and maintained by human hands, many are places waiting for people to return. They are touchstones to past connections to places, a physical manifestation of social relationships and responsibilities. As relatively permanent, albeit submerging and emerging monuments, they are visual cues to places and histories.

Recovering and reinstating traditional systems of coastal ecosystem management will require leadership and collaboration, as well as different land tenure arrangements. Continued research will be needed, using experimental and adaptive methodologies and a range of available technologies, guided by principles of respect and cooperation. Ongoing communication and experiential learning will help to maintain and develop stronger relationships with Indigenous communities and will hopefully help to support decolonization and reconciliation, as well as restoration of landscapes and native species, making our social–ecological systems more resilient in the face of ongoing changes. The natural processes underlying these management systems still exist, and the species involved, from clams, to salmon, to estuarine root vegetables are still key parts of the ecosystems they inhabit. The collaborative work undertaken so far to identify and characterize these management features provides a solid foundation for contemporary Indigenous communities to regain control and use of their traditional food production systems and diverse cultural keystone species, and for ongoing efforts to sustain and enhance the ecosystems on which we all rely.

ACKNOWLEDGMENTS

We extend our deepest gratitude to the Indigenous experts who have kept this critically important knowledge alive and who have shared their experiences and insights with us. Particular thanks go to: Clan Chief Adam Dick (Kwaxsistalla), Kim Recalma-Clutesi (Ogwi'low'qwa), Dr. Daisy Sewid-Smith (Mayanilth), Dr. E. Richard Atleo (Chief Umeek), Chief Earl Maquinna George, Barbara Wilson (Ḵii'iljuus), Joan Morris (Súlhlima), Chief Ron Sam, Elsie Claxton, Dr. Arvid Charlie (Luschiim), John Thomas (Tliishal), Pauline Waterfall (Hilistis), Frank and Kathy Brown, Helen Clifton, Charlie Bob, Johnny Johnson and Emma Reid, Jennifer Carpenter, and Chris Picard. We are also grateful to Dr. Eric Peterson and Christina Munck of the Tula Foundation, for their support over the past five years, to the Social Sciences and Humanities Research Council of Canada, and to the Pierre Elliot Trudeau Foundation. Thank you, too, to our students and colleagues for their contributions; the publications of several are cited here. We extend special thanks to two anonymous reviewers whose comments and suggestions resulted in substantive improvements to our manuscript. Finally, our deep appreciation goes to Phil Levin and Melissa Poe for inviting this contribution to their edited publication.

REFERENCES

Alfred, T., 2009. Colonialism and state dependency. Journal of Aboriginal Health 5, 42–60.

Ames, K.M., 1994. The Northwest Coast: complex hunter-gatherers, ecology, and social evolution. Annual Review of Anthropology 23, 209–229.

Ames, K.M., 1998. Economic prehistory of the northern British Columbia coast. Arctic Anthropology 35, 68–87.

Ames, K.M., Maschner, H.D.G., 1999. Peoples of the Northwest Coast: Their Archaeology and Prehistory. Thames and Hudson Ltd, London.

Anderson, E.N., 1996. Ecologies of the Heart: Emotion, Belief and the Environment. Oxford University Press, New York.

Anderson, E.N., 2014. Caring for Place: Ecology, Ideology, and Emotion in Traditional Landscape Management. Left Coast Press, CA.

Anderson, M.K., 2005. Tending the Wild: Native American Knowledge and Management of California's Natural Resources. University of California Press, Berkeley.

Anderson, M.K., 2009. The Ozette Prairies of Olympic National Park: Their Former Uses & Management. Olympic National Park, Port Angeles, WA.

Anthony, P.J., 1976. Stone fish traps of the Bella Bella region. In: Carlson, R. (Ed.), Current Research Reports. Archaeology Press, Simon Fraser University, Burnaby, BC, pp. 165–173.

Atleo (Chief Umeek), E.R., 2004. Tsawalk: A Nuu-chah-nulth Worldview. UBC Press, Vancouver.

Atleo (Chief Umeek), E.R., 2011. Principles of Tsawalk: An Indigenous Approach to Global Crisis. UBC Press, Vancouver, BC.

Augustine, S., Dearden, P., 2014. Changing paradigms in marine and coastal conservation: a case study of clam gardens in the southern Gulf islands, Canada. The Canadian Geographer 58, 305–314.

Ballard, A.C., 1957. The salmon weir on green river in western Washington. Davidson Journal of Anthropology 3 (1), 37–53.

Berkes, F., 2012. Sacred Ecology: Traditional Ecological Knowledge and Resource Management, third ed. Taylor and Francis, Philadelphia, PA.

Berkes, F., 2015. Coasts for People: Interdisciplinary Approaches to Coastal and Marine Resource Management. Routledge, New York.

Berkes, F., Folke, C., Gadgill, M., 1995. Traditional ecological knowledge, biodiversity, resilience and sustainability. In: Perrings, C.A., Maler, K.G., Folke, C., Holling, C.S., Jansson, B.O. (Eds.), Biodiversity Conservation: Problems and Policies. Springer Science & Business Media B.V., Dordrecht, Netherlands, pp. 281–300.

Boas, F., 1910. Kwakiutl Tales. Columbia University Press, New York.

Boas, F., 1921. Ethnology of the Kwakiutl. Smithsonian Institution, Bureau of American Ethnology, Washington, DC. 35th Annual Report, Parts 1 and 2.

Boas, F., 2002. Indian Myths and Legends from the North Pacific Coast of America: A Translation of Franz Boas's 1895 Edition of Indianische Sagen von der Nord Pacifischen Küste Amerikas (D. Bertz, Trans.). Bouchard, R., Kennedy, D. (Eds.), Talonbooks, Vancouver, BC.

Boas, F., Hunt, G., 1906. Kwakiutl Texts—Second Series, vol. X. Publications of the Jesup North Pacific Expedition, New York. Part 1.

Bouchard, R., Kennedy, D., 1990. Clayoquot Sound Indian Land Use. Report prepared for MacMillan Bloedel Limited, Fletcher Challenge Canada and British Columbia Ministry of Forests.

Bringhurst, R., 2000. A Story as Sharp as a Knife: The Classical Haida Mythtellers and Their World. University of Nebraska Press, Lincoln.

Brown, F., Brown, K., Wilson, B., Waterfall, P., Cranmer Webster, G., 2009. Staying the Course, Staying Alive: Coastal First Nations Fundamental Truths. Biodiversity BC, Victoria.

Butler, V.L., Campbell, S.K., 2004. Resource intensification and resource depression in the Pacific northwest of North America: a zooarchaeological review. Journal of World Prehistory 18 (4), 327–405.

Caldwell, M.E., Lepofsky, D., Combes, G., Washington, M., Welch, J.R., Harper, J.R., 2012. A Bird's eye view of northern Coast Salish intertidal resource management features, southern British Columbia, Canada. The Journal of Island and Coastal Archaeology 7, 219–233.

Campbell, S.K., Butler, V.L., 2010. Archaeological evidence for resilience of Pacific Northwest salmon populations and the socioecological system over the last~7,500 years. Ecology and Society 15 (1), 17.

Cannon, A., Burchell, M., 2009. Clam growth-stage profiles as a measure of harvest intensity and resource management on the central coast of British Columbia. Journal of Archaeological Science 36, 1050–1060.

Cannon, A., Yang, D.Y., 2006. Early storage and sedentism on the Pacific Northwest Coast: ancient DNA analysis of salmon remains from Namu, British Columbia. American Antiquity 71, 123–140.

Carpenter, J., Humchitt, C., Eldridge, M., 2000. Final Report, Fisheries Renewal B.C. Research Reward, Science Council of BC, Reference Number FS99–32 (Unpublished manuscript on file at the Heiltsuk Cultural Education Centre, Bella Bella, B.C.).

Claxton, N.X., 2015. To Fish as Formerly: A Resurgent Journey Back to the Saanich Reef Net Fishery (Ph.D. dissertation). Faculty of Education, University of Victoria, Victoria, BC.

Claxton (YELḰÁT̶E) Sr., E., Elliott (STOL₵EŁ) Sr., J., 1994. Reef Net Technology of the Saltwater People. Saanich Indian School Board, Brentwood Bay, BC.

Cole, D., Chaikin, I., 1990. An Iron Hand upon the People: The Law against the Potlatch on the Northwest Coast. Douglas & McIntyre, Vancouver.

Comberti, C., Thornton, T.F., Wyllie de Echeverria, V., Patterson, T., 2015. Ecosystem services or services to ecosystems? Valuing cultivation and reciprocal relationships between humans and ecosystems. Global Environmental Change 34, 247–262.

Corntassel, J., 2012. Re-envisioning resurgence: indigenous pathways to decolonization and sustainable self-determination. Decolonization: Indigeneity, Education & Society 1 (1), 86–101.

Cuerrier, A., Turner, N.J., Gomes, T., Garibaldi, A., Downing, A., 2015. Cultural keystone places: conservation and restoration in cultural landscapes. Journal of Ethnobiology 35 (3), 427–448.

Cullis-Suzuki, S., Wyllie-Echeverria, S., Dick (Kwaxsistalla), A., Turner, N.J., 2015. Tending the meadows of the sea: a disturbance experiment based on traditional indigenous harvesting of *Zostera marina* L. (Zosteraceae) the southern region of Canada's west Coast. Aquatic Botany 127. http://dx.doi.org/10.1016/j.aquabot.2015.07.001.

Derr, K., 2014. Anthropogenic fire and landscape management on Valdes island, Southwestern BC. Canadian Journal of Archaeology 38 (1), 1.

Deur, D., 2002. Plant Cultivation on the Northwest Coast: A Reassessment. Journal of Cultural Geography 19 (2), 9–35.

Deur, D., 2005. Tending the garden, making the soil: Northwest Coast estuarine gardens as engineered environments. In: Deur, D., Turner, N.J. (Eds.), "Keeping it Living": Traditions of Plant Use and Cultivation on the Northwest Coast of North America. University of Washington Press, Seattle, pp. 296–327.

Deur, D., Dick (Kwaxsistalla), A., Recalma-Clutesi (Ogwi'low'qwa), K., Turner, N.J., 2015a. Kwakwaka'wakw "clam gardens": motive and agency in traditional Northwest Coast mariculture. Human Ecology. http://dx.doi.org/10.1007/s10745-015-9743-3.

Deur, D., Turner, N.J. (Eds.), 2005. Keeping it Living": Traditions of Plant Use and Cultivation on the Northwest Coast of North America. University of Washington Press, Seattle and UBC Press, Vancouver, BC.

Deur, D., Turner, N.J., Dick (Kwaxsistalla), A., Sewid-Smith (Mayanilth), D., Recalma-Clutesi (Ogwi'low'qwa), K., 2013. Subsistence and resistance on the British Columbia coast: Kingcome Village's estuarine gardens as contested space. BC Studies 79, 13–37.

Deur, D., Dick, A., Recalma-Clutesi, K., Turner, N., 2015b. Kwakwaka'wakw "clam gardens". Human Ecology 43, 201–212.

Dillehay, T.D., Ramírez, C., Pino, M., Collins, M.B., Rossen, J., Pino-Navarro, J.D., 2008. Monte Verde: seaweed, food, medicine, and the peopling of South America. Science 320, 784–786.

Drucker, P., 1951. The Northern and Central Nootkan Tribes. Government Printing Office, Washington, DC.

Easton, N., 1985. The Underwater Archaeology of Straits Salish Reefnetting. University of Victoria, Victoria, BC.

Edwards, G.T., 1979. Indian Spaghetti. The Beaver (Autumn), pp. 4–11.

Ellis, D.W., Swan, L., 1981. Teachings of the Tides: Uses of Marine Invertebrates by the Manhousaht People. Theytus Books, Nanaimo, BC.

Ellis, D.W., Wilson, S., 1981. The Knowledge and Usage of Marine Invertebrates by the Skidegate Haida People of the Queen Charlotte Islands. Monograph Series No. 1. Queen Charlotte Islands Museum Society, Skidegate, BC.

Erlandson, J.M., Graham, M.H., Bourque, B.J., Corbett, D., Estes, J.A., Steneck, R.S., 2007. The kelp highway hypothesis: marine ecology, the coastal migration theory, and the peopling of the Americas. The Journal of Island and Coastal Archaeology 2, 161–174.

Erlandson, J.M., 1988. The role of shellfish in prehistoric economies: a protein perspective. American Antiquity 53 (1), 102–109.

Fladmark, K.R., 1979. Routes: alternate migration corridors for early man in north America. American Antiquity 44, 55–69.

Ford, R.I., 1985. Prehistoric Food Production in North America (University of Michigan Museum of Anthropology, Anthropological Papers No. 75). University of Michigan Press, Ann Arbor.

Garibaldi, A., Turner, N.J., 2004. Cultural keystone species: implications for ecological conservation and restoration. Ecology and Society. 9 (3). http://www.ecologyandsociety.org/vol9/iss3/art1.

George, E.M., 2003. Living on the Edge: Nuu-chah-nulth History from an Ahousaht Chief's Perspective. Sono Nis, Winlaw, BC.

Greene, N.A., McGee, D.C., Heitzmann, R.J., 2015. The Comox Harbour fish trap complex: a large-scale, technologically sophisticated intertidal fishery from British Columbia. Canadian Journal of Archaeology 39, 161–212.

Groesbeck, A.S., Rowell, K., Lepofsky, D., Salomon, A.K., 2014. Ancient clam gardens increased shellfish production: adaptive strategies from the past can inform food security today. PLoS One 9, 1–13.

Gruhn, R., 1994. The Pacific Coast route of initial entry: an overview. In: Bonnichsen, R., Steele, D.G. (Eds.), Methods and Theory for Investigating the Peopling of the Americas. Oregon State University Press, Corvallis, pp. 249–256.

Gunther, E., 1926. An analysis of the first salmon ceremony. American Anthropologist 28 (4), 605–617.

Haggan, N., Turner, N., Carpenter, J., Jones, J.T., Mackie, Q., Menzies, C., 2006. 12,000+ Years of Change: Linking Traditional and Modern Ecosystem Science in the Pacific Northwest (Working paper). Fisheries Centre, University of British Columbia, Vancouver, BC.

Harper, J., Haggerty, J., Morris, M., 1995. Broughton Archipelago Clam Terrace Survey. Land Use Coordination Office, BC Ministry of Government Services, Victoria.

Harris, D.C., 2001. Fish, Law, and Colonialism: The Legal Capture of Salmon in British Columbia. University of Toronto Press, Toronto, ON.

Hill-Tout, C., Maud, R., 1978. The Salish People: The Local Contribution of Charles Hill-Tout. Talonbooks, Vancouver, BC.

Hunn, E.N., Johnson, D.R., Russell, P.R., Thornton, T.F., 2003. Huna Tlingit traditional environmental knowledge, conservation, and the management of a "Wilderness" park. Current Anthropology 44, S79–S103.

Jones, R., 1999. Haida Names and Utilization of Common Fish and Marine Mammals. In: Haggan, N., Beattie, A. (Eds.). Haggan, N., Beattie, A. (Eds.), With the Assistance of D. Pauly. 1999. Back to the Future: Reconstructing the Hecate Strait Ecosystem, vol. 7 (3). Fisheries Centre Research Report Series, pp. 39–48.

Jones, J.T., 2002. "We Looked after all the Salmon Streams": Traditional Heiltsuk Cultural Stewardship of Salmon and Salmon Streams: A Preliminary Assessment, Environmental Studies (M.A. thesis). University of Victoria, Victoria, BC.

Joseph, L., 2012. Finding Our Roots: Ethnoecological Restoration of Lhásem (Fritillaria camschatcensis (L.) Ker-gawl), an Iconic Plant Food in the Squamish River Estuary, British Columbia (M.Sc. thesis). University of Victoria, Victoria, BC.

Kennedy, D.I.D., Bouchard, R., 1974. Utilization of Fishes, Beach Foods, and Marine Mammals by the Tl'úhus (Klahoose) Indian People of British Columbia. BC Indian Language Project.

Kennedy, D.I.D., Bouchard, R., 1983. Sliammon Life, Sliammon Lands. Talonbooks, Vancouver, BC.

Kennedy, D.I.D., Bouchard, R., 1990. Bella Coola. In: Suttles, W.P. (Ed.). Suttles, W.P. (Ed.), Handbook of North American Indians, vol. 7. Smithsonian Institution, Washington, DC, pp. 323–329 (Northwest Coast, 323–39).

Krkosek, M., Lewis, M.A., Morton, A., Frazer, L.N., Volpe, J.P., 2006. Epizootics of wild fish induced by farm fish. Proceeds of the National Academy of Science of United States of America 103, 15506–15510.

Langdon, S.J., 2006a. Tidal Pulse Fishing: Selective Traditional Tlingit Salmon Fishing Techniques on the West Coast of the Prince of Wales Archipelago. Traditional Ecological Knowledge and Natural Resource Management.

Langdon, S.J., 2006b. Traditional Knowledge and Harvesting of Salmon by Huna and Hinyaa Tlingit, Final Report (Project 02–104). U.S. Fish and Wildlife Service, Office of Subsistence Management, Fisheries Resource Monitoring Program, Anchorage.

Langdon, S.J., 2007. Sustaining a relationship: inquiry into the emergence of a logic of engagement with salmon among the southern Tlingits. In: Harkin, M.E., Lewis, D.R. (Eds.), Native Americans and the Environment: Perspectives on the Ecological Indian. University of Nebraska Press, Lincoln and London, pp. 233–273.

LeCompte-Mastenbrook, J., 2015. Restoring Coast Salish Foods and Landscapes, a More-than-Human Politics of Place, History and Becoming (Ph.D. dissertation). University of Washington, Seattle, WA.

Lepofsky, D., Caldwell, M., 2013. Indigenous marine resource management on the Northwest Coast of North America. Ecological Processes 2, 12.

Lepofsky, D., Smith, N.F., Cardinal, N., Harper, J., Morris, M., White, E.G., Bouchard, R., Kennedy, D., Salomon, A.K., Puckett, M., Rowell, K., McLay, E.M., 2015. Ancient shellfish mariculture on the Northwest Coast of North America. American Antiquity 80, 236–259.

Lepofsky, D., Lertzman, K., 2008. Documenting ancient plant management in the northwest of North America. Botany 86, 129–145.

Lertzman, K., 2009. The paradigm of management, management systems, and resource stewardship. Journal of Ethnobiology 29 (2), 339–358.

Lightfoot, K.G., Cuthrell, R.Q., Striplen, C.J., Hylkema, M.G., 2013. Rethinking the study of landscape management practices among hunter-gatherers in North America. American Antiquity 78, 285–301.

Losey, R., 2010. Animism as a means of exploring archaeological fishing structures on Willapa Bay, Washington, USA. Cambridge Archaeological Journal 20 (1), 17–32.

Lutz, J., 2008. Makuk: A New History of Aboriginal-White Relations. UBC Press, Vancouver.

MacDonald, G.F., 2006. Coast Tsimshian Pre-contact Economics and Trade: An Archaeological and Ethno-Historic Reconstruction (Metlakatla/Lax Kw'alaams Land Claim File, Submission to Ratliff & Co. by 6347371 Canada Inc.).

Mackie, Q., Fedje, D., McLaren, D., Smith, N., McKechnie, I., 2011. Early environments and Archaeology of coastal British Columbia. In: Bicho, N.F., Haws, J.A., Davis, L.G. (Eds.), Trekking the Shore: Changing Coastlines and the Antiquity of Coastal Settlement. Springer, New York, pp. 51–103.

Matson, R.G., 1992. The evolution of Northwest Coast subsistence. In: Croes, D.R., Hawkins, R.A., Isaac, B.L. (Eds.), Long-term Subsistence Change in Prehistoric North America. JAI Press, Greenwich, Connecticut, pp. 367–428.

Mauss, M., 1990. The Gift: Forms and Functions of Exchange in Archaic Societies (W.D. Halls, Trans.). Routledge, London.

McIlwraith, T.F., 1948. The Bella Coola Indians, 2 vols.. University of Toronto Press, Toronto, ON.

McKechnie, I., 2007. Investigating the complexities of sustainable fishing at a prehistoric village on western Vancouver Island, British Columbia, Canada. Nature Conservation 15, 208–222.

McMillan, A., Hutchinson, I., 2002. When the mountain dwarfs danced: aboriginal traditions of paleoseismic events along the cascadia subduction zone of western North America. Ethnohistory 49, 41–68.

Mobley, C.M., McCallum, W.M., 2001. Prehistoric intertidal fish traps from central Southeast Alaska. Canadian Journal of Archaeology 25, 28–52.

Moore, C., Mason, A., 2011. Demonstration survey of prehistoric reef-net sites with sidescan sonar, near Becher Bay, British Columbia, Canada. International Journal of Nautical Archaeology.

Moss, M.L., 1993. Shellfish, gender, and status on the Northwest Coast: reconciling archaeological, ethnographic, and ethnohistorical records of the Tlingit. American Anthropologist 95, 631–652.

Moss, M.L., 2011. Northwest Coast: Archaeology as Deep History. Society for American Archaeology Press, Washington, DC.

Moss, M.L., 2013. Fishing traps and weirs on the Northwest Coast of North America: new approaches and new insights. In: Menotti, F., O'Sullivan, A. (Eds.), The Oxford Handbook of Wetland Archaeology. Oxford University Press, Oxford, pp. 323–338.

Murray, C.C., Agbayani, S., Ban, N., 2015. Cumulative effects of planned industrial development and climate change on marine ecosystems. Global Ecology and Conservation 4, 110–116.

Peter, D., Shebitz, D., 2006. Historic anthropogenically maintained bear grass savannas of the southeastern Olympic Peninsula. Restoration Ecology 14 (4), 605–615.

Pukonen, J.C., 2008. The λ'aayaʕas Project: Revitalizing Traditional Nuu-chah-nulth Root Gardens in Ahousaht, British Columbia (M.Sc. thesis). University of Victoria, Victoria, BC.

Recalma-Clutesi, K., 2005. Ancient Sea Gardens: Mystery of the Pacific Northwest (Documentary film). Aquaculture Pictures, Toronto, ON.

Reid, B., Bringhurst, R., 1984. The Raven Steals the Light. Douglas & McIntyre, Vancouver.

Satterfield, T., Robertson, L., Turner, N.J., December 2011. Being Gitka'a'ata: A Baseline Report on Gitka'a'ata Ways of Life, a Statement of Cultural Impacts Posed by the Northern Gateway Pipeline, and a Critique of the ENGP Assessment Regarding Cultural Impacts (Submission to the Joint Review Panel for Review of the Enbridge Northern Gateway Project).

Scientific Panel for Sustainable Forest Practices in Clayoquot Sound, 1995. First Nations' Perspectives on Forest Practices in Clayoquot Sound. Report 3. Cortex Consulting and Government of British Columbia, Victoria, BC.

Smith, M.W., 1940. The Puyallup-Nisqually. Columbia University Press, New York.

Smith, B.D., 2011. General patterns of niche construction and the management of 'wild' plant and animal resources by small-scale pre-industrial societies. Philosophical Transactions of the Royal Society B 366 (1566), 836–848. http://dx.doi.org/10.1098/rstb.2010.0253.

Suttles, W.P., 1974. Economic Life of the Coast Salish of Haro and Rosario Straits. Garland Publishing Inc., New York.

Suttles, W., 1987a. Spirit dancing and the persistence of native culture among the Coast Salish. In: Suttles, W. (Ed.), Coast Salish Essays. Talon Books, Burnaby, pp. 199–208.

Suttles, W., 1987b. Part I: models of historic social systems. In: Suttles, W. (Ed.), Coast Salish Essays. Talon Books, Burnaby, pp. 3–66.

Stewart, H., 1977. Indian Fishing. Douglas & McIntyre, Vancouver, BC, and University of Washington Press, Seattle.

Stewart, H., 1984. Cedar: Tree of Life to the Northwest Coast Indians. Douglas & McIntyre, Vancouver, BC.

Tabarev, A.V., 2011. Blessing the salmon: archaeological evidences of the transition to intensive fishing in the final paleolithic, maritime region, Russian far east. In: Bicho, N.F., Haws, J.A., Davis, L.G. (Eds.), Trekking the Shore: Changing Coastlines and the Antiquity of Coastal Settlement. Springer, New York, pp. 105–116.

Thompson, K.-L., Picard, C., 2016. 2015 Intertidal Bivalve Stock Assessment Data Summary Report. Gitga'at Lands and Marine Resources Department, Gitga'at First Nation, Hartley Bay, BC.

Thornton, T.F., 2008. Being and Place Among the Tlingit. University of Washington Press Seattle.

Thornton, T.F., 2015. The ideology and practice of Pacific herring cultivation among the Tlingit and Haida. Human Ecology 43 (2), 213–223.

Thornton, T.F., Moss, M.L., Butler, V.L., Heber, J., Funk, F., 2010. Local and traditional knowledge and the historical ecology of Pacific herring in Alaska. Journal of Ecological Anthropology 14, 81–88.

Thornton, T.F., Deur, D., 2015. Introduction to the special section on marine cultivation among indigenous peoples of the Northwest Coast. Human Ecology 43 (2), 187.

Thornton, T.F., Deur, D., Kitka Sr., H., 2015. Cultivation of salmon and other marine resources on the Northwest Coast of North America. Human Ecology 43 (2), 189–199.

Trant, A.J., Nijland, W., Hoffman, K., Nelson, T., Mathews, D.L., McLaren, D., Starzomski, B.M., 2016. Intertidal resource-use over millennia enhances forest productivity. Nature Communications 7:12491.

Trosper, R.L., 2011. Resilience, Reciprocity and Ecological Economics: Northwest Coast Sustainability. Routledge, London.

Truth and Reconciliation Commission of Canada, 2015. What We Have Learned: Principles of Truth and Reconciliation (Winnipeg, MN) https://ricochet.media/en/844/10-principles-for-truth-and-reconciliation.

Turner, N.J., Loewen, D.C., 1998. The original "free trade": exchange of botanical products and associated plant knowledge in northwestern North America. Anthropologica 40 (1), 49–70.

Turner, N.J., Kuhnlein, H.V., 1982. Two important "root" foods of the Northwest Coast Indians: springbank clover (*Trifolium wormskioldii*) and Pacific silverweed (*Potentilla anserina* ssp. *pacifica*). Economic Botany 36 (4), 411–432.

Turner, N.J., 1999. "Time to burn:" traditional use of fire to enhance resource production by aboriginal peoples in British Columbia. In: Boyd, R. (Ed.), Indians, Fire and the Land in the Pacific Northwest. Oregon State University Press, Corvallis, pp. 185–218.

Turner, N.J., 2003. The ethnobotany of "edible seaweed" (*Porphyra abbottiae* Krishnamurthy and related species; Rhodophyta: Bangiales) and its use by First Nations on the Pacific Coast of Canada. Canadian Journal of Botany 81 (2), 283–293.

Turner, N.J., 2005. The Earth's Blanket: Traditional Teachings for Sustainable Living. Douglas & McIntyre, Vancouver, BC and University of Washington Press, Seattle.

Turner, N.J., 2014. Ancient Pathways, Ancestral Knowledge: Ethnobotany and Ecological Wisdom of Indigenous Peoples of Northwestern North America, 2 vols.. McGill-Queens University Press, Montreal.

Turner, N.J., Berkes, F., 2006. Coming to Understanding: Developing Conservation through Incremental Learning. In: Berkes, F., Turner, N.J. (Eds.)Berkes, F., Turner, N.J. (Eds.), Developing Resource Management and Conservation, Special Issue, Human Ecology, vol. 34 (4), pp. 495–513.

Turner, N.J., Berkes, F., Stephenson, J., Dick, J., 2013a. Blundering intruders: multi-scale impacts on Indigenous food systems. Human Ecology 41 (4), 563–574.

Turner, N.J., Bhattacharyya, J., 2016. Salmonberry bird and goose woman: birds, plants and people in Indigenous Peoples' lifeways in northwestern North America. Journal of Ethnobiology 36 (4), 717–745.

Turner, N.J., Clifton, H., 2006. The forest and the seaweed': Gitga'at seaweed, traditional ecological knowledge and community survival. In: Menzies, C.R. (Ed.), Traditional Ecological Knowledge and Natural Resource Management. University of Nebraska Press, Lincoln, pp. 65–86.

Turner, N.J., Cocksedge, W., 2001. Aboriginal use of non-timber forest products in northwestern North America. Journal of Sustainable Forestry 13 (3–4), 31–58.

Turner, N.J., Deur, D., Lepofsky, D., 2013b. Plant management systems of British Columbia's first peoples. In: Turner, N.J., Lepofsky, D. (Eds.)Turner, N.J., Lepofsky, D. (Eds.), Ethnobotany in British Columbia: Plants and People in a Changing World. Special Issue, BC Studies, vol. 179, pp. 107–133.

Turner, N.J., Gregory, R., Brooks, C., Failing, L., Satterfield, T., 2008. From invisibility to transparency: identifying the implications (of invisible losses to First Nations communities). Ecology and Society 13 (2), 7. [online] http://www.ecologyandsociety.org/vol13/iss2/art7/.

Turner, N.J., Hebda, R.J., 2012. Saanich Ethnobotany: Culturally Important Plants of the WSÁNEĆ (Saanich) People of Southern Vancouver Island. Royal BC Museum, Victoria.

Turner, N.J., Peacock, S., 2005. 'Solving the perennial paradox': ethnobotanical evidence for plant resource management on the Northwest Coast. In: Deur, D., Turner, N.J. (Eds.), "Keeping it Living": Traditions of Plant Use and Cultivation on the Northwest Coast of North America. University of Washington Press, Seattle, and UBC Press, Vancouver, pp. 101–150.

Turner, N.J., Robinson, C., Robinson, G., Eaton, B., 2012. 'To feed all the people': Lucille Clifton's Fall feasts for the Gitga'at community of Hartley Bay, British Columbia. In: Quinlan, M., Lepofsky, D. (Eds.), Explorations in Ethnobiology: The Legacy of Amadeo Rea. Society of Ethnobiology, Department of Geography, University of North Texas, Denton, pp. 322–363.

Turner, N.J., Smith, R.Y., Jones, J.T., 2005. 'A fine line between two Nations': ownership patterns for plant resources among Northwest coast indigenous peoples—implications for plant conservation and management. In: Deur, D., Turner, N.J. (Eds.), "Keeping it Living": Traditions of Plant Use and Cultivation on the Northwest Coast of North America. University of Washington Press, Seattle, and UBC Press, Vancouver, pp. 151–180.

Turner, N.J., Thomas (Tl'iishal), J., Carlson, B.F., Ogilvie, R.T., 1983. Ethnobotany of the Nitinaht Indians of Vancouver Island. British Columbia Provincial Museum, Victoria.

Turner, N.J., Turner, K.L., 2008. 'Where our women used to get the food': cumulative effects and loss of ethnobotanical knowledge and practice. Botany 86, 103–115.

United Nations, 2007. Declaration on the Rights of Indigenous Peoples. http://www.un.org/esa/socdev/unpfii/documents/DRIPS_en.pdf.

Waldbusser, G.G., Powell, E.N., Mann, R., 2013. Ecosystem effects of shell aggregations and cycling in coastal waters: an example of Chesapeake Bay oyster reefs. Ecology 94 (4), 895–903.

Weiser, A., Lepofsky, D., 2009. Ancient land use and management of Ebey's Prairie, Whidbey island, Washington. Journal of Ethnobiology 29 (2), 184–212.

Williams, J., 2006. Clam Gardens: Aboriginal Mariculture on Canada's West Coast. New Star Books, Vancouver.

Wilson (Kii7iljus), B.J., Harris, H., 2005. Tllsda Xaaydas K'aaygang.nga: long, long ago Haida ancient stories. In: Fedje, D.W., Mathewes, R.W. (Eds.), Haida Gwaii: Human History and Environment from the Time of Loon to the Time of the Iron People. UBC Press, Vancouver, pp. 117–139.

Woods, D.J., Woods, D., 2005. Ancient Sea Gardens: Mystery of the Pacific Northwest. DVD. Directed by Aaron Szimanski. Aquaculture Pictures, Toronto.

Xanius White, E.A.F., 2006. Heiltsuk Stone Fish Traps: Products of My Ancestor's Labour (M.A. thesis). Simon Fraser University, Burnaby, BC.

Chapter 10

Blurred Lines: What's a Non-native Species in the Anthropocene Ocean?

Isabelle M. Côté
Simon Fraser University, Burnaby, BC, Canada

INTRODUCTION

People have been moving marine species around for centuries, if not millennia. As early as 1245, some 250 years before Columbus voyage to the New World, the soft-shelled clam *Mya arenaria* was transported from its native northwest Atlantic shores to northern Jutland in Denmark (Petersen et al., 1992), making this the oldest recorded introduction of a marine species beyond its natural range. The bivalve is now widespread across Europe (Strasser, 1999). The Vikings were, of course, not the first long-distance seafarers. As far back as 3000 BCE, trading vessels sailed between the Red Sea, Arabian Sea, and Persian Gulf (Boivin and Fuller, 2009), and a short time later, Malayo-Polynesians were undertaking sea journeys of thousands of kilometers, island-hopping from Taiwan to Hawaii (Bellwood, 2007). These early vessels could have carried marine organisms with ballast in their holds or on their wooden hulls into new areas, sometimes instantly altering the trajectory of millions of years of evolution in ways that remain undetected today. The frequency of human-aided biological introductions in the sea has increased exponentially over the past two centuries as trade and travel have become globalized (Ruiz et al., 2000), such that most marine communities now host at least some novel components (Molnar et al., 2008).

Accelerating introductions are now occurring against a background of multiple pressures on the ocean. The effects of overexploitation, eutrophication, pollution, and habitat degradation (Halpern et al., 2008) are increasingly compounded by rising atmospheric carbon dioxide (CO_2), which is increasing ocean temperatures and lowering pH (IPCC, 2013). Several of these pressures are acting directly or indirectly to enhance or depress various aspects of invasion success (Carlton, 2000). For example, overfishing of planktivorous fishes in the Black Sea in the early 1990s resulted in an abundance of zooplankton

Conservation for the Anthropocene Ocean. http://dx.doi.org/10.1016/B978-0-12-805375-1.00010-6

that fueled blooms of comb jelly *Mnemiopsis leidyi*, recently introduced from the western Atlantic (Daskalov et al., 2007). Similarly, hypoxia induced by eutrophication is more tolerable to non-native mud snails in California than to their native counterparts, favoring the establishment of the invaders in stressed estuaries (Byers, 2000). While climate change, in particular warming sea temperature, might often give non-native species a physiological or ecological advantage over native species (e.g., Stachowicz et al., 2002; Sorte et al., 2010a), it is also generating a new means of reorganizing marine communities by shifting native species out of their recent historical bounds (Walther et al., 2009). Marine species track the speed at which temperature is changing, shifting north or south, deeper or shallower as local climate velocity dictates (Pinsky et al., 2013). However, not all species respond in the same way to the speed of climate change (Poloczanska et al., 2013); hence species assemblages—and all of the interspecific interactions they entail—are becoming increasingly dynamic.

Therefore, there are two types of species out of place in the Anthropocene ocean: those stemming from introductions, that is human-aided translocations that have often occurred across geographical barriers and over vast distances; and those stemming from range expansions, which occur naturally as a result of changing environmental conditions (Carlton, 1987; Walther et al., 2009). I will refer to the former as "non-native" species and latter as "range-shifting" species, applying the term "invasive" to those species in either category that exert substantial impact on native biota, economic values, or human health (Davis and Thompson, 2001). In this chapter, I examine the evidence for the extent and impacts of non-native marine species and then compare it to what we know about range-shifting species. I then review the arguments made in the debate about whether we should worry about non-native species, suggest that the debate should shift to a clear definition of what a non-native species is, which can be operationalized for conservation management, and conclude that the decision to manage non-native species must be made in relation to the risk posed by these species compared to that posed by other threats to the marine environment.

SPECIES INTRODUCTIONS: EXTENT AND EVIDENCE FOR IMPACTS IN THE MARINE ENVIRONMENT

Non-native species have been introduced to virtually every reach of the coastal ocean. Molnar et al. (2008) mapped the distribution of 329 non-native species in nearshore environments and found at least one non-native species in 194 of 232 of the world's marine ecoregions. The absence of non-native species in the 38 remaining ecoregions reflected a lack of data rather than true absence (Molnar et al., 2008). It is therefore fair to think that with complete information, we would see non-native marine species in all regions of the ocean (Fig. 10.1).

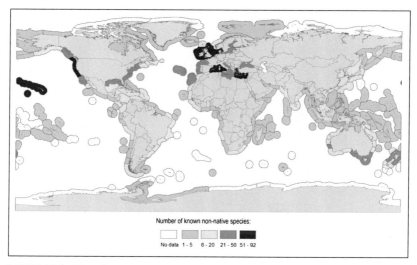

FIGURE 10.1 Map of the number of non-native marine species across coastal ecoregions of the world. Non-native species refer to both introduced and range-shifting species, although the former predominated in this database. The color of ecoregions becomes more intense as the number of non-native species recorded increases. There are no records of non-native marine species in the *white areas. Figure modified from Molnar, J.L., Gamboa, R.L., Revenga, C., Spalding, M.D., 2008. Assessing the global threat of invasive species to marine biodiversity. Frontiers in Ecology and the Environment 6, 485–492.*

There are now, in 2016, nearly seven times more species ($n = 2205$) in the World Register of Marine Species than Molnar et al. mapped in 2008 (Pagad et al., 2016, Fig. 10.2). This tally is still likely to be an underestimate since marine invasions are generally less studied than invasions in other realms (Thomaz et al., 2015), at least some non-native species are currently classed as cryptogenic (i.e., species whose origin is currently unclear; Carlton, 1996), and large numbers of introduced microbes and small invertebrates continue to be overlooked. For example, vertebrates account for more than one-fifth of all non-native species—roughly the same proportion as mollusks (Fig. 10.2), despite the fact that there are twice as many marine mollusks than marine vertebrate species. Côté and Bruno (2015) found a similar pattern on coral reefs. Their near-exhaustive list of non-native taxa recorded from coral reefs included only 57 species, of which 65% were fishes. This bias is almost certainly because fishes are the most easily identifiable and thoroughly monitored taxonomic group on coral reefs.

Non-native species can have a range of effects on their host communities. The most notorious are the deleterious impacts of non-native invasive species. It is beyond the scope of this chapter to review these negative effects in detail; readers are directed to general overviews of these impacts (e.g., Bax et al., 2003; Williams and Smith, 2007; Paolucci et al., 2013; Katsanevakis et al., 2014; Davidson et al., 2015). However, through interspecific interactions

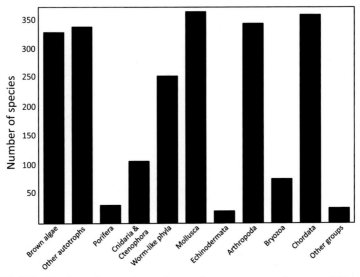

FIGURE 10.2 Numbers of non-native marine species across taxonomic groups. "Other auto-trophs" includes red algae, green algae, and vascular plants. "Worm-like phyla" includes the Annelida, Platyhelminthes, Nematoda, Nemertea, Chaetognatha, and Sipuncula. "Other groups" includes the Brachiopoda, Chaetognatha, Entoprocta, Rotifera, Bacteria, and Fungi. *Data derived from Pagad, S., Hayes, K., Katsanevakis, S., Costello, M.J., 2016. World Register of Introduced Marine Species (WRIMS).* http://www.marinespecies.org/introduced.

such as competition, predation, and facilitation, some marine invasive species have caused major declines in abundance and diversity of native competitors or prey, altered the structure of recipient communities, reduced ecosystem services, and caused economic hardship to people that depend on marine resources (Fig. 10.3).

The key word in the previous sentence is "some." Ecologists know a fair amount of the harmful effects of a handful of non-native marine species—the "poster children" of marine invasions (Fig. 10.3)—and not very much about the rest. Davidson et al. (2015), for example, noted that only 10% of currently known non-native macroalgal species have been the subject of impact studies. I conducted a search on the Web of Science (to 2016) and found that a similar pattern holds for the most common groups of non-native marine animals (i.e., cnidarians and ctenophores, annelids, mollusks, crustaceans, and fishes; Fig. 10.4), but the numbers are even more biased than it appears. For arthropods, for example, more than half (53%) of the studies of non-native species are on just two species: the European green crab *Carcinus maenas* (Fig. 10.3) and the Asian shore crab *Hemigrapsus sanguineus*. Red lionfish *Pterois volitans* (Fig. 10.3) alone accounts for one-third of studies on non-native chordates. Moreover, much of what we do know of negative impacts of marine invasive species has been criticized as stemming from anecdotal observations or unmanipulated studies with low power (Katsanevakis et al., 2014; Davidson et al., 2015).

FIGURE 10.3   Four "poster children" of extremely damaging marine invasions. Two effective space competitors: (A) the "killer alga" *Caulerpa taxifolia* and (B) the golden star tunicate *Botryllus schlosseri*. Two predators implicated in significant native prey declines: (C) the European green crab *Carcinus maenas* and (D) the red lionfish *Pterois volitans*. *Photo credits: (A) National Oceanic and Atmospheric Administration—The Southwest Regional Office of the National Marine Fisheries Service, (B) Norah Brown, (C) and (D) Isabelle M. Côté.*

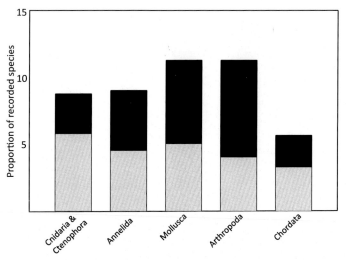

FIGURE 10.4   Proportion of recorded non-native species of major groups of marine animals for which there are studies of ecology, evolution, or behavior (*black bars*) and for which there are studies of ecological impact (*gray bar*). The proportions stem from a search of relevant publications on the Web of Science (to 2016). The total number of non-native species in each group was derived from the World Register of Marine Introduced Species (Pagad et al., 2016).

It is also difficult to pinpoint invaders as the sole culprit of the ultimate detrimental impact: the extinction of native species. Gurevitch and Padilla (2004), for example, found that non-native invasive species directly affected 87 of 737 marine species threatened with extinction and placed on the global IUCN Red List of 2003. Most of the affected species were seabirds. More importantly, all critically endangered species impacted by non-natives (1 mammal and 14 birds) were also affected by other factors such as exploitation or habitat loss, prompting Gurevitch and Padilla to reject the notion that non-native invasive species are a major cause of extinctions, not just in the sea but on land too. Examination of the factors contributing to *actual* marine extinctions suggests that this is correct, at least for now. Some 133 local, regional, or global extinctions of marine populations have been documented so far, and of these, only five have been attributed wholly ($n=3$; a seabird, a mud snail, and a mussel) or in part ($n=2$; a seabird, and a fish) to invasions (Dulvy et al., 2003). The vast majority of marine extinctions across all three geographic scales are caused by exploitation and habitat loss (Fig. 10.5).

All of this does not mean that non-native marine species never have negative impacts. Some of them clearly do. However, some also have positive effects on native populations—for example by providing habitat for the settlement of native species, thereby increasing local abundance and/or diversity (e.g., Bulleri et al., 2006; Gribben et al., 2013; Thyrring et al., 2013)—or on ecosystem function—for example, by altering nutrient cycles to enhance carbon sequestration (e.g., Tait et al., 2015) or providing enhanced coastal protection

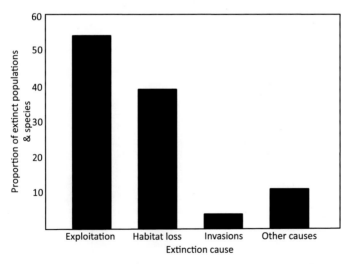

FIGURE 10.5 The probable causes of local, regional, and global extinctions of marine populations and species. "Other causes" include pollution, climate change, and disease. The proportions of all extinctions sum to more than 100% because some extinctions were attributed to more than one cause. *Data from Dulvy, N.K., Sadovy, Y., Reynolds, J.D., 2003. Extinction vulnerability in marine populations. Fish and Fisheries 4, 25–64.*

(e.g., Troost, 2010). Moreover, the same non-native species can have positive effects on some species and negative effects on other species within the same ecological community (e.g., Gribben et al., 2013; Thomsen et al., 2014). The bottom line is that ecological effects, whether positive, negative, or negligible, have not been assessed for the vast majority of non-native marine species (Katsanevakis et al., 2014; Davidson et al., 2015).

SHIFTING RANGES: "NATIVE" INVADERS?

The imprint of anthropogenic climate change on marine life is now pervasive. Marine species have been responding to warming ocean temperatures and lower pH by changing their phenology, abundance, calcification rates, demography, and distribution. In the most comprehensive review to date (1735 observations of 857 species), Poloczanska et al. (2013) found that more than 80% of the observed responses of marine species were consistent with the expected impacts of climate change. Distribution ranges are particularly responsive. Highly mobile or dispersive pelagic organisms, such as phytoplankton, zooplankton, and fishes, show very large range-shift speeds (142–470 km per decade; Poloczanska et al., 2013), although these rates are only half as fast as the average rates of spread of non-native species during colonization (Sorte et al., 2010b). As expected, most shifts are in poleward directions, though some species seek cooler, deeper water instead (Perry et al., 2005), and some show unexpected longitudinal rather than latitudinal shifts that are best explained by local climate velocities (Pinsky et al., 2013; Pinsky, this volume). Most importantly in the context of invasions, the leading (colder) edges of ranges of marine species have been shifting by ~70 km per decade, on average, while the trailing (warmer) edges have been contracting more slowly, by ~15 km per decade (Poloczanska et al., 2013). The result is overall range expansions, but also novel species mixing as lower-latitude species race toward the poles and higher-latitude species are slower to pull their trailing limit away from warming conditions.

The ranges of marine species are thus shifting quickly in the Anthropocene ocean, but range shifts (or "native invaders," sensu Carey et al., 2012) and human-mediated introductions differ in many ways. Ricciardi (2006) and Sorte et al. (2010b) provide complementary overviews of some of the key differences. Here, I focus on three differences that should affect impacts on recipient communities. One important distinction is the spatial scale of species movement. In the case of introductions, species might have traveled very large distances from their native range, crossing continents and oceans on motorized vectors, whereas in range shifts, dispersal is shorter and at the mercy of natural barriers and climate velocity. One consequence of this spatial difference is that introduced species are more likely to be inserted into communities that are phylogenetically dissimilar from their native assemblages. Evolutionarily distinct invaders are expected to have a larger impact on recipient communities because they have left behind the predators, parasites, and competitors that limited their populations at home (Colautti et al., 2004). They can also be more effective predators on naïve prey that have not coevolved with them (Salo et al., 2007; Sih et al., 2010). On the other hand, shifting species might be especially well equipped to deal with new enemies that are closely related to those from their original ranges, so biotic resistance might be expected to be stronger toward non-native than range-shifting species (Levine et al., 2004).

A second key difference is in the types of species involved in movement. Most species in any given marine assemblage will respond to warming temperatures, albeit with some species-specific differences. In contrast, human-assisted invaders are an extremely nonrandom segment of biodiversity (Ricciardi, 2006). They typically have traits associated with specific human interests, for example species that can be cultured (e.g., oysters; Ruesink et al., 2005) or kept in aquaria (Padilla and Williams, 2004; Semmens et al., 2004), or traits that allow them to survive inadvertent long-distance journeys during unintentional human-mediated transport. If these traits are also associated with the ability to survive in novel habitats, compete or prey upon native species, and spread, then the impact of introduced species might be expected to be greater than that of range-shifting species.

Finally, while range shifts occur more or less irrespective of habitat quality, introductions happen particularly frequently in areas that are likely to be disturbed. Shipping and aquaculture are the two main vectors of marine introductions (Bax et al., 2003; Molnar et al., 2008). Shipping, in particular, is associated with the development of extensive infrastructure in centers of high human population density, with accompanying high levels of pollution, eutrophication, and (over)fishing. Ecosystems stressed by multiple pressures might be more likely to be invaded, either because propagule pressure is chronically high in these areas or because native species diversity, which could provide biotic resistance (Levine et al., 2004), has been eroded (Occhipinti-Ambrogi and Savini, 2003; Wonham and Carlton, 2005; Williams and Smith, 2007). Hence, non-native species might have an easier time becoming established than range-shifting species, though the impact of invaders might be expected to be lower in more disturbed habitats where stressful conditions overtake biotic interactions as the dominant drivers of community structure (Tamburello et al., 2015).

It is difficult to test these predictions because we still know relatively little about the impact of range-shifted marine species. Sorte et al. (2010b) found evidence of community- or ecosystem-level effects for 8 of 129 shifting species. There was no information for the remaining 121 species. A review of publications since 2008 (the last year of Sorte et al. literature search) added eight more papers and five new systems (Table 10.1). The shifting species competed with, parasitized, grazed, added nutrients to, or preyed upon various components of their new habitats. The effects in all but one case study were negative, but given how few studies exist, the conclusion that these effects are comparable in frequency and magnitude to those of introduced species is premature (Sorte et al., 2010b). Impact studies were dubbed the next frontier of

TABLE 10.1 Range-Shifting Marine Species With Their Observed Ecological Effects

Species	Group	Ecological Effect	Mechanism	Sources
Perkinsus marinus	Protist	↑ oyster mortality	Parasitism	Ford and Smolowitz (2007)
Amphistegina spp.	Protist	↓ foraminifera richness change in sediment composition	Competition	Mouanga and Langer (2014)
Fucus serratus	Alga	↓ cover of other seaweeds	Competition	Arrontes (2002)
Lamineria ochroleuca	Alga	↑ primary productivity ↑ gastropod abundance	Competition	Smale et al. (2015)
Patella ulyssiponensis	Gastropod	↓ seaweed cover	Herbivory	O'Connor and Crowe (2005)
"Lottia" depicta	Gastropod	↓ seagrass growth and biomass	Herbivory	Jorgensen et al. (2007) Zimmerman et al. (1996)
Petrolisthes armatus	Crab	↓ microalgal biomass	Herbivory	Hollebone and Hay (2007)
Jasus lalandii	Lobster	↓ grazers ↑ macroalgal cover ↓ encrusting algae	Predation	Haley et al. (2011) Blamey and Branch (2012)
Centrostephanus rodgersii	Urchin	↓ seaweed biomass	Herbivory	Ling (2008)
		↓ abalone abundance	Competition	Strain et al. (2013)
		↑ barren depth range	Herbivory	Perkins et al. (2015)

Continued

TABLE 10.1 Range-Shifting Marine Species With Their Observed Ecological Effects—cont'd

Species	Group	Ecological Effect	Mechanism	Sources
Dosidicus gigas	Squid	↓ hake (fish) abundance	Predation	Zeidberg and Robison (2007)
Siganus luridus and *Siganus rivulatus*	Fish	↓ benthic biomass, species richness, and algal canopy	Herbivory	Vergés et al. (2014)
Lutjanus griseus and *Lutjanus synagris*	Fish	No effect on growth and abundance of pinfish	–	Gericke et al. (2014)
Larus delawarensis	Bird	↑ local nitrate ↑ local phosphorus ↑ cover of non-native plants	Nutrient	Hogg and Morton (1983)

Case studies with sources prior to 2009 were derived from Sorte et al. (2010b).

climate change research in 2008, on the eve of an important American election (Kintisch, 2008). It seems that nearly a decade later, that frontier is still on the horizon.

THE DEBATE OVER NON-NATIVE SPECIES: GOOD, BAD, OR IS THIS NOW THE WRONG QUESTION?

Since as early as the late 19th century, opinions about non-native species have been deeply polarized (Simberloff, 2003). Elton's (1958) seminal book on invasion ecology eventually catalyzed a new field of applied ecological endeavor that has focused largely on measuring and predicting the threats posed by invaders. This focus, as well as some of the evocative terminology used (e.g., "invasion," "colonization," "resistance," "eradication," "meltdown"), has led to a perceived generalization that native species are "good" and non-native species are "bad" (e.g., Sagoff, 1999; Slobodkin, 2001; Brown and Sax, 2004). This position, attributed to those who favor attempts to control non-native species, has been variously labeled biased, alarmist, and even racist, nativist, and xenophobic (reviewed by Simberloff, 2003).

The last decade or so has seen multiple attempts to paint non-native species with more discriminating hues. Some authors have highlighted the potential conservation value of non-native species in terms of providing habitat or food for rare species, serving as functional substitutes for extinct taxa, delivering desirable ecosystem functions such as pollination and biocontrol, and giving rise to novel, beneficial evolutionary lineages and interactions (e.g., Schlaepfer et al., 2011). Others, echoing Elton (1958), have called for shifting the concern from "Where does a species originate?" to "What is its impact?" (e.g., Davis et al., 2011; Thompson, 2014).

It is likely that the "good versus bad" dichotomy—however unproductive—will linger because our understanding of the effect of non-native species is based on studies of a very small fraction of these species. For many reasons (including funding), ecologists tend to study non-native species that have visible, negative effects, and the benefits of non-native species are largely underreported (Bonnano, 2016). Both trends cause a potentially biased perception. We also worry about the invasion debt—the fact that the impacts of non-native species might remain negligible for a long time, until their populations explode unpredictably and their effects are felt (Simberloff et al., 2013; Simberloff, 2014). These uncertainties make the notions that deliberate introductions of beneficial, non-native species should sometimes be done for conservation (Schlaepfer et al., 2011), that we should embrace novel ecosystems (Davis, 2000) or to "learn to love" non-native species if they are not a medical or economic problem (Sagoff, 1999), seem imprudent at best and wreckless at worst.

The clear imprint of climate change, especially in the ocean, now calls for a shift in the debate to what a non-native species is. Species transported by humans to locations far beyond their native ranges are clearly not native to their new environment. However, the distribution ranges of most species have changed over time. As continents drifted, dispersal barriers disappeared, currents altered, and the climate changed, species boundaries have expanded, shrunk, or relocated. On both evolutionary and ecological timescales, extraordinary dispersal events—think of organisms floating over long distances on rafts of macroalgae—have been relatively rare, but still frequent enough to account for many natural range expansions in marine species (Thiel and Haye, 2006). The range shifts currently observed in the Anthropocene ocean could therefore be seen as largely natural phenomena. So should a species whose range has shifted northward because of warming temperatures be considered a non-native species in its new northern home? Is it more or less non-native than a species introduced to the same locality by human means? And what if this species was introduced centuries ago to a more southern area and has only recently expanded its range in response to ocean warming? Does that make it a native or a non-native species in the north, or something else altogether (Fig. 10.6A)?

There is an urgent need to provide some clarity to these thorny questions, which are not just of academic interest but have important management consequences too. The definitions in the ecological literature of what constitute

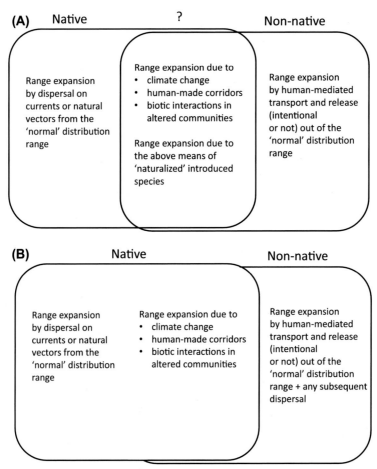

FIGURE 10.6 Conceptual diagrams depicting (A) the current ambiguity in defining and identifying the status of marine species that are out of their natural ("normal") range; and (B) a resolution of the ambiguity, as proposed by Gilroy et al. (2016), achieved by defining non-native status based on direct human agency.

native and non-native species are ambiguous (Usher, 2000; Crees and Turvey, 2015; Gilroy et al., 2016). This scientific vagueness has crept into national and international legislation, potentially leading to diametrically different policy and management action for range-shifting species in different countries (Gilroy et al., 2016).

The role played by humans in the arrival of a species in a new location should be the most critical aspect of the definition of non-nativeness (Gilroy et al., 2016). Humans can either play no role (e.g., when dispersal is by means of currents), a direct role (e.g., by transporting and/or releasing a species out of its usual range, either intentionally or inadvertently), or an indirect role when species move in response to human-driven environmental change, e.g., climate change, biotic interactions in altered communities, or corridors created between previously unconnected water bodies (Hulme, 2015). Non-native status should be reserved exclusively for species introduced via direct human transport pathways. This also applies to populations established through direct human transport that have subsequently spread and dispersed into new areas. In contrast, all species moving out of their native range via natural dispersal under indirect human influence, including those driven by habitat change and other infrastructure developments, should be considered "native" in their newly colonized ranges (Fig. 10.6B). This simple definition avoids any weakening of protection for potentially threatened species that depend on expanding their ranges to persist in the face of environmental change. At the same time, problematic shifting-range species can still be controlled to mitigate their undesired impacts, in the same way as "pest" species are, without having to invoke and defend a "non-native" label.

This simple definition facilitates policy articulation and implementation, but a few important gray areas still remain. For example, natural expansions from the current natural ranges of species that have been introduced elsewhere by human agency might challenge policy makers and managers. Moreover, populations that are reintroduced after being extirpated should technically be considered non-native, although in practice, they would be seen as native (i.e., the "restored natives" of Crees and Turvey, 2015). The same might apply to populations that are translocated out of their current range in assisted migration efforts, although in these cases, the specter of invasive introductions will loom large (Hulme, 2015). Cryptogenic species pose the greatest challenge to implementation of the framework. Such species are sometimes numerous. In the San Francisco Bay, for example, 37% of the 336 species deemed to be definitively or potentially non-native were cryptogenic (Carlton, 1996). The effort required to resolve the ambiguity of the origin of these species would be enormous. The precautionary principle suggests that we should consider them as non-native, at the risk of potentially targeting a native species for control or eradication.

In conclusion, species out of place will prevail in the Anthropocene ocean. From a management perspective, little can be done to limit range shifts. While many countries have made some strides in legislating and implementing measures to curb the rate of new marine species introductions (e.g., Firestone and Corbett, 2005; Sambrook et al., 2014), we are a long way from meeting the global target of having "invasive alien species and pathways identified and prioritized, priority species controlled and eradicated, and measures in place to manage pathways to prevent their introduction and establishment" by 2020 (Aichi target 9, CBD, 2014). However, in the cold light of day, we have to ask whether tackling non-native species is the best use of our limited conservation resources. Even if marine non-native species are as bad as we think they could be, or will be in the near future, the threat they pose pales in comparison to the impacts of rapidly accelerating global climate change. Therefore, it seems prudent to weigh the risks imposed by non-native species against those of other environmental stressors when deciding whether or not to take action against a given non-native species. In some or perhaps many cases, diverting resources needed to substantially reduce CO_2 emissions or fishing effort and allocating them toward marine non-native control or eradication might be analogous to rearranging deck chairs on a sinking ship.

ACKNOWLEDGMENTS

Thank you to Fraser Januchowski-Hartley for creating Fig. 10.1, to Beth Sanderson and Katie Barnas for constructive criticisms, and to the anonymous reviewer who urged me to go out on a limb and suggested the provocative last sentences of this chapter.

REFERENCES

Arrontes, J., 2002. Mechanisms of range expansion in the intertidal brown alga *Fucus serratus* in northern Spain. Marine Biology 141, 1059–1067.

Bax, N., Williamson, A., Aguero, M., Gonzalez, E., Geeve, W., 2003. Marine invasive alien species: a threat to global biodiversity. Marine Policy 27, 313–323.

Bellwood, P., 2007. Prehistory of the Indo-Malaysian Archipelago, third ed. ANU E Press, Canberra, Australia. 384 pp.

Blamey, L.K., Branch, G.M., 2012. Regime shift of a kelp-forest benthic community induced by an 'invasion' of the rock lobster *Jasus lalandii*. Journal of Experimental Marine Biology and Ecology 420–421, 33–47.

Boivin, N., Fuller, D.Q., 2009. Shells middens, ships and seeds: exploring coastal subsistence, maritime trade and the dispersal of domesticates in and around the ancient Arabian peninsula. Journal of World Prehistory 22, 113–180.

Bonanno, G., 2016. Alien species: to remove or not to remove? That is the question. Environmental Science & Policy 59, 67–73.

Brown, J.H., Sax, D.F., 2004. An essay on some topics concerning invasive species. Austral Ecology 29, 530–536.

Bulleri, F., Airoldi, L., Branca, G.M., Abbiati, M., 2006. Positive effects of the introduced green alga, *Codium fragile* ssp. *tomentosoides*, on recruitment and survival of mussels. Marine Biology 148, 1213–1220.

Byers, J.B., 2000. Differential susceptibility to hypoxia aids estuarine invasion. Marine Ecology Progress Series 203, 123–132.

Carey, M.P., Sanderson, B.L., Barnas, K.A., Olden, J.D., 2012. Native invaders: emerging challenges for conservation. Frontiers in Ecology and the Environment 10, 373–381.

Carlton, J.T., 1987. Patterns of transoceanic marine biological invasions in the Pacific Ocean. Bulletin of Marine Science 41, 452–465.

Carlton, J.T., 1996. Biological invasions and cryptogenic species. Ecology 77, 1653–1655.

Carlton, J.T., 2000. Global change and biological invasions in the oceans. In: Mooney, H.A., Hobbs, R.J. (Eds.), Invasive Species in a Changing World. Island Press, Washington, D.C, pp. 31–53.

Colautti, R.I., Ricciardi, A., Grigorovich, I.A., MacIsaac, H.J., 2004. Is invasion success explained by the enemy release hypothesis? Ecology Letters 7, 721–733.

Convention on Biological Diversity, 2014. Global Biodiversity Outlook 4. Secretariat of the Convention on Biological Diversity, Montréal.

Côté, I.M., Bruno, J.F., 2015. Impacts of invasive species on coral reef fishes. In: Mora, C. (Ed.), Ecology of Fishes on Coral Reefs. Cambridge University Press, pp. 154–165.

Crees, J.J., Turvey, S.T., 2015. What constitutes a 'native' species? Insights from the quaternary faunal record. Biological Conservation 186, 143–148.

Daskalov, G.M., Grishin, A.N., Rodionov, S., Mihneva, V., 2007. Trophic cascades triggered by overfishing reveal possible mechanisms of ecosystem regime shifts. Proceedings of the National Academy of Sciences of the United States of America 104, 10518–10523.

Davidson, A.D., Campbell, M.L., Hewitt, C.L., Schaffelke, B., 2015. Assessing the impacts of non-indigenous marine macroalgae: an update of current knowledge. Botanica Marina 58, 55–79.

Davis, M.A., Thompson, K., 2001. Invasion terminology: should ecologists define their terms differently than others? No, not if we want to be of any help. Bulletin of the Ecological Society of America 82, 206.

Davis, M.A., Chew, M.K., Hobbs, R.J., Lugo, A.E., Ewel, J.J., Vermeij, G.J., Brown, J.H., Rosenzweig, M.L., Gardener, M.R., Carroll, S.P., Thompson, K., Pickett, S.T.A., Stromberg, J.C., Del Tredici, P., Suding, K.N., Ehrenfeld, J.G., Grime, J.P., Mascaro, J., Briggs, J.C., 2011. Don't judge species on their origins. Nature 474, 153–154.

Davis, M.A., 2000. "Restoration" – a misnomer? Science 287, 1203.

Dulvy, N.K., Sadovy, Y., Reynolds, J.D., 2003. Extinction vulnerability in marine populations. Fish and Fisheries 4, 25–64.

Elton, C.S., 1958. The Ecology of Invasions by Animals and Plants. Methuen and Co., Ltd., London, UK.

Firestone, J., Corbett, J.J., 2005. Coastal and port environments: international legal and policy responses to reduce ballast water introductions of potentially invasive species. Ocean Development and International Law 36, 291–316.

Ford, S.E., Smolowitz, R., 2007. Infection dynamics of an oyster parasite in its newly expanded range. Marine Biology 151, 119–133.

Gericke, R.L., Heck Jr., K.L., Fodrie, F.J., 2014. Interactions between northern-shifting tropical species and native species in the northern Gulf of Mexico. Estuaries and Coasts 37, 952–961.

Gilroy, J.J., Avery, J.D., Lockwood, J.L., 2016. Seeking international agreement on what it means to be "native". Conservation Letters. http://dx.doi.org/10.1111/conl.12246.

Gribben, P.E., Byers, J.E., Wright, J.T., Glasby, T.M., 2013. Positive versus negative effects of an invasive ecosystem engineer on different components of a marine ecosystem. Oikos 122, 816–824.

Gurevitch, J., Padilla, D.K., 2004. Are invasive species a major cause of extinction? Trends in Ecology and Evolution 19, 470–474.

Haley, C., Blamey, L.K., Atkinson, L.J., Branch, G.M., 2011. Dietary change of the rock lobster *Jasus lalandii* after an 'invasive' geographic shift: effects of size, density and food availability. Estuarine and Coastal Shelf Science 93, 160–170.

Halpern, B.S., Walbridge, S., Selkoe, K.A., Kappel, C.V., Micheli, F., D'Agrosa, C., Bruno, J.F., Casey, K.S., Ebert, C., Fox, H.E., Fujita, R., Heinemann, D., Lenihan, H.S., Madin, E.M.P., Perry, M.T., Selig, E.R., Spalding, M., Steneck, R., Watson, R., 2008. A global map of human impact on marine ecosystems. Science 319, 948–952.

Hogg, E.H., Morton, J.K., 1983. The effects of nesting gulls on the vegetation and soil of islands in the Great Lakes. Canadian Journal of Botany 61, 3240–3254.

Hollebone, A.L., Hay, M.E., 2007. Population dynamics of the non-native crab *Petrolisthes armatus* invading the South Atlantic Bight at densities of thousands m^{-2}. Marine Ecology Progress Series 336, 211–223.

Hulme, P.E., 2015. Invasion pathways at a crossroad: policy and research challenges for managing alien species introductions. Journal of Applied Ecology 52, 1418–1424.

IPCC, 2013. Climate change 2013: the physical science basis. In: Stocker, T.F., Qin, D., Plattner, G.-K., Tignor, M., Allen, S.K., Boschung, J., Nauels, A., Xia, Y., Bex, V., Midgley, P.M. (Eds.), Contribution of Working Group I to the Fifth Assessment Report of the Intergovernmental Panel on Climate Change. Cambridge University Press, Cambridge, United Kingdom and New York, NY, USA. 1535 pp.

Jorgensen, P., Ibarra-Obando, S.E., Carriquiry, J.D., 2007. Top-down and bottom-up stabilizing mechanisms in eelgrass meadows differentially affected by coastal upwelling. Marine Ecology Progress Series 333, 81–93.

Katsanevakis, S., Wallentinus, I., Zenetos, A., Leppakoski, E., Çinar, M.E., Ozturk, B., Grabowski, M., Golani, D., Cardoso, A.C., 2014. Impacts of invasive alien marine species on ecosystem services and biodiversity: a pan-European review. Aquatic Invasions 9, 391–432.

Kintisch, E., 2008. Impacts research seen as next climate frontier. Science 322, 182–183.

Levine, J.M., Adler, P.B., Yelenik, S.G., 2004. A meta-analysis of biotic resistance to exotic plant invasions. Ecology Letters 7, 975–989.

Ling, S.D., 2008. Range expansion of a habitat-modifying species leads to loss of taxonomic diversity: a new and impoverished reef state. Oecologia 156, 883–894.

Molnar, J.L., Gamboa, R.L., Revenga, C., Spalding, M.D., 2008. Assessing the global threat of invasive species to marine biodiversity. Frontiers in Ecology and the Environment 6, 485–492.

Mouanga, G.H., Langer, M.R., 2014. At the front of expanding ranges: shifting community structures at amphisteginid species range margins in the Mediterranean Sea. Nues Jahrbuch für Geologie und Palaontologie-Abhandlungen 271, 141–150.

O'Connor, N.E., Crowe, T.P., 2005. Biodiversity loss and ecosystem functioning: distinguishing between number and identity of species. Ecology 86, 1783–1796.

Occhipinti-Ambrogi, A., Savini, D., 2003. Biological invasions as a component of global change in stressed marine ecosystems. Marine Pollution Bulletin 46, 542–551.

Padilla, D.K., Williams, S.L., 2004. Beyond ballast water: aquarium and ornamental trades as sources of invasive species in aquatic ecosystems. Frontiers in Ecology and the Environment 2, 131–138.

Pagad, S., Hayes, K., Katsanevakis, S., Costello, M.J., 2016. World Register of Introduced Marine Species (WRIMS). http://www.marinespecies.org/introduced.

Paolucci, E.M., MacIsaac, H.J., Ricciardi, A., 2013. Origin matters: alien consumers inflict greater damage on prey populations than do native consumers. Diversity and Distributions 19, 988–995.

Perkins, N.R., Hill, N.A., Foster, S.D., Barrett, N.S., 2015. Altered niche of an ecologically significant urchin species, *Centrostephanus rodgersii*, in its extended range revealed using an Autonomous Underwater Vehicle. Estuarine and Coastal Shelf Science 155, 56–65.

Perry, A.L., Low, P.J., Ellis, J.R., Reynolds, J.D., 2005. Climate change and distribution shifts in marine fishes. Science 308, 1912–1915.

Petersen, K.S., Rasmussen, K.L., Heinemeier, J., Rud, N., 1992. Clams before Columbus? Nature 359, 679.

Pinsky, M.L., Worm, B., Fogarty, M.J., Sarmiento, J.L., Levin, S.A., 2013. Marine taxa track local climate velocities. Science 341, 1239–1242.

Poloczanska, E.S., Brown, C.J., Sydeman, W.J., Kiessling, W., Schoeman, D.S., Moore, P.J., Brander, K., Bruno, J.F., Buckley, L.B., Burrows, M.T., Duarte, C.M., Halpern, B.S., Holding, J., Kappel, C.V., O'Connor, M.I., Pandolfi, J.M., Parmesan, C., Schwing, F., Thompson, S.A., Richardson, A.J., 2013. Global imprint of climate change on marine life. Nature Climate Change 3, 919–925.

Ricciardi, A., 2006. Are modern biological invasions an unprecedented form of global change? Conservation Biology 21, 329–336.

Ruesink, J.L., Lenihan, H.S., Trimble, A.S., Heiman, K.W., Micheli, F., Byers, J.E., Kay, M.C., 2005. Introduction of non-native oysters: ecosystem effects and restoration implications. Annual Review of Ecology and Systematics 36, 643–689.

Ruiz, G.M., Fofonoff, P.W., Carlton, J.T., Wonham, M.J., Hines, A.H., 2000. Invasion of coastal marine communities in North America: apparent patterns, processes, and biases. Annual Review of Ecology and Systematics 31, 481–531.

Sagoff, M., 1999. What wrong with exotic species? Reports of the Institute of Philosophy and Public Policy 19, 16–23.

Salo, P., Korpimäki, E., Banks, P.B., Nordström, M., Dickman, C.R., 2007. Alien predators are more dangerous than native predators to prey populations. Proceedings of the Royal Society London B: Biological Sciences 274, 1237–1243.

Sambrook, K., Holt, R.H.F., Sharp, R., Griffith, K., Roche, R.C., Newstead, R.G., Wyn, G., Jenkins, S.R., 2014. Capacity, capability and cross-border challenges associated with marine eradication programmes in Europe: the attempted eradication of an invasive non-native ascidian, *Didemnum vexillum* in Wales, United Kingdom. Marine Policy 48, 51–58.

Schlaepfer, M.A., Sax, D.F., Olden, J.D., 2011. The potential conservation value of non-native species. Conservation Biology 25, 428–437.

Semmens, B.X., Buhle, E.R., Salomon, A.K., Pattengill-Semmens, C.V., 2004. A hotspot of non-native marine fishes: evidence for the aquarium trade as an invasion pathway. Marine Ecology Progress Series 266, 239–244.

Sih, A., Bolnick, D.I., Luttbeg, B., Orrock, J.L., Peacor, S.D., Pintor, L.M., Preisser, E., Rehage, J.S., Vonesh, J.R., 2010. Predator–prey naïveté, antipredator behavior, and the ecology of predator invasions. Oikos 119, 610–621.

Simberloff, D., Martin, J.L., Genovesi, P., Maris, V., Wardle, D.A., Aronson, J., Courchamp, F., Galil, B., García-Berthou, E., Pascal, M., Pyšek, P., Sousa, R., Tabacchi, E., Vilà, M., 2013. Impacts of biological invasions: what's what and the way forward. Trends in Ecology & Evolution 26, 8–66.

Simberloff, D., 2003. Confronting introduced species: a form of xenophobia? Biological Invasions 5, 179–192.

Simberloff, D., 2014. Biological invasions: what's worth fighting and what can be won? Ecological Engineering 65, 112–121.

Slobodkin, L.B., 2001. The good, the bad and the reified. Evolutionary Ecology Research 3, 1–13.

Smale, D.A., Wernberg, T., Yunnie, A.L.E., Vance, T., 2015. The rise of *Laminaria ochroleuca* in the Western English Channel (UK) and comparisons with its competitor and assemblage dominant *Laminaria hyperborean*. Marine Ecology 36, 1033–1044.

Sorte, C.J.B., Williams, S.L., Zerebecki, R.A., 2010a. Ocean warming increases threat of invasive species in a marine fouling community. Ecology 91, 2198–2204.

Sorte, C.J.B., Williams, S.L., Carlton, J.T., 2010b. Marine range shifts and species introductions: comparative spread rates and community impacts. Global Ecology and Biogeography 19, 303–316.

Stachowicz, J.J., Terwin, J.R., Whitlatch, R.B., Osman, R.W., 2002. Linking climate change and biological invasions: ocean warming facilitates nonindigenous species invasions. Proceedings of the National Academy of Sciences of the United States of America 99, 15497–15500.

Strain, E.M.A., Johnson, C.R., Thomson, R.J., 2013. Effects of a range-expanding sea urchin on behaviour of commercially fished abalone. PLoS One 8, e73477.

Strasser, M., 1999. *Mya arenaria* – an ancient invader of the North Sea coast. Helgoländer Meeresuntersuchungen 52, 309–324.

Tait, L.W., South, M.W., Lilley, S.A., Thomsen, M.S., Schiel, D.R., 2015. Assemblage and understory carbon production of native and invasive canopy-forming macroalgae. Journal of Experimental Marine Biology and Ecology 469, 10–17.

Tamburello, L., Maggi, E., Benedetti-Cecchi, L., Bellistri, G., Rattray, A.J., Ravaglioli, C., Rindi, L., Roberts, J., Bulleri, F., 2015. Variation in the impact of non-native seaweeds along gradients of habitat degradation: a meta-analysis and an experimental test. Oikos 124, 1121–1131.

Thiel, M., Haye, P.A., 2006. The ecology of rafting in the marine environment. III. Biogeographical and evolutionary consequences. Oceanography and Marine Biology Annual Review 44, 323–429.

Thomaz, S.M., Kovalenko, K.E., Havel, J.E., Kats, L.B., 2015. Aquatic invasive species: general trends in the literature and introduction to the special issue. Hydrobiologia 746, 1–12.

Thompson, K., 2014. Where Do Camels Belong? The Story and Science of Invasive Species. Profile Books, London, UK. 272 pp.

Thomsen, M.S., Byers, J.E., Schiel, D.R., Bruno, J.F., Olden, J.D., Wernberg, T., Silliman, B.R., 2014. Impacts of marine invaders on biodiversity depend on trophic position and functional similarity. Marine Ecology Progress Series 495, 39–47.

Thyrring, J., Thomsen, M.S., Wernberg, T., 2013. Large-scale facilitation of a sessile community by an invasive habitat-forming snail. Helgoland Marine Research 67, 789–794.

Troost, K., 2010. Causes and effects of a highly successful marine invasion: case-study of the introduced Pacific oyster *Crassostrea gigas* in continental NW European estuaries. Journal of Sea Research 64, 145–165.

Usher, M.B., 2000. The nativeness and non-nativeness of species. Watsonia 23, 323–326.

Vergés, A., Tomas, F., Cebrian, E., Ballesteros, E., Kizilkaya, A., Dendrinos, P., Karamanlidis, A.A., Spiegel, S., Sala, E., 2014. Tropical rabbitfish and the deforestation of a warming temperate sea. Journal of Ecology 102, 1518–1527.

Walther, G.R., Roques, A., Hulme, P.E., Sykes, M.T., Pysek, P., Kuhn, I., Zobel, M., Bacher, S., Botta-Dukat, Z., Bugmann, H., Czucz, B., Dauber, J., Hickler, T., Jarosik, V., Kenis, M., Klotz, S., Minchin, D., Moora, M., Nentwig, W., Ott, J., Panov, V.E., Reineking, B., Robinet, C., Semenchenko, V., Solarz, W., Thuiller, W., Vila, M., Vohland, K., Settele, J., 2009. Alien species in a warmer world: risks and opportunities. Trends in Ecology & Evolution 24, 686–693.

Williams, S.L., Smith, J.E., 2007. A global review of the distribution, taxonomy, and impacts of introduced seaweeds. Annual Review of Ecology and Systematics 38, 327–359.

Wonham, M.J., Carlton, J.T., 2005. Trends in marine biological invasions at local and regional scales: the Northeast Pacific Ocean as a model system. Biological Invasions 7, 369–392.

Zeidberg, L.D., Robison, B.H., 2007. Invasive range expansion by the Humboldt squid, *Dosidicus gigas*, in the eastern North Pacific. Proceedings of the National Academy of Sciences of the United States of America 104, 12948–12950.

Zimmerman, R.C., Kohrs, D.G., Alberte, R.S., 1996. Top-down impact through a bottom-up mechanism: the effect of limpet grazing on growth, productivity and carbon allocation of *Zostera marina* L. (eelgrass). Oecologia 107, 560–567.

Chapter 11

Can Ecosystem Services Make Conservation Normal and Commonplace?

Kai M.A. Chan, Paige Olmsted, Nathan Bennett, Sarah C. Klain, Elizabeth A. Williams
University of British Columbia, Vancouver, BC, Canada

CONSERVATION AS NORMAL AND COMMONPLACE

What would constitute success for the conservation movement? Meeting the Convention on Biological Diversity Aichi targets? That is, protecting 17% of all terrestrial and inland water areas, and 10% of coastal and marine areas by 2020? One might argue that such targets, despite their apparent specificity, would constitute success only given many caveats and conditions (e.g., that the protected areas are representative sites connected meaningfully in networks, that they are not already too degraded and are strongly protected from ongoing stressors). Perhaps such targets are neither necessary nor sufficient for conservation success.

If, on the other hand, protecting biodiversity were both normal and commonplace, many conservationists might agree that conservation success was virtually assured. In this chapter, we argue that achieving *normality*—that is, widespread practice with some kind of moral tinge, even if implicit—is a necessary criterion for success. Currently, economic activity is prioritized as somehow more essential to human well-being than the ecosystems on which they rely, artificially separating economic activity and human well-being from the natural environment (Levin, 1996; Rees, 1998; Daly, 2005). As long as protecting biodiversity is seen as a luxury or elitist endeavor, and most economic activity can proceed apace without actively accounting for and mitigating impacts on biodiversity, ecosystems will continue to be buffeted by massive, unprecedented, and growing pressures. The cumulative impacts of such pressures can only be large. We are, after all, in the Anthropocene (Steffen et al., 2007; Dirzo et al., 2014).

Conservation must account for and enlist the billions of people who regularly consume products made from raw materials that are often extracted from ecosystems and the Earth's crust thousands of kilometers away, then processed into components and assembled in other distant locations. In the context of this global economy, we see conservation as no less than the internalization of externalities throughout global supply chains. To succeed in the Anthropocene, conservation must go far beyond protected areas and endangered species in places we know; it will require the proactive mitigation of impacts on biodiversity and ecosystem processes in all manner of economic activity, anywhere (Ehrlich and Pringle, 2008). By our view, it must also attend to the human consequences of these ecological effects and to issues of justice via the production and distribution of goods and ills and the processes involved (Schlosberg, 2004; Chan and Satterfield, 2013). Put this way, success seems a distant possibility.

Normalizing this kind of conservation is a bold aspiration and not yet a goal that unites the conservation movement. Although conservationists might generally agree with such an objective, most are preoccupied with protecting particular cherished species and places from the negative consequences of economic activity. We must fight the fires. But the global aggregation of such local "fire-fighting" does not equate to charting an economic transformation that prevents such impacts and neither does it create a society that takes responsibility for them. Fire-fighting is not fire-proofing.

In this chapter, we consider what conservation fire-proofing or normalization might entail. We first consider the rise of ecosystem services as a field, and the hope that it might make conservation commonplace, amidst concerns that embracing ecosystem services might undermine. We then consider what it might mean to normalize conservation, the extent to which this lofty goal is even an active project within the conservation movement, and the role of ecosystem services in these efforts to date. Although we conclude that ecosystem services have played at best a marginal role in norm change thus far, we argue that it presents two kinds of unheralded promises to contribute to this important project, via broadly shared values of justice and equity, and the identity-forming role that cultural ecosystem services have historically played. Highlighting relevant emerging institutions for ecosystem services, we propose an agenda for scaling these up and including a broad base of corporations and citizens in doing so, in part via a new initiative called CoSphere (a community of small-planet heroes). We close by considering the kinds of governance challenges that any such initiative must address to normalize conservation.

ECOSYSTEM SERVICES: THE GREAT GREEN HOPE?

The rapid rise in popularity of the ecosystem services concept stems partly from the hope that it might normalize conservation. The argument can be phrased as follows: given the underpinning role of biodiversity in the production of benefits that are directly connected to people's daily lives, ecosystem services might embed conservation activities in a wide range of decisions where biodiversity per se lacked traction (Daily, 1997; Balvanera et al., 2001).

Indeed, the Economist's first major coverage of ecosystem services embodied such logic (Economist, 2005): if people could profit from ecosystem-service protection, well, that was the kind of concept that could rescue environmentalism (from the "death" of the movement that had been declared months earlier; Shellenberger and Nordhaus, 2004). Although this media attention followed closely after the publication of the Millennium Ecosystem Assessment (2005), the Economist articles primarily covered corporate action to pay for conservation that mitigated risks to future business. The focus on reducing risks and expanding business opportunities appealed most to corporate actors (World Resources Institute (WRI) et al., 2008).

Meanwhile, in government circles, ecosystem service approaches appeared to gain traction based on the logic that explicit valuation of ecosystem services would enable cost–benefit analyses to assist policy making and provide other processes to "mainstream" conservation in decision making (TEEB, 2009; Daily et al., 2011).

Similarly, nongovernmental organizations saw great potential for the ecosystem services concept to inspire new members and donors to conservation causes. Thus, conservation nongovernmental organizations (NGOs) have been among the greatest popularizers of the concept, including through planning (Chan et al., 2006, 2011b), on-ground programs (Goldman-Benner et al., 2012), and also monetary valuations of nature's benefits to people (Naidoo and Ricketts, 2006; Turner et al., 2007; Naidoo et al., 2008).

With ecosystem services, conservation was no longer just for those who cared about the protection of rare species or actively sought to expand protected areas. The ecosystem services concept created the link to help people understand that anyone who wants to drink clean water or breathe clean air should care about environmental protection, both in remote areas and also close to home.

EXPANDING THE TENT OR WELCOMING THE WOLF?

The concept of ecosystem services is at the heart of the controversy over "new conservation." Proponents of "new conservation" argue that old methods were failing, requiring the experimentation with and adoption of approaches that speak to the masses (Marris, 2011; Kareiva et al., 2012; Kareiva, 2014; Marris, 2014). Whereas many see nature's benefits to people as a strong basis for conservation (Marvier and Wong, 2012), the relevance of intrinsic values of biodiversity is the subject of considerable debate. Moreover, at least in concept, the idea of ecosystem services might provide many firms with diverse reasons for engaging in conservation, both to manage risks and realize new opportunities (World Resources Institute (WRI) et al., 2008).

Of the numerous fierce critiques of this utilitarian turn to modern conservation, several are pertinent here: the contentions that (1) engaging with corporations and neoliberal market-based schemes associated with ecosystem services naively imagines that it can reverse the very forces that pose the greatest risks to nature (Foster, 2002; Peterson et al., 2010); (2) including instrumental motivations associated with ecosystem services will narrow the range of appropriate motivations (Norton and Noonan, 2007; Vatn, 2010; Doak et al., 2015) and detract from "real" conservation of biodiversity (McCauley, 2006; Soulé, 2013); (3) partnering with powerful industry forces will distort conservation agendas away from the most pressing matters, such as the structure of the economy (Igoe et al., 2010); (4) management for ecosystem services will not necessarily yield the biodiversity protection that conservationists seek, despite frequent claims that ecosystem services are ultimately dependent on biodiversity (Chan et al., 2007; Vira and Adams, 2009; Norgaard, 2010); (5) valuation, economic metaphors, and market-based mechanisms may foster the commodification of nature and perhaps thereby undermine conservation agendas (Gómez-Baggethun and Ruiz-Pérez, 2011; Luck et al., 2012) and social justice (e.g., by obscuring the unequal power relations implicit in production) (Kosoy and Corbera, 2010).

Controversy over these issues has persisted even though the field of ecosystem services (which includes a suite of related terms studied in diverse fields, including "environmental services") is actually much broader than the narrow economic–ecological framing that arguably popularized the idea over the past two decades. Recently, much broader conceptions of ecosystem services have received recognition in academic and policy circles, for example, steering toward a more inclusive terminology to represent diverse ways of knowing (Díaz et al., 2015), and acknowledging a broader diversity of value concepts [IPBES (Intergovernmental Science-Policy Platform on Biodiversity and Ecosystem Services), 2015; Chan et al., 2016]. It is possible that if the field of ecosystem services had matured earlier in these ways, it would not have generated the same kind of controversy. Conversely, it is also possible that the field's strong ecological–economic and neoliberal focus generated the attention of government agencies and multinational corporations that was necessary for the field to blossom at all.

Although the earlier debates are lively and acrimonious, we know of scarce empirical examination of the net effect of ecosystem services research and practice on "core" conservation activities and their normalization. One important study, however, found that ecosystem services projects conducted by the Nature Conservancy brought in new funding from an expanded set of funders and that they were just as likely to include or create protected areas as projects without explicit ecosystem-services language (Goldman et al., 2008). Not only is there reason to believe that uptake of ecosystem services valuation in decision making is limited (Honey-Rosés and Pendleton, 2013; Förster et al., 2015; Martinez-Harms et al., 2015), there is a notable absence of empirical work on this topic (Laurans et al., 2013). There is some evidence that implicit ecosystem-services framings do not crowd out diverse metaphors for human relationships with nature (that questions about benefits associated with ecosystems elicit a wide variety of views of nature, well beyond a factory model of production) (Klain et al., 2014; Gould et al., 2015). Furthermore, ecosystem-services applications are often sought without the request for monetary metrics (Ruckelshaus et al., 2015). Overall, has the concept of ecosystem services expanded the scope for conservation without diluting conservation's central objective of biodiversity protection? Empirically, it is too early to say.

NORM CHANGE IN CONSERVATION

Social scientists, in diverse fields including anthropology, sociology, and psychology, have extensively explored factors that contribute to or impede widespread environmental attitudes and actions. These include, for example, connection to nature, sense of place, social identity, and social networks (e.g., Gifford, 2011). Although abundant knowledge exists to aid the project of norm change for conservation, the fields of conservation and ecosystem services have given scant attention to the challenge of norm change. How might conservation engage with norm change more explicitly, and how might this relate to ecosystem services? Drawing on the earlier social science research, we identify four gaps in conservation practice and outline how a new approach might foster broad-scale norm change.

The first gap is that conservation science and practice has focused on behavior, through a variety of efforts that do not easily scale up to norm change. The conservation community has increasingly recognized that conservation is fundamentally a human challenge, fueling the recent rise of interest in the application of behavioral economics to conservation (e.g., Brondízio et al., 2010; Muradian et al., 2013), and in the emerging field of conservation marketing (e.g., Veríssimo, 2013; Wright et al., 2015). Behavioral economic approaches often "nudge" behavior change by making the desired behavior easier and the default option for a specific situation, minimizing inconvenience and mental effort (Thaler and Sunstein, 2008). However, these behavioral approaches fall short of a norm-change agenda unless they engage values or social sanctions (the "moral tinge"), which are crucial for norm enforcement (Fehr and Fischbacher, 2004). According to psychologist Cialdini (2003), norms are the values, customs, traditions, and other "cultural products" that represent individuals' basic knowledge of what others do and what they think they should do. From a social psychology perspective, norms are not behavior per se but rather "mental representations of appropriate behavior" (Aarts and Dijksterhuis, 2003). In contrast, some sociological perspectives emphasize the role of formal and informal social sanctions in maintaining norms by rewarding or punishing behavior (suggesting sanctions via external social pressure can be more powerful than individual beliefs; Elster, 2000; Hechter and Opp, 2001). Stemming from sociology, a social practice perspective explicitly accounts for the centrality of norms (including institutions and personal habits) in constraining and determining actions, which these scholars refuse to term "behavior" to explicitly de-emphasize the role of individual choice (Shove, 2010). Regardless of these differing emphases, norm change can be much more powerful than mere behavior change alone because norms apply to diverse contexts (Reno et al., 1993).

Second, conservation has suffered from focusing unduly on those who are already motivated by environmental causes. As norms establish what is done routinely and what is seen as morally requisite, they can contribute powerfully to both environmental successes and failures, depending on the perceived status quo (Gifford, 2011). Norms, such as those for proenvironmental behavior, often have prescriptive force only for those with a shared identity (Lapinski and Rimal, 2005) and often help to shape a group's identity (Feldman, 1984). These points are crucial when one considers the kinds of behavior-change interventions that are popularly alluded to in contemporary conservation. Many campaigns and actions appear to have appeal and uptake only among die-hard environmentalists, emphasizing the distinction between environmentalists and everyone else. Such segmented normalization might impede broad uptake of certain practices, leaving nonenvironmentalists to reason: "Environmentalists ought to do X," "I don't want to do X/X is too onerous," and therefore, "I am not an environmentalist." Conversely, the alternative reasoning is also true: "Environmentalists do X," "I am not an environmentalist (nor do I want to be), therefore, I will not do X." Following this logic, we might understand why even people with environmental values may not develop the norms, behavior, and shared identity of environmentalists or conservationists. Namely, when actions are perceived to be too onerous or undesirable, and/or individuals do not wish to identify as environmentalists, they may not ascribe to these group norms. Recent research suggests that negative stereotypes of environmentalists are common and may be associated with a reduced desire to affiliate with them and to espouse the behaviors they promote (Bashir et al., 2013). Furthermore, when people anticipate feeling morally judged by groups who deviate from the status quo, they tend to derogate these groups rather than adopt the norms and behaviors common to the group (Minson and Monin, 2012). Thus, rather than making conservation part of everyday thinking and action, some efforts for civic behavior change may actually impede widespread normalization.

As an illustrative example, all of the authors have experience with seafood purchasing campaigns and one of us (KC) routinely asks audiences about their experiences with them. Audiences relate strongly—as evidenced by familiar laughs and nodding heads—to the discomfort of purchasing sustainable seafood. Asking restaurant servers detailed and apparently nitpicky questions about the source of particular seafood product, amid several trips back to the kitchen to ask the chef, can be socially awkward and potentially disruptive. The problem here is that this action—as but one example of what is seen in some circles as being expected of good environmental citizens—brings embarrassment and so impedes normalization [group norms are generally behaviors that *avoid* embarrassing interpersonal experiences (Feldman, 1984), not facilitate them]. And even where there are norms, the norm-activation model suggests that anticipated feelings of pride help to activate those norms in the form of behavior and anticipated guilt impedes activation (Onwezen et al., 2013). Only for a committed few is it more embarrassing to be caught consuming Patagonian toothfish or farmed Atlantic salmon than it is to bewilder one's server with questions she or he cannot answer. Thus, the gulf between environmentalists and the rest of the population widens.

A third gap is that although some environmental practices and values have extended beyond special interest groups to become normalized in general populations, these successes are generally quite limited in their scope and applicability. For example, reusable shopping bags used to be normal only for die-hard environmentalists, in part thanks to popular NGO campaigns depicting the devastating effects of plastic on seabirds, sea turtles, and other marine life. Reusing shopping bags has become normal for lots of people in many places (Clapp and Swanston, 2009), owing both to campaigns encouraging environmental citizenship and fiscal incentive policies (Dobson, 2007). Two other conservation marketing examples highlighted by Wright et al. (2015) demonstrate the effect of increased public awareness on conservation behaviors: a seafood purchasing campaign involving a London luxury department store, Selfridges; and concerns for marine mammals in captivity promulgated by a team of NGOs and the movie Blackfish. The seafood purchasing campaign sought behavior change resulting in consumers purchasing more sustainable seafood. The Blackfish campaign sought changes in behavior (public attendance at orca shows), policy (a California bill banning captive orcas), and practice (changes to SeaWorld practices). Both were important and successful, raising awareness, educating people, and yielding changes in how some people buy seafood and in how SeaWorld housed orcas (SeaWorld subsequently announced it was phasing out captive breeding, such that the current generation of orcas will be the last). However, such successes are very limited in scope and so may contribute little to making conservation—writ large—normal.

Fourth, conservation has often sought normalization by leading with legislation, which can yield unintended consequences for conservation attitudes by overly restricting agency. Laws and policies can contribute eventually to actions being seen as morally requisite, which is crucial for normalization. But when a law or policy heavily burdens some people without their voluntary involvement and/or a broad base of support, it can impede positive norm change by provoking resentment and perverse behaviors. For all the great outcomes that have come from the US Endangered Species Act (Goble et al., 2005; Scott et al., 2006), it provides a prime example of this by burdening landowners found to host endangered species with large restrictions on land-use, which in turn incentivizes a perverse behavior called "shoot, shovel, shut up" (Polasky et al., 1997). Personal agency appears to be key to conservation-friendly sentiments: engaging landowners in conservation in a voluntary capacity has yielded large changes in compliance and attitudes (Wilcove and Lee, 2004). In contrast to the Endangered Species Act, the aforementioned bill banning keeping orcas in captivity was an example of first establishing the moral norm among the populace and then achieving legal progress. "Growing your people" (Heath and Heath, 2010)—in this case via cultivating conservation-supporting identities (even without self-identification as environmentalists)—is important for achieving policy or institutional change supported by a critical mass of individuals (Burstein, 2003).

Considering these gaps, one might argue that what conservation needs to achieve widespread norm change is to give a wide diversity of people the opportunity to enact—conspicuously—a broadly shared value via a sweeping set of conservation-relevant actions that are easy, enjoyable, and not too expensive. By rooting the actions in values, and making them conspicuous (for reinforcing social sanctions), we might move beyond behavior to the realm of norms. By connecting to broadly shared values, and making the actions easy, enjoyable, and not too expensive, we might reach beyond the environmental movement to broader publics. By connecting a sweeping set of conservation-relevant actions under the same umbrella, we might include a broad scope of conservation challenges. And by giving people options, we might "grow our people" and avoid the pitfalls of solutions that overly restrict agency. Before we consider what kind of a program might simultaneously achieve these various requirements, we must first consider the role of ecosystem services.

Of the normalizing conservation successes we know, precious few have stemmed from ecosystem services research and practice. Most seem to stem from a strong principle- or virtue-based position, such as the recent popular views strongly opposed to trophy hunting, notwithstanding apparent contributions from trophy hunting activities to conservation funding (Chan, 2015; Di Minin et al., 2016). Other examples of emerging normalization in conservation include (as earlier) the view that it is wrong for retailers to sell seafood that catches sea turtles, dolphins, and seabirds; that it is wrong to keep intelligent and emotionally complex animals in small tanks; or, in the case of the many successful Rare Pride campaigns, that a particular endangered animal is a cherished component of a place and the responsibility of its citizens (Jenks et al., 2010). Pollution-motivated laws and policies, such as the US Clean Water Act, which requires the protection and compensatory restoration of wetlands, have impacted industry practice but with only limited consequences for conservation. For the most part, ecosystem services research and practice has steered clear of such moral territory (but see Luck et al., 2009; Chan et al., 2011a, 2012b; Luck et al., 2012; Jax et al., 2013). We see this largely uncharted moral territory related to ecosystem services as a key opportunity because the ecosystem services concept offers a critical benefit in the building of norms for conservation.

THE UNCELEBRATED PROMISE OF ECOSYSTEM SERVICES: 1. JUSTICE

Ecosystem services research and practice has widely overlooked principle-based arguments for conservation. Although intrinsic values in nature are widely recognized (e.g., the rights of nonhuman species to exist), but often seen as secondary to human considerations (Marvier and Wong, 2012), the ecosystem services concept explicitly connects the environment to human consequences (Daily, 1997). Because these human consequences can be distributed fairly or unfairly, ecosystem services connect environmental issues to strong normalizing concepts such as justice and equity (Rawls, 1971). Yet, the ecosystem services field has largely aligned itself with a strictly utilitarian logic (but see Sikor, 2013), which is largely silent on justice considerations (Dean Moore and Russell, 2009; Chan and Satterfield, 2013). The alignment of ecosystem services with utilitarian logic likely stems from the key role of economics in the field's origin (Gómez-Baggethun et al., 2010), but the time is ripe for a broadening in this moral framing (Luck et al., 2012; Jax et al., 2013; Chan et al., 2016).

The moral and normative promise of the ecosystem services concept is that it extends "responsibility," a powerful moral concept, through space and time and across environmental contexts. Responsibility is inherent in the Golden Rule (do unto others as you would have them do unto you—were you in their position) (Hare, 1991), a foundation for a wide array of moral frameworks. All people rely on and benefit from ecosystem services, albeit some much more directly than others (Bawa and Gadgil, 1997). Most of us would not want these services degraded, as it routinely occurs every day via pressures associated with human consumption of various goods and services, including extraction, production, transportation, and so on (Millennium Ecosystem Assessment, 2005). It follows then that most of us would recognize a responsibility to use ecosystem services thoughtfully and sustainably as an extension of our responsibilities toward other people. Current supply chains, one might argue, are unjust in their propagation of environmental impacts via pollution, habitat loss, and degradation, with limited opportunity for consideration or mitigation. One meaningful step to restore justice would be mitigating our impacts on ecosystem services to reduce disproportionate impact on impoverished and disempowered peoples (Bawa and Gadgil, 1997; Martinez-Alier, 2002).

Recognizing and leveraging the relevance of ecosystem services research for justice does *not* imply resorting to a new manifestation of moral suasion (behavior change via moral argument). Our position here is rather that ecosystem services have crucial justice dimensions and that these might be leveraged into broadly shared changes in behavior or "social practice" by creating the institutions and infrastructure that enable actions that align with those core values (Shove, 2010). In the final section, we discuss the institutions and infrastructure that might enable consumers, corporations, and producers to mitigate impacts on ecosystem services, in line with environmental dimensions of justice. First, we consider a second important lever provided by ecosystem services—this time associated not with the concept but rather the services themselves.

THE PROMISE OF ECOSYSTEM SERVICES: 2. VALUE FORMATION

Ecosystem services offer a second major leverage point for making conservation normal, in that cultural ecosystem services contribute crucially to the formation of relationships with "nature" and to values about those relationships (relational values) (Chan et al., 2016). A purely instrumental relationship with nature is unlikely to inspire concern for distant people and places. More likely, concern for conservation will stem from a citizen's knowledge of, attachment to, and identification with places/ecosystems/species and associated people.

Cultural ecosystem services, as "ecosystems' contributions to the non-material benefits (e.g., capabilities and experiences) that arise from human–ecosystem relationships" (also see Church et al., 2011; Chan et al., 2012b), are a primary conduit for establishing and renewing strong relationships with nature (Martín-López et al., 2012; Plieninger et al., 2013; Klain et al., 2014; Chan and Satterfield, 2016). Arguably, it is the nonmaterial experience of eco-systems (a cultural service) that propels people to seek out further experiences, even if a provisioning service is also implicated (e.g., provision of fish for subsistence, recreational, or commercial catch) (Chan et al., 2012a). From these repeated interactions arise capabilities (e.g., boat-handling, navigation, reading the weather, knowledge of submerged rocks and currents, cleaning and cook-ing fish), norms, and identities (Chan and Satterfield, 2016). Surely it is such lived experiences, norms, and identities that provide a substantial basis to care *for* places, and *about* other places (Norton and Hannon, 1997; Scannell and Gifford, 2010a,b; Chan et al., 2016).

Arguably, these tangible and intangible relationships with nature are essen-tial to value formation. Given the demographic changes of the 21st century, the majority of human beings live in cities and have experienced a dramatic rise in the substitution of virtual (screen-based) experiences for outdoor ones (but see McLain et al., 2014; Poe et al., 2014; Chan et al., 2016; Chan and Satterfield, 2016). Louv's (2008) reference to the "last child in the woods" is no joke. Meaningful experiences in nature during childhood and over time, espe-cially experiences intertwined with important people in our lives, can greatly influence our values and relationships with nature as adults, and these values can inspire conservation action (Clayton and Myers, 2015; Poe et al., 2016). Accordingly, ecosystem services are central to the project of normalizing con-servation because they are crucial to the formation of environmental identi-ties (including values) upon which conservation relies. Efforts to mainstream conservation must therefore foster the kind of contact with ecosystem services (particularly cultural ones) that fosters attachment and broader environmental identities.

INNOVATIONS AND INSTITUTIONS FOR CONSERVATION IN THE ANTHROPOCENE

Earlier we argue that a justice-based ecosystem services concept holds promise as an additional moral basis for conservation—that we ought to protect and restore the ecosystems upon which we all rely. To move from a principled idea to norm change will require enacting the idea via practical solutions from bold individuals and organizations. Particularly important are the actions of businesses, universities, governments, and religious institutions that have demonstrated a willingness to make principled purchasing and investments (or divestments). Also crucial are affluent individuals (anyone in the middle and upper classes) whose consumption has disproportionate impact and who have the capacity to pay to mitigate their impacts (via improved practices and offsets). Here we briefly discuss existing mechanisms that integrate consideration for ecosystem service impacts, including certification, community supported fisheries, and payments for ecosystem services. The key question is how new institutions might combine aspects of these strategies and amplify their application at broader scales. Toward this end, we introduce CoSphere, a new initiative aimed at spreading norms of environmental responsibility that exemplifies our vision.

Promising Institutions for Ecosystem Services

Apart from those relying entirely and directly on ecosystems for subsistence, we all participate in market economies rooted largely in unsustainable use and degradation of oceans and other ecosystems. Several existing strategies disrupt traditional supply chains to engage consumers and address particular environmental concerns. We highlight a few important examples.

Certification: Certification systems provide one means of reducing environmental degradation, via third party evaluation of certain production practices, but challenges remain. As one example, Marine Stewardship Council certified products, like products labeled under other certification systems, are available in some places for some product types, for an additional cost (often a considerable barrier). They likely confer reduced impacts on ecosystems and ecosystem services, but important residual impacts undoubtedly remain (Jacquet et al., 2010b). Furthermore, certification systems present consumers with a plethora of choices and information that most people are not equipped or interested to deconstruct (Jacquet et al., 2010a).

Community Supported Fisheries: Based on the logic that overexposure and vulnerability to risk prevents individuals and companies from adopting new approaches that might yield environmental gains (Holzmann and Jørgensen, 2001), new strategies like community supported fisheries are appearing to help fishers manage risks, following the community supported agriculture model (Brinson et al., 2011). As a conservation strategy, it is indirect, providing a guaranteed revenue stream via membership fees collected at the outset of the season. This certainty and timing of funds may help fishers (especially small-scale ones, who may lack access to capital) invest in repairs, new gear, upgrades, and otherwise undertake sustainable practices including avoiding by-catch and targeting appropriate species. In practice, community supported fisheries can help reduce greenhouse gas emissions and they may help address other environmental problems (McClenachan et al., 2014).

Payments for Ecosystem Services (PES): PES programs are institutional-ized payments to ecosystem service providers to improve or reduce impacts on ecosystem services (Wunder, 2006; Jack et al., 2008; Schomers and Matzdorf, 2013; Wunder, 2013). Existing designs have yielded mixed results (Kinzig et al., 2011; Naeem et al., 2015) and suffered from a wide range of critiques (Muradian et al., 2013; Pascual et al., 2014; Chan et al., 2017), but some of us have argued elsewhere that a subtle but important redesign of PES could largely address most of the largest critiques (Chan et al., 2017). For example, by chang-ing the program framing (to be a reward for leadership in land management) and by offering only a co-pay, programs could avoid crowding out existing motiva-tions for conservation and distribute responsibilities more evenly so as to avoid "paying the polluter" (for pollution reductions that could be seen as extensions of preexisting responsibilities). And by paying for a wide range of management actions, programs can enlist (rather than squash) landowner agency and creativ-ity (Chan et al., 2017). Despite perceptions of widespread application of PES, PES programs have spread slowly in some contexts (Wunder, 2013).

Insofar as certification, community-supported agriculture/fisheries, and PES programs could contribute to the maintenance of sustainable ocean- and land-based livelihoods (by providing an additional revenue stream), an expansion of these initiatives could foster the continued exposure to cultural ecosystem services that build key capacities for using nature and attachments to it (Chan and Satterfield, 2016).

Novel and Scaled-Up Institutions

To attain broad normalization of conservation, a crucial step is to enable all those who are complicit in environmental degradation to mitigate their impacts via mechanisms that are conspicuous, easy, enjoyable, and not too expensive.

Because many of these impacts are entailed by purchases of products and services, such mitigation opportunities would ideally be linked to purchases [via payments commensurate with the embodied impacts; probably as an opt-out to signal a default and norm of paying (Heath and Heath, 2010; Kahneman, 2013)]. Imagine walking into a grocery store and being able to buy a wide vari-ety of seafood products, knowing that coupled with your purchase of the shrimp (for example) would be a payment that funds fishers and conservationists to prevent by-catch and restore the species and habitats harmed in fishing.

Some of the authors are working to create and propagate such mitigation opportunities via CoSphere (a Community of Small Planet Heroes, Ecologically Regenerating Economies). This fledgling organization is building critical insti-tutional infrastructure for sustainability and an online community to enable consumers and organizations to improve their environmental impact. When all members of the supply chain, in partnership, can take shared responsibility for their negative impacts, pledge their commitment to improve impacts, and take action to achieve such improvements, we can initiate a movement toward mar-ket transformation (see Fig. 11.2).

CoSphere's developers hope that by enabling individuals and firms to have net-positive impacts on our planet's biodiversity and ecosystem services, it can unleash stewardship values across the supply chain. These positive impacts would be achieved via contributions to conservation and stewardship programs directly in the form of creative financing, as described earlier, payment for eco-system services programs (Chan et al., 2017), and impact investing, including social impact bonds (SIBs; see in the following).

CoSphere would be akin to a system of offsetting for biodiversity and eco-system services, with crucial caveats. That is, it would learn from failures in carbon offsets (Bumpus and Liverman, 2008; Corbera et al., 2009; Wittman and Caron, 2009; Wittman et al., 2015)—that have arguably given offsets a bad name—and ongoing lessons about biodiversity offsets (BBOP, 2012a,b; Miller et al., 2015; Moreno-Mateos et al., 2015; Tallis et al., 2015). In particular, four points are key. First, offset transactions would be negotiated by an arms-length third-party organization (Fox et al., 2006), not open to trading as are commodi-ties, so as to avoid speculation and profiteering. Second, payments would go primarily or exclusively to on-ground conservation, restoration, and improved resource management, rather than to companies who manage to find room under artificial "caps" as in a cap-and-trade model. Third, payments would be in pro-portion to impacts caused, to send a price signal that would incentivize supply chains to reduce impacts. Fourth, messaging would target earnest responsible citizens and organizations to convey clearly that CoSphere is aimed at market transformation (rather than a means to justify consumption).

Existing pathways to support sustainability build on consumer demand for spe-cific products, whereas transformation and norm change requires innovation at the system level. CoSphere aims for system-level transformation in several respects. First, by rooting a conspicuous, easy, and enjoyable action (paying for mitigation) in a broadly shared value, it intends to take advantage of positive social sanctions toward norm change. Second, by inviting citizens and corporations alike, it hopes to enable consumers to pressure retailers to mitigate impacts, retailers to require suppliers, and so on down supply chains (and also back up, e.g., via progressive retailers marketing their commitment to consumers). Third, by providing funds directly to on-ground conservation and stewardship, including fishers, farmers, etc., this model would foster the ongoing development of connections to the ocean and the land upon which environmental identities are based. In the following we describe how regenerative renewable energy and new investment strategies pres-ent two potentially impactful pathways to increase visibility and personal connec-tions to ecosystem services, while supporting their restoration.

A likely impediment to the normalization of conservation stems from the ever-present cognitive dissonance for consumers who desire to propel environmen-tal solutions but know that chosen solutions cause other environmental harms. Renewable energy infrastructure, for example, has been opposed in many places, in part due to its impacts on ecosystem services and biodiversity. Specifically, whereas offshore wind turbines could contribute substantially toward meeting electricity demand in numerous coastal countries (Jacobson et al., 2015), they pose a threat to some species of seabirds and migratory birds. Perhaps innovation in the energy sector could tap into latent willingness to pay for biodiversity-friendly renewable energy development, mitigating the impacts of this otherwise "clean" technology. Klain et al. (2017) document high willingness to pay—over $35/month/person—for offshore wind farms that enhance species diversity and abundance via artificial reefs at the base of the turbines. This willingness to pay for biodiversity friendly energy could possibly fund not only the deployment of artificial reefs in association with offshore wind farms, but also seabird nesting habitat protection, conservation in migratory flyways, and other conservation and restoration activities. No energy is entirely free of negative impact, but CoSphere would aim to enable consumer purchases that bundle renewable energy with the onsite or offsite mitigation of residual impacts.

For conservation to be broadly normal, citizens ought to be able to express their conservation values through their investment choices without great difficulty or cost. For example, CoSphere intends to enable conservation impact investing (Olmsted et al., 2017) via a model based on SIBs. SIBs are a relatively new financial mechanism currently being rolled out in several UK and US locations and gaining interest and traction with other regional and federal governments globally (Warner, 2013). Briefly, they are a contract between the private and public sectors where private capital is used to fund a social program, with the government committing to pay back the investor with a return when the investment results in improved social outcomes and public sector savings (Fig. 11.1). Application in the environmental sector is limited to infrastructure and energy projects; conservation projects are less likely to generate returns and so it is more difficult to follow the traditional SIB model in which cost savings are recouped (Olmsted, 2016). We see potential for a similar model, however, that includes upfront support from a foundation or government to match crowd-funded community contributions. We refer to this arrangement as a "stewardship bond" because the finances would be raised primarily by those who live or operate in a region, with the explicit intent that this support would foster a sense of stewardship that previously was latent (such that investors would be truly "invested" in the region).

Because stewardship bonds would not be a traditional donation, and there is the opportunity for (modest) financial gain, they may increase the financial support for conservation projects while attracting new people who would not otherwise participate, broadening the tent on the path to normalization (Olmsted, 2016). Furthermore, the modest nature of the potential financial gain is both intentional and instrumental, for several reasons. First, CoSphere would be intended for investors with genuine concern for local outcomes, not those seeking to "make a quick buck." Second, a modest payoff should be sufficient such that diverse investors (including stakeholder groups that may not often collaborate—for example, the fishing industry, local property owners, other industry and retailers operating in the region, and conservation NGOs) see the benefit in collaborating to achieve the ecosystem service protection targets. The

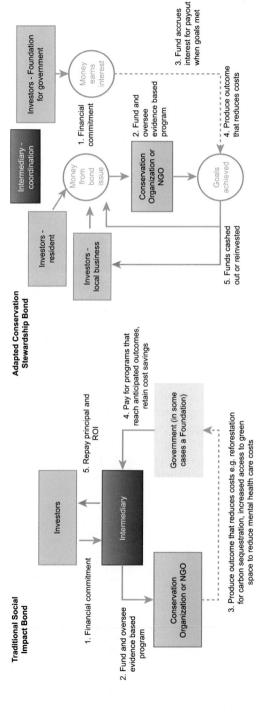

FIGURE 11.1 Schematic representing traditional social impact bonds and an adaptation of these for conservation to enable expression of conservation values via investments to restore ecosystems and ecosystem services. In the traditional model, an upfront financial commitment facilitates a conservation effort that results in a cost savings. This savings enables a return on investment paid to the investor after the agreed term. The adapted model does not require explicit cost savings. For example, crowd funding from residents and local businesses fund a wetland restoration. Matching foundation funds are set aside to generate a small percentage of return. If the conservation goals are met, supporters would have the option to "cash out" or reinvest to projects as in a revolving fund.

incentive to collaborate is enhanced because all investors would receive the payoff on the same schedule. Third, modest gains should lead investors to feel that they could have maximized returns elsewhere. Because accepting lower returns only seems rational if a person is committed to the cause, knowledge of opportunity costs ought to foster the avoidance of cognitive dissonance (Festinger, 1962; Cooper, 2007) to consolidate local stewardship values and identities.

Good Governance and Institutions That Fit

Controversies about due process and human casualties pose a major risk to the widespread normalization of conservation, as illustrated by the Death of Environmentalism (Shellenberger and Nordhaus, 2004), conservation refugees (Dowie, 2005), and Chapin's "challenge to conservationists" (2004). Ecosystem services are intended to address human well-being in conservation, but the risk remains that lofty ideas are applied in inappropriate and even harmful ways on the ground, perhaps especially with innovative "big picture" conservation solutions. For example, the negative impacts of and inappropriate processes through which well-intentioned and globally accepted models of conservation and ecosystem services (e.g., terrestrial and marine protected areas; carbon markets) are well documented (West and Brockington, 2006; Bumpus and Liverman, 2008; Corbera et al., 2009; Wittman and Caron, 2009; Bennett and Dearden, 2014; Franco et al., 2014; Wittman et al., 2015). Persistent critiques have led some to question the effectiveness of capital-driven conservation programs (Brockington et al., 2008; Brockington and Duffy, 2011). Yet, we argue that in the friction between global conservation programs and local practices there is the opportunity to create institutions that are legitimate, enabling and sufficiently flexible to fit diverse contexts. In institutions intended to normalize conservation, we would highlight two key strategies to overcome the challenges discussed earlier.

First, the decision-making bodies and the processes through which decisions are made need to employ good governance to ensure the acceptability and legitimacy of the initiative. Thus good governance is crucial not only for its own sake, but also for practical or instrumental reasons. The concept of "good" governance suggests that there is a right way (and hence, a wrong way) to do things. Normative ideals—such as inclusiveness, participation, deliberation, vision, voice, equity, transparency, and accountability—are key considerations (Lockwood et al., 2010; Turner et al., 2014). Local expectations or perceptions about who is at the table and the way that decisions are made about programmatic priorities, actions to be taken, or projects to be funded can determine whether local people will support initiatives (Bennett, 2016). These literatures suggest that decision-making bodies would need to incorporate local people, and their voices and visions, in transparent deliberation processes. Mechanisms would need to be in place to ensure those making decisions are accountable for social and ecological outcomes. "Good governance" sounds obvious; achieving it in reality (not just rhetoric—Igoe, 2004) is another matter.

Second, the way that decisions are made and actions are employed need to be flexible enough so that they can be adapted to "fit" different contexts (Young, 2002; Epstein et al., 2015; Guerrero et al., 2015). Many top-down conservation initiatives exclude local voices, values, needs, and preexisting governance arrangements (Smith et al., 2009; Bennett et al., 2015). Yet, when conservation programs do not align with local social, cultural, economic, or governance contexts, they are unlikely to be deemed legitimate or adopted broadly (Hoffman, 2009; Ban et al., 2013). Without appropriate safeguards, corporate/private partners or even consumers in market-oriented conservation models such as CoSphere might drive the agenda and propose actions for local implementation with little knowledge of what would actually work. How can we ensure that this does not occur? Program infrastructure would need to ensure that governance and decision-making processes are adaptable and flexible enough to allow for these contextual differences, thus enabling locally appropriate, acceptable, and effective solutions (Barrett et al., 2005; Folke et al., 2005; Ostrom et al., 2007).

Integrating these considerations in the context of initiatives such as CoSphere (which is but one example of the ideas introduced here), we see the following needs and solutions. To foster institutions that fit, institutions would need to be nested across scales, with a guiding principle of subsidiarity (i.e., decision-making authority delegated to the lowest level possible). At CoSphere, a global body would guide, enable, and support regional or national bodies that would be entrusted with making decisions. These local decision-making bodies, made up of selected experts (i.e., scientists, practitioners, policy makers, and community representatives), would be charged with crowd-sourcing proposals (to be funded through a kind of reverse auction, Chan et al., 2017), analyzing their effectiveness and negotiating across scales to select projects that are situated in context and that align with conservation and development priorities and locally salient ecosystem services (locally, nationally, and globally) (Fig. 11.2 depicts this, without the scalar complexity). To further ensure local accountability, regional conservation priorities would be derived from value-focused deliberative multistakeholder processes (Espinosa Romero, 2010; Gregory et al., 2012). All told, the challenge is to embed good governance and participatory processes in a space (environmental management) too often occupied by a combination of highly constrained or even dysfunctional governmental agencies and private corporate control (Dauvergne and Lister, 2013).

CONCLUSION

Conservation is not yet normal, and yet it is an attainable goal. By leveraging the implicit potential of ecosystem services to put a human face on environmental change, conservation can legitimately link environmental concerns to social justice. We see hope, including via CoSphere, that conservation can address the five major concerns regarding ecosystem services we identified (see "Expanding the Tent or Welcoming the Wolf?"), (1) enlisting powerful economic forces

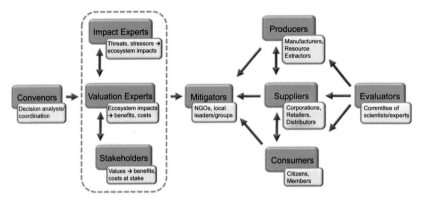

FIGURE 11.2 A schematic representing the institutional arrangements envisioned for CoSphere (a Community of Small-Planet Heroes, Ecologically Regenerating Economies) to make it easy, enjoyable, and not too expensive for organizations and individuals to mitigate impacts on biodiversity and ecosystem services. On the left side, the conveners bring together a variety of local experts and stakeholders in a structured participatory process to prioritize the kinds of mitigation actions that would be most appropriate. On the right side, evaluators help producers, suppliers, and consumers identify kinds of impacts needing mitigation in different regions; based on this, CoSphere members (individuals and organizations) contribute toward funding for conservation.

without succumbing to them, (2) "crowding *in*" inherent motivations for conservation, (3) addressing the structure of the economy in a subtle, inconspicuous way (by popularizing the internalization of ecological externalities), (4) protecting biodiversity as a target alongside ecosystem services, and (5) actively contributing to conservation values and social justice by funding ocean- and land-based people to continue traditional livelihoods and evolve them further toward sustainability. Pursuing this with a combination of incentive programs, finance schemes, and adaptable "good" governance institutions that engender not only behavior change but also value change, conservation could become the new normal—both morally necessary and commonplace.

TAKING ACTION, SPREADING NORMS

The example mentioned earlier of CoSphere is but one potential mechanism to spread conservation norms. Achieving transformational change will by definition require a groundswell of support that could not simply stem from one or two initiatives. Consequently, to support any readers who would like to spread conservation norms in their workplaces, families, and communities, we offer the following as starting points (while naturally welcoming any multitude of other strategies).

- Join and/or build a conservation initiative around an issue that already has community support to leverage existing areas of focus (e.g., local ecosystem services), connecting explicitly to stewardship and responsibility.
- Frame messages to encourage connections between the environment/ environmental values and broader values such as justice and equality, including via ecosystem services.

- Engage stakeholders early in the process of a new initiative to promote buy-in and ownership via fair and inclusive processes.
- Set great examples of actions in accordance with values—social norms are powerful and people tend to follow the lead of those around them.
- Provide pathways to change supply chains and economic structures—via actions that are easy, fun, and broadly inviting, to help generate critical mass beyond self-identified environmentalists.

REFERENCES

Aarts, H., Dijksterhuis, A., 2003. The silence of the library: environment, situational norm, and social behavior. Journal of Personality and Social Psychology 84 (1), 18.

Balvanera, P., Daily, G.C., Ehrlich, P.R., Ricketts, T.H., Bailey, S.A., Kark, S., Kremen, C., Pereira, H., 2001. Conserving biodiversity and ecosystem services. Science. 291 (5511), 2047. http://www.sciencemag.org/cgi/content/full/291/5511/2047.

Ban, N.C., Mills, M., Tam, J., et al., 2013. Towards a social-ecological approach for conservation planning: embedding social considerations. Frontiers in Ecology and the Environment 11, 194–202. http://dx.doi.org/10.1890/110205.

Barrett, C.B., Lee, D.R., McPeak, J.G., 2005. Institutional arrangements for rural poverty reduction and resource conservation. World Development. 33 (2), 193–197. http://www.sciencedirect.com/science/article/pii/S0305750X04001895.

Bashir, N.Y., Lockwood, P., Chasteen, A.L., Nadolny, D., Noyes, I., 2013. The ironic impact of activists: negative stereotypes reduce social change influence. European Journal of Social Psychology 43 (7), 614–626. http://dx.doi.org/10.1002/ejsp.1983.

Bawa, K.S., Gadgil, M., 1997. Ecosystem services in subsistence economies and conservation of biodiversity. In: Daily, G.C. (Ed.), Nature's Services: Societal Dependence on Natural Ecosystems. Island Press, Washington, DC, pp. 295–310.

Business and Biodiversity Offsets Programme (BBOP), 2012a. Resource Paper: Limits to What Can Be Offset. BBOP, Washington, DC, p. 39. http://www.forest-trends.org/documents/files/doc_3128.pdf.

Business and Biodiversity Offsets Programme (BBOP), 2012b. Standard on Biodiversity Offsets. Forest Trends, Washington, DC. http://www.forest-trends.org/documents/files/doc_3078.pdf.

Bennett, N.J., Dearden, P., 2014. Why local people do not support conservation: community perceptions of marine protected area livelihood impacts, governance and management in Thailand. Marine Policy. 44 (0), 107–116. http://www.sciencedirect.com/science/article/pii/S0308597X13001711.

Bennett, N.J., Govan, H., Satterfield, T., 2015. Ocean grabbing. Marine Policy. 57 (0), 61–68. http://www.sciencedirect.com/science/article/pii/S0308597X15000755.

Bennett, N.J., 2016. Using perceptions as evidence to improve conservation and environmental management. Conservation Biology 30 (3), 582–592. http://dx.doi.org/10.1111/cobi.12681.

Brinson, A., Lee, M.-Y., Rountree, B., 2011. Direct marketing strategies: the rise of community supported fishery programs. Marine Policy. 35 (4), 542–548. http://www.sciencedirect.com/science/article/pii/S0308597X11000157.

Brockington, D., Duffy, R., 2011. Capitalism and Conservation. Wiley. https://books.google.ca/books?id=WDHZNEWqSQcC.

Brockington, D., Duffy, R., Igoe, J. (Eds.), 2008. Nature Unbound: The Past, Present and Future of Protected Areas. Earthscan, London. http://books.google.ca/books?id=M0wOJwAACAAJ.

Brondízio, E.S., Gatzweiler, F.W., Zografos, C., Kumar, M., Jianchu, X., McNeely, J., Kadekodi, G.K., Martinez-Alier, J., 2010. Socio-cultural context of ecosystem and biodiversity valuation. In: Kumar, P. (Ed.), The Economics of Ecosystems and Biodiversity: Ecological and Economic Foundations. Earthscan, London and Washington, pp. 150–181. http://www.teebtest. org/2013-08-30_archive/wp-content/uploads/2013/04/D0-Chapter-4-Socio-cultural-context-of-ecosystem-and-biodiversity-valuation.pdf.

Bumpus, A.G., Liverman, D.M., 2008. Accumulation by decarbonization and the governance of carbon offsets. Economic Geography 84 (2), 127–155. http://dx.doi.org/10.1111/j.1944-8287.2008. tb00401.x.

Burstein, P., 2003. The impact of public opinion on public policy: a review and an agenda. Political Research Quarterly. 56 (1), 29–40. http://prq.sagepub.com/content/56/1/29.abstract.

Chan, K.M.A., Satterfield, T., 2013. Justice, equity, and biodiversity. In: Levin, S.A. (Ed.), The Encyclopedia of Biodiversity. Elsevier Ltd., Oxford, pp. 434–441. http://store.elsevier.com/ Encyclopedia-of-Biodiversity/isbn-9780123847195/.

Chan, K.M.A., Satterfield, T., 2016. Managing cultural ecosystem services for sustainability. In: Potschin, M., Haines-Young, R., Fish, R., Turner, R.K. (Eds.), Routledge Handbook of Ecosystem Services. Routledge, London and New York, pp. 343–358.

Chan, K.M.A., Shaw, M.R., Cameron, D.R., Underwood, E.C., Daily, G.C., 2006. Conservation planning for ecosystem services. PLoS Biology. 4 (11), 2138–2152. http://biology.plosjournals. org/perlserv/?request=get-document&doi=10.1371/journal.pbio.0040379.

Chan, K.M.A., Pringle, R.M., Ranganathan, J., et al., 2007. When agendas collide: human welfare and biological conservation. Conservation Biology. 21 (1), 59–68. http://www.blackwell-synergy.com/doi/abs/10.1111/j.1523-1739.2006.00570.x.

Chan, K.M.A., Goldstein, J., Satterfield, T., Hannahs, N., Kikiloi, K., Naidoo, R., Vadeboncoeur, N., Woodside, U., 2011a. Cultural services and non-use values. In: Kareiva, P., Tallis, H., Ricketts, T.H., Daily, G.C., Polasky, S. (Eds.), Natural Capital: Theory & Practice of Mapping Ecosystem Services. Oxford University Press, Oxford, UK, pp. 206–228. http://www.oup.com/us/catalog/ general/subject/Economics/Policy/?view=usa&sf=toc&ci=9780199588992.

Chan, K.M.A., Hoshizaki, L., Klinkenberg, B., 2011b. Ecosystem services in conservation planning: targeted benefits or co-benefits/costs? PLoS One 6 (9), e24378. http://dx.doi.org/10.1371/ journal.pone.0024378.

Chan, K.M.A., Guerry, A., Balvanera, P., et al., 2012a. Where are 'cultural' and 'social' in ecosystem services: a framework for constructive engagement. BioScience 6 (8), 744–756. http:// dx.doi.org/10.1525/bio.2012.62.8.7.

Chan, K.M.A., Satterfield, T., Goldstein, J., February 2012b. Rethinking ecosystem services to better address and navigate cultural values. Ecological Economics. 74, 8–18. http://www.sciencedirect.com/science/article/pii/S0921800911004927.

Chan, K.M.A., Balvanera, P., Benessaiah, K., et al., 2016. Why protect nature? Rethinking values and the environment. Proceedings of the National Academy of Sciences of the United States of America. 113 (6), 1462–1465. http://www.pnas.org/content/113/6/1462.full.

Chan, K.M.A., Anderson, E., Chapman, M., Jespersen, K., Olmsted, P., 2017. Payments for ecosystem services: rife with problems and potential—for transformation towards sustainability. Ecological Economics (in press).

Chan, K.M.A., 2015. Trophy Hunting: A Bugbear for Christy Clark. Vancouver Observer, Vancouver. http://www.vancouverobserver.com/opinion/trophy-hunting-bugbear-christy-clark.

Chapin, M., November/December 2004. A Challenge to Conservationists. World Watch, pp. 17–31. http://www.eldis.org/static/DOC18110.htm.

Church, A., Burgess, J., Ravenscroft, N., et al., 2011. Cultural Services. The UK National Ecosystem Assessment Technical Report. UK National Ecosystem Assessment. UNEP-WCMC, Cambridge, UK, pp. 633–692. http://uknea.unep-wcmc.org/LinkClick.aspx?fileticket =QLgsfedO70I%3d&tabid=82.

Cialdini, R.B., 2003. Crafting normative messages to protect the environment. Current Directions in Psychological Science. 12 (4), 105–109. http://cdp.sagepub.com/content/12/4/105.abstract.

Clapp, J., Swanston, L., 2009. Doing away with plastic shopping bags: international patterns of norm emergence and policy implementation. Environmental Politics 18 (3), 315–332. http://dx.doi.org/10.1080/09644010902823717.

Clayton, S., Myers, G., 2015. Conservation Psychology: Understanding and Promoting Human Care for Nature. Wiley. https://books.google.ca/books?id=6IxxBgAAQBAJ.

Cooper, J., 2007. Cognitive Dissonance: 50 Years of a Classic Theory. SAGE Publications. http://books.google.com/books?id=yKi2cLshWiAC.

Corbera, E., Estrada, M., Brown, K., 2009. How do regulated and voluntary carbon-offset schemes compare? Journal of Integrative Environmental Sciences 6 (1), 25–50. http://dx.doi.org/10.1080/15693430802703958.

Daily, G.C., Kareiva, P.M., Polasky, S., Ricketts, T.H., Tallis, H., 2011. Mainstreaming natural capital into decisions. In: Kareiva, P., Tallis, H., Ricketts, T.H., Daily, G.C., Polasky, S. (Eds.), Natural Capital: Theory & Practice of Mapping Ecosystem Services. Oxford University Press, Oxford, UK, pp. 3–14. http://www.oup.com/us/catalog/general/subject/Economics/Policy/?view w=usa&sf=toc&ci=9780199588992.

Daily, G.C. (Ed.), 1997. Nature's Services: Societal Dependence on Natural Ecosystems. Island Press, Washington, DC. http://www.amazon.ca/gp/product/1559634766/sr=1-1/qid=1155595726/ref=sr_1_1/702-8872424-7903261?ie=UTF8&s=books.

Daly, H.E., 2005. Economics in a full world. Scientific American. 293 (3), 100. http://search.ebscohost.com/login.aspx?direct=true&db=aph&AN=17836492&site=ehost-live.

Dauvergne, P., Lister, J., 2013. Eco-business: A Big-Brand Takeover of Sustainability. MIT Press. http://books.google.ca/books?id=6PlaIDc1d38C.

Dean Moore, K., Russell, R., 2009. Toward a new ethic for the oceans. In: McLeod, K., Leslie, H. (Eds.), Ecosystem-Based Management for the Oceans. Island Press, Washington, DC, pp. 324–340. http://books.google.com/books?id=yn4mL6u35tMC&source=gbs_navlinks_s.

Di Minin, E., Leader-Williams, N., Bradshaw, C.J.A., 2016. Banning trophy hunting will exacerbate biodiversity loss. Trends in Ecology & Evolution 31 (2), 99–102. http://dx.doi.org/10.1016/j.tree.2015.12.006.

Díaz, S., Demissew, S., Joly, C., et al., June 2015. The IPBES Conceptual framework – connecting nature and people. Current Opinion in Environmental Sustainability. 14, 1–16. http://www.sciencedirect.com/science/article/pii/S187734351400116X.

Dirzo, R., Young, H.S., Galetti, M., Ceballos, G., Isaac, N.J.B., Collen, B., 2014. Defaunation in the anthropocene. Science. 345 (6195), 401–406. http://www.sciencemag.org/content/345/6195/401.abstract.

Doak, D.F., Bakker, V.J., Goldstein, B.E., Hale, B., 2015. What is the future of conservation? In: Wuerthner, G., Crist, E., Butler, T. (Eds.), Protecting the Wild. Springer, pp. 27–35.

Dobson, A., 2007. Environmental citizenship: towards sustainable development. Sustainable Development 15 (5), 276–285. http://dx.doi.org/10.1002/sd.344.

Dowie, M., 2005. Conservation refugees. Orion Magazine. http://www.oriononline.org/pages/om/05-6om/Dowie.html.

Economist, T., 2005. Are you being served? The Economist 2005, 76–78.

Ehrlich, P.R., Pringle, R.M., 2008. Where does biodiversity go from here? A grim business-as-usual forecast and a hopeful portfolio of partial solutions. Proceedings of the National Academy of Sciences of the United States of America. 105 (Suppl. 1), 11579–11586. http://www.pnas.org/content/105/suppl.1/11579.abstract.

Elster, J., 2000. Social norms and economic theory. In: Crothers, L., Lockhart, C. (Eds.), Culture and Politics: A Reader. Palgrave Macmillan US, New York, pp. 363–380. http://dx.doi.org/10.1007/978-1-349-62397-6_20.

Epstein, G., Pittman, J., Alexander, S.M., et al., 2015. Institutional fit and the sustainability of social–ecological systems. Current Opinion in Environmental Sustainability 14, 34–40. http://dx.doi.org/10.1016/j.cosust.2015.03.005.

Espinosa Romero, M.J., 2010. Towards Ecosystem-Based Management: Integrating Stakeholder Values in Decision-Making and Improving the Representation of Ecosystems in Ecosystem Models. RMES. Vancouver, BC, University of British Columbia. M.Sc http://hdl.handle.net/2429/28127.

Fehr, E., Fischbacher, U., 2004. Social norms and human cooperation. Trends in Cognitive Sciences. 8 (4), 185–190. http://www.sciencedirect.com/science/article/pii/S1364661304000506.

Feldman, D.C., 1984. The development and enforcement of group norms. Academy of Management Review. 9 (1), 47–53. http://amr.aom.org/content/9/1/47.abstract.

Festinger, L., 1962. A Theory of Cognitive Dissonance. Stanford University Press. http://books.google.com/books?id=voeQ-8CASacC.

Folke, C., Hahn, T., Olsson, P., Norberg, J., 2005. Adaptive governance of social-ecological systems. Annual Review of Environment and Resources. 30 (1), 441–473. http://www.annualreviews.org/doi/abs/10.1146/annurev.energy.30.050504.144511.

Förster, J., Barkmann, J., Fricke, R., et al., 2015. Assessing ecosystem services for informing land-use decisions: a problem-oriented approach. Ecology and Society. 20 (3). http://www.ecologyandsociety.org/vol20/iss3/art31/.

Foster, J.B., 2002. Ecology Against Capitalism. Monthly Review Press. https://books.google.ca/books?id=rqpGBAAAQBAJ.

Fox, J., Daily, G.C., Thompson, B.H., Chan, K.M.A., Davis, A., Nino-Murcia, A., 2006. Conservation banking. In: Scott, J.M., Goble, D.D., Davis, F.W. (Eds.), The Endangered Species Act at Thirty: Conserving Biodiversity in the Human-Dominated Landscape. Island Press, Washington, DC, pp. 228–243.

Franco, J., Buxton, N., Vervest, P., Feodoroff, T., Pedersen, C., Reuter, R., Barbesgaard, M.C., 2014. The Global Ocean Grab: A Primer. Economic Justice Program, Transnational Institute. http://www.tni.org/briefing/global-ocean-grab-primer-0.

Gifford, R., 2011. The dragons of inaction: psychological barriers that limit climate change mitigation and adaptation. American Psychologist. 66 (4), 290–302. http://dx.doi.org/10.1037/a0023566.

Goble, D.D., Scott, J.M., Davis, F.W., 2005. The Endangered Species Act at Thirty. Renewing the Conservation Promise, vol. 1. Island Press, Washington, DC. https://books.google.ca/books?id=vEh7m1k5WMwC.

Goldman, R.L., Tallis, H., Kareiva, P., Daily, G.C., 2008. Field evidence that ecosystem service projects support biodiversity and diversify options. Proceedings of the National Academy of Sciences of the United States of America 105 (27), 9445–9448. http://dx.doi.org/10.1073/pnas.0800208105.

Goldman-Benner, R.L., Benitez, S., Boucher, T., Calvache, A., Daily, G., Kareiva, P., Kroeger, T., Ramos, A., 2012. Water funds and payments for ecosystem services: practice learns from theory and theory can learn from practice. Oryx 46 (01), 55–63. http://dx.doi.org/10.1017/S0030605311001050.

Gómez-Baggethun, E., Ruiz-Pérez, M., 2011. Economic valuation and the commodification of eco-system services. Progress in Physical Geography. 35 (5), 613–628. http://ppg.sagepub.com/content/35/5/613.abstract.

Gómez-Baggethun, E., de Groot, R., Lomas, P.L., Montes, C., 2010. The history of ecosystem services in economic theory and practice: from early notions to markets and payment schemes. Ecological Economics. 69 (6), 1209–1218. http://www.sciencedirect.com/science/article/B6VDY-4XXM2HP-1/2/262d5bd14da586b14dfa2e2590ac30fd.

Gould, R.K., Klain, S.C., Ardoin, N.M., Satterfield, T., Woodside, U., Hannahs, N., Daily, G.C., Chan, K.M., 2015. A protocol for eliciting nonmaterial values using a cultural ecosys-tem services frame. Conservation Biology. 29 (2), 575–586. http://onlinelibrary.wiley.com/doi/10.1111/cobi.12407/full.

Gregory, R., Failing, L., Harstone, M., Long, G., McDaniels, T., 2012. Structured Decision Making: A Practical Guide to Environmental Management Choices. John Wiley & Sons, Inc., Hoboken, NJ. http://books.google.ca/books?id=pU8-YgEACAAJ.

Guerrero, A.M., Bodin, Ö., McAllister, R.R.J., Wilson, K.A., 2015. Achieving social-ecological fit through bottom-up collaborative governance: an empirical investigation. Ecology and Society. 20 (4). http://www.ecologyandsociety.org/vol20/iss4/art41/.

Hare, R.M., 1991. Universal prescriptivism. In: Singer, P. (Ed.), A Companion to Ethics. Blackwell Publishers Inc., Oxford, UK, pp. 451–463.

Heath, C., Heath, D., 2010. Switch: How to Change Things When Change Is Hard. Crown Publishing Group, New York. http://books.google.ca/books?id=QgzBqhbdlvUC.

Hechter, M., Opp, K.D., 2001. Social Norms. Russell Sage Foundation. https://books.google.ca/books?id=uPiFAwAAQBAJ.

Hoffman, D., 2009. Institutional legitimacy and co-management of a marine protected area: implementation lessons from the case of Xcalak Reefs National Park, Mexico. Human Organization. 68 (1), 39–54. http://www.sfaajournals.net/doi/abs/10.17730/humo.68.1.28gw1106u131143h.

Holzmann, R., Jørgensen, S., 2001. Social risk management: a new conceptual framework for social protection, and beyond. International Tax and Public Finance 8 (4), 529–556. http://dx.doi.org/10.1023/A:1011247814590.

Honey-Rosés, J., Pendleton, L.H., 2013. A demand driven research agenda for ecosystem services. Ecosystem Services. http://dx.doi.org/10.1016/j.ecoser.2013.04.007.

Igoe, J., Neves, K., Brockington, D., 2010. A spectacular eco-tour around the historic bloc: theoris-ing the convergence of biodiversity conservation and capitalist expansion. Antipode 42 (3), 486–512. http://dx.doi.org/10.1111/j.1467-8330.2010.00761.x.

Igoe, J., 2004. Disciplining democracy: development discourses and good governance in Africa. The International Journal of African Historical Studies. 37 (3), 577. https://www.questia.com/library/journal/1P3-806342301/disciplining-democracy-development-discourses-and.

IPBES (Intergovernmental Science-Policy Platform on Biodiversity and Ecosystem Services), 2015. Preliminary Guide Regarding Diverse Conceptualization of Multiple Values of Nature and Its Benefits, Including Biodiversity and Ecosystem Functions and Services (Deliverable 3 (d)), p. 95. https://www.researchgate.net/publication/271529734_Preliminary_guide_regard-ing_diverse_conceptualization_of_multiple_values_of_nature_and_its_benefits_including_biodiversity_and_ecosystem_functions_and_services.

Jack, B.K., Kousky, C., Sims, K.R.E., 2008. Designing payments for ecosystem services: lessons from previous experience with incentive-based mechanisms. Proceedings of the National Academy of Sciences of the United States of America 105 (28), 9465–9470. http://dx.doi.org/10.1073/pnas.0705503104.

Jacobson, M.Z., Delucchi, M.A., Bazouin, G., et al., 2015. 100% clean and renewable wind, water, and sunlight (WWS) all-sector energy roadmaps for the 50 United States. Energy & Environmental Science 8 (7), 2093–2117. http://dx.doi.org/10.1039/C5EE01283J.

Jacquet, J., Hocevar, J., Lai, S., Majluf, P., Pelletier, N., Pitcher, T., Sala, E., Sumaila, R., Pauly, D., 2010a. Conserving wild fish in a sea of market-based efforts. Oryx. 44 (1), 45–56. http://journals.cambridge.org/production/action/cjoGetFulltext?fulltextid=6829476.

Jacquet, J., Pauly, D., Ainley, D., Holt, S., Dayton, P., Jackson, J., 2010b. Seafood stewardship in crisis. Nature 467 (7311), 28–29. http://dx.doi.org/10.1038/467028a.

Jax, K., Barton, D.N., Chan, K.M.A., et al., 2013. Ecosystem services and ethics. Ecological Economics. 93 (0), 260–268. http://www.sciencedirect.com/science/article/pii/S0921800913002073.

Jenks, B., Vaughan, P.W., Butler, P.J., 2010. The evolution of rare pride: using evaluation to drive adaptive management in a biodiversity conservation organization. Evaluation and Program Planning. 33 (2), 186–190. http://www.sciencedirect.com/science/article/pii/S0149718909000755.

Kahneman, D., 2013. Thinking, Fast and Slow, Farrar, Straus and Giroux. http://books.google.ca/books?id=iBs9uAAACAAJ.

Kareiva, P., Lalasz, R., Marvier, M., 2012. Conservation in the anthropocene: beyond solitude and fragility. Breakthrough Journal (Winter). http://thebreakthrough.org/index.php/journal/past-issues/issue-2/conservation-in-the-anthropocene.

Kareiva, P., 2014. New conservation: setting the record straight and finding common ground. Conservation Biology 28 (3), 634–636. http://dx.doi.org/10.1111/cobi.12295.

Kinzig, A.P., Perrings, C., Chapin, F.S., Polasky, S., Smith, V.K., Tilman, D., Turner, B.L., 2011. Paying for ecosystem services—promise and peril. Science. 334 (6056), 603–604. http://www.sciencemag.org/content/334/6056/603.short.

Klain, S., Satterfield, T., Chan, K.M.A., 2014. What matters and why? Ecosystem services and their bundled qualities. Ecological Economics. 107, 310–320. http://www.sciencedirect.com/science/article/pii/S0921800914002730.

Klain, S.C., Satterfield, T., Chan, K.M.A., 2017. Quantifying public support for ecologically regenerative renewable energy: high willingness to pay for offshore wind farms with biodiversity benefits (in preparation).

Kosoy, N., Corbera, E., 2010. Payments for ecosystem services as commodity fetishism. Ecological Economics. 69 (6), 1228–1236. http://www.sciencedirect.com/science/article/pii/S0921800909004510.

Lapinski, M.K., Rimal, R.N., 2005. An explication of social norms. Communication Theory 15 (2), 127–147. http://dx.doi.org/10.1111/j.1468-2885.2005.tb00329.x.

Laurans, Y., Rankovic, A., Billé, R., Pirard, R., Mermet, L., 2013. Use of ecosystem services economic valuation for decision making: questioning a literature blindspot. Journal of Environmental Management. 119, 208–219. http://www.sciencedirect.com/science/article/pii/S0301479713000285.

Levin, S.A., 1996. Economic growth and environmental quality. Ecological Applications. 6 (1), 12. http://www.jstor.org/stable/2269538.

Lockwood, M., Davidson, J., Curtis, A., Stratford, E., Griffith, R., 2010. Governance principles for natural resource management. Society & Natural Resources 23 (10), 986–1001. http://dx.doi.org/10.1080/08941920802178214.

Louv, R., 2008. Last Child in the Woods: Saving Our Children from Nature-Deficit Disorder. Algonquin Books, Chapel Hill, NC. http://richardlouv.com/last-child-purchase.

Luck, G.W., Chan, K.M.A., Fay, J.P., 2009. Protecting ecosystem services and biodiversity in the world's watersheds. Conservation Letters 2, 179–188. http://dx.doi.org/10.1111/j.1755-263X.2009.00064.x.

Luck, G., Chan, K.M.A., Eser, U., Gómez-Baggethun, E., Matzdorf, B., Norton, B., Potschin, M., 2012. Ethical considerations in on-ground applications of the ecosystem services concept. BioScience 62 (12), 1020–1029. http://dx.doi.org/10.1525/bio.2012.62.12.4.

Marris, E., 2011. Rambunctious Garden: Saving Nature in a Post-wild World. Bloomsbury Publishing. http://books.google.ca/books?id=vW_bWhe5rwIC.

Marris, E., 2014. 'New conservation' is an expansion of approaches, not an ethical orientation. Animal Conservation. http://dx.doi.org/10.1111/acv.12129 n/a–n/a.

Martinez-Alier, J., 2002. The Environmentalism of the Poor: A Study of Ecological Conflicts and Valuation. Edward Elgar Publishing, Northampton, MA. http://books.google.com/books?id=JR7_onvkj8UC&source=gbs_navlinks_s.

Martinez-Harms, M.J., Bryan, B.A., Balvanera, P., Law, E.A., Rhodes, J.R., Possingham, H.P., Wilson, K.A., 2015. Making decisions for managing ecosystem services. Biological Conservation. 184, 229–238. http://www.sciencedirect.com/science/article/pii/S0006320715000452.

Martín-López, B., Iniesta-Arandia, I., García-Llorente, M., et al., 2012. Uncovering ecosystem service bundles through social preferences. PLoS One 7 (6), e38970. http://dx.doi.org/10.1371/journal.pone.0038970.

Marvier, M., Wong, H., 2012. Resurrecting the conservation movement. Journal of Environmental Studies and Sciences 2 (4), 291–295. http://dx.doi.org/10.1007/s13412-012-0096-6.

McCauley, D.J., 2006. Selling out on nature. Nature. 443 (7107), 27–28. http://www.nature.com/nature/journal/v443/n7107/pdf/443027a.pdf.

McClenachan, L., Neal, B.P., Al-Abdulrazzak, D., Witkin, T., Fisher, K., Kittinger, J.N., 2014. Do community supported fisheries (CSFs) improve sustainability? Fisheries Research. 157, 62–69. http://www.sciencedirect.com/science/article/pii/S0165783614000988.

McLain, R.J., Hurley, P.T., Emery, M.R., Poe, M.R., 2014. Gathering "wild" food in the city: rethinking the role of foraging in urban ecosystem planning and management. Local Environment 19 (2), 220–240. http://dx.doi.org/10.1080/13549839.2013.841659.

Millennium Ecosystem Assessment, 2005. Ecosystems and Human Well-Being: Synthesis. Island Press, Washington, DC. http://www.millenniumassessment.org/documents/document.356.aspx.pdf.

Miller, K., Trezise, J., Kraus, S., Dripps, K., Evans, M., Gibbons, P., Possingham, H.P., Maron, M., 2015. The development of the Australian environmental offsets policy: from theory to practice. Environmental Conservation First View 1–9. http://dx.doi.org/10.1017/S037689291400040X.

Minson, J.A., Monin, B., 2012. Do-gooder derogation: disparaging morally motivated minorities to defuse anticipated reproach. Social Psychological and Personality Science. 3 (2), 200–207. http://spp.sagepub.com/content/3/2/200.abstract.

Moreno-Mateos, D., Maris, V., Béchet, A., Curran, M., 2015. The true loss caused by biodiversity offsets. Biological Conservation. 192, 552–559. http://www.sciencedirect.com/science/article/pii/S0006320715300665.

Muradian, R., Arsel, M., Pellegrini, L., et al., 2013. Payments for ecosystem services and the fatal attraction of win-win solutions. Conservation Letters 6 (4), 274–279. http://dx.doi.org/10.1111/j.1755-263X.2012.00309.x.

Naeem, S., Ingram, J.C., Varga, A., et al., 2015. Get the science right when paying for nature's services. Science. 347 (6227), 1206–1207. http://www.sciencemag.org/content/347/6227/1206.short.

Naidoo, R., Ricketts, T.H., 2006. Mapping the economic costs and benefits of conservation. PLoS Biology. 4 (11), 2153–2164. http://biology.plosjournals.org/perlserv/?request=get-document&doi=10.1371%2Fjournal.pbio.0040360.

Naidoo, R., Balmford, A., Costanza, R., Fisher, B., Green, R.E., Lehner, B., Malcolm, T.R., Ricketts, T.H., 2008. Global mapping of ecosystem services and conservation priorities. Proceedings of the National Academy of Sciences of the United States of America. 105 (28), 9495–9500. http://www.pnas.org/content/105/28/9495.abstract.

Norgaard, R.B., 2010. Ecosystem services: from eye-opening metaphor to complexity blinder. Ecological Economics 69 (6), 1219–1227. http://dx.doi.org/10.1016/j.ecolecon.2009.11.009.

Norton, B.G., Hannon, B., 1997. Environmental values: a place-based theory. Environmental Ethics 19 (3), 227–245. http://dx.doi.org/10.5840/enviroethics199719313.

Norton, B.G., Noonan, D., 2007. Ecology and valuation: big changes needed. Ecological Economics 63 (4), 664–675. http://dx.doi.org/10.1016/j.ecolecon.2007.02.013.

Olmsted, P., Honey-Rosés, J., Satterfield, T., Chan, K.M.A., 2017. Conservation impact investing: promise, pitfalls, and pathways forward (submitted for publication).

Olmsted, P., 2016. Social Impact Investing and the Changing Face of Conservation Finance. IUCN, p. 32. https://www.iucn.org/sites/dev/files/pdf_final_social_impact_investing.pdf.

Onwezen, M.C., Antonides, G., Bartels, J., 2013. The Norm Activation Model: an exploration of the functions of anticipated pride and guilt in pro-environmental behaviour. Journal of Economic Psychology. 39, 141–153. http://www.sciencedirect.com/science/article/pii/S0167487013000950.

Ostrom, E., Janssen, M.A., Anderies, J.M., 2007. Going beyond panaceas. Proceedings of the National Academy of Sciences of the United States of America. 104, 15176–15178. http://www.pnas.org/content/104/39/15176.

Pascual, U., Phelps, J., Garmendia, E., Brown, K., Corbera, E., Martin, A., Gomez-Baggethun, E., Muradian, R., 2014. Social equity matters in payments for ecosystem services. BioScience. 64 (11), 1027–1036. http://bioscience.oxfordjournals.org/content/64/11/1027.abstract.

Peterson, M.J., Hall, D.M., Feldpausch-Parker, A.M., Peterson, T.R., 2010. Obscuring ecosystem function with application of the ecosystem services concept. Conservation Biology 24 (1), 113–119. http://dx.doi.org/10.1111/j.1523-1739.2009.01305.x.

Plieninger, T., Bieling, C., Ohnesorge, B., Schaich, H., Schleyer, C., Wolff, F., 2013. Exploring futures of ecosystem services in cultural landscapes through participatory scenario development in the Swabian Alb, Germany. Ecology and Society. 18 (3). http://www.ecologyandsociety.org/vol18/iss3/art39/.

Poe, M.R., LeCompte, J., McLain, R., Hurley, P., 2014. Urban foraging and the relational ecologies of belonging. Social & Cultural Geography 15 (8), 901–919. http://dx.doi.org/10.1080/14649365.2014.908232.

Poe, M.R., Donatuto, J., Satterfield, T., 2016. "Sense of place": human wellbeing considerations for ecological restoration in Puget Sound. Coastal Management 1–18. http://dx.doi.org/10.1080/08920753.2016.1208037.

Polasky, S., Doremus, H., Rettig, B., 1997. Endangered species conservation on private land. Contemporary Economic Policy. 15 (4), 66–76. http://onlinelibrary.wiley.com/doi/10.1111/j.1465-7287.1997.tb00490.x/abstract.

Rawls, J., 1971. A Theory of Justice. The Belknap Press of Harvard University Press, Cambridge, MA.

Rees, W.E., 1998. How should a parasite value its host? Ecological Economics 25 (1), 49–52. http://dx.doi.org/10.1016/S0921-8009(98)00015-9.

Reno, R.R., Cialdini, R.B., Kallgren, C.A., 1993. The transsituational influence of social norms. Journal of Personality and Social Psychology. 64 (1), 104. https://www.researchgate.net/profile/Robert_Cialdini/publication/232604958_The_transsituational_influence_of_social_norms/links/0a85e53b2ea8be1795000000.pdf.

Ruckelshaus, M., McKenzie, E., Tallis, H., et al., 2015. Notes from the field: lessons learned from using ecosystem service approaches to inform real-world decisions. Ecological Economics. 115, 11–21. http://www.sciencedirect.com/science/article/pii/S0921800913002498.

Scannell, L., Gifford, R., 2010a. Defining place attachment: a tripartite organizing framework. Journal of Environmental Psychology. 30 (1), 1–10. http://www.sciencedirect.com/science/article/pii/S0272494409000620.

Scannell, L., Gifford, R., 2010b. The relations between natural and civic place attachment and pro-environmental behavior. Journal of Environmental Psychology. 30 (3), 289–297. http://www.sciencedirect.com/science/article/pii/S0272494410000198.

Schlosberg, D., 2004. Reconceiving environmental justice: global movements and political theories. Environmental Politics 13 (3), 517–540.

Schomers, S., Matzdorf, B., 2013. Payments for ecosystem services: a review and comparison of developing and industrialized countries. Ecosystem Services. 6 (0), 16–30. http://www.sciencedirect.com/science/article/pii/S221204161300003X.

Scott, J.M., Goble, D.D., Davis, F.W., 2006. The Endangered Species Act at Thirty. Conserving Biodiversity in Human-Dominated Landscapes, vol. 2. Island Press, Washington, DC. https://books.google.ca/books?id=luJtr72vqi0C.

Shellenberger, M., Nordhaus, T., 2004. The Death of Environmentalism: Global Warming Politics in a Post-environmental World. http://www.thebreakthrough.org.

Shove, E., 2010. Beyond the ABC: climate change policy and theories of social change. Environment and Planning A. 42 (6), 1273–1285. https://blog.itu.dk/hest/files/2012/10/shove_abc.pdf.

Sikor, T., 2013. The Justices and Injustices of Ecosystem Services. Taylor & Francis, New York, NY. http://books.google.ca/books?id=dd0cAAAAQBAJ.

Smith, R.J., Verissimo, D., Leader-Williams, N., Cowling, R.M., Knight, A.T., 2009. Let the locals lead. Nature 462 (7271), 280–281. http://dx.doi.org/10.1038/462280a.

Soulé, M., 2013. The "new conservation," Conservation Biology 27 (5), 895–897. http://dx.doi.org/10.1111/cobi.12147.

Steffen, W., Crutzen, P.J., McNeill, J.R., 2007. The anthropocene: are humans now overwhelming the great forces of nature. AMBIO: A Journal of the Human Environment 36 (8), 614–621. http://dx.doi.org/10.1579/0044-7447(2007)36%5B614:TAAHNO%5D2.0.CO;2.

Tallis, H., Kennedy, C.M., Ruckelshaus, M., Goldstein, J., Kiesecker, J.M., 2015. Mitigation for one & all: an integrated framework for mitigation of development impacts on biodiversity and ecosystem services. Environmental Impact Assessment Review. 55 (0), 21–34. http://www.sciencedirect.com/science/article/pii/S0195925515000566.

TEEB, 2009. The Economics of Ecosystems and Biodiversity for National and International Policy Makers. United Nations Environment Programme, Wesseling, Germany. http://www.teebweb.org/ForPolicymakers/tabid/1019/language/en-US/Default.aspx.

Thaler, R.H., Sunstein, C.R., 2008. Nudge: Improving Decisions about Health, Wealth, and Happiness. Yale University Press. http://books.google.ca/books?id=dSJQn8egXvUC.

Turner, W.R., Brandon, K., Brooks, T.M., Costanza, R., da Fonseca, G.A.B., Portela, R., 2007. Global conservation of biodiversity and ecosystem services. BioScience 57 (10), 868–873. http://dx.doi.org/10.1641/B571009.

Turner, R.A., Fitzsimmons, C., Forster, J., Mahon, R., Peterson, A., Stead, S.M., 2014. Measuring good governance for complex ecosystems: perceptions of coral reef-dependent communities in the Caribbean. Global Environmental Change. 29, 105–117. http://www.sciencedirect.com/science/article/pii/S0959378014001447.

Vatn, A., 2010. An institutional analysis of payments for environmental services. Ecological Economics. 69 (6), 1245–1252. http://www.sciencedirect.com/science/article/pii/S0921800909004674.

Veríssimo, D., 2013. Influencing human behaviour: an underutilised tool for biodiversity management. Conservation Evidence 10, 29–31.

Vira, B., Adams, W.M., 2009. Ecosystem services and conservation strategy: beware the silver bullet. Conservation Letters 2 (4), 158–162. http://dx.doi.org/10.1111/j.1755-263X.2009.00063.x.

Warner, M.E., 2013. Private finance for public goods: social impact bonds. Journal of Economic Policy Reform 16 (4), 303–319. http://dx.doi.org/10.1080/17487870.2013.835727.

West, P., Brockington, D., 2006. An anthropological perspective on some unexpected consequences of protected areas. Conservation Biology. 20 (3), 609–616. http://www.blackwell-synergy.com/doi/abs/10.1111/j.1523-1739.2006.00432.x.

Wilcove, D.S., Lee, J., 2004. Using economic and regulatory incentives to restore endangered species: lessons learned from three new programs. Conservation Biology. 18 (3), 639–645. http://www.blackwell-synergy.com/doi/abs/10.1111/j.1523-1739.2004.00250.x.

Wittman, H.K., Caron, C., 2009. Carbon offsets and inequality: social costs and co-benefits in Guatemala and Sri Lanka. Society & Natural Resources 22 (8), 710–726. http://dx.doi.org/10.1080/08941920802046858.

Wittman, H., Powell, L.J., Corbera, E., 2015. Financing the agrarian transition? The clean development mechanism and agricultural change in Latin America. Environment and Planning A. 47 (10), 2031–2046. http://epn.sagepub.com/content/47/10/2031.abstract.

World Resources Institute (WRI), World Business Council on Sustainable Development (WBCSD), Meridian Institute, 2008. The Corporate Ecosystem Services Review: Guidelines for Identifying Business Risks and Opportunities Arising from Ecosystem Change. WBCSD Publications. 37+vi http://www.wbcsd.org/Plugins/DocSearch/details.asp?DocTypeId=25&ObjectId=Mjg5NjQ.

Wright, A.J., Veríssimo, D., Pilfold, K., et al., 2015. Competitive outreach in the 21st century: why we need conservation marketing. Ocean & Coastal Management. 115, 41–48. http://www.sciencedirect.com/science/article/pii/S0964569115001829.

Wunder, S., 2006. Are direct payments for environmental services spelling doom for sustainable forest management in the tropics? Ecology and Society. 11 (2). http://www.ecologyandsociety.org/vol11/iss2/art23/.

Wunder, S., 2013. When payments for environmental services will work for conservation. Conservation Letters 6 (4), 230–237. http://dx.doi.org/10.1111/conl.12034.

Young, O.R., 2002. The Institutional Dimensions of Environmental Change: Fit, Interplay, and Scale. MIT Press, Cambridge, MA. https://books.google.ca/books?id=jd8rD4gEJLQC.

Chapter 12

Beyond Privatization: Rethinking Fisheries Stewardship and Conservation in the North Pacific

Rachel Donkersloot[1], Courtney Carothers[2]
[1]Alaska Marine Conservation Council, Anchorage, AK, United States; [2]University of Alaska Fairbanks, Anchorage, AK, United States

INTRODUCTION

The conservation community has increasingly embraced privatization as a means to promote environmental goals. This so-called neoliberal conservation situates property rights and free market exchange at the heart of environmental governance (Igoe and Brockington, 2007). In the case of fisheries, individual transferable quotas (ITQs) are imagined to be secure property rights that confer economic incentives for owners of fishery access rights (quota holders) to fish sustainably. Many international conservation organizations, such as the Environmental Defense Fund, are vocal proponents of ITQs, or catch shares, for conservation based on this logic:

> Catch shares "right the ship." With a secure share of the catch, there is no pressure or need to race for fish. And with a clear stake in the overall health and sustainability of the fishery, fishermen's incentives change from maximizing volume to maximizing value. Fishermen no longer become fierce competitors but are now inspired to collaborate as environmental stewards of the resource their livelihood depends on. This type of cooperation is almost unheard of in non-catch share fisheries where competition—not communication—is the rule… Evidence shows that catch shares overcome the "tragedy of the commons" by providing a clear economic rationale for conserving resources.
>
> EDF (2016)

The logic underpinning such a theoretical link—that secure private property rights make owners care better for the resource for the long term—is faulty for several reasons (Bromley, 2015; Macinko and Bromley, 2004, 2002). First, although ITQs often function like private property rights, they are rarely legally defined as such, but usually as revocable access privileges (Shotton, 2001; Hannesson, 2006; Abbott et al., 2010). Second, even if ITQs are fully privatized and monetized, market incentives and economic pressures may make degradation of a fishery financially rational (Sumaila, 2010), as Acheson (2006) describes for privately owned forests in Maine. Third, as we discuss in the following, leasing of access rights means that in many fisheries, the "owners" of access rights are not active fishermen directly engaged in harvesting the resource.

Nevertheless *catch shares for conservation* has become a potent narrative perpetuated in academia (e.g., Costello et al., 2016, 2008) and popular media (e.g., Rowley, 2016; NPR, 2015; Economist, 2008). A series of researchers have challenged the methodology and conclusions drawn by a well-cited paper that links catch shares with the prevention of fisheries "collapse" (Costello et al., 2008). These studies note that biological tools such as setting an appropriate total allowable catch for fisheries limits overharvesting better than the implementation of catch shares, a primarily allocative tool designed to promote economic efficiencies (e.g., Acheson et al., 2015; Melnychuk et al., 2012; Essington et al., 2012; Essington, 2010; Branch, 2009; Chu, 2009; Ban et al., 2008). In this chapter, we challenge the claim that the privatization of a public resource leads to an enhanced conservation ethic and suggest a need for more holistic approaches to managing fisheries as complex socio-ecological systems. We highlight potential community-oriented alternatives as fishery conservation solutions that do not come at the expense of fishing communities, rural livelihoods, and future generations.

The story about catch shares "righting" the tragedy of the commons obscures not only the historical and contemporary successes in managing common pool resources (Ostrom, 1990), but also the ways in which catch shares as quasi-private property rights are creating the "tragic commons" (Chambers and Carothers, 2017) or the "tragedy of the commodity" (e.g., Longo et al., 2015; Carothers, 2010; McCay, 2004). It is clear that the commodification of fishery access rights is remaking fishery systems with largely negative impacts to small-scale fishermen, non-owners, young and new fishery entrants, and rural and indigenous communities (e.g., Carothers and Chambers, 2012; Olson, 2011; Knapp and Lowe, 2007; McCay, 2004). ITQ programs create a host of equity issues and contribute to the alienation of fishing rights from longstanding fishing communities and cultures (e.g., Donkersloot and Carothers, 2016; Carothers, 2015, 2010; Pinkerton and Davis, 2015; McCormack, 2012; Pinkerton and Edwards, 2009). As well, the positioning of catch shares as the prevailing answer to fishery conservation problems tends to overshadow the point that catch shares are primarily a tool to promote economic efficiencies and maximize aggregate fleet-wide profits, typically with little concern for distribution of that wealth. Sumaila (2010, p. 1) calls attention to this point noting that "if economic efficiency were the only concern of fisheries management, then ITQs would be a great tool for achieving management objectives…, but fisheries management is not about economic efficiency alone. It is about conserving the resources, preserving the ecosystems that support the resources through time, and ensuring equity and social justice in the use of these resources."

The purpose of this chapter is twofold. For one, we challenge the underlying assumptions driving fishery privatization processes, paying particular attention to the validity of the ownership-promotes-stewardship thesis. As part of this effort, we draw attention to the mixed conservation outcomes of catch share programs and consider whether the benefits associated with slowing down the "race for fish"—often attributed to the creation of alienable property rights—might be achieved in ways that do not require allocating access rights in perpetuity to the current generation of harvesters. Second, we argue that conservation solutions that create social inequities and alienate local resource users from the resource base are unsustainable and run counter to fisheries management goals of social and economic sustainability. We draw on contemporary case studies from the North Pacific to highlight complex environmental and equity concerns in fisheries managed under various forms of catch shares. We argue that the outflow of fishing rights from fishery-dependent communities, now a predictable outcome of ITQ management, is antithetical to the goals of resource governance and fishery conservation today, including fishing community stability and the sustained participation of fishing communities. We suggest that rather than fully alienable private property rights that serve to sever relationships between people, places, and resources, we must consider place-based fishing livelihoods and human–environment connections as fundamental to the sustainability of healthy social–ecological systems. Central here is the need for alternative constructions of stewardship to better inform fisheries management.

DEBUNKING THE INDIVIDUAL OWNERSHIP-PROMOTES-STEWARDSHIP THESIS

The ownership-promotes-stewardship thesis conjures up powerful imagery of a derelict commons driven to collapse by the inability of competitive, self-interested actors to limit use of a common resource. From these assumptions, the creation of property rights signals a promising shift in behavior. Empowered with newly acquired assets, resource users are incentivized as owners to act as better stewards of the resource for their long-term interest and economic gain (Arnason, 2012; Grafton et al., 2006). Environmental Defense Fund's *Catch Share Design Manual* unpacks these underlying assumptions more completely noting that "by allocating participants a secure [share] of the catch, catch share programs give participants a long-term stake in the fishery and tie their current behavior to future outcomes. This security provides a stewardship incentive for fishermen that was previously missing or too uncertain to influence their behavior toward long-term conservation" (Bonzon et al., 2013, p. 2).

Catch share programs come in a diversity of forms and have been designed to address a number of management objectives including increasing economic efficiencies, reducing overcapitalization, reducing bycatch, extending fishing seasons, reducing market gluts, and improving at-sea safety. Nevertheless, it is their potential to solve conservation problems that is frequently singled out. Catch shares are increasingly heralded as a necessary incentive needed to improve resource conservation. As de facto property rights assumed to inspire environmental stewardship, catch shares have been embraced by fishing nations around the world. There are an estimated 250 fisheries globally managed under some form of an ITQ system today (Chu, 2009). In the United States, the rise of the "privatize or perish" message peaked in 2009 when the National Oceanic and Atmospheric Administration (NOAA) explicitly promoted catch shares as a desired management tool and encouraged "the voluntary use of well-designed catch share programs in appropriate fisheries to help rebuild and sustain fisheries and support fishermen, communities and vibrant working waterfronts" (NOAA Fisheries Service, 2010, p. 1). Endorsement of catch shares as a cure-all contrasts sharply with a growing body of literature documenting the undesirable social costs of catch share programs and importantly, a lack of empirical evidence in support of ITQs as contributing to enhanced stewardship ethics (e.g., Pinkerton and Davis, 2015; Van Putten et al., 2014; Reedy and Maschner, 2014; Carothers and Chambers, 2012; Pinkerton and Edwards, 2009; Lowe and Carothers, 2008; Gilmour et al., 2012).

ITQs, Conservation, and Stewardship

Despite the fact that catch shares are primarily a tool to promote economic efficiencies (Pinkerton, 2013; Sumaila, 2010; McCay, 2004), proponents of privatizing fisheries access frequently assert that catch shares are linked to increased resource conservation and stewardship outcomes. The linkage between ITQs and resource conservation is not as straightforward as proponents contend. Acheson et al. (2015, p. 7) offer a succinct review of literature examining the effects of ITQs on fish stocks and conclude "the best evidence strongly suggests the effects are mixed."

> *Chu (2009) points out that in a study of 20 stocks where biomass changes were analyzed, there was an improvement in 12 stocks after the advent of an ITQ program. The other eight (40%) continued to decline... [The] same lack of consistent results was reported by Branch (2009), who analyzed the effect of ITQs as reported in 227 peer reviewed papers... Thirty-five papers or 15% of the total reported on the biological effects of ITQs. The results were mixed, with 60% of these reporting a positive effect, while 23% reported a negative effect, and another 14% reported a mixed effect... Essington (2010) assessed the effect of implementing catch shares on certain indicators of conservation, including biomass, fishing effort, and discards. With the exception of a decline in discard rate, he is unable to see any significant change in these indicators following implementation of ITQs.*
>
> Acheson et al. (2015, p. 7)

Other researchers demonstrate the mixed success of ITQ management regimes to provide consistent conservation outcomes (e.g., Melnychuk et al., 2012; Thébaud et al., 2012; Sumaila, 2010; Gibbs, 2009). In a comprehensive analysis of how catch shares affect fished populations in 84 catch share and 140 reference fisheries, Essington et al. (2012, p. 8) found that "many of the elements of the fishing systems—including the economic and social systems—that promoted overexploitation prior to catch shares largely persisted after catch shares were implemented."

The link between privatized access and stewardship ethics and behaviors is also not straightforward. Some studies have documented how the incentives embedded in ITQs have actually led to unsustainable fishing practices (e.g., Chambers and Carothers, 2017; McCay, 2004; Copes and Pálsson, 2000), including higher discard rates. Rieser et al. (2013) examined how ownership incentives contribute to enhanced stewardship in catch share programs in place in Alaska and New Zealand trawl fisheries where quota holders worked with fishery managers to create bottom trawl closures. In the Alaska example, the outcome resulted in a closure of nearly 1 million km^2 of seafloor, effectively freezing the trawl footprint. However, as the authors note, only 10% of the closed area was considered fishable due to ocean depths (greater than 1000 m) and other factors, and some of the most important coral and sponge habitat areas identified remained unprotected in areas which remained open to trawling (Rieser et al., 2013, p. 77). In this way, the authors differentiate between the "creation of a perception of habitat stewardship" in these fisheries and the adoption of industry-supported measures for "protection of seafloor habitat that can be characterized as responsible stewardship" (Rieser et al., 2013, p. 82). Elsewhere, examining the theoretical relationship between the implementation of ITQs and changes in environmental stewardship of fishers, Van Putten et al. (2014, p. 5) note, "there is no evidence available to indicate that this environmental stewardship has changed as a consequence of fisheries management changes."

Prevalent shifts in patterns of quota ownership and the leasing of access rights raise further questions about the conservation logic driving the shift toward fisheries privatization. Of the many undesirable social consequences of catch share programs documented around the world, quota leasing practices, marked by high lease fees and absentee ownership, are among the most disconcerting (NPFMC, 2015, p. 50; Pinkerton and Edwards, 2009). Consolidation of fishing vessels and concentration of quota ownership and wealth are well documented and often intended outcomes of catch share programs designed to reduce overcapitalization (Knapp and Lowe, 2007; Knapp, 2006). Such shifts in (absentee and in some cases corporate) ownership complicate the supposed connection between ownership rights and stewardship due to the leasing of access rights, which means that in many fisheries "owners" are not the same individuals harvesting the fish. ITQ systems are creating a new class of fishermen, likened to "sharecroppers" in feudal systems (see van der Woo, 2013), who must now rent the right to fish from quota holders who stay ashore and collect fishery earnings as absentee owners (Van Putten et al. (2014,; Pinkerton and Edwards, 2009; Helgason and Palsson, 1997; Eythórsson, 1996). These types of arrangements are at odds with the underlying arguments driving the ownership-promotes-stewardship thesis. Given the complex relationships in fishery systems and the trends noted earlier, we resist the oversimplification that the privatization of a public resource uniformly leads to enhanced conservation; rather we point to the importance of understanding how other fishery management design features, such as setting appropriate total catch limits, monitoring, enforcement, improved communication, and others related to slowing down the race for fish, contribute to ensuring sustainable fishing practices. This will be discussed further in the following.

COMMUNITIES, CONSERVATION, AND CATCH SHARES: EXAMPLES FROM THE NORTH PACIFIC

The North Pacific is a region recognized as a global leader in managing sustainable fisheries and developing innovative and community-oriented management models. It is also a site of powerful examples of how catch shares are remaking fishery systems, and how they can—sometimes predictably, sometimes unexpectedly—create inequities and impact the sustainability of socio-ecological systems by turning the right to fish into a tradable commodity.

There are currently five catch share programs in place in the North Pacific, with another potential program under development. Only one of these programs was developed to specifically address conservation concerns. The Bering Sea non-pollock groundfish trawl fishery (often referred to as the "Amendment 80" fleet) was rationalized in 2008 to improve bycatch reduction and accountability among the fleet targeting Atka mackerel, Pacific ocean perch, and three flatfish species. Other programs including the Gulf of Alaska Rockfish Program (2010), Bering Sea Aleutian Island (BSAI) crab rationalization program (2005), American Fisheries Act pollock cooperatives program (1999), and halibut and sablefish IFQ program (1995) were created to address management concerns of overcapitalization, allocation disputes, US ownership requirements, safety and derby style fishing marked by short seasons, loss of product/quality and the "race for fish" (Fina, 2011, p. 165).

The Halibut IFQ Fishery

The halibut IFQ fishery is an insightful starting point to consider the complex and shifting dynamic between catch shares, conservation, and communities in the North Pacific.

The US North Pacific halibut fishery shifted to ITQ management along with the sablefish fishery in 1995. This fishery has long been used as an example of the economic efficiencies and improvements in safety that can accompany ITQ management (e.g., Matulich and Clark, 2003; Hartley and Fina, 2001). Short and chaotic fishing openings that produced a glut of product on the market transitioned into a 9-month fishing season following ITQ implementation, with increases in halibut value for a harvesting fleet that consolidated by over 50% (NOAA Fisheries Service, 2016; Herrmann and Criddle, 2006). However, despite design parameters that attempted to maintain diversity in the fleet, such as owner-on-board provisions for the next generation of quota holders and restrictions on quota transfers between vessel class sizes, disproportionate social impacts were felt by crew members, skippers, small-scale fishermen, and rural, primarily indigenous communities post-ITQ implementation (Carothers, 2013; NPFMC, 2016). For example, since the program was implemented there has been a 57% decrease in the number of residents in small, mostly indigenous Gulf of Alaska communities who hold halibut quota (NOAA Fisheries Service, 2014). Equity concerns were raised and the North Pacific Fishery Management Council responded by implementing a community purchase program in 2004 that enables small communities to collectively purchase halibut and sablefish quota. This program has yet to lead to any significant reallocation of halibut quota to affected communities (Carothers, 2011; NPFMC, 2016; NPFMC, 2010b; Langdon, 2008). This is due in part to the very high cost of purchasing halibut fishing rights and lack of available quota shares for sale (Carothers, 2011).

In recent years, the halibut fishery has struggled with uncertainty over the health of the resource. Under ITQ management, the total allowable catch (TAC) over the past decade has decreased substantially due to declining stock abundance of harvestable halibut (i.e., exploitable biomass). This decline is linked to significant reductions in female spawning biomass and decreasing size at age (Stewart et al., 2014). The International Pacific Halibut Commission (IPHC), which manages Pacific halibut stocks in waters off Alaska, British Columbia, and the west coast of the United States, estimates a 66% decline in catch rates in the directed halibut fishery in Alaska from 2000 to 2013 (Stewart et al., 2014). In 2015, conservation concerns over the halibut resource culminated into crisis when the IPHC recommended another 60% reduction from 2014 harvest levels. As catch limits plummeted due to declining stock abundance of harvestable halibut, equity concerns emerged over what amounted to a dramatic reallocation of the halibut resource in the North Pacific as bycatch for Bering Sea groundfish fisheries.

Halibut Bycatch in the Bering Sea Groundfish Fisheries

Bering Sea groundfish fisheries, and their associated halibut bycatch, are managed by the North Pacific Fishery Management Council (NPFMC). Prior to 2015, on average, Bering Sea groundfish fisheries took around 5 million pounds of halibut a year as bycatch. Because the NPFMC manages halibut bycatch with a fixed hard cap that is not indexed to abundance levels, the IPHC must deduct the previous year's halibut bycatch before setting annual catch limits in the directed halibut fishery throughout Alaska. Because bycatch limits do not shift with abundance levels, in recent years, as the halibut stock has declined, bycatch has become the primary source of halibut mortality in the Bering Sea. The impacts of halibut bycatch mortality are felt well beyond the Bering Sea, which serves as nursery grounds for stocks which embark on lifelong migrations as far south as Oregon, but the situation is particularly problematic for small Bering Sea communities dependent on halibut harvested around the Pribilof Islands and along the mainland coast of western Alaska in management Area 4CDE (see Fig. 12.1).

In 2015, the IPHC recommended a harvest limit of 370,000 pounds for Area 4CDE, amounting to a 71% cut from the 2014 limit of 1.29 million pounds (see Fig. 12.2). IPHC estimates showed that at the 2015 projected harvest level, bycatch would account for 93% of all halibut removals in the Bering Sea. Such a stark disparity in allocation of the resource would effectively eliminate the directed halibut fishery in this area. This, paired with the failure of the NPFMC to take action to reduce halibut bycatch at a December 2014 meeting, prompted all six of the Alaska members of the 11 (voting) member NPFMC to request the National Marine Fisheries Service Assistant Administrator for an emergency 33% reduction in Bering Sea halibut bycatch in 2015. The response to the emergency regulation request recommended the IPHC "provide adequate harvest opportunities" for Area 4CDE halibut fishermen in 2015 without a reduction in bycatch levels in groundfish fisheries. The recommendation was based on the "potentially serious socio-economic impacts of a low catch limit" for the remote fishery-dependent communities in the region while recognizing the "extensive new efforts being taken by the [groundfish] fleet to further minimize bycatch" in 2015.fn11[1] The Amendment 80 fleet in particular has made a concerted effort to address bycatch problems through innovative measures and experimentation, such as deck sorting initiatives and handling practices, technological advances, use of excluder devices, and improved communication among vessels, tools that are made more successful in the cooperative management system. In the end, the IPHC exceeded their own recommended catch rate for 2015 and maintained the 2014 harvest level of 1.29 million pounds for Area 4CDE. At the June 2015 meeting, the NPFMC recommended new halibut bycatch limits in the Bering Sea which would ultimately reduce bycatch in the Amendment 80 fleet by 17% relative to 2014 levels. Though important, this action falls short of the reduction needed to support the directed catch limit of 1.29 million pounds that the IPHC estimates to be equivalent to a reduction in the halibut bycatch limit of approximately 41% (IPHC, 2015). The North Pacific Fishery Management Council is currently working toward developing additional conservation tools in Bering Sea groundfish fisheries, including abundance-based halibut bycatch caps.

---

1. Letter from Eileen Sobeck, NMFS Assistant Administrator for Fisheries, to Bruce Leaman, IPHC Executive Director, dated January 20, 2015.

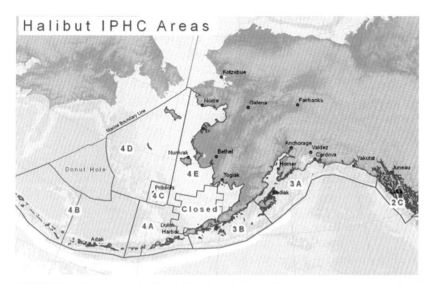

FIGURE 12.1 International Pacific Halibut Commission halibut management areas. *Reproduced from NOAA, 2015. Transfer Report, Changes Under Alaska's Halibut IFQ Program, 1995 Through 2014. p. 3. Available at: https://alaskafisheries.noaa.gov/sites/default/files/reports/halibut-transfer-frpt2015.pdf#page=17&zoom=auto,-73,620.*

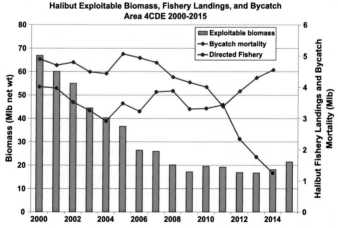

FIGURE 12.2 Trends in halibut exploitable biomass, bycatch mortality, and directed halibut fish-ery catch limits in IPHC regulatory area 4CDE, 2000–15. *Reproduced from IPHC, 2015. IPHC letter to NPFMC of May 26, 2015. June 2015 meeting, Agenda Item C2 BSAI PSC Limits. Available at: http://www.iphc.int/documents/bycatch/IPHC2NPFMC_PSClimitJune2015b.pdf.*

This reallocation of the halibut resource is a complex example occurring across two management bodies (IPHC and NPFMC) governing small-scale and industrial fisheries with varying degrees of political power, and managed under two different catch share programs. We draw on it here to illustrate the siloed and cascading effects of catch shares as a management tool within and across these fisheries, including their limitations in responding to environmental change and related community sustainability concerns. In the Gulf of Alaska, the privatization of access to the halibut fishery has redistributed access rights and severely limited local access for rural and indigenous communities and new entrants (Donkersloot and Carothers, 2016; Carothers, 2011, 2013). In the Bering Sea, the shrinking halibut resource has been disproportionately allocated to larger and industrial fisheries operating offshore contributing to great instability, hardship, and social and economic inequities in the halibut fisheries (commercial and subsistence) and fishing communities. In this way, although an ITQ system in the halibut fishery has served to extend the fishing season and improve product quality, it has not been able to create the economic stability or ensure the overall health of the directed halibut fishery because it is nested within larger political and power structures which prioritize the needs of competing fishery systems. Even more concerning is the fact that, although the fleet has not been reaching historic bycatch levels and has been operating below the bycatch cap for some years, halibut bycatch in the Bering Sea groundfish fisheries rose steadily between 2011 and 2014 despite the presence of a form of catch shares as a management tool (IPHC, 2015). This increase occurred while the directed halibut fishery experienced rapid declines in catch limits. This example demonstrates how fishery management tools are embedded in complex political processes with complex political outcomes (Donkersloot, 2016). It also draws attention to the need for additional and alternative management tools in both reducing bycatch and ensuring fishing community participation and stability and suggests that quasi-private property rights alone are inadequate in achieving conservation goals and solving the complex problems facing fisheries and fishing communities today (Dietz et al., 2003).

MOVING PAST COMMODITIZED FISHING RIGHTS TOWARD OTHER DESIGN PRINCIPLES

In the North Pacific, various forms of catch share programs have been implemented as a solution to a range of economic and environmental concerns (Fina, 2011). These programs have achieved some management objectives including improving economic efficiencies, product quality and safety, but they have also resulted in a number of undesirable social consequences and inequities evidenced in place-based fishing communities and dynamics within and across North Pacific fisheries. In the context of fishery conservation, privatization has not consistently produced the desired conservation outcomes. Instead, the earlier examples point to the importance of specific provisions for achieving conservation gains including but not limited to, catch limits, abundance-based bycatch caps, monitoring and enforcement, information sharing, and improved fleet coordination. Some of these features, especially information sharing and cooperation among vessels, have been identified as benefits of catch share programs because catch shares end the race for fish. Is it possible to reimagine fisheries management regimes in ways that draw upon these conservation tools without commodifying and privatizing the right to access fishery resources? Again, the North Pacific offers a compelling example.

A new management structure is currently under development for the Gulf of Alaska groundfish trawl fishery. The new program is driven by the need to provide the trawl fleet with the tools to better address mounting bycatch concerns. Catch shares have been identified as a preferred tool for bycatch management in the discussion to date but the Council process has been wrought with conflict and disagreement over whether catch shares are the right way forward for all stakeholders including vessel owners, crew, processors, and communities (Donkersloot, 2016). In October 2015, the Commissioner of the Alaska Department of Fish and Game, who represents the State of Alaska on the NPFMC, introduced an alternative program structure with the aim of enabling bycatch reductions while avoiding the creation of new economic assets that result from privatized and transferrable quota. The Council is also considering a novel allocation of quota to a community entity as a means to mitigate some of the potential impacts of the program on Gulf of Alaska fishing communities (Donkersloot, 2016). It is too early to tell which direction the Council will take in designing a new management structure. In the meantime, the Gulf of Alaska trawl fleet is operating under a voluntary cooperative structure to better communicate information and adapt to the recent implementation of bycatch caps, including first-ever caps on Chinook salmon bycatch and a 15% reduction in the halibut bycatch limit in Gulf of Alaska trawl fisheries.

Taken in sum, the successes and challenges in the North Pacific compel us to ask whether the allocation of quasi-private property rights is the best mechanism to manage our fisheries. What is happening in the Gulf of Alaska suggests that there are potential alternatives to addressing conservation concerns that allow for a more equitable allocation of fishery resources without transforming the right to fish into a tradable commodity with windfall gains to initial recipients (see also Eythórsson, 2016; Foley et al., 2013, 2015). Considering equity in fisheries, management becomes more imperative in light of the rise in corporate and processor-owned quota share that has been described as a new tragedy in fisheries created under rights-based management regimes (Donkersloot, 2016; Olson, 2011; Pauly, 2008; Dewees, 1989).

RETHINKING STEWARDSHIP—FROM COMMODITY TO COMMUNITY

Fishery managers in the North Pacific have become increasingly aware of the ways in which catch share programs negatively impact communities, crew, and new entrants (Carothers, 2011; Reedy and Maschner, 2014; Knapp, 2006; Knapp and Lowe, 2007; Lowe and Carothers, 2008), as well as the widespread difficulty of reversing the inequitable outcomes of ITQ regimes (Chambers and

Carothers, 2017; Copes and Pálsson, 2000). For example, the high cost of quota coupled with high leasing fees and absentee ownership in Bering Sea crab and North Pacific halibut and sablefish fisheries continues to impede upward mobility of crew members in these fisheries (Szymkowiak and Himes-Cornell, 2015; NPFMC, 2010a). North Pacific fishery managers have implemented a number of community-oriented provisions to mitigate these types of impacts including crew shares, community quotas, and consolidation caps, among others. These provisions aim to ensure that fishery conservation solutions do not come at the expense of fishing communities, rural livelihoods, and future generations. Some of these provisions have been more successful than others. For example, the Western Alaska Community Development Quota (CDQ) program, first implemented in 1992, is often considered the "crown jewel" in fisheries management models while the potential for the Community Quota Entity program in place in the Gulf of Alaska remains largely unrealized (Carothers, 2011). The CDQ program was created in 1992 as part of the rationalization of Bering Sea pollock fishery. At the time, the program allocated 7.5% of the pollock resource to Western Alaska communities, many of which are economically disadvantaged, geographically isolated, and largely Alaska Native. Today the CDQ program has grown to include a 10% allocation of all BSAI quotas for groundfish, halibut, and crab. The program embeds resource wealth from Bering Sea offshore fisheries in Western Alaska communities (see also Foley et al., 2013, 2015). CDQ entities use royalties from these fisheries to advance regional economic development through investments in local industry, part ownership of the offshore vessels, infrastructure, and education. In the context of halibut, the CDQ program also provides real fishing opportunity for CDQ residents who actively fish the CDQ allocation. In contrast, the Community Quota Entity (CQE) program did not allocate quota to communities, but allows eligible communities to purchase quota to lease to resident fishermen. The high cost of halibut quota coupled with sharp declines in the TAC means that many CQE communities cannot afford to take advantage of the program. For example, the price of halibut quota has increased from around $10.00 per pound in 1995 to more than $50.00 per pound in 2016.[2]

On a broader scale, the marginalization of place and community-based livelihoods in fisheries privatization processes has resulted in a more explicit articulation of social goals in US fisheries policy. National Standard 8 of the Magnuson-Stevens Fisheries Conservation and Management Act (MSA) pushes fisheries management policy beyond the narrow scope of environmental sustainability to recognize community sustainability as central to fisheries management. Specifically, it instructs policy makers to "take into account the importance of fishery resources to fishing communities in order to (A) provide for sustained participation of such communities, and (B) to the extent practicable, minimize adverse impacts on such communities." The 2006 MSA

---

2. See NPFMC (2016, p. 100) and http://www.alaskabroker.com/listings/halquota.html#quota_three.

reauthorization reaffirms the importance of place in fisheries management with the inclusion of language authorizing mechanisms to distribute fishing privileges to communities [see Sections 303A(c)(3)]. Community held fishing rights have been described as the "next step in the evolution of catch shares" (see Donkersloot, 2016) and "appear to have been driven by Congress' interest in supporting small-scale and community-based operations" given the tendency for these to be disproportionately negatively impacted by limited access privilege programs (Stoll and Holliday, 2014, p. 2). The opening up of the narrow corridors of catch share programs to include recognition of the importance of sustaining communities and community-based livelihoods in fisheries allocation regimes is a meaningful contribution to public policy. The authorization of allocations to community entities in particular is an implicit challenge to the argument that private property rights are the solution to fishery conservation problems. The move toward community held fishing rights helps to highlight the difference between the actual tool of assigning a portion of the total catch and the ideology of "privatize or perish" that insists on giving away fisheries resource wealth to the current generation of harvesters. Community allocations suggest that responsible stewardship is possible without quasi-private property rights.

CONCLUDING THOUGHTS

If we view marine conservation as stewarding sustainable human-marine connections, ITQs become antithetical to conservation because they function as a mechanism for the alienation of local fishing rights embedded in place. The widespread privatization of access rights to fish is creating a crisis of social and cultural sustainability in our world's fisheries systems. This crisis is largely obscured by the hopeful rhetoric that continues to tout catch shares as the remedy for the problems facing our fisheries today.

Rather than a theoretical link between private property rights and stewardship, we emphasize the need to widen the conceptual framing of stewardship to recognize (1) the social dimensions of fishery systems and (2) the role of communities, values, and institutions (in promoting responsible fishing behavior) beyond the narrow confines of private property rights. Environmental stewardship can be defined as a set of values that individuals, communities, and cultures draw upon to form their relationships and interactions with the environment and its resources (Van Putten et al., 2014, p. 3). The ownership-promotes-stewardship thesis tends to simplify these complex values and institutions, replacing this diversity with universalist assumptions that all fishermen act selfishly in their own self-interest and the private property rights are the means necessary to incentivize long-term self-interest. The work of Mansfield (2011) and other political ecologists helps to reframe fishery conservation problems—not as an outcome of lack of property rights—but rather as a product of a global political economy that has fostered the industrialization of fishing fleets, large subsidization of fisheries development, and highly unequal systems of fisheries extraction from the global South for consumption in the global North. These industrializing processes are central to our crisis of ecological sustainability of fishery stocks (Donkersloot and Menzies, 2015; Carothers and Chambers, 2012; Mansfield, 2011). As we have seen here, a central dimension of these processes is the uneven power relations at play in fisheries development (Donkersloot, 2016; Donkersloot and Menzies, 2015; Jentoft, 2007). We cannot ignore the role of power in shaping fishery systems in the North Pacific, or the role that catch shares play in the consolidation of power.

As researchers, policy makers, and communities we must look beyond privatization to articulate alternative constructions of stewardship as place-based, multi-generational, and inclusive of attachments to place, among other things. Creating space to allow for alternative constructions of stewardship, which acknowledge people–place–resource connections, to meaningfully inform fisheries management decisions is a step toward empowering communities, independent fishing operations, and small-scale fishing operations to address fishery conservation problems in ways that do not threaten their sustainability. "Righting" the tragedy of fisheries today means approaching fisheries as integrated social–ecological systems, ensuring equity in the distribution of fishery access and benefits, and resituating communities and people–place–resource connections at the center of conservation solutions.

REFERENCES

Abbott, J.K., Garber-Yonts, B., Wilen, J.E., 2010. Employment and remuneration effects of IFQs in the Bering Sea/Aleutian Islands crab fisheries. Marine Resource Economics 25 (4), 333–354.

Acheson, J., Apollonio, S., Wilson, J., 2015. Individual transferable quotas and conservation: a critical assessment. Ecology and Society 20 (4) art. 7.

Acheson, J., 2006. Institutional failure in resource management. Annual Review of Anthropology 35, 117–134.

Arnason, R., 2012. Property rights in fisheries: how much can individual transferable quotas accomplish? Review of Environmental Economics and Policy 6 (2), 217–236.

Ban, N., Blight, N., Foster, S., Morgan, S., Donnell, K., 2008. Pragmatism before prescription for managing global fisheries. Frontiers in Ecology and Environment 6 (10).

Bonzon, K., McIlwain, K., Strauss, C.K., Van Leuvan, T., 2013. Catch Share Design Manual, Volume 1: A Guide for Managers and Fishermen, second ed. Environmental Defense Fund.

Branch, T.A., 2009. How do individual transferable quotas affect marine ecosystems? Fish and Fisheries 10, 39–57.

Bromley, D.W., 2015. Correcting the whimsies of U.S. fisheries policy. Choices 30 (4). Available at: http://choicesmagazine.org/choices-magazine/submitted-articles/correcting-the-whimsies-of-us-fisheries-policy.

Carothers, C., Chambers, C., 2012. Fisheries privatization and the remaking of fishery systems. Environment and Society: Advances in Research 3, 39–59.

Carothers, C., 2010. Tragedy of commodification: transitions in Alutiiq fishing communities in the Gulf of Alaska. Maritime Studies (MAST) 90 (2), 91–115.

Carothers, C., 2011. Equity and access to fishing rights: exploring the community quota program in the Gulf of Alaska. Human Organization 70 (3), 713–723.

Carothers, C., 2013. A survey of halibut IFQ holders: market participation, attitudes, and impacts. Marine Policy 38, 515–522. http://dx.doi.org/10.1016/j.marpol.2012.08.007.

Carothers, C., 2015. Fisheries privatization, social transitions, and well-being in Kodiak, Alaska. Marine Policy 61, 313–322. http://dx.doi.org/10.1016/j.marpol.2014.11.019.

Chambers, C., Carothers, C., 2017. Thirty years after privatization: a survey of Icelandic small-boat fishermen. Marine Policy. http://dx.doi.org/10.1016/j.marpol.2016.02.026. (in press).

Chu, C., 2009. Thirty years later: the global growth of ITQs and their influence on stock status in marine fisheries. Fish and Fisheries 10, 217–230.

Copes, Pálsson, 2000. Challenging ITQs: legal and political action in Iceland, Canada and Latin America. In: IIFET 2000 Proceedings, pp. 1–6.

Costello, C., Gaines, S.D., Lynham, J., 2008. Can catch shares prevent fisheries collapse? Science 321 (5896), 1678–1681.

Costello, C., Ovando, D., Clavelle, T., Strauss, C.K., Hilborn, R., Melnychuk, M.C., Branch, T.A., Gaines, S.D., Szuwalski, C.S., Cabral, R.B., Rader, D.N., Leland, A., 2016. Global fishery prospects under contrasting management regimes. Proceedings of the National Academy of Sciences of the United States of America 113 (18), 5125–5129.

Dewees, C.M., 1989. Assessment of the implementation of individual transferable quotas in New Zealand's inshore fishery. North American Journal of Fisheries Management 9 (2), 131–139. http://dx.doi.org/10.1577/1548-8675(1989)009%3C0131:AOTIOI%3E2.3.CO;2.

Dietz, T., Ostrom, E., Stern, P., 2003. The struggle to govern the commons. Science 302, 1907–1912.

Donkersloot, R., Carothers, C., 2016. The graying of the Alaskan fishing fleet. Environment: Science and Policy for Sustainable Development 58 (3), 30–42.

Donkersloot, R., Menzies, C., 2015. Place-based fishing livelihoods and the global ocean: the Irish pelagic fleet at home and abroad. Maritime Studies 14 (20). http://dx.doi.org/10.1186/s40152-015-0038-5.

Donkersloot, R., 2016. Considering Community Allocations in the Emerging Gulf of Alaska Catch Share Program. Marine Policy 74, 300–308 (special issue).

Economist, September 18, 2008. Fishing and Conservation: A Rising Tide. Available at: http://www.economist.com/node/12253181.

Environmental Defense Fund (EDF), 2016. How Catch Shares Work: A Promising Solution. https://www.edf.org/oceans/how-catch-shares-work-promising-solution.

Essington, T.E., Melnychuk, M.C., Branch, T.A., Heppell, S.S., Jensen, O.P., Link, J.S., Martell, S.J.D., Parma, A.M., Pope, J.G., Smith, A.D.M., 2012. Catch shares, fisheries, and ecological stewardship: a comparative analysis of resource responses to rights-based policy instrument. Conservation Letters 5, 186–195.

Essington, T.E., 2010. Ecological indicators display reduced variation in North American catch share fisheries. Proceedings of the National Academy of Sciences of the United States of America 107 (7), 754–759.

Eythórsson, E., 1996. Theory and practice of ITQs in Iceland privatization of common fishing rights. Marine Policy 20 (3), 269–281.

Eythórsson, E., 2016. A milder version of ITQs? Post-ITQ provisions in Norway's fisheries. In: Cullenberg, P. (Ed.), Fisheries Access for Alaska—Charting the Future: Workshop Proceedings. Alaska Sea Grant, University of Alaska Fairbanks, AK-SG-16-02, Fairbanks, pp. 145–148. http://dx.doi.org/10.4027/faacfwp.

Fina, M., 2011. Evolution of catch share management: lessons from catch share management in the North Pacific. Fisheries 36 (4), 164–177.

Foley, P., Mather, C., Neis, B., 2013. Fisheries Allocation Policies and Regional Development: Successes from the Newfoundland and Labrador Shrimp Fishery. The Harris Centre, Memorial University.

Foley, P., Mather, C., Neis, B., 2015. Governing enclosure for coastal communities: social embeddedness in a Canadian shrimp fishery. Marine Policy 61, 390–400.

Gibbs, M.T., 2009. Individual transferable quotas and ecosystem based fisheries management: it's all in the T. Fish and Fisheries 10 (4), 470–474. http://dx.doi.org/10.1111/j.1467-2979.2009.00343.x.

Gilmour, P.W., Day, R.W., Dwyer, P.D., 2012. Using private rights to manage natural resources: is stewardship linked to ownership? Ecology and Society 17 (3), 1. http://dx.doi.org/10.5751/ES-04770-170301.

Grafton, R.Q., Arnason, R., Bjorndal, T., Campbell, D., Campbell, H.F., Clark, C.W., Connor, R., Dupont, D.P., Hannesson, R., Hilborn, R., Kirkley, J.E., Kompas, T., Lane, D.E., Munro, G.R., Pascoe, S., Squires, D., Steinshamn, S.I., Turris, B.R., Weninger, Q., 2006. Incentive-based approaches to sustainable fisheries. Canadian Journal of Fisheries and Aquatic Sciences 63 (3), 699–710.

Hannesson, R., 2006. The Privatization of the Oceans. MIT Press, Cambridge.

Hartley, M., Fina, M., 2001. Changes in fleet capacity following the introduction of individual vessel quotas in the Alaskan Pacific halibut and sablefish fishery. In: Shotton, R. (Ed.), Case Studies on the Effects of Transferable Fishing Rights on Fleet Capacity and Concentration of Quota Ownership. FAO, Rome.

Helgason, A., Pálsson, G., 1997. Contested commodities: the moral landscape of modernist regimes. Journal of the Royal Anthropological Institute 3, 451–471.

Herrmann, M., Criddle, K., 2006. An econometric market model for the Pacific halibut fishery. Marine Resource Economics 21 (2), 129–158.

Igoe, J., Brockington, D., 2007. Neoliberal conservation: a brief introduction. Conservation and Society 5 (4), 432–449.

IPHC, 2015. IPHC letter to NPFMC of May 26, 2015. June 2015 meeting, Agenda Item C2 BSAI PSC Limits. Available at: http://www.iphc.int/documents/bycatch/IPHC2NPFMC_PSClimitJune2015b.pdf.

Jentoft, S., 2007. In the power of power: understanding aspects of fisheries and coastal management. Human Organization 66 (4), 426–437.

Knapp, G., Lowe, M., 2007. Economic and Social Impacts of BSAI Crab Rationalization on the Communities of King Cove, Akutan and False Pass Report prepared for Aleutians East Borough, City of King Cove ISER Publication, University of Alaska, Anchorage. Available at: http://www.iser.uaa.alaska.edu/people/knapp/personal/pubs/Knapp_&_Lowe_AEB_Crab_Rationalization_Final_Report_November_2007.pdf.

Knapp, G., 2006. Economic Impacts of BSAI Crab Rationalization on Kodiak Fishing Employment and Earnings and Kodiak Businesses. A Preliminary Analysis. ISER Publication, University of Alaska, Anchorage. Available at: http://www.iser.uaa.alaska.edu/people/knapp/personal/Knapp_Kodiak_Crab_Rationalization_Preliminary_Report.pdf.

Langdon, S.J., 2008. The community quota program in the Gulf of Alaska: a vehicle for Alaska native village sustainability? In: Lowe, M., Carothers, C. (Eds.), Enclosing the Fisheries: People, Places, and Power: American Fisheries Society Symposium 68. American Fisheries Society, Bethesda, MD, pp. 155–194.

Longo, S.B., Clausen, R., Clark, B., 2015. The Tragedy of the Commodity: Oceans, Fisheries and Aquaculture. Rutgers University Press.

Lowe, M., Carothers, C. (Eds.), 2008. Enclosing the Fisheries: People, Places, and Power. American Fisheries Society, Symposium 68. Bethesda, MD.

Macinko, S., Bromley, D., 2002. Who Owns America's Fisheries? Island Press, Washington, DC.

Macinko, S., Bromley, D., 2004. Property and fisheries for the twenty-first century: seeking coherence from legal and economic doctrine. Vermont Law Review 28 (3), 623–661.

Mansfield, B., 2011. 'Modern' industrial fisheries and the crisis of overfishing. In: Peet, R., Robbins, P., Watts, M. (Eds.), Global Political Ecology. Routledge, New York, pp. 84–99.

Matulich, S.C., Clark, M., 2003. North Pacific halibut and sablefish IFQ policy design: quantifying the impacts on processors. Marine Resource Economics 149–166.

McCay, B.J., 2004. ITQs and community: an essay on environmental governance. Review of Agricultural and Resource Economics 33 (2), 162–170.

McCormack, F., 2012. The reconstitution of property relations in New Zealand fisheries. Anthropological Quarterly 85 (1), 171–201.

Melnychuk, M.C., Essington, T.E., Branch, T.A., Heppell, S.S., Jensen, O.P., Link, J.S., Smith, A.D., 2012. Can catch share fisheries better track management targets? Fish and Fisheries 13, 267–290.

NOAA Fisheries Service, 2010. NOAA Catch Share Policy. Silver Spring, MD. Available at: http://www.nmfs.noaa.gov/sfa/management/catch_shares/about/documents/noaa_cs_policy.pdf.

NOAA Fisheries Service, 2014. Report on Holdings of Individual Fishing Quota (IFQ) by Residents of Selected Gulf of Alaska Fishing Communities 1995–2014. https://alaskafisheries.noaa.gov/sites/default/files/reports/ifq_community_holdings_95-14.pdf.

NOAA Fisheries Service, 2016. Changes in Halibut Quota Share Holdings between Issuance and Currently Issued. Juneau, Alaska https://alaskafisheries.noaa.gov/sites/default/files/reports/16ifqqscompare.pdf.

NPFMC, December 28, 2010a. Executive Summary: 5 year Review of the Crab Rationalization Program for Bering Sea Aleutian Island Crab Fisheries, pp. 1–24 Available at: http://www.npfmc.org/wp-content/PDFdocuments/catch_shares/Crab/5yearSummaryCrab911.pdf.

NPFMC, 2010b. Review of the Community Quota Entity (CQE) Program under the Halibut/Sablefish IFQ Program. Final report. Available at: http://www.npfmc.org/wp-content/PDFdocuments/halibut/CQEreport210.pdf.

NPFMC, April 2015. Workplan for the 10-year Review of the Bering Sea/Aleutian Island Crab Rationalization Program Report to the SSC.

NPFMC, October 2016. Twenty Year Review of the Halibut/Sablefish Individual Fishing Quota Management Program. Agenda Item C6.

NPR, November 4, 2015. The Less Deadly Catch. Planet Money: Episode 661 Available at: http://www.npr.org/sections/money/2015/11/04/454698093/episode-661-the-less-deadly-catch.

Olson, J., 2011. Understanding and contextualizing social impacts from the privatization of fisheries: an overview. Ocean and Coastal Management 54 (5), 353–363.

Ostrom, E., 1990. Governing the Commons: The Evolution of Institutions for Collective Action. Cambridge University Press.

Pauly, D., 2008. Agreeing with Daniel Bromley. Maritime Studies 6 (2), 27–28.

Pinkerton, E., Davis, R., 2015. Neoliberalism and the politics of enclosure in North American small-scale fisheries. Marine Policy 61, 312–313.

Pinkerton, E., Edwards, D., 2009. The elephant in the room: the hidden costs of leasing individual transferable fishing quotas. Marine Policy 33, 707–713.

Pinkerton, E., 2013. Alternatives to ITQs inequity-efficiency-effectiveness trade-offs: how the lay-up system spread effort in the BC halibut fishery. Marine Policy 42, 5–13.

Reedy, K., Maschner, H., 2014. Traditional foods and corporate controls: networks of household access to key marine species in southern Bering sea villages. Polar Record 50, 364–378.

Rieser, A., Watling, L., Guinotte, J., 2013. Trawl fisheries, catch shares and the protection of benthic marine ecosystems: has ownership generated incentives for seafloor stewardship? Marine Policy 40, 75–83.

Rowley, S., April 19, 2016. How Dwindling Fish Stocks Got a Reprieve. New York Times. Available at: http://opinionator.blogs.nytimes.com/2016/04/19/how-dwindling-fish-stocks-got-a-reprieve/?_r=0.

Shotton, R. (Ed.), 2001. Case Studies on the Allocation of Transferable Quota Rights in Fisheries. FAO, Rome. Technical Paper No. 411.

Stewart, I.J., Martell, S., Leaman, B.M., Webster, R.A., Sadorus, L.L., June 2014. Report to the North Pacific Fishery Management Council on the Status of Pacific Halibut in the Bering Sea and Aleutian Islands and the Impacts of Prohibited Species Catch.

Stoll, J.S., Holliday, M.C., 2014. The Design and Use of Fishing Community and Regional Fishery Association Entities in Limited Access Privilege Programs. U.S. Dept. of Commer., NOAA. NOAA Technical Memorandum NMFS-F/SPO-138. Available at: http://spo.nmfs.noaa.gov/tm/.

Sumaila, U.R., 2010. A cautionary note on individual transferable quotas. Ecology and Society 15 (3), 36–43.

Szymkowiak, M., Himes-Cornell, A., 2015. Towards individual-owned and owner-operated fleets in the Alaskan halibut and sablefish IFQ program. Maritime Studies 14 (1), 1–19.

Thébaud, O., Innes, J., Ellis, N., 2012. From anecdotes to scientific evidence? A review of recent literature on catch share systems in marine fisheries. Frontiers in Ecology and the Environment 10 (8), 433–437. http://dx.doi.org/10.1890/110238.

Van der Woo, L., January 8, 2013. Sharecroppers of the Sea. Seattle Weekly News. Available at: http://www.seattleweekly.com/2013-01-09/news/sharecroppers-of-the-sea/.

Van Putten, I., Boschetti, F., Fulton, E.A., Smith, A.D.M., Thebaud, O., 2014. Individual transferable quota contribution to environmental stewardship: a theory in need of validation. Ecology and Society 19 (2), 35. http://dx.doi.org/10.5751/ES-06466-190235.

Chapter 13

Addressing Socioecological Tipping Points and Safe Operating Spaces in the Anthropocene

Benjamin S. Halpern[1,2]
[1]UC Santa Barbara, Santa Barbara, CA, United States; [2]Imperial College London, Ascot, United Kingdom

INTRODUCTION

Humans have been altering the planet for millennia, often in dramatic ways. We started driving species extinct at least a 1000 years ago, and the rate has been accelerating, especially in the last century (Barnosky et al., 2011; Ceballos et al., 2015; Dirzo et al., 2014; McCauley et al., 2015). Concomitantly, we have altered the appearance and function of ecosystems as human populations have grown, spread, inhabited, and cultivated larger and larger swaths of land (Rockstrom et al., 2009b). Similar changes came much later in the ocean, but many signs point to an emerging global imprint of humanity on the oceans (Halpern et al., 2008, 2015; Jackson et al., 2001; McCauley et al., 2015). We have entered the era of the Anthropocene (Crutzen, 2002; Lewis and Maslin, 2015; Waters et al., 2016).

One of the biggest concerns about the Anthropocene is that we may be pushing the planet and its nature-based support systems past sustainable thresholds. To ensure a viable future, we likely need to rein ourselves in to remain within, or perhaps get back to, a "safe operating space" for the planet and humanity. At least that is how many scientists, and some policy makers, describe the concern (Dearing et al., 2014; Hughes et al., 2013; Rockstrom et al., 2009a). But what do those words really mean, and what do they mean for how we should manage our planet? In this chapter, I argue that we still know relatively little that is practical about these key ecosystem concepts, creating significant uncertainty in how best to set management strategies for getting close to, but not crossing, these planetary thresholds. It is an unsettling

Conservation for the Anthropocene Ocean. http://dx.doi.org/10.1016/B978-0-12-805375-1.00013-1

uncertainty—the stakes are incredibly high if we get things wrong. As a result, management will likely be most effective and efficient if focused on managing for this uncertainty, rather than trying to remove it in the quest for predicting where and when these thresholds occur.

The idea of a tipping point is intuitive, the proverbial sense that there is a "straw that breaks the camel's back" in natural systems. Social sciences have used the idea of tipping points for decades to describe, for example, shifts in cultural norms or the composition of neighborhoods, and the term was popularized by Malcolm Gladwell's book *The Tipping Point*. In natural systems, tipping points are generally defined as rapid, nonlinear change in ecosystem condition in response to the intensity of a human or natural driver (Selkoe et al., 2015). The theory of ecosystem thresholds, or ecosystem tipping points, and the inherent resilience in ecosystems (or erosion thereof) that helps avoid (or cause) these threshold responses emerged in recent decades (Folke et al., 2004; Holling, 1973; Lewontin, 1969; Scheffer et al., 2001; Scheffer and Carpenter, 2003). In particular, this theoretical research has focused on describing and understanding how (or if) hysteresis can lead to alternate stable states in ecosystems, and whether early warning indicators exist that can signal the onset of a tipping point. The theory in part grew out of observing dramatic changes in natural systems when, for example, harvest of a top predator causes a trophic cascade of changes in a land- or seascape (Estes et al., 1998; Frank et al., 2005; Myers et al., 2007) or overexploitation of a key species collapses that species, with collateral consequences for the ecosystem (Ferretti et al., 2010; Frank et al., 2011; Rasher et al., 2013; Steneck and Wahle, 2013). Classic examples in the ocean include extirpation of sea otters along the US West Coast that led to sea urchin population explosions that in turn grazed down kelp forests (Estes and Palmisano, 1974; Filbee-Dexter and Scheibling, 2014), and overharvest of cod in the Gulf of Maine that flipped the entire ecosystem to a lobster-dominated one, with dramatic socioeconomic implications for the region (Frank et al., 2005; Steneck et al., 2011; Steneck and Wahle, 2013). Natural processes can also flip systems into a different state, whether through climate variability (e.g., oceanographic cycles) or boom and bust population cycles (Pershing et al., 2015; Ware and Thomson, 2005). Dramatic shifts in pelagic food webs, often driven by changes in primary productivity that respond to decadal-scale oscillations in regional climate, are common examples of these kinds of shifts (Richardson and Schoeman, 2004; Ware and Thomson, 2005).

Ecosystems that have more redundancy of key functional components, or that have more abundant populations of the full suite of native species, are generally more resilient to various pressures and likely able to rebound after experiencing the pressures (Worm et al., 2006; Worm and Duffy, 2003). For example, healthy coral reefs experiencing modest fishing pressure and little else tend to be able to recover quickly from bleaching events or storm damage because the full suite of native species are present and sufficiently abundant to repopulate and/or recover from the disturbance (Hughes et al., 2003). The erosion of this resilience increases uncertainty in what future ecosystems will look like. In other words, intact natural systems are less likely to tip into alternate states. But intact does not mean pristine; people can harvest species and modify habitats and still allow for ecosystem resilience, as long as those activities are done in fully sustainable ways, that is, in ways that do not compromise the integrity and resilience of the species or the ecosystem in which it lives. Ecosystem-based harvest of fish stocks is a clear example of sustained resilience (when harvest is truly sustainable), and modest nutrient pollution into a system can leave ecosystems largely intact.

Ecosystem tipping points garner a lot of scientific and public attention because they typically represent dramatic shifts in what people get out of the system. Whether it is the loss of a key food resource due to a stock collapse, change in the aesthetics and functioning of an ecosystem following the population explosion of an invasive species, or loss of a habitat and its associated species due to coastal development, people often directly experience the implications of ecosystem tipping points (Biggs et al., 2012; Graham et al., 2013; Lebel et al., 2006). For example, the collapse of the California abalone fishery and near extinction of the species in the 1990s led to complete closure of the commercial fishery along the entire coast in 1997 (Hobday et al., 2001; Murray et al., 1999).

Another reason for increasing research focus and policy discussions about ecosystem tipping points is that the planet seems to be experiencing more and more of them. Examples exist from around the world in almost every type of habitat (Kelly et al., 2014; oceantippingpoints.org/our-work/management-practice). The increasing human population in the coming decades suggests ecosystem tipping points will become even more common.

The concept (and reality) of ecosystem tipping points suggests that if we stay within threshold values, the things we care about and want from ecosystems are "safe." Indeed, the concept of a "safe operating space" for the planet emerged from the theory of threshold responses in ecosystems (Hughes et al., 2013; Rockstrom et al., 2009b). This safe operating space is defined by the threats or pressures of human activities, such as climate change, land-system change, ocean acidification, and freshwater use; their impact on key biological resources or benefits we derive from the planet; and threshold values of impact that create the boundaries of where humanity should operate (Rockstrom et al., 2009a). Initial estimates suggest we have exceeded the boundaries for at least two of these attributes, biochemical flows of nitrogen and phosphorous and the integrity of genetic diversity, and are approaching the limit for one or two others (Rockstrom et al., 2009b; Steffen et al., 2015). The question remains how many can be exceeded before the planet fully tips into an unrecognizable (and likely uninhabitable) place.

Although these ideas of "ecosystem tipping points" and "safe operating space" are compelling ways to conceptualize and approach managing the planet, at least two key hurdles exist that challenge efforts to operationalize the concepts. First, the coupled social–ecological systems that define ecosystem state and influence if and how ecosystems "tip" are complex enough to defy predictability in most cases. Second, even if we improve our ability to describe and predict ecosystem tipping points, we always face the challenge of not being able to disprove the null hypothesis, of not knowing if management effort was actually necessary (or efficient) for avoiding a tipping point. Maybe the system would not have tipped even in the absence of our efforts. The rest of the chapter builds the case for these arguments and their implications, but first dives deeper into how socioecological tipping points work, and why it is so important to still try to find, describe, and predict them, despite these challenges.

SOCIOECOLOGICAL TIPPING POINTS

Initially, research on ecosystem tipping points focused on their ecological causes and consequences, with tipping points often described in different terms. In particular, research on fisheries stock dynamics (and collapses) and trophic cascades defined our understanding of how species and ecological communities changed in dramatic (i.e., nonlinear) ways under increasing human pressure. Both cases provide important lessons for what we do and can know, what is difficult to know, and what these lessons mean for management.

Fisheries science is implicitly defined by its quest to find, anticipate, and manage for tipping points. Maximum sustainable yield, a common objective of fisheries management, is the point beyond which a fishery tips into an unsustainable level (assuming fishing pressure remains the same) and that can ultimately lead to stock collapse. Globally, nearly half of all fisheries are collapsed or predicted to be overfished (Pauly and Zeller, 2016; SOFIA, 2014), in large part because they are data limited, making it difficult to know how to set appropriate yield targets and to motivate adherence to and enforcement of uncertain targets (Bentley and Stokes, 2009; Costello et al., 2012). Even some stocks with data-rich monitoring are overfished, emphasizing the challenges in translating science into actionable policy. For example, Atlantic Bluefin tuna and some stocks of Pacific Bluefin tuna are heavily overfished despite massive monitoring programs (and an international agency dedicated to their management). For the many well-managed stocks, precautionary buffers (i.e., basing targets on optimal rather than maximum yield) are needed to manage uncertainty in stock assessments, despite substantial monitoring programs. The need for such precautionary buffers has recently been extended to the idea of "pretty good yield," intended to avoid recruitment overfishing (Rindorf et al., 2016). Alaskan groundfish are well known for being sustainably managed, in large part because precautionary, ecosystem-based harvest targets are used (Witherell et al., 2000).

Another main area of research on ecosystem tipping points comes from the trophic cascade literature. Overharvest of predators and/or excess pollution into systems has caused hundreds of examples of ecosystems tipping into an alternate state (Daskalov et al., 2007; Frank et al., 2005; Shurin et al., 2002). Until recently, these trophic cascades were not explicitly described as ecosystem tipping points; they are now recognized as some of the best and most dramatic examples. In many cases, these trophic cascades result from changes in the abundance of keystone species, that is, species that play a disproportionate role in structuring an ecosystem. The iconic example of this is when otter extirpation along the US west coast caused the shift from kelp-dominated to urchin-dominated rocky reefs (Estes and Duggins, 1995; Estes and Palmisano, 1974; Estes et al., 1978; Filbee-Dexter and Scheibling, 2014). Species can also play a keystone role in human cultures, such that shifts in the species' abundance can cause tipping points in human communities (Garibaldi and Turner, 2004).

Research and management addressing invasive species and restoration are two additional disciplines that have addressed tipping points extensively, although often not framed explicitly this way. The most pernicious cases of invasive species completely alter the appearance and function of an ecosystem and essentially cannot be removed—the system has tipped past a critical threshold into a true alternate stable state. For example, the invasive seaweed *Caulerpa* spp. have completely overrun reefs throughout the Mediterranean (and many other places around the world), changing species composition and ecosystem function (Williams and Smith, 2007). Restoration, on the other hand, is the active manipulation of a species or ecosystem with the aim to recover to a previous (or new) state, which usually requires pushing the system past a tipping point in the opposite direction of ecosystem degradation. The theory and practice of how and where to do restoration have provided many lessons on the challenges of reversing tipping points. For example, the inertia (or hysteresis) within degraded ecosystems often requires far more effort to restore ecosystem function than was used to degrade it (Selkoe et al., 2015).

In the last decade or so, the science of ecological tipping points has advanced the underlying theory about how tipping points work and highlighted the potential for tipping points to lead to alternate stable regimes (Scheffer et al., 2001; Scheffer and Carpenter, 2003). More recently, the science has been placed within a growing awareness of the intertwined nature of human and ecological systems: the idea of the ecosystem that includes people and nature. This recognition is commonly called ecosystem-based management or ecosystem-based fisheries management and has led to a new emphasis on coupled human–natural systems science (McLeod and Leslie, 2009). In this frame, emphasis is given to how human values and behaviors interact with ecological processes. With respect to ecosystem tipping points, these feedbacks mean that human behavior can amplify or dampen potential ecological tipping points, and ecological tipping points can initiate or inhibit social tipping points (e.g., Travis et al., 2014). For example, when harvest of a stock makes it less abundant and thus harder to catch, fishermen may need to fish even harder to catch enough to remain profitable, and the negative feedback continues until the stock collapses and the fishery gets shut down.

These interactions between people and nature are varied and potentially very complex. For example, rarity of a species can make people want to increase their efforts to protect it or create perverse incentives to harvest it even more because it is more valuable (Courchamp et al., 2006; Gault et al., 2008; Hall et al., 2008). Cultural values (e.g., is protection or economic value more important) and cultural context (e.g., can fishermen easily move into other sectors, do cultural practices depend on a species) can make the same ecological situation play out in completely different ways. Invasive species that alter ecosystem structure and function can be seen as a nuisance, unless the species is economically valuable and then people may embrace the change (e.g., the introduction and invasion of red king crab into Norwegian coastal waters; Jorgensen and Nilssen, 2011). Different people and groups within the same community can have very different values and incentives, and who holds power can change outcomes. The implication of this complexity is that to fully understand and predict socioecological tipping points one has to fully understand the dynamics of both the natural and human systems.

The most dramatic examples of coupled system tipping points are perhaps the most worrying ones, where social tipping points lead to ecological ones and vice versa. If a human community is dependent on a few key fish stocks, stock collapses can lead to loss of a substantial proportion of local jobs and an entire sector of the economy and even cause dramatic changes to a local culture (Hamilton and Butler, 2001; Hamilton and Haedrich, 1999; Perry et al., 2011). The overharvest of cod in Newfoundland, Canada, and its ultimate collapse in 1992, and the subsequent upheaval of coastal fishing communities, is a dramatic example of such coupled tipping points. Similarly, sudden shifts in cultural diet preferences can lead to an emerging fishery that gets rapidly depleted, causing dramatic ecological shifts (Levin and Dufault, 2010). The sudden rise in popularity of blackened redfish in the 1980s (attributed to the famous New Orleans chef Paul Prudhomme) and Chilean seabass in the 1990s (attributed to marketers rebranding the Antarctic toothfish with a tastier name) led both species to quickly become threatened.

Yet there are lots of reasons that social and ecological tipping points would not be synched in space or time, or even connected. For example, incremental and linear declines in the abundance of key species within a natural system may eventually lead to the population being too small to be commercially viable for harvest, collapsing an economic sector and potentially changing a culture even though the natural system persists without dramatic change. Conversely, incremental increases in coastal development within an estuary may push the system to a break point where key species can no longer persist, yet the human community is essentially unaffected. The development of San Francisco bay and its ultimate impact on delta smelt populations is a good example of this.

Ultimately, it is the social impacts of tipping points that managers, and in general the public, care most about. How many jobs will be lost due to ecosystem tipping points, and how much economic revenue will be lost? How will cultural values based on a particular ecosystem state or abundance of key species be affected if the system shifts? How will human health and nutrition be impacted by changing ecosystem state?

THE QUEST TO FIND TIPPING POINTS

It seems obvious why resource managers and society in general would want to know if and where ecosystem tipping points will occur. The ecological, economic, and social costs of crossing a tipping point can be profound. Some notable examples include the collapse of cod in Newfoundland, Canada (Hamilton and Butler, 2001; Hamilton and Haedrich, 1999), the shift of many Caribbean coral reefs to an algal-dominated state (Gardner et al., 2003; Karr et al., 2015; Pandolfi et al., 2005), and sequential collapse of fisheries in the Black Sea (Daskalov et al., 2007).

The last few decades have seen important advances in the description, explanation, and prediction of ecosystem tipping points. With respect to description, we now have documented examples from hundreds of cases around the world from nearly every marine system (Rocha et al., 2015; Scheffer et al., 2001; Shurin et al., 2002), giving insight into how common ecosystem tipping points may be and suggesting that no system is immune. With respect to explanation, research has identified several key variables that can explain the occurrence, nature, and magnitude of tipping points, in particular, overharvest that can trigger trophic cascades (e.g., Casini et al., 2009; Daskalov et al., 2007) and shifts in environmental variables (e.g., temperature, dissolved oxygen, salinity) that move a system beyond tolerance thresholds for key species (e.g., Hunsicker et al., 2016). For example, climate-driven ocean warming along the west coast of Australia tipped a temperate kelp forest ecosystem into a warm seaweed turf/coral ecosystem in just a few years (Wernberg et al., 2016). With respect to prediction, recent advances have identified several potential early warning signs of the onset of a regime shift, including things such as increased temporal or spatial variance in the abundance of key species (sometimes termed "flickering") and slower return rates in ecosystem conditions following a disturbance event (called "critical slowing down"; Carpenter et al., 2011; Dakos et al., 2012a,b; Scheffer et al., 2009).

Fisheries science is perhaps the most advanced field in studying and managing for tipping points, although it is often not framed as such. Over the past decades, there have been hundreds of stock collapses, providing a host of lessons learned and insight into how to avoid stock collapses, and in some cases how to better predict where and when future stock collapses may occur (Burgess et al., 2013). Probably the most important of these lessons has been to adjust harvest targets away from "maximum sustainable yield" to a more precautionary "optimal sustainable yield" that acknowledge the difficulties and uncertainties of knowing, and hitting, a target precisely. Predicting stock collapses is also relatively straightforward (when harvest rate gets too high, the stock will tip toward collapse), and so measures of known harvest and estimates of unreported catch can provide strong early warning indicators of impending collapse. Unfortunately, it is difficult and costly to collect effort and catch data in many places around the world.

All of these examples and associated research are building our capacity to predict tipping points. The value of this line of work seems obvious enough to almost not need stating. If we can improve our ability to understand and predict tipping points, and ultimately be fully able to do so, it would be like knowing ahead of time that the stock market was about to crash—you can plan for it and ideally make necessary changes to avoid it. It seems worth doubling down on our efforts to understand and predict ecosystem tipping points.

THE CHALLENGE IN PREDICTING TIPPING POINTS

Here is where I argue that maybe we should not do that. We need to know a little about tipping points, but not everything, to do efficient and effective management. Instead of trying to perfectly predict ecosystem tipping points, the primary focus for science and management should be on managing with uncertainty rather than removing it, focusing, in particular, on building ecosystem resilience and adaptive capacity. This argument rests on three salient attributes of current tipping point science and management and the overall implications of them.

First, we still know relatively little about ecosystem tipping points. Despite all of the science referenced earlier, it remains very difficult to anticipate and predict precisely in any given location when a species' population trend will tip it toward extinction, what exact level of harvest to allow to maximize catch while maintaining sustainable fish stocks, or how much cumulative pressure on ecosystems is allowable to keep an ecosystem within a safe operating space. Major research and monitoring efforts could help alleviate this challenge in the long run but are not likely to solve the problem. It will be very difficult to develop strong, general predictability that works in every situation. Knowing that ecosystem tipping points exist and what causes them in general may be sufficient; a quest to precisely find them so that we can push resource exploitation levels right up against them without consequence seems likely to fail.

Second, and related, it is incredibly data intensive to understand and predict ecosystem tipping points. Fisheries science demonstrates the reality of this challenge. Well-managed stocks that have stayed within a safe operating space are almost exclusively those informed by very large monitoring and assessment efforts. These assessments usually cost a lot. For example, on the west coast of the United States, the process of collecting data for surveys, staffing vessel monitoring programs, and conducting stock assessments for the roughly 20 regional groundfish species with formal assessments costs millions of dollars each year. Monitoring and measuring sufficient components of an ecosystem to be able to anticipate broader tipping points, across multiple scales and in different ecosystems and locations around the world, would cost exponentially more, almost certainly a prohibitively large amount.

Finally, and perhaps most perniciously, we will always be faced with the problem of not knowing if monitoring efforts (and cost) intended to identify and avoid a specific tipping point were worth it if the ecosystem never goes past the tipping point. Maybe the ecosystem would not have tipped in the absence of such monitoring and management, in which case those efforts would have been poorly spent; demands for efficient (i.e., least cost) management would suggest not doing the costly monitoring. In contrast, if great expense is made to anticipate tipping points and yet the system tips anyway, for unknown or unanticipated reasons, then the management failed at great cost. The value of information gained by monitoring and studying a system that ultimately tips is of course relevant and important, but for reasons of efficiency, and likely effectiveness, it may be better to save those expenses, allocate modest resources to basic monitoring that can help indicate general ecosystem trajectories, and manage with more precaution.

Given these challenges, it seems a better strategy to focus on building resilience and adaptive capacity into systems, knowing that tipping points will likely occur at some point, and to manage conservatively to minimize the chance of crossing a tipping point. Indeed, building adaptive capacity into ecosystems through actions that, for example, maintain or enhance species functional diversity (Gunderson, 2000) or strengthen human social networks and economic flexibility (Cinner et al., 2015) not only helps ecosystems respond to tipping points when they occur but may also reduce the occurrence of them (Folke et al., 2004). To manage in this way still requires keeping tipping points in mind, leveraging tipping point science, but it places emphasis on reducing the need to worry about them by taking actions designed to reduce their likelihood. It is a bit like setting the cruise control on your car below the posted speed limit versus keeping your foot on the gas and hoping it does not get too "heavy" when you pass a hidden police officer. This approach to management is essentially embodied in best-practice fisheries management, where typically limited monitoring data inform estimates of maximum sustainable yield, and target yields are set at a buffered value below that maximum to account for inherent and difficult-to-manage uncertainty in how the fishery will actually unfold the following year. We need to find ways to further streamline collection of "minimum necessary information" to inform where thresholds likely exist, while simultaneously moving management as much as possible away from trying to get close to these thresholds and toward a precautionary approach that focuses on building resilience and adaptive capacity.

An emphasis on maintaining and building resilience would focus management on actions that protect or enhance functional diversity and redundancy, both in ecological and human communities. Protecting habitat-forming species because of the unique functional role they play (i.e., maintaining functional diversity) or restoring herbivorous coral reef fish to increase the number and abundance of species present that provide that key ecological function (i.e., increasing functional redundancy) are both examples of managing with ecosystem resilience in mind. In essence, a focus on resilience acknowledges that pressures and disturbances to the system will continue and may increase over time and that to keep an ecosystem in its current state requires building the capacity to recover in the face of these disturbances. The aim is not to find and describe the tipping point and then push up against it, but instead to assume it exists and take action to make it irrelevant.

In contrast, an emphasis on adaptive capacity acknowledges that tipping points are inevitable, at least to some extent, and that management actions should therefore focus on helping promote the ability for ecological and human communities to transition to the new ecosystem configuration after a tipping point occurs, or at a minimum to be able to cope with the altered configuration until management actions can be enacted to reverse the system back to its original state. Adaptive capacity in ecological communities arises from community elements that increase resilience (Gunderson, 2000), whereas human adaptive capacity often relates to variables such as occupational diversity and mobility, social networks and capital, and financial variables such as access to credit (e.g., Cinner et al., 2015). Setting and adjusting targets for such objectives often benefit from comanagement strategies (Gelcich et al., 2010). Management actions aimed at supporting or improving these dimensions of ecological and human communities will be useful even if, and perhaps especially if, nothing is known about an ecosystem's tipping point(s).

CONCLUSIONS

It is hard to know exactly what the planet will look like in the future, but it is certain that it will look different than it did in the past or than it does today. And a world in which we push systems outside their safe operating space is certain to experience very dramatic changes. Where, how often, and to what extent these changes will happen remains uncertain.

The science of ecosystem tipping points has advanced significantly and rapidly in the past decade, helping reduce some of the uncertainty in predicting tipping points, and providing initial guidelines and estimates for and leading indicators of impending shifts in ecosystem state. These are undeniably exciting advances. Yet, we have a very long way to go before the theory and results will be specific enough to guide management actions aimed at pushing up against, but not exceeding, ecosystem tipping points. Ultimately, then, we sit at a management tipping point: do we significantly increase funding and research to identify where tipping points are in every possible location and context so that we can maximize our use of the ecosystem while still avoiding tipping points, or do we start managing our planet using a precautionary approach so that we do not need to know precisely where they are? Fisheries science realized the value of precautionary buffers long ago; when implemented well, they largely work. Policy and management at all levels—local to international—need to embrace the idea of well-informed ecosystem-based precautionary buffers.

Practically, this "embrace" will still require monitoring efforts and scientific research aimed at understanding the dynamics of ecosystem tipping points. However, more effort needs to focus on managing for the large amount of uncertainty that remains. Where uncertainty is large, managers should set targets well below values perceived as possible. Put the burden of proof on users of the ecosystem to demonstrate that their actions will *not* create ecological or social tipping points, rather than on managers that an action *will* cause one. Such changes in strategy represent a substantial shift in current ideology and management practice. If we are unable to make this management tipping point happen, though, we seem poised to exit the planetary safe operating space and tip into an unrecognizable, and potentially uninhabitable, global state.

ACKNOWLEDGMENTS

Thanks to Courtney Scarborough and two anonymous reviewers for helpful comments on earlier drafts.

REFERENCES

Barnosky, A.D., Matzke, N., Tomiya, S., Wogan, G.O.U., Swartz, B., Quental, T.B., Marshall, C., McGuire, J.L., Lindsey, E.L., Maguire, K.C., Mersey, B., Ferrer, E.A., 2011. Has the Earth's sixth mass extinction already arrived? Nature 471, 51–57.

Bentley, N., Stokes, K., 2009. Contrasting paradigms for fisheries management decision making: how well do they serve data-poor fisheries? Marine and Coastal Fisheries 1, 391–401.

Biggs, R., Schlueter, M., Biggs, D., Bohensky, E.L., BurnSilver, S., Cundill, G., Dakos, V., Daw, T.M., Evans, L.S., Kotschy, K., Leitch, A.M., Meek, C., Quinlan, A., Raudsepp-Hearne, C., Robards, M.D., Schoon, M.L., Schultz, L., West, P.C., 2012. Toward principles for enhancing the resilience of ecosystem services. In: Gadgil, A., Liverman, D.M. (Eds.), Annual Review of Environment and Resources, vol. 37, pp. 421–448.

Burgess, M.G., Polasky, S., Tilman, D., 2013. Predicting overfishing and extinction threats in multispecies fisheries. Proceedings of the National Academy of Sciences of the United States of America 110, 15943–15948.

Carpenter, S.R., Cole, J.J., Pace, M.L., Batt, R., Brock, W.A., Cline, T., Coloso, J., Hodgson, J.R., Kitchell, J.F., Seekell, D.A., Smith, L., Weidel, B., 2011. Early warnings of regime shifts: a whole-ecosystem experiment. Science 332, 1079–1082.

Casini, M., Hjelm, J., Molinero, J.-C., Lovgren, J., Cardinale, M., Bartolino, V., Belgrano, A., Kornilovs, G., 2009. Trophic cascades promote threshold-like shifts in pelagic marine ecosystems. Proceedings of the National Academy of Sciences of the United States of America 106, 197–202.

Ceballos, G., Ehrlich, P.R., Barnosky, A.D., García, A., Pringle, R.M., Palmer, T.M., 2015. Accelerated modern human–induced species losses: entering the sixth mass extinction. Science Advances 1.

Cinner, J.E., Huchery, C., Hicks, C.C., Daw, T.M., Marshall, N., Wamukota, A., Allison, E.H., 2015. Changes in adaptive capacity of Kenyan fishing communities. Nature Climate Change 5, 872–876.

Costello, C., Ovando, D., Hilborn, R., Gaines, S.D., Lester, S.E., 2012. Status and solutions for the world's unassessed fisheries. Science 338, 517–520.

Courchamp, F., Angulo, E., Rivalan, P., Hall, R.J., Signoret, L., Bull, L., Meinard, Y., 2006. Rarity value and species extinction: the anthropogenic Allee effect. PLoS Biology 4, 2405–2410.

Crutzen, P.J., 2002. Geology of mankind. Nature 415, 23.

Dakos, V., Carpenter, S.R., Brock, W.A., Ellison, A.M., Guttal, V., Ives, A.R., Kefi, S., Livina, V., Seekell, D.A., van Nes, E.H., Scheffer, M., 2012a. Methods for detecting early warnings of critical transitions in time series illustrated using simulated ecological data. PLoS One 7.

Dakos, V., van Nes, E.H., D'Odorico, P., Scheffer, M., 2012b. Robustness of variance and autocorrelation as indicators of critical slowing down. Ecology 93, 264–271.

Daskalov, G.M., Grishin, A.N., Rodionov, S., Mihneva, V., 2007. Trophic cascades triggered by overfishing reveal possible mechanisms of ecosystem regime shifts. Proceedings of the National Academy of Sciences of the United States of America 104, 10518–10523.

Dearing, J.A., Wang, R., Zhang, K., Dyke, J.G., Haberl, H., Hossain, M.S., Langdon, P.G., Lenton, T.M., Raworth, K., Brown, S., Carstensen, J., Cole, M.J., Cornell, S.E., Dawson, T.P., Doncaster, C.P., Eigenbrod, F., Floerke, M., Jeffers, E., Mackay, A.W., Nykvist, B., Poppy, G.M., 2014. Safe and just operating spaces for regional social-ecological systems. Global Environmental Change-Human and Policy Dimensions 28, 227–238.

Dirzo, R., Young, H.S., Galetti, M., Ceballos, G., Isaac, N.J.B., Collen, B., 2014. Defaunation in the Anthropocene. Science 345, 401–406.

Estes, J.A., Duggins, D.O., 1995. Sea otters and kelp forests in Alaska – generality and variation in a community ecological paradigm. Ecological Monographs 65, 75–100.

Estes, J.A., Palmisano, J.F., 1974. Sea otters – their role in structuring nearshore communities. Science 185, 1058–1060.

Estes, J.A., Smith, N.S., Palmisano, J.F., 1978. Sea otter predation and community organization in western Aleutian Islands, Alaska. Ecology 59, 822–833.

Estes, J.A., Tinker, M.T., Williams, T.M., Doak, D.F., 1998. Killer whale predation on sea otters linking oceanic and nearshore ecosystems. Science 282, 473–476.

Ferretti, F., Worm, B., Britten, G.L., Heithaus, M.R., Lotze, H.K., 2010. Patterns and ecosystem consequences of shark declines in the ocean. Ecology Letters 13, 1055–1071.

Filbee-Dexter, K., Scheibling, R.E., 2014. Sea urchin barrens as alternative stable states of collapsed kelp ecosystems. Marine Ecology Progress Series 495, 1–25.

Folke, C., Carpenter, S., Walker, B., Scheffer, M., Elmqvist, T., Gunderson, L., Holling, C.S., 2004. Regime shifts, resilience, and biodiversity in ecosystem management. Annual Review of Ecology Evolution and Systematics 35, 557–581.

Frank, K.T., Petrie, B., Choi, J.S., Leggett, W.C., 2005. Trophic cascades in a formerly cod-dominated ecosystem. Science 308, 1621–1623.

Frank, K.T., Petrie, B., Fisher, J.A.D., Leggett, W.C., 2011. Transient dynamics of an altered large marine ecosystem. Nature 477, 86–89.

Gardner, T.A., Cote, I.M., Gill, J.A., Grant, A., Watkinson, A.R., 2003. Long-term region-wide declines in Caribbean corals. Science 301, 958–960.

Garibaldi, A., Turner, N., 2004. Cultural keystone species: implications for ecological conservation and restoration. Ecology and Society 9.

Gault, A., Meinard, Y., Courchamp, F., 2008. Consumers' taste for rarity drives sturgeons to extinction. Conservation Letters 1, 199–207.

Gelcich, S., Hughes, T.P., Olsson, P., Folke, C., Defeo, O., Fernandez, M., Foale, S., Gunderson, L.H., Rodriguez-Sickert, C., Scheffer, M., Steneck, R.S., Castilla, J.C., 2010. Navigating transformations in governance of Chilean marine coastal resources. Proceedings of the National Academy of Sciences of the United States of America 107, 16794–16799.

Graham, N.A.J., Bellwood, D.R., Cinner, J.E., Hughes, T.P., Norstrom, A.V., Nystrom, M., 2013. Managing resilience to reverse phase shifts in coral reefs. Frontiers in Ecology and the Environment 11, 541–548.

Gunderson, L.H., 2000. Ecological resilience—in theory and application. Annual Review of Ecology and Systematics 31, 425–439.

Hall, R.J., Milner-Gulland, E.J., Courchamp, F., 2008. Endangering the endangered: the effects of perceived rarity on species exploitation. Conservation Letters 1, 75–81.

Halpern, B.S., Walbridge, S., Selkoe, K.A., Kappel, C.V., Micheli, F., D'Agrosa, C., Bruno, J.F., Casey, K.S., Ebert, C., Fox, H.E., Fujita, R., Heinemann, D., Lenihan, H.S., Madin, E.M.P., Perry, M.T., Selig, E.R., Spalding, M., Steneck, R., Watson, R., 2008. A global map of human impact on marine ecosystems. Science 319, 948–952.

Halpern, B.S., Frazier, M., Potapenko, J., Casey, K.S., Koenig, K., Longo, C., Lowndes, J.S., Rockwood, R.C., Selig, E.R., Selkoe, K.A., Walbridge, S., 2015. Spatial and temporal changes in cumulative human impacts on the world's ocean. Nature Communications 6.

Hamilton, L.C., Butler, M.J., 2001. Outport Adaptations: social indicators through Newfoundland's cod crisis. Research in Human Ecology 8, 1–11.

Hamilton, L.C., Haedrich, R.L., 1999. Ecological and population changes in fishing communities of the North Atlantic Arc. Polar Research 18, 383–388.

Hobday, A.J., Tegner, M.J., Haaker, P.L., 2001. Over-exploitation of a broadcast spawning marine invertebrate: decline of the white abalone. Reviews in Fish Biology and Fisheries 10, 493–514.

Holling, C.S., 1973. Resilience and stability of ecological systems. Annual Review of Ecology and Systematics 4, 1–23.

Hughes, T.P., Baird, A.H., Bellwood, D.R., Card, M., Connolly, S.R., Folke, C., Grosberg, R., Hoegh-Guldberg, O., Jackson, J.B.C., Kleypas, J., Lough, J.M., Marshall, P., Nystrom, M., Palumbi, S.R., Pandolfi, J.M., Rosen, B., Roughgarden, J., 2003. Climate change, human impacts, and the resilience of coral reefs. Science 301, 929–933.

Hughes, T.P., Carpenter, S., Rockstrom, J., Scheffer, M., Walker, B., 2013. Multiscale regime shifts and planetary boundaries. Trends in Ecology & Evolution 28, 389–395.

Hunsicker, M.E., Kappel, C.V., Selkoe, K.A., Halpern, B.S., Scarborough, C., Mease, L., Amrhein, A., 2016. Characterizing driver-response relationships in marine pelagic ecosystems for improved ocean management. Ecological Applications 26, 651–663.

Jackson, J.B.C., Kirby, M.X., Berger, W.H., Bjorndal, K.A., Botsford, L.W., Bourque, B.J., Bradbury, R.H., Cooke, R., Erlandson, J., Estes, J.A., Hughes, T.P., Kidwell, S., Lange, C.B., Lenihan, H.S., Pandolfi, J.M., Peterson, C.H., Steneck, R.S., Tegner, M.J., Warner, R.R., 2001. Historical overfishing and the recent collapse of coastal ecosystems. Science 293, 629–638.

Jorgensen, L.L., Nilssen, E.M., 2011. The invasive history, impact and management of the red king crab paralithodes camtschaticus off the coast of Norway. In: Galil, B.S., Clark, P.F., Carlton, J.T. (Eds.), In the Wrong Place—Alien Marine Crustaceans: Distribution, Biology and Impacts, pp. 521–536.

Karr, K.A., Fujita, R., Halpern, B.S., Kappel, C.V., Crowder, L., Selkoe, K.A., Alcolado, P.M., Rader, D., 2015. Thresholds in Caribbean coral reefs: implications for ecosystem-based fishery management. Journal of Applied Ecology 52, 402–412.

Kelly, R.P., Erickson, A.L., Mease, L.A., 2014. How not to fall off a cliff, or, using tipping points to improve environmental management. Ecology Law Quarterly 41, 843–886.

Lebel, L., Anderies, J.M., Campbell, B., Folke, C., Hatfield-Dodds, S., Hughes, T.P., Wilson, J., 2006. Governance and the capacity to manage resilience in regional social-ecological systems. Ecology and Society 11.

Levin, P.S., Dufault, A., 2010. Eating up the food web. Fish and Fisheries 11, 307–312.

Lewis, S.L., Maslin, M.A., 2015. Defining the Anthropocene. Nature 519, 171–180.

Lewontin, R.C., 1969. The meaning of stability. Brookhaven Symposia in Biology 22, 13–24.

McCauley, D.J., Pinsky, M.L., Palumbi, S.R., Estes, J.A., Joyce, F.H., Warner, R.R., 2015. Marine defaunation: animal loss in the global ocean. Science 347.

McLeod, K.L., Leslie, H.M., 2009. Ecosystem-Based Management for the Oceans. Island Press, Washington, DC.

Murray, S.N., Ambrose, R.F., Bohnsack, J.A., Botsford, L.W., Carr, M.H., Davis, G.E., Dayton, P.K., Gotshall, D., Gunderson, D.R., Hixon, M.A., Lubchenco, J., Mangel, M., MacCall, A., McArdle, D.A., Ogden, J.C., Roughgarden, J., Starr, R.M., Tegner, M.J., Yoklavich, M.M., 1999. No-take reserve networks: sustaining fishery populations and marine ecosystems. Fisheries 24, 11–25.

Myers, R.A., Baum, J.K., Shepherd, T.D., Powers, S.P., Peterson, C.H., 2007. Cascading effects of the loss of apex predatory sharks from a coastal ocean. Science 315, 1846–1850.

Pandolfi, J.M., Jackson, J.B.C., Baron, N., Bradbury, R.H., Guzman, H.M., Hughes, T.P., Kappel, C.V., Micheli, F., Ogden, J.C., Possingham, H.P., Sala, E., 2005. Are US coral reefs on the slippery slope to slime? Science 307, 1725–1727.

Pauly, D., Zeller, D., 2016. Catch reconstructions reveal that global marine fisheries catches are higher than reported and declining. Nature Communications 7.

Perry, R.I., Ommer, R.E., Barange, M., Jentoft, S., Neis, B., Sumaila, U.R., 2011. Marine social-ecological responses to environmental change and the impacts of globalization. Fish and Fisheries 12, 427–450.

Pershing, A.J., Alexander, M.A., Hernandez, C.M., Kerr, L.A., Le Bris, A., Mills, K.E., Nye, J.A., Record, N.R., Scannell, H.A., Scott, J.D., Sherwood, G.D., Thomass, A.C., 2015. Slow adaptation in the face of rapid warming leads to collapse of the Gulf of Maine cod fishery. Science 350, 809–812.

Rasher, D.B., Hoey, A.S., Hay, M.E., 2013. Consumer diversity interacts with prey defenses to drive ecosystem function. Ecology 94, 1347–1358.

Richardson, A.J., Schoeman, D.S., 2004. Climate impact on plankton ecosystems in the Northeast Atlantic. Science 305, 1609–1612.

Rindorf, A., Cardinale, M., Shepherd, S., De Oliveira, J.A.A., Hjorleifsson, E., Kempf, A., Luzenczyk, A., Miller, C., Miller, D.C.M., Needle, C.L., Simmonds, J., Vinther, M., 2016. Fishing for MSY: using "pretty good yield" ranges without impairing recruitment. ICES Journal of Marine Science.

Rocha, J.C., Peterson, G.D., Biggs, R., 2015. Regime shifts in the Anthropocene: drivers, risks, and resilience. PLoS One 10.

Rockstrom, J., Steffen, W., Noone, K., Persson, A., Chapin III, F.S., Lambin, E., Lenton, T.M., Scheffer, M., Folke, C., Schellnhuber, H.J., Nykvist, B., de Wit, C.A., Hughes, T., van der Leeuw, S., Rodhe, H., Sorlin, S., Snyder, P.K., Costanza, R., Svedin, U., Falkenmark, M., Karlberg, L., Corell, R.W., Fabry, V.J., Hansen, J., Walker, B., Liverman, D., Richardson, K., Crutzen, P., Foley, J., 2009a. Planetary boundaries: exploring the safe operating space for humanity. Ecology and Society 14.

Rockstrom, J., Steffen, W., Noone, K., Persson, A., Chapin III, F.S., Lambin, E.F., Lenton, T.M., Scheffer, M., Folke, C., Schellnhuber, H.J., Nykvist, B., de Wit, C.A., Hughes, T., van der Leeuw, S., Rodhe, H., Sorlin, S., Snyder, P.K., Costanza, R., Svedin, U., Falkenmark, M., Karlberg, L., Corell, R.W., Fabry, V.J., Hansen, J., Walker, B., Liverman, D., Richardson, K., Crutzen, P., Foley, J.A., 2009b. A safe operating space for humanity. Nature 461, 472–475.

Scheffer, M., Carpenter, S.R., 2003. Catastrophic regime shifts in ecosystems: linking theory to observation. Trends in Ecology & Evolution 18, 648–656.

Scheffer, M., Carpenter, S., Foley, J.A., Folke, C., Walker, B., 2001. Catastrophic shifts in ecosystems. Nature 413, 591–596.

Scheffer, M., Bascompte, J., Brock, W.A., Brovkin, V., Carpenter, S.R., Dakos, V., Held, H., van Nes, E.H., Rietkerk, M., Sugihara, G., 2009. Early-warning signals for critical transitions. Nature 461, 53–59.

Selkoe, K.A., Blenckner, T., Caldwell, M.R., Crowder, L.B., Erickson, A.L., Essington, T.E., Estes, J.A., Fujita, R.M., Halpern, B.S., Hunsicker, M.E., Kappel, C.V., Kelly, R.P., Kittinger, J.N., Levin, P.S., Lynham, J.M., Mach, M.E., Martone, R.G., Mease, L.A., Salomon, A.K., Samhouri, J.F., Scarborough, C., Stier, A.C., White, C., Zedler, J., 2015. Principles for managing marine ecosystems prone to tipping points. Ecosystem Health and Sustainability 1, 17.

Shurin, J.B., Borer, E.T., Seabloom, E.W., Anderson, K., Blanchette, C.A., Broitman, B., Cooper, S.D., Halpern, B.S., 2002. A cross-ecosystem comparison of the strength of trophic cascades. Ecology Letters 5, 785–791.

SOFIA, 2014. The state of world fisheries and aquaculture. In: Food and Agriculture Organization of the United Nations (Rome, Italy).

Steffen, W., Richardson, K., Rockstrom, J., Cornell, S.E., Fetzer, I., Bennett, E.M., Biggs, R., Carpenter, S.R., de Vries, W., de Wit, C.A., Folke, C., Gerten, D., Heinke, J., Mace, G.M., Persson, L.M., Ramanathan, V., Reyers, B., Sorlin, S., 2015. Planetary boundaries: guiding human development on a changing planet. Science 347.

Steneck, R.S., Wahle, R.A., 2013. American lobster dynamics in a brave new ocean. Canadian Journal of Fisheries and Aquatic Sciences 70, 1612–1624.

Steneck, R.S., Hughes, T.P., Cinner, J.E., Adger, W.N., Arnold, S.N., Berkes, F., Boudreau, S.A., Brown, K., Folke, C., Gunderson, L., Olsson, P., Scheffer, M., Stephenson, E., Walker, B., Wilson, J., Worm, B., 2011. Creation of a gilded trap by the high economic value of the Maine lobster fishery. Conservation Biology 25, 904–912.

Travis, J., Coleman, F.C., Auster, P.J., Cury, P.M., Estes, J.A., Orensanz, J., Peterson, C.H., Power, M.E., Steneck, R.S., Wootton, J.T., 2014. Integrating the invisible fabric of nature into fisheries management. Proceedings of the National Academy of Sciences of the United States of America 111, 581 4644–4646, 2013.

Ware, D.M., Thomson, R.E., 2005. Bottom-up ecosystem trophic dynamics determine fish production in the northeast Pacific. Science 308, 1280–1284.

Waters, C.N., Zalasiewicz, J., Summerhayes, C., Barnosky, A.D., Poirier, C., Galuszka, A., Cearreta, A., Edgeworth, M., Ellis, E.C., Ellis, M., Jeandel, C., Leinfelder, R., McNeill, J.R., Richter, D.d., Steffen, W., Syvitski, J., Vidas, D., Wagreich, M., Williams, M., An, Z., Grinevald, J., Odada, E., Oreskes, N., Wolfe, A.P., 2016. The Anthropocene is functionally and stratigraphically distinct from the Holocene. Science 351, 137.

Wernberg, T., Bennett, S., Babcock, R.C., de Bettignies, T., Cure, K., Depczynski, M., Dufois, F., Fromont, J., Fulton, C.J., Hovey, R.K., Harvey, E.S., Holmes, T.H., Kendrick, G.A., Radford, B., Santana-Garcon, J., Saunders, B.J., Smale, D.A., Thomsen, M.S., Tuckett, C.A., Tuya, F., Vanderklift, M.A., Wilson, S., 2016. Climate-driven regime shift of a temperate marine ecosystem. Science 353, 169–172.

Williams, S.L., Smith, J.E., 2007. A global review of the distribution, taxonomy, and impacts of introduced seaweeds. Annual Review of Ecology Evolution and Systematics 327–359.

Witherell, D., Pautzke, C., Fluharty, D., 2000. An ecosystem-based approach for Alaska groundfish fisheries. Ices Journal of Marine Science 57, 771–777.

Worm, B., Duffy, J.E., 2003. Biodiversity, productivity and stability in real food webs. Trends in Ecology & Evolution 18, 628–632.

Worm, B., Barbier, E.B., Beaumont, N., Duffy, J.E., Folke, C., Halpern, B.S., Jackson, J.B.C., Lotze, H.K., Micheli, F., Palumbi, S.R., Sala, E., Selkoe, K.A., Stachowicz, J.J., Watson, R., 2006. Impacts of biodiversity loss on ocean ecosystem services. Science 314, 787–790.

Section III

Conservation in the
Anthropocene in Practice

Chapter 14

Stakeholder Participation in Marine Management: The Importance of Transparency and Rules for Participation

Christine Röckmann[1], Marloes Kraan[1], David Goldsborough[2],
Luc van Hoof[1]
[1]Wageningen Marine Research, Den Helder, The Netherlands; [2]VHL University of Applied Sciences, Leeuwarden, The Netherlands

INTRODUCTION

Conserving nature requires the management of people and managing together with people, because nature itself can only seldomly be forced to comply with a conservation plan. As noted already in 1973 by Rittel and Webber, environmental management and conservation issues are essentially social problems, characterized by high uncertainties and high stakes of those involved. Such "wicked problems" cannot be solved objectively or neutrally by scientific textbook knowledge, and "it makes no sense to talk about 'optimal solutions' to social problems unless severe qualifications are imposed first. Even worse, there are no 'solutions' in the sense of definitive and objective answers" (Rittel and Webber, 1973). In other words, the solution to complex conservation problems is not singular in either design or implementation (Röckmann et al., 2015).

Instead, highly uncertain and important conservation challenges call for holistic approaches, where all stakeholders—decision makers, scientists, and other actors, such as industry—participate in conservation planning and implementation (Dankel et al., 2012; Röckmann et al., 2012; van der Sluijs et al., 2008; Wilson, 2009). The importance of stakeholder participation has been widely acknowledged (Dreyer and Renn, 2011; Ehler and Douvere, 2006; Flannery and Ó Cinnéide, 2012; Jentoft and Chuenpagdee, 2009; Linke and Jentoft, 2012; Pomeroy and Douvere, 2008; Reed, 2008). Participation—"the cornerstone of democracy" (Arnstein, 1969)—implies the involvement of user groups in the decision-making and implementation process (Röckmann et al., 2015).

Conservation for the Anthropocene Ocean. http://dx.doi.org/10.1016/B978-0-12-805375-1.00014-3
289

A stakeholder is anybody who has, or feels to have, a stake, that is, any kind of interest, in the question/problem under consideration and related activities. The main motivations for (increased) stakeholder involvement and reported successes were synthesized by Röckmann et al. (2015) and Reed (2008).

> *"Participation can [...] bring additional knowledge and values into decision-making in order to make better decisions (Badalamenti et al., 2000; Renn, 2008), provide greater legitimacy (Raakjaer Nielsen and Mathiesen, 2003; Raakjaer Nielsen and Vedsmand, 1995), increase trust (de Vos and Mol, 2010; Luoma and Löfstedt, 2007; Munton, 2003; Renn and Levine, 1990; Young et al., 2013), enhance compliance (Christie, 2011; Christie et al., 2009; Jentoft, 2000), and reduce the intensity of conflict (Young et al., 2013). An improved overall process quality can result in increased management efficiency, equity, sustainability, reduction of administration and enforcement costs (Raakjaer Nielsen and Vedsmand, 1995), making the management not only more legitimate, salient, credible, but also enforceable and realistic (Craye et al., 2005; de Vos and van Tatenhove, 2011; Fiorino, 1990; Leslie and McLeod, 2007; Renn, 2008; Tallis et al., 2010; van der Sluijs, 2002; Wilson, 2009)"*

Röckmann et al. (2015, p. 157)

The earlier grants the benefits of participation to managers, but the participants themselves can also perceive advantages. Reed (2008, p. 2420) states that when considering local interests and concerns at an early stage, for example, to inform project design, the likelihood increases that local needs and priorities are successfully met. Irvin and Stansbury (2004) argue that participation enables stakeholders to gain some control over the policy process, which can lead to more appropriate policy and implementation decisions. The latter is clearly demonstrated in fisheries comanagement arrangements in the Netherlands: government and fisheries organizations jointly develop and implement social–economic policies, and organized interests do not have to lobby but are welcome partners at the table (van Hoof, 2010).

One premise of effective and sustainable environmental governance is that it draws from the iterative process of adaptive management. Costanza et al. (1998) concluded, for example, that the "key to achieving sustainable governance of the oceans is an integrated (across disciplines, stakeholder groups, and generations) approach based on the paradigm of 'adaptive management' whereby policy making is an iterative experiment acknowledging uncertainty, rather than a static answer."

Despite the promises, involving stakeholders can be challenging, time consuming, and expensive. Factors influencing the outcomes of a participatory process include the following: the willingness of stakeholders to interact with each other, level of transparency, availability of resources (e.g., financial, time), trust, state of knowledge, and shared language. Moreover, it is crucial to address equity issues such as which actors can participate, both in terms of who is invited to the table and also who can afford to sit at the table in the policy process and to what extent (Röckmann et al., 2015). If poorly handled, participatory processes can result in counterproductive negative consequences (e.g., erosion of trust between involved stakeholders and end of cooperation) (Johnson and van Densen, 2007; Reed et al., 2009). Röckmann et al. (2015, p.156) argue that "each individual situation requires context-specific trade-offs between ecological, economic and social sustainability criteria, based on an understanding of the institutional and political setting, local dynamics and context-dependent cultural constructs of the environment (Campbell et al., 2009; Christie, 2011)." Hence, the design and the expected outcomes of a participatory process are highly context specific and dependent on the way in which the coordinating party grants stakeholders a degree of ownership over the process.

Marine management and conservation is inherently political, as it deals with access to resources and hence making hard choices (Kooiman, 2005). At the same time, marine management depends on a high input of (scientific) knowledge due to the complexities and high uncertainties of the many issues to be addressed (e.g., indeterminate living and nonliving marine resources, and often intractable human activities at sea). Practitioners' knowledge is thus important as an additional source of knowledge. It can be unlocked when bringing stakeholders to the table. The integration of fishers' knowledge in management and science has been studied intensively (see Hind (2015) for an historic account of fishers' knowledge research), can take many forms, and can be integrated at different degrees (Stephenson et al., 2016). Knowledge sharing is, therefore, an important feature of participatory processes. But, a second important feature is negotiation. Based on experiences with participatory processes and, in particular, our analysis of the two case studies described later, we hypothesize that in practice, processes usually aim mutually exclusively at either the sharing of knowledge (to better inform a decision making process) or the sharing of power by negotiating (to jointly arrive at a decision). Participatory processes become messy as soon as the two aims are mixed, when focus is lacking, and when there is no clarity about participants' decision-making power. The challenge is to separate knowledge sharing and negotiating as much as possible. To avoid frustration and stakeholder fatigue in a participatory process it is a prerogative to be transparent about decision making, as well as roles, responsibilities, and mandate.

Here we argue that the increased integration of stakeholder knowledge in marine management requires clear objectives and a clear role for stakeholders in decision making. Making hard choices in marine management depends not only on scientific and local knowledge, but also must reflect societal values, which often requires trade-offs (Weible and Sabatier, 2005). In addition, we emphasize the importance of understanding underlying, often implicit, world views and strongly held ideas of the different stakeholders involved (Kraan, 2009), as they can fundamentally shape a process and debate (Kooiman and Jentoft, 2009). For example, different actor groups held different beliefs regarding the role of science in management, the seriousness of various marine problems, and interpretations of the scientific uncertainty concerning the implementation of Marine Protected Areas (MPAs) in California (Weible, 2008). According to Weible (2007) public policy controversies are driven more by value differences than by technical deficiencies.

This chapter discusses best practices and challenges with stakeholder participation. It draws from existing literature on theories of stakeholder participation (with a focus on knowledge sharing, negotiation, transparency, and rules for participation in decision making). The chapter then examines two different cases of stakeholder participation in applied marine science and management of the North Sea. In the description of the cases, we specifically draw attention to the specifics that shape the participatory process, following Kooiman and Jentoft (2009) and the work of Weible and Sabatier (2005) and Weible (2007, 2008). These cases provide the empirical foundation for summarizing lessons learned in the challenges and success of stakeholder participation in marine conservation.

KEY FEATURES AND BEST PRACTICES OF STAKEHOLDER PARTICIPATION

Levels of Participation

The extent of stakeholder participation varies. Arnstein (1969) described a gradated typology of eight levels of participation ranging from "nonparticipation" of citizens up to full "citizen power" in decision making. Similar gradations of participation have been described by others (Mathbor, 2008; Raakjaer Nielsen and Vedsmand, 1995; Tosun, 1999); they basically follow Arnstein'sladder of participation, starting at the lowest participation level with a centralistic top-down, government-based model that "enables power holders to educate citizens" (i.e., not allowing any participation); to gradually increasing degrees of stakeholder participation via instructing, consulting, cooperating, advising, being informed; and to self-management by user groups (cf. Röckmann et al., 2015, pp. 157–158). Similarly, Kooiman and Bavinck (2005) distinguish three ideal modes of governance, transferable to levels of participation: hierarchical governance, self-governance, and cogovernance. Often mixes of these modes occur, such as in fisheries governance systems (Kooiman and Bavinck, 2005).

Best Practices

The quality of participatory management outcomes in a context of complex, dynamic environmental problems is strongly dependent on the nature of the stakeholder process. Reed (2008) identifies eight best practices that improve the quality and effectiveness of stakeholder participation.

1. Stakeholder participation needs to be underpinned by a philosophy that emphasizes empowerment, equity, trust, and learning.
2. Where relevant stakeholders should be involved as early as possible and throughout the process.

3. A systematic stakeholder analysis should be carried out to ensure representative involvement of those stakeholders relevant to the environmental management question.
4. Clear objectives for the participatory process need to be agreed among stakeholders at the outset.
5. Methods should be selected and tailored to the decision-making context, considering the objectives, type of participants, and appropriate level of engagement.
6. Highly skilled facilitation is essential.
7. Local and scientific knowledge should be integrated (to provide a more comprehensive understanding of complex and dynamic socioecological systems and processes).
8. Participation needs to be institutionalized (creating organizational cultures that can facilitate processes where goals are negotiated and outcomes are necessarily uncertain).

Participation as a Shared Value

A philosophy of empowerment, equity, trust, and learning (the first best practice) provides the most crucial basis for any participatory process. Here, understanding the underlying values held by the different stakeholder groups is important for building trust and shared decision making. This foundational practice increases the governability of a system, that is, the capability to control/manage (Kooiman et al., 2008). Note that governability is the ability to govern a marine socioecological system (e.g., a fishery). Such systems are divers, complex and dynamic, and hence, governability depends on many factors. Moreover, even if principles and values are not shared, it is important that participants are aware of the differences (Mahon et al., 2011). Insights in how participants' underlying principles and values compare, why, and how they are intertwined in the debate may facilitate mutual understanding and increase trust, transparency, and clarity about process goals, rules, and participants' roles, responsibilities, mandates, and expectations.

Participatory stakeholder processes for environmental management and nature conservation are diverse. It is crucial to identify the shared expectations of the level of participation by stakeholders in the particular management context, together with creating transparency regarding the roles, responsibilities, and decision-making power of stakeholders participating in the particular context.

CASE STUDIES IN STAKEHOLDER PARTICIPATION FROM THE NORTH SEA

Two different cases of stakeholder participation in applied marine science and management are compared using the eight best practices mentioned earlier. The examples illustrate common dilemmas that arise with participation in the real world: confusion and lack of transparency over the levels of participation; failure to initially agree and clarify the participatory philosophy; decisions subject to political will; and diverse and hidden stakeholder perceptions and values.

The examples are from two stakeholder participation processes in the designation of Marine Protected Areas (MPAs) under EU legislation in the North Sea: one international European case ("Dogger Bank Case") and one national Dutch case (Dutch Marine Strategy).

METHOD

Our analyses are based on applied research carried out in several projects.

In the international Dogger Bank case, researchers facilitated and then assessed the process of preparing an international transboundary fisheries management plan for the Dogger Bank (MASPNOSE, 2012). Additionally, the Dogger Bank served as a case to test an extensive maritime spatial planning (MSP) and governance analysis framework (Goldsborough, 2013). The role and uptake of science were evaluated based on 10 interviews (Kraan and Pastoors, 2014).

The Dutch Marine Strategy case is based on participant observation during meetings organized in 2015 by the Dutch Ministry of Infrastructure and Environment in the stakeholder process related to the designation of MPAs at the Frisian Front and Central Oyster grounds in the North Sea. In addition, the fishers' knowledge that formed the basis of the fishery sector's zoning proposal was compared with scientific data and documented in a letter to the Ministry. Observations and results have not been published yet.

For both cases, we briefly describe the background, including the set-up and intention for creating a participatory process; we then evaluate how participation played out in practice.

Case One: International Process to Prepare Management Measures for the Special Areas of Conservation on the Dogger Bank

Background

The Dogger Bank is a large sand bank in the middle of the North Sea, slightly covered by seawater permanently. It stretches over parts of the Exclusive Economic Zones of four EU countries [the United Kingdom (UK), the Netherlands (NL), Germany (DE), and Denmark (DK)]. The UK, DE, and NL proposed parts of the Dogger Bank as Special Areas of Conservation (SACs) under the EU Habitat Directive. To comply with the European Habitat Directive legal obligations, the three countries had to identify specific conservation objectives for their national SACs and then develop suitable management measures to meet the conservation objectives. The proposed management measures to protect the sand bank habitat were to prohibit seafloor bottom impacting fisheries in parts of the SACs with a fisheries management plan. Goldsborough (2013) described in detail the international process, serving as empirical basis for our analysis.

There was no consensus, not between the Member States or between the different stakeholders, on which are the most sensitive parts of the Dogger Bank that need protection (NSAC, 2012). Each country has set different national priorities with slightly differing SAC conservation objectives, and partly conflicting economic interests. These relate to the different sectors: flatfish fisheries in the UK and NL; sandeel fishery in DK and wind energy in the UK. All management actions focused on limiting bottom impacting fisheries.

Set-up and Intention of the Participatory Process

An international MSP process was initiated by the Dutch authorities during a conference of the Dutch FIMPAS project (FIsheries Measures in Marine Protected Areas; https://noordzee.wordpress.com/2009/11/01/project-fimpas-official-summary/; Goldsborough, 2013). The main intention of this initiative was to facilitate alignment between DE, NL, and the UK concerning the SAC management objectives and measures for this valuable European sand bank habitat. The involved Member States agreed that cross-border management of this shared sand bank required a joint approach.

Stakeholders were involved at the initial FIMPAS conference through the already existing stakeholder platform, the North Sea Regional Advisory Council (NSAC). Advisory councils, consisting of the fishing industry and nongovernmental organizations (NGOs), evolved from the Regional Advisory Councils introduced in the 2002 reform of the European Common Fisheries Policy and have been established to provide advice and greater stakeholder involvement in fisheries management at the regional level (CEC, 2004). The international MSP process for the Dogger Bank triggered the establishment of an intergovernmental Dogger Bank Steering Group (DBSG), consisting of representatives of governmental authorities of the four Dogger Bank countries (UK, NL, DE, DK) and the European Commission. The International Council for the Exploration of the Sea (ICES) supported the process. The DBSG was tasked with the coordination of the cross-border process, dealing with competence struggles between national and European legislation.

Initially, the ultimate objective was to jointly propose a cross-border fisheries management plan, in close interaction with the regional fisheries stakeholder platform NSAC, and hence there was a clear focus on negotiation and less so on sharing knowledge. Later on, once the intergovernmental steering group was established, the regional fisheries stakeholder platform, the NSAC, held an active observer status with two seats. In a first step, the DBSG formally invited the NSAC to suggest zoning proposals for a fisheries management plan for the combined Dogger Bank SAC area. This invitation was not accompanied by any terms of reference, as the DBSG had not defined them yet. Furthermore, the DBSG itself did not participate in this stakeholder-led process.

In addition, another, parallel, preparatory action on MSP (MASPNOSE project: MAritime Spatial Planning in the North Sea) facilitated and supported the NSAC in their Dogger Bank zoning exercises, with the clear goal of sharing knowledge. This MASPNOSE project actually enabled the participatory process by contributing the funding that was crucial for the NSAC to accept the invitation from the DBSG. The NSAC formed a Focus Group, involving participants from the UK, NL, and DK fishing sectors (the DE sector declined), NGOs, the wind sector, and the MASPNOSE team. The subsequent participatory process covered about 1 year (MASPNOSE, 2012).

Realization of the Participatory Process

The participatory process led by the stakeholders and the MASPNOSE project was highly appreciated by the stakeholders. Key achievements were the sharing of knowledge and the building of trust, as stakeholders got to know each other and each other's perspectives and became increasingly aware of all the stakes, including different positions of the Member States in the DBSG (Goldsborough, 2013; MASPNOSE, 2012; NSAC, 2012).

In contrast, the overarching intergovernmental DBSG process lacked clarity, transparency, joint political will, commitment, and direction for negotiating a joint, collaborative management approach between the Dogger Bank Member States (NSAC, 2012; Goldsborough, 2013). In the initial 8 months, no stakeholder was involved, no terms of reference were in place and little communication with the outside world occurred. DBSG workshops were initially held behind closed doors, leading to speculations by the stakeholders that were excluded from this process. In its final position paper, the collaborating stakeholders (i.e., the NSAC) identified the "lack of a more joined up approach of all parties," the lack of relevant stakeholders, notably windfarm developers, the lack of clarity in the Terms of Reference as "factors which hindered the delivery of a consensual result" from the outset (NSAC, 2012, pp. 15–16).

Once the DBSG opened up for stakeholder observers (NSAC), and DBSG members attended the NSAC workshops, it became easier for all involved stakeholders to understand the various positions and procedures; the focus further shifted from knowledge sharing to negotiation. Nonetheless, the main drawback was the opaqueness concerning the goal of the broader participatory process, because the intergovernmental DBSG process lacked an official mandate for negotiating and agreeing on a joint MSP on the Dogger Bank. Transparency concerning roles and responsibilities of stakeholders, mainly in the DBSG (less so in the NSAC) was lacking (NSAC, 2012; Goldsborough, 2013).

In summary, disagreement between the countries and conflicting interests concerning proposed management measures have to date hampered the development of a joint decision not to mention implementation. Sectoral as well as national interests on the different national parts of the Dogger Bank have not fully converged. Differing legal obligations on the various institutional levels impede a straight-forward implementation of one joint, cross-border management plan.

Neither the intergovernmental DBSG nor the involved stakeholders have a mandate to take decisions. The lack of international consent and joint political direction has triggered stakeholder frustration and demotivation. In the meantime, a joint and cross-border management plan for the Dogger Bank SACs is in its final stages but it has not yet been approved and implemented. The adoption of this overall plan remains the competency of the European Commission (EC), but each individual Member State has to approve (Goldsborough, 2013; IJlstra, 2013).

Case Two: Participatory Process to Implement the Dutch Marine Strategy

Background

The Dutch government developed a Marine Strategy in 2012 (Rijksoverheid, 2012). The Marine Strategy is the Dutch plan to achieve/maintain Good Environmental Status in its marine environment by 2020 and is thereby linked to the EU Marine Strategy Framework Directive (MSFD). One of the ambitions of the Dutch government is to protect 10%–15% of the Dutch part of the North Sea from bottom impact. As 8.5% of the Dutch part of the North Sea has already been appointed (mostly under N2000 legislation), the Dutch government designated two areas as exploratory areas for introducing additional spatial measures for protection of the sea floor ecosystem: the Central Oyster Grounds and Frisian Front. The remaining 1.5%–6% of the Dutch part of the North Sea to be protected is to be found in these deep zones with muddy habitat, which are not yet part of the cluster of Dutch MPAs (Rijksoverheid, 2014).

The hydrographic and ecological conditions drive particularly high productivity and biodiversity in the Frisian Front (Lindeboom et al., 2015). Within the exploratory area, a spatial plan had to be developed, indicating how large the closures needed to be to meet the Dutch protection requirement. Also, the exact spatial boundaries needed to be determined.

Set-up, Intention and Realization of the Participatory Process

The government explicitly aimed at making choices by balancing economic and environmental costs and benefits. The ideal picture would be finding one or several areas, which if closed, would minimally affect the fishing sector, yet would contribute significantly to the conservation objectives. The Dutch government commissioned research to three national research institutes [Wageningen Marine Research (WMR), Wageningen Economic Research (WEcR), Royal Netherlands Institute for Sea Research (NIOZ)] and set up a stakeholder process with all relevant stakeholders, that is, the environmental NGOs (North Sea Foundation, World Wildlife Fund (WWF), Greenpeace) and the Dutch fishing sector.

With the aim to gather the best information available from all relevant stakeholders, the Dutch government initiated a stakeholder process, inviting the fishing sector, environmental NGOs, and scientists to various information sharing sessions (e.g., professionally facilitated "mapping table sessions"). All events served to share knowledge, ideas, and stakes to feed into the process of MPA proposal development. Undoubtedly, stakeholders joined with their own agendas but due to the set-up of the meetings, sharing information was key. Grouped around the mapping tables, everyone pitched in to discuss, for instance, the theoretical (scientists) and visible (fishers) characteristics of the dynamics of the Frisian Front. Alongside the stakeholder process, the government commissioned supporting research (such as a societal costs and benefits analysis) to also advise on the optimal protected areas based on the best available socioeconomic and ecological science (Oostenbrugge et al., 2015).

Both the fishing sector and NGOs were asked to develop proposals for spatial plans within the areas they considered best suited for this. The fishing sector's proposal was based on the minimum percentage of area protection (1.5%), whereas the NGOs proposal was built around the maximum (6.5%). Additionally, the fishing sector proposed areas outside of the official exploratory area, with the explanation that the initially designated study area had been much larger and that according to the Dutch Marine Strategy, the boundaries of the study area were not cast in stone (Rijksoverheid, 2014). This illustrates how the information sharing intent of the stakeholder process that focused on discussing the characteristics of an area, gradually shifted to a process with more emphasis on negotiation, as both stakeholder parties made use of the space to maneuver (an area between 1.5% and 6%) based on their competing proposals (i.e., the soft boundaries).

The fisheries sector's proposal had been developed in a participatory way. A scientist working for the sector together with two fishers developed the proposal, after having consulted other fishers active in the area. The sector considered the joint proposal a major achievement, as it represented a compromise resulting from negotiations between all involved fishers, whereby some proposed closure areas were more important catch grounds for one than the other fishery. The fisheries sector felt that this proposal was their best possible offer.

In order to fully understand the proposal's value, the fishers' underlying principles, values, worldviews, and specific context need to be understood and explained. It is sensitive for fishers to share knowledge on location and quality of fishing grounds with colleagues and with managers. Their knowledge is their capital. Fishers are cooperative, but they are also competitors (van Ginkel, 2009); usually there is no reason to help a competitor by sharing your own knowledge on fishing grounds, which "makes your colleagues smarter than they already are." Similarly, fishers hesitate to point out their key fishing grounds to managers and other stakeholders, as they are afraid that this sensitive information could be used against them. Additionally, fishers do not easily "give up" or "hand in" an area as being "not interesting" and hence a potential area for closure, as they have a strong conception of the sea as being dynamic. An area can quickly change in characteristics overnight depending on many aspects. Hence, giving up a "not interesting" area today may exclude its use in the future. Also, fishers have a strong sense of freedom. Selecting areas they "do not need" is unlikely, especially in a context with competing claims at sea (e.g., offshore wind farms) and they see their freedom of movement already affected.

ANALYSIS AND DISCUSSION

An overview of how the two case studies compare with the eight key features of participation (Reed, 2008) is presented in Table 14.1.

The analysis highlights that despite representative stakeholder involvement (3), appropriate use of participatory methods (5), skilled process facilitation,

TABLE 14.1 Implementation of Key Features in the Two Participatory Case Studies

	Case Study	International Dogger Bank SACs	Dutch Marine Strategy
	Involved stakeholders	• DBSG: governmental representatives of EU member states • NSAC: fishing industry, NGOs • Scientists • EU • ICES	• Fisheries sector • NGOs • Scientists • Ministry responsible for fisheries and nature conservation; ministry responsible for MSFD implementation
1	Stakeholder participation needs to be underpinned by a philosophy that emphasises empowerment, equity, trust, and learning.	The intergovernmental DBSG consulted stakeholders and officially invited them to present a joint proposal for zonation. Focus was on a joint stakeholder action. An explicit philosophy on participation was lacking: Stakeholders' mandate was unclear. There was no empowerment of stakeholders to join the decision-making process.	Stakeholders were consulted to share knowledge and insights and invited to present individual proposals for zonation. Consultation was explicitly communicated as the underpinning philosophy. The stakeholders' roles were clearly communicated, the mandate was clear.
2	Stakeholder participation should be considered as early as possible and throughout the process	Partly: The NSAC was formally invited to contribute to the DBSG process early on, but accepted as observers to the intergovernmental process only after a year.	Ok: Stakeholders were formally involved from the beginning.

Continued

TABLE 14.1 Implementation of Key Features in the Two Participatory Case Studies—cont'd

Case Study		International Dogger Bank SACs	Dutch Marine Strategy
3	Relevant stakeholders need to be analyzed and represented systematically	Ok: Stakeholders were involved through the existing regional stakeholder platform NSAC.	Ok: NGOs and all relevant fishing representatives could participate.
4	Clear objectives for the participatory process need to be agreed among stakeholders at the outset	No: The involvement of stakeholders was only a formal invitation to contribute to the process without specified ToRs.	Ok: It was made clear that stakeholders would be consulted and that the government would make a final decision on the zonation.
5	Methods selected and tailored to the decision-making context, considering the objectives, type of participants and appropriate level of engagement	NSAC ok: Information sharing sessions DBSG: Not possible to evaluate as the process was not open and transparent and no documentation is available.	Ok: Interactive sessions were organized to share knowledge. As a follow up sector and NGO were individually asked to propose zoning scenarios.
6	Highly skilled facilitation is essential	Ok: Professional facilitation was provided by the MASPNOSE project for the NSAC and by ICES for the DBSG.	Ok: Professional facilitation was provided by independent experts.
7	Local and scientific knowledge should be integrated	Ok: Knowledge was provided by industry, NGOs and science and was then integrated for the joint development of the zoning proposals.	Ok: Fishers' knowledge was compared with scientific knowledge.
8	Participation needs to be institutionalized	Partly: The DBSG as well as the NSAC are officially institutionalized bodies. However, the NSAC's participatory role in the governmental DBSG process was limited due to the lack of official mandate to be involved in the decision making process.	OK: The government institutionalized participation by officially organizing participatory processes. Participation was limited to consultation.

and good knowledge integration (7), a participatory process can still be unsuccessful, as illustrated in the Dogger Bank case. The Dogger Bank process has not yet resulted in a final management decision, although stakeholders had unanimously presented zoning proposals already in 2012 (NSAC, 2012). This is due to a lack of empowerment (1), a delay of involvement (2), a lack of clarity and transparency about roles (1), a lack of institutionalized procedures (8), a lack of clear objectives (4), and a lack of mandates (1,4,8).

Both the Dogger Bank case and the Dutch Marine Strategy case illustrate the importance of transparency about the underpinning philosophy of empowerment in marine management, that is, process goals, roles and responsibilities, mandate, as well as a common understanding of context and being explicit about different perceptions between the different stakeholders. In both cases, the embedded processes were more successful than the overarching processes. So when the NSAC organized their own process it functioned better than that of the DBSG. The same holds true for the fishers in the Dutch Marine Strategy case: They succeeded in agreeing on their own spatial plan proposal. In these smaller and well-defined subgroups, the process participants knew each other and agreed on the procedure and the importance of the process.

In the Dogger Bank case, the intergovernmental group officially invited stakeholders to suggest zoning proposals. The rules of the game, stakeholders' roles, responsibilities, and timelines for participation in the intergovernmental process remained unclear, though. Terms of Reference were formulated by the intergovernmental group and provided to the stakeholders only late in the process. The intergovernmental group mainly worked behind closed doors, focusing on political interpretations of the legal obligation to implement EU policy, and thus, essentially on negotiation. And despite the establishment of the intergovernmental DBSG, ultimately the decision-making process remained with the individual member states and their individual responsibility to implement European nature legislation nationally. In contrast to the intergovernmental DBSG, the NSAC itself organized a clear and transparent process, starting out by agreeing on scope and procedure, selecting, and agreeing on issues and tasks that would be dealt with in the group and with the available resources. Moreover, a shared understanding was reached on the goal of nature conservation and principles for fisheries management on the Dogger Bank. The NSAC process could thus focus on practical implementation of policy rather than on policy interpretation only. Although the different stakeholder groups (i.e., industry and environmental NGOs) did not arrive at one joint proposal to the governmental group, they did succeed in agreeing on a fundamental understanding of the conservation task, sharing their knowledge and views and integrating the different knowledge bases to collaborate and actually provide joint knowledge and advice to the official intergovernmental decision-making process (NSAC, 2012). A stalemate between sector and NGOs was thus reverted into cooperation toward providing joint advice, accompanied by two zoning proposals. The only difference between the two proposals was the percentage of area to be protected (NSAC, 2012). In line with Costanza et al. (1998), Kooiman and Bavinck (2005), and Mahon et al. (2009), basic agreement on high-level norms, values, and principles helped stimulate and foster realization of day-to-day operations in the participatory process.

In the Dutch stakeholder process for the designation of additional protected areas under the MSFD, it was communicated transparently from the beginning that the participatory process focused on stakeholder consultation and sharing information and that a final decision would be taken by the government. The process was set-up to share knowledge but gradually evolved in a process where negotiation set in, as both fishers and NGOs took advantage of the bandwidth to push for different sizes and boundaries of the protected area. One of the side-effects of this was that the fishing sector proposal was perceived as driven by stakes rather than by knowledge. However, taking into account fishers' general aversion to area closures, their proposal comprised more than only a strategic step in a negotiation process. It was a step toward cooperating in developing a joint proposal for closure.

CONCLUSION

Marine management is inherently political and relies on scientific knowledge and a broad range of expertise. Practitioners' knowledge is increasingly important to marine management and to expand the knowledge base. Participatory processes can also be a space for negotiating diverse conservation values. Problems emerge when the process and objectives of stakeholder participation are ambiguous, and when participation is driven by cross-purposes of knowledge sharing or negotiation. Therefore, the challenge is to separate as much as possible these two participatory aspects. Ultimately, stakeholders' role in decision making needs to be clear and transparent for greatest success.

The underpinning philosophy of empowerment, equity, trust, and learning frames the basis and sets the pace for participatory processes. This chapter has highlighted the importance of transparency about the level of decision-making power in marine management, which includes clarifying process goals, roles, and responsibilities, mandate, as well as developing a common understanding of the decision context and the diverse values among different stakeholders. It should be clear which actor has decision power and when decisions will be taken. In addition, especially when based on different guiding principles, the success of the planning process is highly dependent on the degree of transparency and trust between participants.

Moreover, it is essential that all parties involved in the process have commitment and resources available to take on their role in the process. Despite representative stakeholder involvement, appropriate use of participatory methods, skilled process facilitation, and good knowledge integration, a participatory process can still not be successful if an underpinning participatory philosophy and clear objectives are lacking, participation is delayed and not well institutionalized.

Different stakeholder groups have their own set of ideas, experiences, and beliefs. These differences influence how stakeholders participate in participatory processes. It is recommended to share the different contexts of departure at the beginning of a participatory process. Also it should be made explicit who are considered stakeholders in the issue at hand, which of these groups can participate and in which form (rules of the game), and who decides on all of this, in short: who is the owner of the participatory process.

In summary, our analyses illustrate that a participatory process can still not be successful if an underpinning participatory philosophy and clear objectives are lacking, participation is delayed and not well institutionalized. Clarity is needed about the participatory philosophy and process objective. The goal can be sharing knowledge or negotiating a decision. The increased need of stakeholder knowledge requires clarity about which of the two is driving the process. Rules of the game, including roles, responsibilities, and mandate need to be clear to all participants from the beginning.

REFERENCES

Arnstein, S.R., 1969. A ladder of citizen participation. Journal of the American Planning Association 35, 216–224.

Badalamenti, F., Ramos, A.A., Voultsiadou, E., Sánchez Lizaso, J.L., D'Anna, G., Pipitone, C., Mas, J., Fernandez, J.A.R., Whitmarsh, D., Riggio, S., 2000. Cultural and socio-economic impacts of Mediterranean marine protected areas. Environmental Conservation 27, 110–125.

Campbell, L.N., Gray, N.J., Hazen, E.L., Shackeroff, J.M., 2009. Beyond baselines: rethinking priorities for ocean conservation. Ecology and Society 14.

CEC, 2004. Council decision 2004/585/EC of 19 July 2004 establishing regional advisory councils under the common fisheries policy. In: Commission C.o.t.E. 2004/585/EC.

Christie, P., 2011. Creating space for interdisciplinary marine and coastal research: five dilemmas and suggested resolutions. Environmental Conservation 38, 172–186.

Christie, P., Pollnac, R.B., Oracion, E.G., Sabonsolin, A., Diaz, R., Pietri, D., 2009. Back to basics: an empirical study demonstrating the importance of local-level dynamics for the success of tropical marine ecosystem-based management. Coastal Management 37, 349–373.

Costanza, R., Andrade, F., Antunes, P., den Belt, M.V., Boersma, D., Boesch, D.F., Catarino, F., Hanna, S., Limburg, K., Low, B., Molitor, M., Pereira, J.G., Rayner, S., Santos, R., Wilson, J., Young, M., 1998. Principles for sustainable governance of the oceans. Science 281, 198–199.

Craye, M., Funtowicz, S., van der Sluijs, J.P., 2005. A reflexive approach to dealing with uncertainties in environmental health risk science and policy. International Journal of Risk Assessment and Management 5, 216–236.

Dankel, D.J., Aps, R., Padda, G., Röckmann, C., van der Sluijs, J.P., Wilson, D.C., Degnbol, P., 2012. Advice under uncertainty in the marine system. ICES Journal of Marine Science 69, 3–7.

de Vos, B.I., Mol, A.P.J., 2010. Changing trust relations within the Dutch fishing industry: the case of National Study Groups. Marine Policy 34, 887–895.

de Vos, B.I., van Tatenhove, J.P.M., 2011. Trust relationships between fishers and government: new challenges for the co-management arrangements in the Dutch flatfish industry. Marine Policy 35, 218–225.

Dreyer, M., Renn, O., 2011. Participatory approaches to modelling for improved learning and decision-making in natural resource governance: an editorial. Environmental Policy and Governance 21, 379–385.

Ehler, C., Douvere, F., 2006. Visions for a Sea Change, Report of the First International Workshop on Marine Spatial Planning. UNESCO, Paris.

Fiorino, D.J., 1990. Citizen participation and environmental risk: a survey of institutional mechanisms. Science, Technology & Human Values 15, 226–243.

Flannery, W., Ó Cinnéide, M., 2012. Deriving lessons relating to marine spatial planning from Canada's Eastern Scotian Shelf integrated management initiative. Journal of Environmental Policy & Planning 14, 97–117.

Goldsborough, D., 2013. Governance Analysis, WP6. Case Study: Dogger Bank. A Case Study Report for Work Package 6 of the MESMA Project (www.mesma.org). 45 pp. MESMA Deliverable 6.1, Appendix A 7.3. pp. 50–95. http://mesma.org/default.asp?ZNT=S0T1O-1P171.

Hind, E.J., 2015. A review of the past, the present, and the future of fishers' knowledge research: a challenge to established fisheries science. ICES Journal of Marine Science 72, 341–358.

IJlstra, T., 2013. The international Dogger Bank steering group. In: Presented at the NSRAC Executive Committee Meeting, Amsterdam 28 February 2013. http://nsrac.org/wp-content/uploads/2013/01/International-Dogger-Bank-Steering-Group-Ton-Ijstra-presentation1.ppt.

Irvin, R.A., Stansbury, J., 2004. Citizen participation in decision making: is it worth the effort? Public Administration Review 64, 55–65.

Jentoft, S., 2000. Legitimacy and disappointment in fisheries management. Marine Policy 24, 141–148.

Jentoft, S., Chuenpagdee, R., 2009. Fisheries and coastal governance as a wicked problem. Marine Policy 33, 553–560.

Johnson, T.R., van Densen, W.L.T., 2007. Benefits and organization of cooperative research for fisheries management. ICES Journal of Marine Science 64, 834–840.

Kooiman, J., 2005. Fish for Life: Interactive Governance for Fisheries. Leiden University Press.

Kooiman, J., Bavinck, M., 2005. The governance perspective. In: Jan, K., Maarten, B., Jentoft, S., Pullin, R. (Eds.), Fish for Life. Interactive Governance for Fisheries. Amsterdam University Press, Amsterdam, p. 16.

Kooiman, J., Bavinck, M., Chuenpagdee, R., Mahon, R., Pullin, R., 2008. Interactive governance and governability: an introduction. Journal of Transdisciplinary Environmental Studies 7, 1–11.

Kooiman, J., Jentoft, S., 2009. Meta-governance: values, norms and principles, and the making of hard choices. Public Administration 87, 818–836.

Kraan, M., Pastoors, M., 2014. The role of science in Marine Policy making: reflecting on the role of IMARES as boundary organisation in applied marine research projects. In: Van Hoof (Ed.), Zee op Zicht: Inzicht; Een zoektocht naar een integraal afwegingskader voor het gebruik van de zee. IMARES rapport. , pp. 36–47. http://library.wur.nl/WebQuery/wurpubs/fulltext/332224.

Kraan, M.L., 2009. (Ph.D.). Creating Space for Fishermen's Livelihoods. Anlo-ewe Beach Seine Fishermen's Negotiations for Livelihood Space within Multiple Governance Structures in Ghana, vol. 19. African Studies Centre. African Studies Collection, Leiden.

Leslie, H.M., McLeod, K.L., 2007. Confronting the challenges of implementing marine ecosystem-based management. Frontiers in Ecology and the Environment 5, 540–548.

Lindeboom, H., Rijnsdorp, A.D., Witbaard, R., Slijkerman, D., Kraan, M., 2015. Het Ecologisch Belang Van Het Friese Front. IMARES report C137/15A. http://edepot.wur.nl/370466.

Linke, S., Jentoft, S., 2012. A communicative turnaround: shifting the burden of proof in European Fisheries Governance. Marine Policy.

Luoma, S.N., Löfstedt, R.E., 2007. Contaminated salmon and the Public's trust. Environmental Science & Technology 41, 1811–1814.

Mahon, R., Fanning, L., McConney, P., 2009. A governance perspective on the large marine ecosystem approach. Marine Policy 33, 317–321.

Mahon, R., Fanning, L., McConney, P., 2011. Principled ocean governance for the wider Caribbean region. In: Towards marine ecosystem-based management in the Wider Caribbean, pp. 27–37.

MASPNOSE, 2012. MASPNOSE Deliverable D1.2: Report on Cross-border Maritime Spatial Planning in Two Case Studies. http://www.wur.nl/upload_mm/7/6/2/92fbfd4c-5b01-4e8e-9a82-de877fa6d515_MASPNOSE%20D1.2%20MSP%20in%20case%20studies.pdf.

Mathbor, G.M., 2008. Chapter 6: A Typology of Community Participation, Effective Community Participation in Coastal Development. Lyceum Books, Inc., p. 144.

Munton, R., 2003. Deliberative democracy and environmental decision making. In: Berkhout, F., Scoones, I., Leach, M. (Eds.), Negotiating Change: Advances in Environmental Social Science. Edward Elgar, Camberley, pp. 63–80.

NSAC, 2012. Final Position Paper April 2012. Fisheries Management in Relation to Nature Conservation for the Combined Area of 3 National Natura 2000 Sites (SACs) on the Dogger Bank. http://nsrac.org/wp-content/uploads/2012/07/NSRAC-1112-7-2012-04-09-Dogger-Bank-SACs-Position-Paper-FINAL.pdf.

Oostenbrugge, H.v., Slijkerman, D., Hamon, K., Bos, O., Machiels, M., van de Valk, O., Hintzen, N., Bos, E., van der Wal, J.T., Coolen, J., 2015. A Cost Benefit Analysis. LEI Wageningen UR report Wageningen, December 2015.

Pomeroy, R., Douvere, F., 2008. The engagement of stakeholders in the marine spatial planning process. Marine Policy 32, 816–822.

Raakjaer Nielsen, J., Mathiesen, C., 2003. Important factors influencing rule compliance in fisheries lessons from Denmark. Marine Policy 27, 409–416.

Raakjaer Nielsen, J., Vedsmand, T., 1995. Fisheries Co-management: An Alternative Strategy in Fisheries – Cases from Denmark – an Issue Paper for the OECD Study on the Efficient Management of Living Marine Resources. OECD, Paris, France, pp. 1–33.

Reed, M.S., 2008. Stakeholder participation for environmental management: a literature review. Biological Conservation 141, 2417–2431.

Reed, M.S., Graves, A., Dandy, N., Posthumus, H., Hubacek, K., Morris, J., Prell, C., Quinn, C.H., Stringer, L.C., 2009. Who's in and why? A typology of stakeholder analysis methods for natural resource management. Journal of Environmental Management 90, 1933–1949.

Renn, O., 2008. Risk Governance. Coping with Uncertainty in a Complex World. Earthscan, London.

Renn, O., Levine, D., 1990. Credibility and Trust in Risk Communication. Springer.

Rijksoverheid, October 2012. Marine Strategy for the Netherlands Part of the North Sea 2012–2020, Part 1. Published by Ministry of Infrastructure and the Environment in Cooperation with Ministry of Economic Affairs the Hague. The Netherlands https://www.noordzeeloket.nl/en/Images/Marine%20Strategy%20for%20the%20Netherlands%20part%20of%20the%20North%20Sea%202012-2020%2C%20Part%201_683.pdf.

Rijksoverheid, 2014. Marine Strategy for the Dutch Part of the North Sea 2012–2020 Part 3. MSFD Programme of Measures Appendix 5 to the National Water Plan 2016–2021. December 2015. Joint Publication of Ministry of Infrastructure and the Environment in Cooperation with Ministry of Economic Affairs the Hague, The Netherlands. https://www.noordzeeloket.nl/en/Images/NL%20Marine%20Strategy%20part%203%20English%20translation_5022.pdf.

Rittel, H.W.J., Webber, M.M., 1973. Dilemmas in a general theory of planning. Policy Sciences 4, 155–169.

Röckmann, C., Ulrich, C., Dreyer, M., Bell, E., Borodzicz, E., Haapasaari, P., Hauge, K.H., Howell, D., Mäntyniemi, S., Miller, D., Tserpes, G., Pastoors, M., 2012. The added value of participatory modelling in fisheries management – what has been learnt? Marine Policy 36, 1072–1085.

Röckmann, C., van Leeuwen, J., Goldsborough, D., Kraan, M., Piet, G., 2015. The interaction triangle as a tool for understanding stakeholder interactions in marine ecosystem based management. Marine Policy 52, 155–162.

Stephenson, R.L., Paul, S., Pastoors, M.A., Kraan, M., Holm, P., Wiber, M., Mackinson, S., Dankel, D.J., Brooks, K., Benson, A., 2016. Quo Vadimus. Integrating fishers' knowledge research in science and management. ICES Journal of Marine Science. http://dx.doi.org/10.1093/icesjms/fsw025.

Tallis, H., Levin, P.S., Ruckelshaus, M., Lester, S.E., McLeod, K.L., Fluharty, D.L., Halpern, B.S., 2010. The many faces of ecosystem-based management: making the process work today in real places. Marine Policy 34, 340–348.

Tosun, C., 1999. Towards a typology of community participation in the tourism development process. Anatolia 10, 113–134.

van der Sluijs, J.P., 2002. A way out of the credibility crisis of models used in integrated environmental assessment. Futures 34, 133–146.

van der Sluijs, J.P., Petersen, A.C., Janssen, P.H.M., Risbey, J.S., Ravetz, J.R., 2008. Exploring the quality of evidence for complex and contested policy decisions. Environmental Research Letters 3, 024008.

van Ginkel, R., 2009. Braving Troubled Waters: Sea Change in a Dutch Fishing Community (Texel) and Beyond. Amsterdam University Press, Amsterdam.

van Hoof, L., 2010. Co-management: an alternative to enforcement? ICES Journal of Marine Science 67.

Weible, C.M., 2007. An advocacy coalition framework approach to stakeholder analysis: understanding the political context of California marine protected area policy. Journal of public Administration Research and Theory 17, 95–117.

Weible, C.M., 2008. Caught in a maelstrom: implementing California marine protected areas. Coastal Management 36, 350–373.

Weible, C.M., Sabatier, P.A., 2005. Comparing policy networks: marine protected areas in California. Policy Studies Journal 33, 181–201.

Wilson, D.C., 2009. The Paradoxes of Transparency. Science and the Ecosystem Approach to Fisheries Management in Europe. Amsterdam University Press, Amsterdam.

Young, J.C., Jordan, A., Searle, K.R., Butler, A., Chapman, D.S., Simmons, P., Watt, A.D., 2013. Does stakeholder involvement really benefit biodiversity conservation? Biological Conservation 158, 359–370.

Chapter 15

Marine Conservation as Complex Cooperative and Competitive Human Interactions

Xavier Basurto[1], Esther Blanco[2,3], Mateja Nenadović[1], Björn Vollan[3,4]
[1]Duke University, Beaufort, NC, United States; [2]Indiana University, Bloomington, IN, United States; [3]Innsbruck University, Innsbruck, Austria; [4]University of Marburg, Marburg, Germany

INTRODUCTION

Around the world, the promise and potential of marine protected areas (MPAs) as effective conservation tools has made them the preferred approach to protect biodiversity (Morton-Lefévre, 2014). Although understanding of MPAs' impacts on biological systems has made important strides (Mumby et al., 2007; Lester et al., 2009; Selig and Bruno, 2010), understanding of social impacts remains limited in comparison (Cinner, 2007; Mascia et al., 2010; Pollnac et al., 2010; Fox et al., 2012; Halpern et al., 2013; Gruby et al., 2015). Yet, the importance of minimizing this deficit is increasingly recognized as necessary to fulfill MPAs' promise as long-term conservation tools (West et al., 2006; Edgar et al., 2014). Sustaining the biological benefits that MPAs can provide to ecosystems and humans in the long term depends on the strength of the civil society valuing MPAs' contributions. Civil societies are made of a complex mix of cooperative and competitive human interactions (Oakerson and Parks, 2011) and in this chapter we extend this view to marine conservation interventions and particularly to the study of the effects of MPAs in fishing communities. We argue that the science of MPAs can benefit from insights from the cooperation, competition, and collective action literature. Understandings of how MPAs change stakeholder complex cooperative and competitive interactions could provide insights about their likelihood for long-term conservation success. For instance, our findings from Mexico show statistically higher levels of cooperation and hypercompetition in MPA communities than in non-MPA sites, where hypercompetition constitutes an antisocial form of competition. Interestingly, this antisocial behavior does not seem to discourage the capacity of members of the fishing community to cooperate and engage in successful collective action,

Conservation for the Anthropocene Ocean. http://dx.doi.org/10.1016/B978-0-12-805375-1.00015-5

307

which is essential in people's ability to organize for the provision of public goods or sustainable consumption of common-pool resources in protected areas (Oakerson and Parks, 2011).

Moreover, individuals behaving as "hypercompetitive co-operators" were found not only among fishers but also among nonfishers in MPA fishing communities. MPAs restrict access to fishing grounds and thus foster competition among fishers, yet MPAs not only affect fishers but also a broader set of stakeholders including tourism and industrial fishing sectors. We suggest that new income opportunities associated to the implementation of MPAs would increase the competitiveness of fishers and nonfishers, whereas concurrently broader social well-being can be strengthened through mutual cooperation, which helps to overcome social cleavages and engage in collective action.

Interest and research on the social impacts of MPAs have increased in the last decade, both theoretically and analytically. Researchers have engaged economic theories (e.g., Smith et al., 2010; Herrera et al., 2016), property rights perspectives (e.g., Mascia and Claus, 2009; Ban et al., 2015), and anthropological (Guerrón-Montero, 2005; Brondo and Woods, 2007) and critical geography approaches (Gruby and Basurto, 2013; Chaigneau and Brown, 2016). Different analytical perspectives such as ethnography have provided nuanced understandings about MPAs relationships with human displacement and conflict (e.g., West et al., 2006) and livelihoods (e.g., Peterson and Stead, 2011). Meta-analyses have been useful to examine MPAs effects on human well-being (e.g., Mascia et al., 2010) and the effectiveness of community-based conservation projects (e.g., Brooks et al., 2012). Experimental approaches have provided insights on societal norms affecting participation in management (Gurney et al., 2016), and large-n surveys have been extensively used to better understand the relationship between socioeconomic and governance characteristics with biophysical conditions (e.g., Cinner et al., 2012) or food security (Darling, 2014).

Our work combines experimental, large-n surveys, and participatory observation to explore the social effects of MPAs on individual and collective behavior in communities affected by MPAs. Lessons learned from our long-term engagement in the region since 1999 (Basurto, 2005; Basurto et al., 2012, 2013; Nenadović et al., 2016; Nenadovic and Epstein, 2016) informed the design and deployment of lab-in-the-field economic experiments with fishers and nonfishers (n = 127) in two MPAs (and two controls) of the Peninsula of Baja California of Northwest Mexico, a recognized region for its global marine biodiversity (Álvarez-Romero et al., 2013). We determined the external validity of our findings through a large-n survey with 71% of all fishers (n = 544, 48% active, Table 15.1) and developed potential explanatory mechanisms through interviews with expert informants (n = 77) in the four most important MPAs in the Peninsula (Fig. 15.1).

Our multimethod approach grounds our interpretation to the local and historical context. We examine the need to attend to the ways in which protected

TABLE 15.1 Characteristics of Our Study Area and Our Surveying Effort

Marine Protected Areas (MPAs)[a]	Localities	Popul.[b]	Data Collection	No. of Employed Enumerators	No. of Identified Fishers[c]	No. of Total Surveys	No. of Surveys With Active Commercial Fishers
UGBR	San Felipe	16,702	Survey	4	197	86	79
	Mulegé	3,821	Experiment	–	–	–	–
LBNP	Loreto	14,724	Survey, experiment	6	229	227	129
	Juncalito	40	Survey				
	Ligüí	203	Survey, experiment				
	Ensenada Blanca	255	Survey, experiment				
ESNP[d]	La Paz	215,178	Survey	6	161	71	63
	Todos Santos	5,148	Experiment	–	–	–	–
CPNP	La Ribera	2,050	Survey, experiment	4	177	160	100
	Cabo Pulmo	50	Survey				
	Los Frailes[e]	9	–				
	Agua Amarga	382	Survey, experiment				
	Boca del Alamo	100	Survey				
Grand totals:					764 (100%)	544 (72.20%)	371 (48.56%)

[a]Official names of each MPA are as follows: UGBR = Upper Gulf Biosphere Reserve; LBNP = Loreto Bay National Park; ESNP = Espiritu Santo National Park; CPNP = Cabo Pulmo National Park. These are four of seven protected areas in the Peninsula of Baja California. The other three include Los Angeles Bay Biosphere Reserve, San Lorenzo Archipelago National Park, and Flora and Fauna Area Cabo San Lucas.
[b]Data from the 2010 population census, Instituto Nacional de Estadística y Geografía.
[c]The total number of fishers (active and inactive) is based on our work. This information is not available through official records.
[d]We focused our surveying effort to identify fishers who exclusively use or used fishing area within the local MPA.
[e]Fishing community of Los Frailes consists of fishers who migrate from other locations within Baja California Sur to this area to fish during a fishing season. Most prominent are fishers from Agua Amarga and Boca del Alamo.

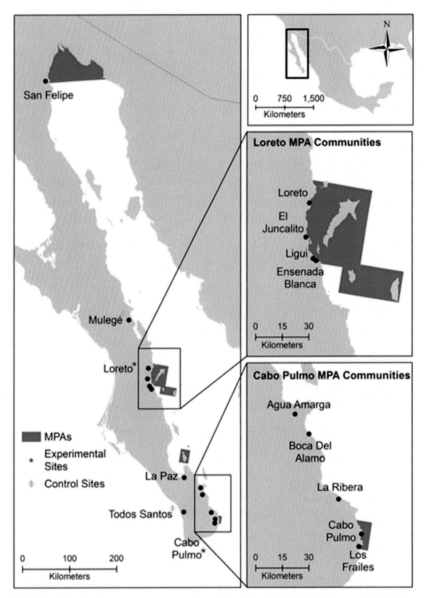

FIGURE 15.1 Study area showing location of marine protected areas, associated fishing communities, and control sites.

areas produce different types of "peoples" (West et al., 2006, p. 251) by assessing whether and how MPAs contribute to the emergence of new patterns of cooperation or competition. We assess the creation of subjects or different types of peoples by measuring changes in competitive and cooperative behavior and particularly focus on the concept of friendly rivalry, which encompasses mutual existence of cooperative and competitive behavior within individual members of a local community. These two types of behavior interweave through interactions with others and the environment through every day activities (McCay, 2002; Noussair et al., 2015), thus shaping communities and fishers' identities (Acheson, 1988; McCormack, 2012). Past studies show that competitiveness or cooperation among fishers emerges in certain environmental conditions as a result of learning and adaptation (Henrich et al., 2004; Leibbrandt et al., 2013). We conceptualize MPAs as a governance intervention that modifies incentives for collective action (i.e., cooperation and competition) by creating new rules, norms, and practices regulating stakeholders' interactions with the marine environment and with each other.

Although both cooperative and competitive interactions form a part of everyday life in fishing communities in what we refer to as friendly rivalry, these interactions need to be differentiated from competition and hypercompetition among nonlocal fishers (e.g., industrial or small-scale) and local fishers, which can trigger open conflict through their perceived destructiveness or illegality. These conflicts have been extensively documented around the world (Phyne, 1990; Schlager and Ostrom, 1992; Berkes et al., 2001; Bavinck, 2005; Pomeroy et al., 2007; DuBois and Zografos, 2012) and our own work shows that our study area is no exception.

MATERIALS AND METHODS

Study Sites

We conducted fieldwork in 12 communities along the Baja California Peninsula (Fig. 15.1 and Table 15.1). Small-scale fisheries are the principal source of income generating in all locations with the exception of La Paz, which is the economic and administrative center of the state of Baja California Sur, and to a smaller extent of Loreto and Todos Santos. Surveys were conducted in 10 communities, all of them associated with a MPA. Experiments were administered in seven communities for a total of nine sessions. Five locations constituted hometowns of fishers in communities influenced by two MPAs, whereas two are hometowns of fishers outside of the influence of MPA authorities. In all locations, participants included fishers and nonfishers (representing other stakeholders, e.g., tourism operators, recreational fishers) living in the community, and each of them made decisions toward a fisher and toward a nonfisher in their session.

Field Experiments

Economic experiments are controlled interactions among individuals based on game theoretic predictions. The use of pecuniary or other material incentives and anonymity makes experiments less prone to hypothetical bias or social-desirability biases than surveys (Smith, 1982; Cardenas and Carpenter, 2008). There is a spectrum of alternative implementations of economic experiments, ranging from laboratory experiments, conducted at universities, having college students as subjects, and subjects being aware of their participation in an experiment, to natural field experiments, where the experiment occurs in the environment where the subjects are naturally undertaking certain tasks and the subjects do not know that they are participants in an experiment (List, 2011). Our application is an "artifactual field experiment," which lies in-between these two extremes. It mimics a lab experiment except that participants are drawn from the setting of interest, and thus they are aware that they are taking part in an experiment and are explicitly told that their experience is for research purposes. Artifactual field experiments have been used extensively for the study of social dilemmas in the management of natural resources, showing its relevance for the study of social norms (see research summarized in Cardenas (2011)), responses to scarcity (Prediger et al., 2014; Blanco et al., 2015; Gatiso et al., 2015; Pfaff et al., 2015), or institutional development (Bouma et al., 2008; Rodriguez-Sickert et al., 2008; Vollan, 2008; Janssen et al., 2013) among others. Moreover, some previous studies have explicitly studied the external validity of artifactual field experiments (Bouma et al., 2008; Rustagi et al., 2010; Gelcich et al., 2013).

We operationalize cooperation experimentally by implementing a public goods game (Isaac et al., 1994) and a modified joy-of-destruction game to measure potential hypercompetition (Abbink and Herrmann, 2011). We define cooperation in the public goods game as the prosocial ability to forgo immediate individual benefit in lieu of collective gain (see discussion in Ledyard (1995); Henrich and Henrich (2007)). Hypercompetition constitutes an extreme, antisocial dimension of competition, where individuals are willing to incur personal costs to destroy other's welfare motivated by a desire to be ahead, focusing on an own advantage in relative payoffs at a cost to absolute payoffs (Jensen, 2010).

We followed a standard implementation of an artifactual field experiment. In particular, each session consisted of three main parts: First, we implemented a standard linear public goods game (Isaac et al., 1984), a straightforward workhorse to measure cooperation. In the two-player version that we implemented, each subject had an initial monetary endowment and could make voluntary contributions to a common fund. For every peso contributed to the common fund by either player we added 50 cents and the total was later evenly divided between the two players, irrespective of the amounts invested by each player. The social optimum for a pair of players was to invest all their endowment in the common fund; however, each of them had incentives to free ride in the investments of the other and not contribute.

Second, a modified joy-of-destruction game (Abbink and Herrmann, 2011) was designed to capture spite or hypercompetitiveness. In this game, players used part of an initial monetary endowment to reduce money from the other player at a personal cost. The money of the other player was not "stolen" but simply destroyed. Reciprocity, inequity aversion, or envy was removed as potential motive for reducing money given this were one-shot games with no reciprocal behavior and equal endowments. Two of the motivations for reducing the other person's money were spite and the desire to be ahead of the other player.

Last, an ex-post socioeconomic survey that included personal demographic characteristics, social norms, opinions regarding MPAs in the past and present, and opinions about local authorities and social organizations was conducted. For an in-depth description of the experimental procedure, including a transcript of the experimental instructions and ex-post questionnaire, see the supplementary materials in Basurto et al. (2016).

The session started with a general description of the structure and context of the activity. We then provided the specific instructions for the first part of the game, including examples. Questions were answered privately and subjects answered a set of quiz questions to check for their level of understanding of the game before making decisions. Answering the quiz questions correctly was not a pre-requisite for continuation in the session, but we used this information to check for the robustness of results when restricting to the participants who answered all questions correctly or when relaxing such requirements. All results are robust. The participants then made their decisions for the first part of the session and we proceeded to present the details of the second part of the session without providing any feedback on the decisions of others for the first part of the game. The development of the second part of the session was equivalent to that of the first. Finally, we implemented the ex-post questionnaire by reading it out loud for participants with mild literacy constraints jointly with instructions on how to record their desired response. The experimenters and field assistants assisted participants with more severe literacy constraints to fill in their decision sheets and questionnaire sheets.

We conducted a series of regression analyses. All results for the effect of MPA remain highly significant to a stepwise inclusion of a rich set of individual control variables obtained from the postexperimental survey (Table 15.2) and are robust to a collection of other robustness checks.

Standardized Survey, Archival Research, Semistructured Interviews, and Participant Observation

We documented fishers' perceptions of MPA effects on their fishing activities and catch, and their ability to work with each other to solve collective-action problems through a standardized survey (n = 544, 48% active), which encompassed 10 localities within the influence of the four MPAs in Baja California (Fig. 15.1 and Table 15.1). We hired local assistants with knowledge of who were fishers in all localities under the influence of MPA regulations. We identified and surveyed 71.20% of all fishers (48.56% active) in our study area. Survey pretests were conducted with 12 fishers before deployment of the instrument.

TABLE 15.2 Description of Variables

Variable	Description	Median (Mean)	Std.	Min/Max	No Obs.	Measuring Unit
Dependent Variables						
Contfish	Percentage contribution to fishers	0.375 (0.411)	0.353	0/1	127	Percentage
Comp.fish	Percentage hypercompetitiveness to fishers	0.05 (0.047)	0.041	0/1	127	Percentage
Contcit	Percentage contribution to nonfisher	0.25 (0.361)	0.329	0/1	127	Percentage
Comp.cit	Percentage hypercompetitiveness to nonfisher	0.05 (0.049)	0.040	0/1	127	Percentage

Variable	Description	Median (Mean)	Std.	Min/Max	No Obs.	Measuring Unit
Explanatory Variables						
Continuous Variables						
Age	Age of participant	37 (39.589)	14.536	18/82	129	Years

Variable	Coding	Mean (SD)		Range	No. Obs.	Units
Locality	No of year a participant is living in this particular locality	28 (30.664)	17.457	1/82	128	Years
Conflict	Number of conflicts participants had with other participants	0 (0.156)	0.539	0/4	128	Number
Closefam	Number of close family members and relatives	1 (2.085)	3.003	0/20	129	Number

Variable	Coding	Frequency in %	No. Obs.	Description
Categorical Variables				
Edu	No education Primary Secondary High school College	9.48 31.90 48.28 1.72 8.62	116	Educational level of participant

Variable	Coding	Frequency in %	No. Obs.	Description
Dichotomous Variables				
Fisher	Fisher Nonfisher	41.86 58.14	129	Participant is a fisher or nonfisher
Marine protected area (MPA)	No Yes	45.74 54.26	129	Participant lives in a fishing village influenced by MPA rules and regulations

Continued

TABLE 15.2 Description of Variables—cont'd

Variable	Coding	Frequency in %	No Obs.	Description
Gender	M F	70.54 29.46	129	Gender of participant
Winning 2[a]	Disagree Agree	60.48 39.52	124	Winning is most important
Society 2[a]	Disagree Agree	52.07 47.93	121	It is not possible to have a good society without competition
Thinkact 2[a]	Disagree Agree	93.80 6.20	129	I usually think before I act
Coopmember	No Yes	71.32 28.68	129	I am a member of a cooperative
W. cooperation 2[a]	Unlikely Likely	64.80 35.20	125	If there was a problem with water supply in your community, how likely do you think is that people would cooperate to solve the problem?

[a]Recoded from a 5-point Likert scale to dummy variables. "Agree" means that the person either "strongly agrees" or "agrees" to the statement, whereas "Disagree" included "don't know," "disagree," and "strongly disagree."

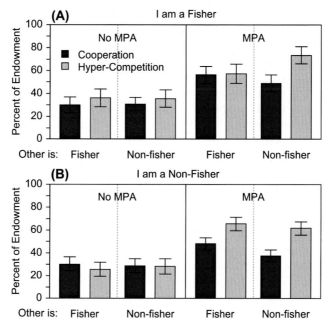

FIGURE 15.2 Higher average cooperation and hypercompetition in marine protected area (MPA) than in non-MPA sites. (A and B) The percentage of endowment used for cooperation and hyper-competition in no-MPA and MPA sites when the individual making a choice to cooperate and/or hypercompete is a fisher (A) and a nonfisher (B). "I am" refers to the participant making the choice depicted and "other is" refers to the participant the "I am" is paired with. Cooperation is defined as the percentage of endowment allocated to a group fund, and hypercompetition is defined as the percentage of endowment allocated to reduction of others' payoff.

Data collection lasted for 13 months between July 2012 and July 2013 and was captured in the field using a Microsoft Access database (version 2010). Back at Duke University, a team of three undergraduates curated all data entries to ensure no capture errors remained prior to data analysis. In our analysis, we only included data from fishers that self-reported to be actively participating in fishing activities at the time the survey was conducted (Fig. 15.4; n=371).

In-depth contextual knowledge of fishing as a social, economic, and cultural activity was gained through overtime archival research, semistructured interviews, and participatory observation in the study area and broader region over more than a decade (Basurto, 2005; Cudney-Bueno and Basurto, 2009; Basurto and Coleman, 2010; Basurto et al., 2013; Leslie et al., 2015; Nenadović et al., 2016; Nenadovic and Epstein, 2016). This understanding was crucial to interpret the experimental and large-n survey findings. For instance, it was through our participatory observation work that we established that cooperative and competitive behavior are interweaved in daily fishing activities, and it is better understood as friendly rivalry and not as a decidedly cooperative or anomalous competitive behavior due to some other exogenous factor.

Our archival research and 77 semistructured interviews sought to understand the history and policy process of establishing MPAs in the study area and targeted key informants such as fisheries leaders, government MPA officials, and non-government organizations' personnel. Interviews were transcribed and systematically coded using the NVivo software (version 10, QSR International) for three main themes: (1) representation of a sociopolitical context during the creation and implementation of the four MPAs, (2) identification of nuances related to the public participation process, and (3) understanding of perceptions of social and ecological impacts attributed to the establishment of the four MPAs.

FINDINGS

Fifteen years after the establishment of the MPAs, there is, on average, higher cooperation and hypercompetition in MPAs than in non-MPAs. This result holds for all combination of pairings between fishers and nonfishers (Fig. 15.2). Subjects made decisions in the experiment being aware of whether the person they interacted with was a fisher or a nonfisher. We observe that fishers making decisions toward others (irrespective of whether those others are fishers or not fishers) are more cooperative and hypercompetitive in MPAs and that the same result holds for behavior of nonfishers.

A substantial portion of subjects in MPA sites (47%) were simultaneously highly hypercompetitive (using above 50% of endowment in the joy-of-destruction game) and highly cooperative (contributing above 50% of endowment in the public goods game) toward fishers. Those "hypercompetitive cooperators" include both fishers (39%) and nonfishers (61%). In non-MPAs we only found "hypercompetitive cooperators" in 8% of the sample. Fig. 15.3 presents, within and outside MPAs, disaggregated percentages of "hypercompetitive cooperators" for decisions made by fishers and nonfishers ("I am") toward another player that is a fisher or a nonfisher ("other is"). Regression analyses on covariates for "hypercompetitive cooperators" show a high tendency to agree with the statement "competition is important in a functioning society," and this statement was not associated with cooperative or hypercompetitive behaviors separately (Basurto et al., 2016).

Although in this study we analyze behavioral actions taken by individual actors in cooperative and hypercompetitive settings, we cautiously extend some of our findings to the community level. Future work with more MPA and non-MPA sites and perhaps various types of MPAs is needed to achieve greater external validity and better understanding of the effect of MPAs on cooperation and hypercompetition.

Were People Already Hypercompetitive Cooperators Before the Establishment of the MPAs?

If this was the case, our reported results would not stem from the creation of the MPA but measure preexisting differences in behavior. Yet, the history of

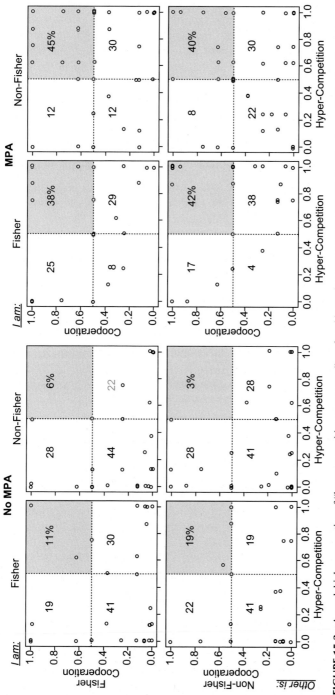

FIGURE 15.3   A much higher proportion of "hypercompetitive cooperators" are found in marine protected area (MPA) sites than in non-MPA sites. Percentage of hypercompetitive cooperators (also highlighted in yellow) are found in the upper right quadrant at no-MPA and MPA sites, and they are defined as those individuals allocating at least 50% of their endowment to the group fund and 50% of their endowment to reducing others' payoff. "I am" refers to the participant making the choice depicted and "other is" refers to the participant the "I am" is paired with.

siting selection criteria for the MPAs in our study indicates that social considerations like social cohesion, conflict, or hypercompetition among community stakeholders did not play a role. Literature reviewing the political process of establishment and implementation of these MPAs establishes that biological considerations were paramount in the designation of MPAs in the study area (Bezaury-Creel, 2005; Peterson, 2011; Rife et al., 2013). Government's focus on biological and ecosystem conservation can be traced back to the early 1990s when Mexico made international commitments to increase the overall area of conserved seascapes (i.e., San Felipe was established in 1993, Cabo Pulmo in 1995, Loreto in 1996, and La Paz in 2007). According to a National Commission of Protected Areas Governmental official who was engaged in the early designation of MPAs in Mexico:

> This was already discussed in Rio (1992) and the international commitments to start allocating spaces for marine conservation, because Mexico almost hadn't had marine protected areas. Very few. In those years most of what we had were 15 million hectares or something like that between marine and terrestrial. There started the idea to protect five or six percent of the national territory.

Interviewee #63.

Furthermore, our ex-post survey data does not reveal any significant differences between participants from MPA and non-MPA sites in observable characteristics, or perceptions on social functioning of their communities (Basurto et al., 2016). The use of biological values as the guiding criteria for the designation of MPAs sites internationally, currently including more than 15,200 sites according to the MPA Atlas Website (www.mpatlas.org), is common practice (Jones, 1994; Fox et al., 2012).

Are Hypercompetitive Cooperators an Isolated Phenomenon?

Can the presence of hypercompetitive cooperators be extrapolated to other MPAs in the study area? The MPAs we studied are typical examples of the most common type of coastal MPAs globally, designated as multiple use areas with no-take core areas outside of which artisanal fishing is permitted (www.mpatlas.org). These MPAs exhibit similar coastal and marine degradation trends as elsewhere (Aburto-Oropeza et al., 2011).

Using in-person survey responses by fishers in four MPAs in Baja California (n = 544), we verified that there were no systematic variations between the two experimental MPA sites and the two nonexperimental MPAs. All MPAs had a considerable impact on the day-to-day activities of fishers (Fig. 15.4A). In all MPAs, there is strong polarization on the view of the impact of conservation on catch (Fig. 15.4B), and most interestingly, 80% of all fishers in each of the four MPAs stated that they are able to solve their problems jointly (Fig. 15.4C). In sum, although the MPAs entailed tangible impacts on fishers, cooperation is high in all four MPAs, which is in line with our experimental results. We also

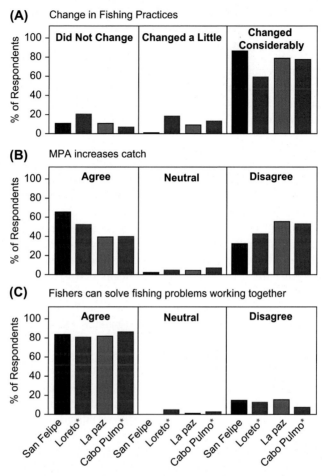

FIGURE 15.4   Fishers' self-reported responses on the experimental and nonexperimental marine protected areas (MPAs) in our study area. (A–C) Fishers' self-reported responses (n = 544) on how MPAs have affected their fishing practices (A), catch (B), and their own capacity to work together to solve fishing problems (C). Experimental MPA sites are marked with an *asterisk*.

have some suggestive evidence that all four MPAs likewise created both winners and losers (Fig. 15.4B) through an increase in income opportunities outside of fishing for some families but not others, either through a direct increase of tourism-related jobs (e.g., diving tourism and recreational fishing in Cabo Pulmo) or through seasonal programs sponsored by MPAs, to provide alternative sources of income. An overtime assessment of the government subsidies awarded to fishers associated with the MPA near La Paz suggests that some of them are more likely than others to successfully renew their grants (Nenadović et al., 2016). Altogether our data suggest an increased social stratification in MPAs, which we argue is a precursor for strengthening competition.

INSTITUTIONAL CHANGE AND SOCIAL DIFFERENTIATION EXPLAIN INCREASED COOPERATION AND HYPERCOMPETITION IN MPAs

Our experimental results show more hypercompetitive cooperators in MPA than in non-MPA fishing communities. Effects are found not only on fishers but also among nonfishers. Thus potential mechanisms need to explain a general pattern of community norms.

Cooperation

Our survey results show that cooperative behavior does not derive from higher fish abundance and catch in MPAs (Fig. 15.4B). Instead semistructured interviews and observations suggest the emergence of norms for cooperation among stakeholders through exposure to trust building exercises or through being subject to government regulations promoting cooperative interactions (Hoffman, 2009). In Loreto, Cabo Pulmo, and La Paz, local and international nongovernmental organizations have engaged stakeholders in trust-building exercises through meetings and activities encouraging participants to work together after the creation of the MPA (Weiant, 2005). For instance, in Loreto, conflict resolution experts engaged stakeholders through trust building exercises described as "a democratic experience in the community [that was] very satisfactory for most stakeholders because it was a very inclusive and transparent process where most stakeholders felt represented" (Eraso et al., 2010, p. 31).

Moreover, in MPAs, fishers were incentivized to organize into fishing cooperatives to gain fishing permits to fish within MPA boundaries (Basurto et al., 2016). Our survey data show that 80% of all co-ops were created after the establishment of the MPAs and 70% of co-op members are related by kin. One of the fishers from Loreto described how new fishing regulations around MPAs encouraged him to start his own family-based fishing cooperative:

> We formed the [fishing cooperative] in 2001 so we could get fishing permits. In 2003 we got our first fishing permit... [It was after the creation of the Loreto MPA] when I started to have problems because I couldn't go out to fish at all without the fishing permit. The inspections started so we began to look for different options. So I talked to the family of my wife and my father-in-law. I told them: "Why don't we organize as a co-op so we can request a fishing permit?" He had his boat, I had mine, his sons fished too. He said: "sure thing." We formed the co-op and we started to request fishing permits.

> Interviewee #56.

State incentives for fishing cooperatives' formation have been a common approach around the world (Jentoft, 1986; Amarasinghe and Bavinck, 2011), and given that often cooperatives are organized around family ties, it is reasonable to expect an increase in cooperative behavior within the fishing fleet, ceteris paribus, given that kinship is thought to be the social basis of cooperation (Henrich and Henrich, 2007).

Hypercompetition

Our study suggests that the willingness to destroy part of the other person's payoff in the experiment is caused by hypercompetition. When subjects were asked for their motives to reduce another subject's income, they stated that they wanted to earn more money than the other person (30%), whereas motives like vengeance (5%), deriving pleasure from harming others (7%), or disliking people (7%) were not named as often. In the same vein, only six people in MPAs and four outside MPAs were stated to have conflicts with other participants in the session, whereas 62% of participants in MPAs and 55% in non-MPAs had at least one family member in the session. Finally, there is a large and significant difference in agreement to the statement "winning is most important to me." Within MPAs, 48% respondents agreed compared to only 29% in non-MPAs. This difference is statistically significant at the 5% level (Fisher exact test; $P = .03$). Thus it seems that preference for being mean or conflictive toward others did not play a major role, whereas being ahead of others in a competition did.

Similar behaviors have been documented from reductions in fishing areas and/or restrictions on fishing gear brought by the implementation of a MPA. For instance, Lédée et al. (2012) observed an increase in competition among commercial fishers operating within the Great Barrier Reef Marine Park after the rezoning of the park that substantially reduced areas where fishing was allowed. Horta e Costa et al. (2013) observed a similar pattern among artisanal fishers over a 3-year period following implementation of Arrábida Marine Park in Portugal and suggested that gear type might also be mediating the level of competition among fishers.

Another potential contributor to hypercompetitive behavior might be attributed to strong social comparisons, where individuals weigh their respective gains and losses in relative instead of in absolute terms (Jensen, 2010). From this perspective, MPAs could be increasing social comparisons in class-based societies like Mexico due to the higher number of stakeholders (e.g., tourism operators, recreational fishers) interacting than in non-MPA sites, where fishers mostly interact only with other fishers. This social differentiation dynamic might have been captured when we asked fishers and nonfishers their perception about the political influence of fishers in their community compared to those individuals that did not fish. Respondents perceived fishers to have less political power in MPAs (28%) than in non-MPAs (56%) ($\chi^2 = 3.6313$, P-value = .01). Such a lack of political influence has also been reported in the case of small-scale fishers from southeast Cebu, Philippines, who felt disfavored in their dealings with government related to the creation of a local MPA (Segi, 2013). In addition, the availability of temporary grants to fishers and nonfishers inside of

MPAs and the evidence suggesting that some individuals get funded repeatedly more often than others (Nenadović et al., 2016) might also be driving this social differentiation dynamics.

Coexistence of Cooperation and Hypercompetition

Most surprising was to find the two extreme opposite behaviors (cooperation and hypercompetition) coexisting at a substantially larger proportion in MPAs than in non-MPA sites, and these extreme behaviors affecting similarly fishers and nonfishers. One might speculate about several possible explanations: Fishing might be so important that norms of behavior spill over to the general population, the general population is also directly affected by the MPA creation or the behavior of fishers and nonfishers is unrelated and there are different mechanisms explaining high competitiveness and high cooperation among these two subpopulations. Our preferred explanation based on our long-term engagement in the region is that MPAs enhance the diversification of market-based activities that increase trust and cooperation among its actors (Henrich et al., 2004) at the same time that market diversification may lead to envy and social stratification, which might spur hypercompetition (Kottak, 1992). The markets are institutions engaged in the exchange of information, goods, or services among many other types of exchanges that can take place among individuals or corporations.

There is significant support in the MPA literature for MPAs' role in promoting economic diversification (i.e., mostly related to tourism) in diverse locations around the world (Leisher et al., 2007; Sørensen and Thomsen, 2009; Brondo and Bown, 2011; Cárcamo et al., 2014). However, income diversification, and to a greater extent, income disparity, among culturally cohesive rural fishing communities, has also been shown to result in envy and group differentiation (Oracion et al., 2005; Brondo and Bown, 2011). In Cabo Pulmo National Park, for instance, various subsidy programs permitted the diversification of alternative economic activities to commercial fishing in the community of Cabo Pulmo. Unlike Cabo Pulmo, neighboring communities did not receive much attention from the government, leading to social comparisons as expressed by this fisher:

> *[While] the community of La Ribera received government support for a collection of garbage along beaches all the major government programs [promoting alternative livelihood opportunities] were lost to the village of Cabo Pulmo. People there get them year after year.*

Fisher #16.

Similar findings have been reported for other MPAs elsewhere. For example, Brondo and Bown (2011) reported that income diversification caused by MPA establishment created social divisions and tensions among small-scale fishing communities in Honduras. In addition, some of the anthropological work from Mexico and elsewhere suggests that market integration or diversification within historically isolated communities may disrupt societal norms and thus trigger envy and even witchcraft (i.e., the evil eye) toward those that become economically more prosperous (Dow, 1981; de Vidas, 2007). From this perspective, social differentiation in regards to the economic and occupational status could have been exacerbated by the implementation of the MPA leading to higher levels of antisocial behavior when compared to non-MPA sites.

Overall, our findings suggest how MPAs seem to be producing particular types of peoples (recall West et al., 2006). Studied MPAs enhanced coupled cooperative and competitive behavior (Brandenburger and Nalebuff, 1997) through the creation of communities of "hypercompetitive cooperators" within MPAs. This view is substantiated by people who stated "competition is important for society" was more likely to be identified as "hypercompetitive cooperators."

It is important to emphasize that cooperation and competition coexist in fishers' daily life, in what we call friendly rivalry; as illustrated by a fisher in our study area who in response to our question "What is the most exciting part of fishing?" said:

> *To be the* best *but also to be* a good friend *to everyone. I never fish by myself and I always try to beat the rest. When all of the boats arrive back on the beach we look to see who has caught the most fish. Sometimes we win by one kilo, sometimes we tie, but it is always exciting to see who won.*

Interviewee #4.

Could friendly rivalry have been magnified through the process of establishment and implementation of MPAs? Our interviews suggest that the establishment of MPAs created new opportunities for cooperation and competition that previously were not available as the number and type of stakeholders increased. In some instances, new alliances were formed to confront a common enemy or threat (Basurto et al., 2016). As one of the tourism developers from Loreto pointed out:

> *One day we had a meeting about the shrimpers [large industrial boats]… Someone said: "Lets put them on the other side of the legal line. Why don't we create a national park, a protected area, where these creatures of god [shrimpers] cannot come in!" Good idea!… We signed a letter including the taxi drivers, everyone supported the petition to make a protected area which would protect our [marine] resources. The government found it very entertaining that a community asked the federal government to create a national park. In this country all national parks were decreed by the government.*

Interviewee #72.

The presence of MPAs changed the balance of winners and losers redefining when, how, and who will be permitted or forbidden to use particular areas within the MPA (Peterson, 2011). Furthermore, the market diversification process that ensued (e.g., employment opportunities in tourism), likely magnified friendly rivalries' desire for competition through an increased desire for social differentiation and relative evaluation among stakeholders. Market diversification also offers new arenas of interaction and opportunities for different work organization, reorganization of social structure, and identity (McCormack, 2012). Such process of institutional change would not surprisingly, fundamentally redefine collective relationships with each other and toward the ocean.

IMPLICATIONS FOR THE SCIENCE ON MPAs

Our findings elicit questions for future research related to the long-term viability of MPAs. For instance, how important are populations of "hypercompetitive cooperators" to MPAs long-term effectiveness? Jensen (2010) argues that hypercompetition by some actors within a cooperative network might be potentially key for its stability. This could be exemplified, for instance, by the willingness of some fishers to speak up against rule breakers, which can ensure individual accountability and overall group cooperation for the maintenance of healthy fish stocks in MPAs. Yet, the distinction between hypercompetitive cooperative behavior and punishment motivated behavior would need to be better established.

MPAs in our study area intensified positive and negative interactions without undermining successful collective action. Social capacity (e.g., trust building activities and conflict resolution skills brought about by civil society groups) played an important role, bringing attention to the role civil society groups play in the maintenance of the success of MPAs. Examining how local or outside civil society groups engage with local inhabitants affected by MPAs and what types of engagements are most effective in building social capacity will continue to be an important area of research on MPA effectiveness.

This study identified some potential mechanisms for the emergence of hypercompetitive behavior. Yet, it is unclear at what point and under what circumstances further intensification of hypercompetition would eventually disrupt overall cooperation. However, this work provides a theoretical and empirical basis to operationalize a way to identify such tipping point, that is, when hypercompetitive behavior might no longer help reinforce cooperation but rapidly unravel into other forms of antisocial behavior like conflict. There are abundant reports within the conservation science literature of conflict among stakeholders related to fisheries and MPAs (see Agardy et al., 2003; Stevenson et al., 2013). Could these cases be theoretically and qualitatively similar to the cases we presented here and yet just be at opposite sides but in the same spectrum of hypercompetition?

Theoretical issues apart, the clear lesson for managers is to pay particular attention to how the creation of MPAs would affect issues of social inequality that could in the longer term affect incentives for cooperation among those most affected by the presence of the MPA.

CONCLUSIONS

Our research approach combined participatory observation, semistructured interviews, surveys, and experiments to analyze the effects of MPAs in fishing communities. These methods rarely enter in conversation with each other particularly in the context of the study of MPAs. Yet, a multimethod approach also provides fertile ground for a broader engagement with social science concepts (Hicks et al., 2016), a key area in the need of expansion to drive the science of MPAs forward. The complexity of issues facing marine conservation in the Anthropocene ensures us that multiple methods will increasingly play an important role toward investigating the effects of MPAs in communities and their potential for long-lasting conservation and human well-being.

ACKNOWLEDGMENTS

We thank Martín Almaráz, Leticia Angulo, Edg,ardo Camacho, Ricardo Chollet, Carmen Cota, Felix Flores, Vismar Flores, Gabriela García, Francisco Gómez, Iván González, Liza Hoos, Heather Leslie, Aldair Murillo, Hervey Murillo, Karla Murillo, Vanessa Murillo, Nicole Naar, Jessica Navarro, Uriel Rubio, Bernardo Sánchez, Edrey Villalejo for field assistance; Hudson Weaver from our partner organization Niparajá for logistical and strategic planning; fishers that participated in the meetings; Katy May, Samantha Emmert, and especially Simone Posch and Peter Zaykoski for data processing and figures. Brian Silliman and Daniel Dunn for comments to previous drafts. Funding support was provided by the Walton Family Foundation (XB), WWF Fuller Fellowship (MN), and Duke University (XB). IRB permit #B0259 Duke University.

REFERENCES

Abbink, K., Herrmann, B., 2011. The moral costs of nastiness. Economic Inquiry 49, 631–633.

Aburto-Oropeza, O., Erisman, B., Galland, G.R., Mascareñas-Osorio, I., Sala, E., Ezcurra, E., 2011. Large recovery of fish biomass in a no-take marine reserve. PLoS One 6, e23601.

Acheson, J.M., 1988. The Lobster Gangs of Maine. University Press New England, Hanover, NH.

Agardy, T., Bridgewater, P., Crosby, M.P., Day, J., Dayton, P.K., Kenchington, R., Laffoley, D., McConney, P., Murray, P.A., Parks, J.E., 2003. Dangerous targets? Unresolved issues and ideological clashes around marine protected areas. Aquatic Conservation: Marine and Freshwater Ecosystems 13, 353–367.

Álvarez-Romero, J.G., Pressey, R.L., Ban, N.C., Torre-Cosío, J., Aburto-Oropeza, O., 2013. Marine conservation planning in practice: lessons learned from the Gulf of California. Aquatic Conservation: Marine and Freshwater Ecosystems 23, 483–505.

Amarasinghe, O., Bavinck, M., 2011. Building resilience: fisheries cooperatives in southern Sri Lanka. In: Jentoft, S., Eide, A. (Eds.), Poverty Mosaics: Realities and Prospects in Small-Scale Fisheries. Springer, New York, NY, pp. 383–406.

Ban, N.C., Evans, L.S., Nenadovic, M., Schoon, M., 2015. Interplay of multiple goods, ecosystem services, and property rights in large social-ecological marine protected areas. Ecology and Society 20.

Basurto, X., Coleman, E., 2010. Institutional and ecological interplay for successful self-governance of community-based fisheries. Ecological Economics 69, 1094–1103.

Basurto, X., Cinti, A., Bourillón, L., Rojo, M., Torre, J., Weaver, A.H., 2012. The emergence of access controls in small-scale fishing commons: a comparative analysis of individual licenses and common property-rights in two Mexican communities. Human Ecology 40, 597–609.

Basurto, X., Bennett, A., Weaver, A.H., Rodriguez-Van Dyck, S., Aceves-Bueno, J.-S., 2013. Cooperative and noncooperative strategies for small-scale fisheries' self-governance in the globalization era: implications for conservation. Ecology and Society 18, 38.

Basurto, X., Blanco, E., Nenadovic, M., Vollan, B., 2016. Integrating simultaneous prosocial and antisocial behavior into theories of collective action. Science Advances 2.

Basurto, X., 2005. How locally designed access and use controls can prevent the tragedy of the commons in a Mexican small-scale fishing community. Society and Natural Resources 18, 643–659.

Bavinck, M., 2005. Understanding fisheries conflicts in the South—a legal pluralist perspective. Society and Natural Resources 18, 805–820.

Berkes, F., Mahon, R., McConney, P., Pollnac, R., Pomeroy, R., 2001. Managing Small-Scale Fisheries: Alternative Directions and Methods. International Development Research Centre, Otawa.

Bezaury-Creel, J.E., 2005. Protected areas and coastal and ocean management in Mexico. Ocean & Coastal Management 48, 1016–1046.

Blanco, E., Lopez, M.C., Villamayor-Tomas, S., 2015. Exogenous degradation in the commons: field experimental evidence. Ecological Economics 120, 430–439.

Bouma, J., Bulte, E., Van Soest, D., 2008. Trust and cooperation: social capital and community resource management. Journal of Environmental Economics and Management 56, 155–166.

Brandenburger, A.M., Nalebuff, B.J., 1997. Co-opetition: A Revolution Mindset that Combines Competition and Cooperation. Currency Doubleday, New York, NY.

Brondo, K.V., Bown, N., 2011. Neoliberal conservation, garifuna territorial rights and resource management in the cayos cochinos marine protected area. Conservation and Society 9, 91.

Brondo, K.V., Woods, L., 2007. Garifuna land rights and ecotourism as economic development in Honduras' cayos cochinos marine protected area. Ecological and Environmental Anthropology 3, 2–18.

Brooks, J.S., Waylen, K.A., Borgerhoff Mulder, M., 2012. How national context, project design, and local community characteristics influence success in community-based conservation projects. Proceedings of the National Academy of Sciences 109, 21265–21270.

Cárcamo, P.F., Garay-Flühmann, R., Squeo, F.A., Gaymer, C.F., 2014. Using stakeholders' perspective of ecosystem services and biodiversity features to plan a marine protected area. Environmental Science & Policy 40, 116–131.

Cardenas, J.C., Carpenter, J., 2008. Behavioural development economics: lessons from field labs in the developing world. The Journal of Development Studies 44, 311–338.

Cardenas, J.C., 2011. Social norms and behavior in the local commons as seen through the lens of field experiments. Environmental and Resource Economics 48, 451–485.

Chaigneau, T., Brown, K., 2016. Challenging the win-win discourse on conservation and development: analyzing support for marine protected areas. Ecology and Society 21.

Cinner, J.E., McClanahan, T.R., MacNeil, M.A., Graham, N.A.J., Daw, T.M., Mukminin, A., Feary, D.A., Rabearisoa, A.L., Wamukota, A., Jiddawi, N., Campbell, S.J., Baird, A.H., Januchowski-Hartley, F.A., Hamed, S., Lahari, R., Morove, T., Kuange, J., 2012. Comanagement of coral reef social-ecological systems. Proceedings of the National Academy of Sciences 109, 5219–5222.

Cinner, J.E., 2007. Designing marine reserves to reflect local socioeconomic conditions: lessons from long-enduring customary management systems. Coral Reefs 26, 1035–1045.

Cudney-Bueno, R., Basurto, X., 2009. Lack of cross-scale linkages reduces robustness of community-based fisheries management. PLoS One 4, e6253.

Darling, E.S., 2014. Assessing the effect of marine reserves on household food security in Kenyan coral reef fishing communities. PLoS One 9, e113614.

de Vidas, A.A., 2007. The symbolic and ethnic aspects of envy among a Teenek community (Mexico). Journal of Anthropological Research 215–237.

Dow, J., 1981. The image of limited production: envy and the domestic mode of production in peasant society. Human Organization 40, 360–363.

DuBois, C., Zografos, C., 2012. Conflicts at sea between artisanal and industrial fishers: intersectoral interactions and dispute resolution in Senegal. Marine Policy 36, 1211–1220.

Edgar, G.J., Stuart-Smith, R.D., Willis, T.J., Kininmonth, S., Baker, S.C., Banks, S., Barrett, N.S., Becerro, M.A., Bernard, A.T.F., Berkhout, J., Buxton, C.D., Campbell, S.J., Cooper, A.T., Davey, M., Edgar, S.C., Forsterra, G., Galvan, D.E., Irigoyen, A.J., Kushner, D.J., Moura, R., Parnell, P.E., Shears, N.T., Soler, G., Strain, E.M.A., Thomson, R.J., 2014. Global conservation outcomes depend on marine protected areas with five key features. Nature 506, 216–220.

Eraso, L.M., Hernández, M., Portilla, J., Espinosa, R., Gómez Chow, L., 2010. Proceso de Revisión del Programa de Manejo de Bahía de Loreto. Módulo 2. Unpublished Report to the National Commission of Protected Areas, Gulf of California.

Fox, H.E., Mascia, M.B., Basurto, X., Costa, A., Glew, L., Heinemann, D., Karrer, L.B., Lester, S.E., Lombana, A., Pomeroy, R., 2012. Reexamining the science of marine protected areas: linking knowledge to action. Conservation Letters 5, 1–10.

Gatiso, T.T., Vollan, B., Nuppenau, E.-A., 2015. Resource scarcity and democratic elections in commons dilemmas: an experiment on forest use in Ethiopia. Ecological Economics 114, 199–207.

Gelcich, S., Guzman, R., Rodríguez-Sickert, C., Castilla, J.C., Cárdenas, J.C., 2013. Exploring external validity of common pool resource experiments: insights from artisanal benthic fisheries in Chile. Ecology and Society 18, 2.

Gruby, R.L., Basurto, X., 2013. Multi-level governance for large marine commons: politics and polycentricity in Palau's protected area network. Environmental Science & Policy 33, 260–272.

Gruby, R.L., Gray, N.J., Campbell, L.M., Acton, L., 2015. Toward a social science research agenda for large marine protected areas. Conservation Letters 9 (3).

Guerrón-Montero, C., 2005. Marine protected areas in Panama: grassroots activism and advocacy. Human Organization 64, 360–373.

Gurney, G., Cinner, J., Sartin, J., Pressey, R., Ban, N., Marshall, N., Prabuning, D., 2016. Participation in devolved commons management: multiscale socioeconomic factors related to individuals' participation in community-based management of marine protected areas in Indonesia. Environmental Science & Policy 61, 212–220.

Halpern, B.S., Klein, C.J., Brown, C.J., Beger, M., Grantham, H.S., Mangubhai, S., Ruckelshaus, M., Tulloch, V.J., Watts, M., White, C., Possingham, H.P., 2013. Achieving the triple bottom line in the face of inherent trade-offs among social equity, economic return, and conservation. Proceedings of the National Academy of Sciences 110, 6229–6234.

Henrich, J., Henrich, N., 2007. Why Humans Cooperate: A Cultural and Evolutionary Explanation. Oxford University Press, New York, NY.

Henrich, J., Boyd, R., Bowles, S., Camerer, C., Fehr, E., Gintis, H., 2004. Foundations of Human Sociality: Economic Experiments and Ethnographic Evidence from Fifteen Small-Scale Societies. Oxford University Press.

Herrera, G.E., Moeller, H.V., Neubert, M.G., 2016. High-seas fish wars generate marine reserves. Proceedings of the National Academy of Sciences 113, 3767–3772.

Hicks, C.C., Levine, A., Agrawal, A., Basurto, X., Breslow, S.J., Carothers, C., Charnley, S., Coulthard, S., Dolsak, N., Donatuto, J., Garcia-Quijano, C., Mascia, M.B., Norman, K., Poe, M.R., Satterfield, T., St Martin, K., Levin, P.S., 2016. Engage key social concepts for sustainability. Science 352, 38–40.

Hoffman, D.M., 2009. Institutional legitimacy and co-management of a marine protected area: implementation lessons from the case of Xcalak Reefs National Park, Mexico. Human Organization 68, 39–54.

Horta e Costa, B., Batista, M.I., Gonçalves, L., Erzini, K., Caselle, J.E., Cabral, H.N., Gonçalves, E.J., 2013. Fishers' behaviour in response to the implementation of a marine protected area. PLoS One 8, e65057.

Isaac, R.M., Walker, J.M., Thomas, S.H., 1984. Divergent evidence on free riding: an experimental examination of possible explanations. Public Choice 43, 113–149.

Isaac, R.M., Walker, J.M., Williams, A.W., 1994. Group size and the voluntary provision of public goods: experimental evidence utilizing large groups. Journal of Public Economics 54, 1–36.

Janssen, M.A., Bousquet, F., Cardenas, J.-C., Castillo, D., Worrapimphong, K., 2013. Breaking the elected rules in a field experiment on forestry resources. Ecological Economics 90, 132–139.

Jensen, K., 2010. Punishment and spite, the dark side of cooperation. Philosophical Transactions of the Royal Society B: Biological Sciences 365, 2635–2650.

Jentoft, S., 1986. Fisheries co-operatives: lessons drawn from international experiences. Canadian Journal of Development Studies 7, 197–209.

Jones, P., 1994. A review and analysis of the objectives of marine nature reserves. Ocean & Coastal Management 24, 149–178.

Kottak, C.P., 1992. Assault on Paradise: Social Change in a Brazilian Village. McGraw-Hill, New York, NY.

Lédée, E.J., Sutton, S.G., Tobin, R.C., De Freitas, D.M., 2012. Responses and adaptation strategies of commercial and charter fishers to zoning changes in the Great Barrier Reef Marine Park. Marine Policy 36, 226–234.

Ledyard, J., 1995. Public goods: a survey of experimental research. In: Kagel, J., Roth, A. (Eds.), Handbook of Experimental Economics. Princeton University Press, Princeton, NJ.

Leibbrandt, A., Gneezy, U., List, J.A., 2013. Rise and fall of competitiveness in individualistic and collectivistic societies. Proceedings of the National Academy of Sciences 110, 9305–9308.

Leisher, C.D., van Beukering, P., Scherl, L.M., 2007. Nature's investment bank: how marine protected areas contribute to poverty reduction. Nature Conservancy p. 43. http://www.nature.org/media/science/mpa_report.pdf.

Leslie, H.M., Basurto, X., Nenadovic, M., Sievanen, L., Cavanaugh, K.C., Cota-Nieto, J.J., Erisman, B.E., Finkbeiner, E., Hinojosa-Arango, G., Moreno-Báez, M., Nagavarapu, S., Reddy, S.M.W., Sánchez-Rodríguez, A., Siegel, K., Ulibarria-Valenzuela, J.J., Weaver, A.H., Aburto-Oropeza, O., 2015. Operationalizing the social-ecological systems framework to assess sustainability. Proceedings of the National Academy of Sciences 112, 5979–5984.

Lester, S.E., Halpern, B.S., Grorud-Colvert, K., Lubchenco, J., Ruttenberg, B.I., Gaines, S.D., Airamé, S., Warner, R.R., 2009. Biological effects within no-take marine reserves: a global synthesis. Marine Ecology Progress Series 384, 33–46.

List, J.A., 2011. Why economists should conduct field experiments and 14 tips for pulling one off. The Journal of Economic Perspectives 25, 3–15.

Mascia, M.B., Claus, C.A., 2009. A property rights approach to understanding human displacement from protected areas: the case of marine protected areas. Conservation Biology 23, 16–23.

Mascia, M.B., Claus, C.A., Naidoo, R., 2010. Impacts of marine protected areas on fishing communities. Conservation Biology 24, 1424–1429.

McCay, B.J., 2002. Emergence of institutions for the commons: contexts, situations, and events. In: Ostrom, E., Dietz, T., Dolsak, N., Stern, P.C., Stonich, S., Weber, E.U. (Eds.), The Drama of the Commons. National Academy Press, Washington, DC, pp. 361–402.

McCormack, F., 2012. Indigeneity as process: Māori claims and neoliberalism. Social Identities 18, 417–434.

Morton-Lefévre, J., 2014. Editorial. Science 346, 525.

Mumby, P.J., Harborne, A.R., Williams, J., Kappel, C.V., Brumbaugh, D.R., Micheli, F., Holmes, K.E., Dahlgren, C.P., Paris, C.B., Blackwell, P.G., 2007. Trophic cascade facilitates coral recruitment in a marine reserve. Proceedings of the National Academy of Sciences 104, 8362–8367.

Nenadovic, M., Epstein, G., 2016. The relationship of social capital and fishers' participation in multi-level governance arrangements. Environmental Science & Policy 61, 77–86.

Nenadović, M., Basurto, X., Weaver, A.H., 2016. Contribution of subsidies and participatory governance to fishers' adaptive capacity. Journal of Environment & Development 25 (4), 1–29.

Noussair, C.N., van Soest, D., Stoop, J., 2015. Punishment, reward, and cooperation in a framed field experiment. Social Choice and Welfare 45, 537–559.

Oakerson, R.J., Parks, R.B., 2011. The study of local public economies: multi-organizational, multi-level institutional analysis and development. Policy Studies Journal 39, 147–167.

Oracion, E.G., Miller, M.L., Christie, P., 2005. Marine protected areas for whom? Fisheries, tourism, and solidarity in a Philippine community. Ocean & Coastal Management 48, 393–410.

Peterson, A.M., Stead, S.M., 2011. Rule breaking and livelihood options in marine protected areas. Environmental Conservation 1–11.

Peterson, N.D., 2011. Excluding to include: (non)participation in Mexican natural resource management. Agriculture and Human Values 28, 99–107.

Pfaff, A., Vélez, M.A., Ramos, P.A., Molina, A., 2015. Framed field experiment on resource scarcity & extraction: path-dependent generosity within sequential water appropriation. Ecological Economics 120, 416–429.

Phyne, J., 1990. Dispute settlement in the newfoundland inshore fishery. Maritime Anthropological Studies (MAST) 3, 88–102.

Pollnac, R., Christie, P., Cinner, J.E., Dalton, T., Daw, T.M., Forrester, G.E., Graham, N.A.J., McClanahan, T.R., 2010. Marine reserves as linked social-ecological systems. Proceedings of the National Academy of Sciences of the United States of America 107, 18262–18265.

Pomeroy, R., Parks, J., Pollnac, R., Campson, T., Genio, E., Marlessy, C., Holle, E., Pido, M., Nissapa, A., Boromthanarat, S., 2007. Fish wars: conflict and collaboration in fisheries management in Southeast Asia. Marine Policy 31, 645–656.

Prediger, S., Vollan, B., Herrmann, B., 2014. Resource scarcity and antisocial behavior. Journal of Public Economics 119, 1–9.

Rife, A.N., Erisman, B., Sanchez, A., Aburto-Oropeza, O., 2013. When good intentions are not enough... insights on networks of "paper park" marine protected areas. Conservation Letters 6, 200–212.

Rodriguez-Sickert, C., Guzmán, R.A., Cárdenas, J.C., 2008. Institutions influence preferences: evidence from a common pool resource experiment. Journal of Economic Behavior & Organization 67, 215–227.

Rustagi, D., Engel, S., Kosfeld, M., 2010. Conditional cooperation and costly monitoring explain success in forest commons management. Science 330, 961.

Schlager, E., Ostrom, E., 1992. Property-rights regimes and natural resources: a conceptual analysis. Land Economics 68, 249–262.

Segi, S., 2013. The making of environmental subjectivity in managing marine protected areas: a case study from Southeast Cebu. Human Organization 72, 336–346.

Selig, E.R., Bruno, J.F., 2010. A global analysis of the effectiveness of marine protected areas in preventing coral loss. PLoS One 5, e9278.

Smith, M.D., Lynham, J., Sanchirico, J.N., Wilson, J.A., 2010. Political economy of marine reserves: understanding the role of opportunity costs. Proceedings of the National Academy of Sciences 107, 18300–18305.

Smith, V.L., 1982. Microeconomic systems as an experimental science. The American Economic Review 923–955.

Sørensen, T.K., Thomsen, L.N., 2009. A comparison of frameworks and objectives for implementation of marine protected areas in Northern Europe and in Southeast Asia. Aquatic Ecosystem Health & Management 12, 258–263.

Stevenson, T.C., Tissot, B.N., Walsh, W.J., 2013. Socioeconomic consequences of fishing displacement from marine protected areas in Hawaii. Biological Conservation 160, 50–58.

Vollan, B., 2008. Socio-ecological explanations for crowding-out effects from economic field experiments in southern Africa. Ecological Economics 67, 560–573.

Weiant, P.A., 2005. A Political Ecology of Marine Protected Areas (MPAs): Case of Cabo Pulmo National Park, Sea of Cortez, Mexico (Ph.D.). University of California, Santa Barbara.

West, P., Igoe, J., Brockington, D., 2006. Parks and peoples: the social impact of protected areas. Annual Review of Anthropology 35, 251–277.

Chapter 16

Transdisciplinary Research for Conservation and Sustainable Development Planning in the Caribbean

Katie K. Arkema[1,2], Mary Ruckelshaus[1,2]

[1]*Stanford University, Stanford, CA, United States;* [2]*University of Washington, Seattle, WA, United States*

> *If you want to go quickly, go alone. If you want to go far, go together.*
>
> African quote

INTRODUCTION

Globally, the human population is fast approaching 10 billion people (UN, 2015), with nearly a third located within 100 km of the sea (Small and Nicholls, 2003; IPCC, 2014). As the list of environmental ills facing ocean and coasts grows longer (Halpern et al., 2008; Worm and Lenihan, 2013; IPCC, 2014), it becomes increasingly important to understand the cumulative effects of anthropogenic stressors and the most promising interventions to bolster ecosystems (Adger et al., 2005; Melillo et al., 2014; Bloomberg et al., 2014). In recent years, scientists and practitioners increasingly have acknowledged and leveraged the two-way interactions between people and nature (Mace, 2014; Levin, 2014), two-way interactions that are brought into stark relief particularly in coastal systems when a fishery closes or a hurricane hits (Allison and Horemans, 2006; Day et al., 2007; Coulthard et al., 2011). By accounting for the ways in which communities depend on ecosystems (MEA, 2005), as well as affect them, the hope is to reach a broader set of actors and to direct investments, planning, and decision making to promote conservation and foster human well-being at the same time (TEEB, 2010; Kinzig et al., 2011; Guerry et al., 2015).

Growing awareness of the benefits that nature provides to people (i.e., ecosystem services, Daily, 1997) has motivated governments, nongovernmental organizations (NGOs), multilateral development banks, major corporations, and other institutions to account for human–environment interactions within large-scale initiatives such as Integrated Coastal Zone Management (ICZM), Coastal and Marine Spatial Planning, Sustainable Development Planning, and payments or subsidies in exchange for ecosystem services (Arkema et al., 2006; Douvere, 2008; Ruckelshaus et al., 2008; McLeod and Leslie, 2009; Lubchenco and Sutley, 2010; He et al., 2014; Ouyang et al., 2016). Several of the recently passed sustainable development goals (SDGs) point to the importance of healthy coastal and ocean ecosystems for social outcomes and poverty alleviation (UNGA, 2015, e.g., SDGs2, 14). Multilateral development banks are pursuing integrated environment–human development strategies such as the Inter-American Development Bank's Biodiversity and Ecosystem Services Program. Nevertheless, there are too few examples of governments and other institutions accounting for natural capital in coastal policy or finance decisions. Innovative work within environmental agencies such as the Coastal Zone Management Authority and Institute (CZMAI) in Belize (see in the following and Arkema et al., 2014, 2015), integrated climate adaptation and watershed planning in coastal California (Langridge et al., 2014), or post-Hurricane Sandy investments in restoration in the northeastern United States (NFWF, 2014) offer promising examples that are not being replicated broadly or rapidly enough.

The persistent marginalization of natural capital in public and private decision making may in part reflect a dearth of actionable, salient science (Clark et al., 2011, 2016). Although use-inspired research is on the rise, the vast majority of academic studies still focus on fundamental rather than applied science (Stokes, 1997; Leford et al., 2015). Not enough time is spent understanding what kind of information decision makers need and what form it should take to change the status quo (Guerry et al., 2015; Clark et al., 2016). This understanding is especially important in developing countries, where links between ecosystem change and vulnerability of human well-being can be acute (MEA, 2005; West and Brockington, 2006; Allison et al., 2009). Another limitation relates to basic scientific understanding. In the decade since the Millennium Ecosystem Assessment (MEA, 2005), science demonstrating the ways in which ecosystems, especially ocean ecosystems, function as life support systems has advanced considerably (Barbier et al., 2011; Guerry et al., 2012; Liquete et al., 2013). Yet there are major gaps in our knowledge and ability to predict how policies and human behavior lead to changes in ecosystem structure and function and how these changes in turn influence human well-being (see in the following and Ferraro et al., 2015; Guerry et al., 2015). Finally, the slow uptake of ecosystem services information into mainstream decision making may reflect the practical difficulties of working across boundaries between institutions such as universities and governments that have their own distinct cultures and reward systems (Stokes, 1997). These challenges point to a need for investigators from diverse disciplines to work closely with leaders in government, business, and civil society to identify the most pressing conservation and development issues, to translate these into novel research programs, and to help integrate new science into everyday decisions (Guerry et al., 2015; Ledford, 2015).

Larger-scale, more complex societal challenges require new ways of working across single disciplines and integrating societal and scientific bodies of knowledge (Kates et al., 2001). A transdisciplinary approach to research provides a framework for doing just that (Lang et al., 2012). Transdisciplinarity is the youngest of the trio of terms that also includes multidisciplinarity and interdisciplinarity (Choi and Pak, 2006). Although definitions differ across fields, in general, multidisciplinary research draws on knowledge from different disciplines but stays within individual disciplinary boundaries. Interdisciplinary research links two or more disciplines by integrating theoretical frameworks, methodology, and the perspectives and skills from several disciplines. Transdisciplinary research creates new conceptual, theoretical, methodological, and translational innovations to societal and related scientific problems concurrently by integrating various societal and scientific bodies of knowledge (Fig. 16.1; Choi and Pak, 2006; Lang et al., 2012).

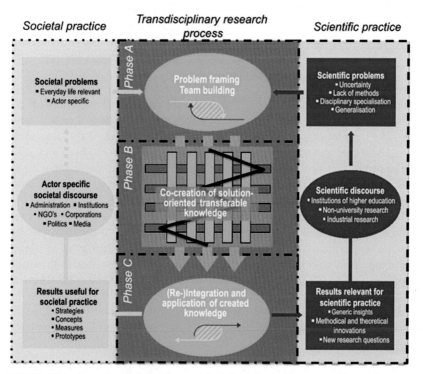

FIGURE 16.1 Conceptual model of a transdisciplinary research process (Lang et al., 2012 Sustainability Science).

Transdisciplinary research agendas increasingly are being recognized as transformative approaches in a variety of fields, from education to health to sustainability science (Choi and Pak, 2006; Brandt et al., 2013). Key arguments for such approaches are first that research on complex problems requires input from various bodies of knowledge to ensure the contribution of information from all relevant disciplines and actor groups (Lang et al., 2012; Brown et al., 2015). Second, research on solution options requires knowledge production beyond problem analysis, as goals, norms, and visions are needed to provide guidance for transition and intervention strategies (Kates et al., 2001; Lang et al., 2012). Third, collaborative efforts between researchers and nonacademic stakeholders increase legitimacy, ownership, and accountability for the problem and the possible solution (Lang et al., 2012; Brown et al., 2015). The presence of these critical elements will dictate how well science contributes to the goals of sustainable development, which are to promote human well-being through integrated stewardship of the earth's environmental, economic, and social assets (UNGA, 2015).

Recent work toward transdisciplinary research in sustainability science suggests that crossing boundaries between science and practice, as well as among academic disciplines, can reap great returns (Brown et al., 2015). Yet navigating these boundaries is not easy and takes far more time than single-discipline research (Van Noorden, 2015). The notion of ecosystem services is potentially a powerful boundary concept (Reyers et al., 2010) for informing transdisciplinary approaches to sustainability because it requires integrating processes and values across disciplines (Tallis and Polasky, 2009). Specific examples of benefits that flow from coastal ecosystems to people (e.g., nursery habitat to support fisheries, coral reefs to attract scuba divers) tend to resonate with the public and decision makers even if the actual term "ecosystem services" does not (The Language of Conservation, 2013; McKenzie et al., 2014; Rosenthal et al., 2014). However, research on ecosystem services also has been criticized for an overly narrow focus on economic valuation, a limited perspective based on western attitudes toward nature, and a lack of societal engagement (Reyers et al., 2010; Seppelt et al., 2011; Laurans et al., 2013; Schoeter et al., 2014). Many studies simply map a suite of services identified by scientists and hand them off to policy makers or managers, without engaging stakeholders who are most closely connected to changes in the natural benefits (reviewed in Mezel and Teng, 2010). Despite these criticisms, societal demand for information about the ways in which ecosystems support economic development and human well-being is growing, as is academic inquiry into processes that underlie the benefits of ecosystems to people (Naidoo et al., 2016) and interdisciplinary collaboration for solving complex challenges (Van Nooden, 2015). These patterns suggest that ecosystem services have the potential to be a transformative tool for sustainability (Reyers et al., 2010; Ruckelshaus et al., 2015), especially if the research agenda can pivot to testing whether what has been proposed to work in theory actually plays out in practice (Lang et al., 2012; Fisher et al., 2013).

In this chapter, we share our experience using transdisciplinary approaches and ecosystem services to inform two government-led spatial planning processes in the Caribbean: ICZM in Belize and Sustainable Development Planning in The Bahamas. We describe the science-policy process in these two countries in light of three important components of transdisciplinarity: (1) solutions-oriented research, (2) coproduction of knowledge, and (3) multiple disciplines (Choi and Pak, 2006; Lang et al., 2012). We highlight challenges and lessons learned for each of these components and end with a discussion of opportunities for leveraging transdisciplinary science of ecosystem services to inform sustainable development.

CASE STUDIES

The Caribbean is an ideal study region for exploring how the science of ecosystem services can be leveraged through transdisciplinary research to advance sustainable development. The economies of many Caribbean countries and well-being of their people are closely tied to natural assets with US$15.7 billion or 4.6% of total gross domestic product coming from the travel and tourism industry (Wilson et al., 2014) and 5% of the population involved in some aspect of the fishing industry (Masters, 2012). Caribbean island and low-lying coastal nations also stand to lose a lot in the face of climate change due to biophysical and sociopolitical vulnerabilities (Allison et al., 2009). Many are pursuing ecosystem-based adaptation strategies that provide multiple cobenefits while reducing risks from coastal hazards (CCCCC, 2009). The Caribbean Community (CARICOM) proved to be a formidable force at the Conference of the Parties (COP21) meeting in Paris in December 2015, pushing for limited global warming to prevent further suffering from climate factors such as storms, sea-level rise, and drought and the collapse of ecosystems that support economic development. Three of the 15 CARICOM states have already developed and approved ICZM plans, whereas another four are in the planning phase (Fig. 16.2). Many of these countries are also pursuing sustainable development initiatives. We have been developing a transdisciplinary research program in the Caribbean that embeds ecosystem service science within stakeholder engagement processes to inform ICZM in Belize and sustainable development in The Bahamas.

Belize

Flanked by the second largest barrier reef in the world, hundreds of cayes, three large atolls, and an extensive system of mangroves and littoral forests, the country of Belize is known for its unique coastal and marine ecosystems. Its highly productive coastal zone is home to a diversity of species, including endangered manatees and sea turtles, as well as 35% of the human population. Fisheries are the basis of livelihoods for many communities and world-renowned snorkeling and diving draw more than 800,000 tourists annually (Cooper et al., 2009). In

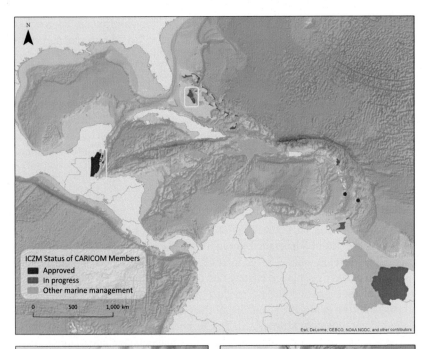

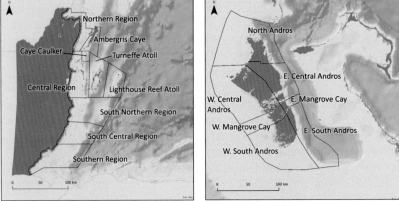

FIGURE 16.2 Integrated Coastal Zone Management (ICZM) Plan status of Caribbean Community (CARICOM) countries (top) where the outlined areas are Belize (bottom left) and Andros Island in The Bahamas (bottom right). *Black lines* and *text* indicate coastal planning regions for the two case studies.

addition to attracting fishers and tourists, coral reefs, mangrove forests, and seagrass beds shield coastal communities from flooding and erosion (Fig. 16.3). Despite the importance of Belize's coast, an increasing diversity and intensity of coastal and ocean sectors, including coastal development, aquaculture, oil, and gas, and marine transportation threaten the very ecosystems that underlie the national economy (http://whc.unesco.org/en/danger/).

FIGURE 16.3 Scuba diver over coral reef in Belize (left; photo credit Melanie McField Marinephotobank). Fishermen and lobster in The Bahamas (top right; photo credit Earthwatch Institute). Mangrove forests fronting coastline in Belize (bottom right; photo credit Nadia Bood).

To address the ad hoc nature of human activities in the coastal zone and ensure sustainable ecosystems for generations to come, The Government of Belize embarked upon a massive effort to design the nation's first ICZM Plan. The 2000 Coastal Zone Management Act called for a plan based on both expert science and local knowledge that would be national in scope but emphasize regional differences. The plan would provide spatially explicit guidance about where and how to engage in ocean and coastal activities to achieve both conservation and development goals. The CZMAI was charged with developing the plan and conducting stakeholder engagement. CZMAI partnered with the authors at the Natural Capital Project (NatCap) in 2010 to use models of ecosystem services to assess multiple tourism, fisheries, and coastal protection objectives under several future scenarios for zoning human activities designed through the process. Through a highly iterative and participatory process of stakeholder engagement and ecosystem modeling, the preferred spatial plan ultimately predicted higher returns in coastal protection and tourism than alternative plans emphasizing either conservation or development alone (Arkema et al., 2015; CZMAI, 2016).

The Bahamas

The Island of Andros lies 40 miles to the west of Nassau, the capital of The Bahamas. Encompassing a land area greater than all other 700 Bahamian islands combined, Andros remains largely undeveloped. Vast mangrove and coppice forests, the third largest coral reef in the world, seagrass beds, sand flats, and a concentrated system of blue holes support the country's commercial and sport-fishing industries, nature-based tourism activities, agriculture, and freshwater resources (Fig. 16.3). Yet many Androsians lack the basic elements of human development, such as essential infrastructure and educational opportunities, to support their livelihoods and those of generations to come. The central challenge confronting The Government of The Bahamas is to design a sustainable development plan that will harness the island's wealth of natural assets without sacrificing the ecosystems that underlie its economy and sustain the well-being of its citizens.

To address this challenge, the Office of the Prime Minister (OPM), with support from the Inter-American Development Bank (IDB), is engaging in an innovative process to design a Sustainable Development Master Plan for Andros Island. The goal is to identify public and private investment opportunities, policy recommendations, zoning guidelines, and other management actions to guide sustainable development of the island over the next 25 years. In consultation with Androsians, government agencies, NGOs, and other stakeholders, OPM is working with NatCap to shape the master plan through providing a preferred island-wide future scenario depicting where human activities and a suite of environmental and social objectives can be harmonized. To do this, we are helping to design several alternative scenarios for future development of Andros based on stakeholders' visions for the future of their island and what benefits they most want to secure. Underpinning and integrating those human development benefits are functioning ecosystems; thus spatial models quantifying ecosystem services information are being used to ensure fisheries, tourism, and coastal resilience objectives are met. The planning process is delivering a community-supported vision within which to embed specific development projects, policies, and investments.

People in countries such as Belize and The Bahamas increasingly realize the interconnections between ecosystem condition and human well-being. Aligning various sectoral interests in envisioning and designing multiobjective, sustainable development plans is one challenge requiring transdisciplinary approaches. The transdisciplinary work does not stop when the plans are complete—more challenges lie ahead in identifying policy and investment mechanisms that ensure plans are implemented and then adapted as lessons emerge. Science can inform such multifaceted problems by both illuminating novel ecological, social, or institutional pathways to consider and also through highlighting solutions under current sociopolitical structures (Folke et al., 2002; Berkes et al., 2003; Folke, 2006; Polasky et al., 2011; Plaganyi et al., 2013; Sellberg et al., 2015).

SOLUTIONS-ORIENTED SCIENCE

For the better part of the last half a century, most of the developed world has viewed scientific inquiry as either basic or applied, presuming that the creativity of basic science would be lost if constrained by the practical (Bush, 1945).

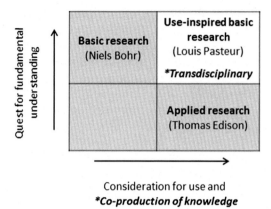

FIGURE 16.4 Pasteur's quadrant with the addition of *Coproduction of knowledge to the *x*-axis to indicate that use-inspired basic research is becoming increasingly *transdisciplinary and involving collaboration among scientists and practitioners, decision makers, and other members of society. *Adapted from Stokes, D.E., 1997. Pasteur's Quadrant: Basic Science and Technological Innovation. Brookings Institution Press, Washington, DC.*

However, this paradigm is finally changing. About the same time that ecologists and economists were beginning to articulate the concept of ecosystem services (Daily, 1997; Costanza et al., 1997) and the role of natural science in sustainable development (Kates et al., 2001; Folke et al., 2002), Donald Stokes published his book, *Pasteur's Quadrant: Basic Science and Technological Innovation* (1996). Stokes argued that rather than a linear progression from basic scientific inquiry eventually leading to applied, science is better considered in two dimensions with the vertical axis representing the quest for fundamental understanding and the horizontal axis representing solutions-oriented research (Fig. 16.4). He identified the top right-hand quadrant as the opportunity for true innovation, emblematic of Louis Pasteur's ground-breaking and use-driven research toward the cause and prevention of disease.

Similar to disease prevention, sustainability is central to societal well-being, broad in scope, and complex, requiring many layers and scales of scientific inquiry. Meeting fundamental human well-being through integrated stewardship of the planet's environmental, economic, and social assets for all people now and in the future is a massive challenge (UNGA, 2015). In the authors' experience, countries involved in national and regional development planning increasingly seek to achieve this goal but lack a road map for doing so; especially one that is tailored to the local social, economic, and environmental context. Here we describe the societal challenge and corresponding solution-driven research questions in Belize and The Bahamas.

We spent months working closely with practitioners and stakeholders in Belize and The Bahamas (see next section on Coproduction of Knowledge) to understand the management challenges and hone the central research questions. In Belize, our science-policy team eventually articulated the following question: Where should the government site zones of human activities to reduce risk to ecosystems and enhance the benefits they provide to people (Arkema et al., 2014, 2015)? This question reflects CZMAI's mandate to develop a spatially explicit plan for managing a suite of ocean and coastal activities, such as infrastructure development, transportation, oil and gas exploration, dredging, and fishing, to ensure the long-term health of ecosystems that underlie the economy and support human well-being (CZMAI, 2016). Our research question also requires scientific advancement. The cumulative influence of multiple stressors on coastal and marine ecosystems was, and continues to be, an active area of scientific inquiry (Halpern et al., 2008; Patrick et al., 2010; Hobday et al., 2011; Samhouri and Levin, 2012; Arkema et al., 2014; Mills et al., 2015; Yates et al., 2015). Similarly, there is much to be learned about relationships between change in ecosystem health and change in human well-being (Carpenter et al., 2006; Guerry et al., 2015). Models that incorporate social, economic, and ecological factors to assess trade-offs among multiple ocean benefits were only recently developed for coastal and marine systems (Watts et al., 2009; White et al., 2012; Guerry et al., 2012; Arkema et al., 2015). Finally, few projects have applied ecosystem services theory in coastal and marine ecosystems. Our work embedding ecosystem services within a stakeholder engagement process in Belize was an opportunity to study a coupled social–ecological system through the lens of the government's ICZM process (Arkema et al., 2015; CZMAI, 2016). In the next two sections, we elaborate on how embedding our science in this process led to the second and third hallmarks of transdisciplinarity: coproduction of knowledge through collaboration among multiple disciplinary experts.

When we first began working in The Bahamas, we wondered if the project would be repetitive of our work in Belize, as the island country is another low-lying coastal nation in the same region of the world. However, we soon realized the challenge was different. Rather than collaborating with a natural resource agency as we did in Belize, our primary partner in The Bahamas is the OPM. Further, the primary funder for the development planning process is a multilateral development bank that is interested in connecting the development plan with subsequent loans for implementation. Our research in The Bahamas focuses on the following question: What public and private investments should be made—and where—to enhance food and water security, coastal resilience, transportation and connectivity, livelihoods and income inequality, and education and capacity building? The Bahamas focus is on the set of specific projects needed to achieve human development goals within the context of a broad framework for sustainable development, rather than asking how to zone ocean sectors and human activities to balance conservation and coastal development goals. Thus we have encountered a more explicit desire from our partners in The Bahamas to understand and capture change in human well-being and livelihoods as a result of management interventions and project investments.

In our experience, a natural capital approach can help to find synergies in management objectives and minimize negative trade-offs in solutions. Interestingly, in The Bahamas, topline objectives may not appear to be "environmental" at first glance. Only when looking at how to achieve sustainable livelihoods, secure infrastructure, and food and water security does the prominent role of ecosystems as underpinning those objectives become clear. This highlights an important lesson for the sustainability science and ocean conservation communities. There are several ways to talk about the importance of ecosystems and natural capital. Ecosystem health may not be the stakeholders' first explicit worry. But by working together, both scientists and practitioners can better understand the critical roles ecosystems play in underpinning human development goals that coastal communities care about most.

COPRODUCTION OF KNOWLEDGE

Coproduction of knowledge and participatory processes are key to successful coastal planning and sustainable development (Clark et al., 2011; Schultz et al., 2015; Posner et al., 2016). Engagement processes that include a variety of actors, foster active participation, prioritize information exchange, and emphasize transparency are more likely to be supported by stakeholders, meet management objectives, and fulfill economic, societal, and conservation goals (Cash et al., 2003). The same is true for the science needed to inform sustainable development in coastal systems. Coproduction of knowledge between scientists, stakeholders, and policy makers increases the chances that scientific results will be salient, credible, and legitimate (Lee, 1993; UNEP & IOC UNESCO, 2009; Clark et al., 2011; McKenzie et al., 2014; Posner et al., 2016). Participatory processes are critical for eliciting information from diverse sources—both scientific and societal—to form a complete picture and the building blocks for transformative solutions (Lang et al., 2012; Clark et al., 2016).

Despite growing awareness among academics that coproduction of knowledge and participatory processes offer potential for both scientific advancement and societal benefit, much of the scientific community has little training or practical experience engaging stakeholders or working with policy makers, resource managers, or other practitioners (Clark et al., 2016). Academics are trained to identify research questions based on gaps in scientific understanding through the peer-review literature, direct observation, and interaction with other scientists. However, posing and tackling research questions that fall in Pasteur's quadrant necessarily involve collaboration between producers and users of science (see * indicating the authors' additions to Stoke's original diagram, Fig. 16.4). This collaborative process involves listening to the needs of practitioners, continually adjusting research agendas to ensure their salience and relevance to the local context, eliciting feedback from a diversity of actors, and developing a suite of both scientific (i.e., peer review papers) and societal (i.e., management plans, blogs, news articles, and community presentations) outputs that leverage local technologies and institutions. Transdisciplinary science has a key role to play in this process. So does explicit definition of roles and boundaries (Reyers et al., 2010; Lang et al., 2012). For example, stakeholders define objectives and alternative futures; scientists help spatially quantify and visualize consequences so that stakeholders and policy makers can make informed decisions. This all takes time.

Our work in Belize and The Bahamas is the outcome in each place of a unique collaboration between scientists and managers to coproduce information about ecosystem services that integrates stakeholder interests, values, and local knowledge into comprehensive sustainable development and coastal zone management plans. Together our team of scientists and practitioners codeveloped an iterative process (Rosenthal et al., 2014; Ruckelshaus et al., 2015) through which we engaged Belizeans and Bahamians from a variety of stakeholder groups to (1) identify and elicit feedback on planning objectives; (2) collect a variety of information including anecdotal knowledge, spatial, and nonspatial data from government agencies, and satellite imagery; (3) design several future scenarios based on climate projections and stakeholder and policy-maker values and visions for the future; (4) use open-source software to understand how decisions that are made now will affect shared ecosystem service values in the future; (5) incorporate feedback from stakeholders and results from models into revised development scenarios; and (6) build consensus around a preferred coastal zoning scheme (Belize) and sustainable development scenario (The Bahamas).

In both countries, we engaged multiple groups of people in different ways at various stages throughout the process. First, close and consistent communication among the core team of partners (e.g., international and local academic institutions, governments, multilateral development banks, consulting firms) formed the basis of our participatory approach. This group met remotely for one hour every week over the course of the whole, multiyear engagement to ensure regular communication. Collaboration during these meetings shaped decisions at every step of the process. For example, in The Bahamas the core team identified a planning horizon of 2040 that would align with the National Development Plan and serve as the timeline for the scientific analysis. We helped to figure out how to best capture Androsian visions for the future in different scenario storylines, and decided which data sources would best serve as inputs into the models for ecosystem services. We used informal agenda setting that incorporated science and policy needs and emphasized developing relationships among members of the core team. These remote meetings were co-led by the policy and science leads. Second, the formation and engagement of advisory groups was built into the process and involved representatives from key ministries, the private sector, and NGO communities. The advisory groups helped to confirm critical decisions, provide feedback, and elicit broad buy-in throughout the process. Third, the core team planned for and conducted stakeholder engagement. In Belize, CZMAI held 50 public and Coastal Advisory Committee meetings throughout the nine coastal planning

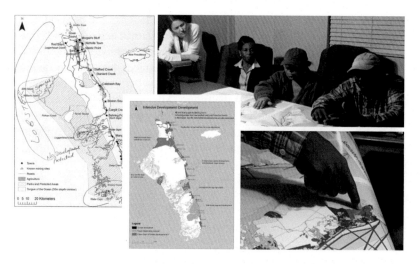

FIGURE 16.5 Scenario design through stakeholder engagement. Stakeholders drew on maps (left) to illustrate where human activities occur now and where they would like them to occur in the future. Over 150 hand-drawn maps from 250 stakeholders were digitized and coalesced to represent different sectors of human use, for example, development (middle). In subsequent meetings, stakeholders provided feedback on proposed future scenarios (right).

regions over a 6-year period. In The Bahamas, we engaged stakeholders from each of the four districts on Andros Island through more than 25 public meetings, open houses, one-on-one visits to homes and businesses, and targeted meetings with local government officials and school groups. The greatest challenge was prioritizing how to use the time with stakeholders given the numbers of people to reach, budget, and capacity.

Over the course of work in each country, we asked the core team, advisory committees, and stakeholders several key questions. First, we asked about the challenges and issues people face in their communities. In The Bahamas, responses ranged from flooding during large storm events to the quality of roads and reliability of flights, accessibility of health clinics, and emigration patterns of young people in search of employment. We used the responses to these questions to understand what issues the management plans needed to address. We then asked what people valued most. On Andros, for example, residents told us about the bonefish that support a highly lucrative sport-fishing tourism sector, the profitable land crab that "put our children through college," and the unique island lifestyle. We used these responses to understand shared stakeholder objectives and which ecosystem services we needed to account for in our analysis of alternative management scenarios. We then asked, what human activities and ocean sectors are prominent on the land- and seascape?, where are these activities located now?, and where do people anticipate, or desire, that they be located in the future? We collected this information through discussion, in hand-written notes from participants, as global positioning system reference points, and through hand-drawn maps (Fig. 16.5).

We grouped different desires and recommendations into several future development scenarios and mapped the location of human activities and climate drivers for each distinct scenario. We then used a suite of models for ecosystem services to understand how climate and the management decisions made today would influence the things that stakeholders told us they valued most (several of which flow from ecosystems) in the future. After vetting with the core team to identify metrics for the ecosystem services that would resonate with stakeholders and how to clearly communicate unexpected trade-offs and synergies, we returned to the advisory committees and stakeholders with maps of human activities and a suite of ecosystems services, both now and under several future management scenarios (Fig. 16.5). We asked the stakeholders if the scenarios reflected the alternative futures they described to us and whether the results from the analysis of ecosystem services aligned with anticipated outcomes or revealed new insights. Our aim at this stage was to understand stakeholder reactions to the results, find out whether we were missing key information, and whether consensus was forming around a particular plan.

We then used the initial results from the models and stakeholder engagement to revise the preferred management scenario for each case. In Belize, by accounting for spatial variation in the impacts of coastal and ocean activities on benefits that ecosystems provide to people, our models allowed stakeholders and policymakers to refine zones of human use (Arkema et al., 2014). The final version of the preferred plan improved expected coastal protection by >25% and more than doubled the revenue from fishing compared with earlier versions based on stakeholder preferences alone (Arkema et al., 2015). Better understanding of how human activities cumulatively shape ecosystem benefits through models encouraged stakeholders to suggest alternative zoning options and also to provide more information that we used to improve modeling and scenario development. In The Bahamas, stakeholders flagged several changes, including the need to assess through scenario modeling a major new mining project that had just been proposed by an international corporation in the northern part of the island. In addition, results from the ecosystem service models suggested locations to prioritize the removal of invasive *Casuarina* trees to reduce coastal erosion and enhance native nursery habitat for fish—a relationship not known to all stakeholders. In the social–ecological systems of both Belize and The Bahamas, knowledge and decision making continually reshaped one another in a process that reflects "coproduction" of knowledge for sustainable development in coastal systems (Jasanoff, 2004; Reyers et al., 2015).

HARNESSING MULTIPLE DISCIPLINES

To achieve sustainable development and conservation of coastal systems that foster societal, environmental, and economic prosperity, scientists from diverse disciplines, including those holding local knowledge in less formal forms, are increasingly working together (Vanoorden, 2015). Societal problems and policy or financial solutions are rarely informed by confining knowledge produced in isolation through separate disciplines. A diversity of approaches is needed to identify and solve the complex issues facing coastal ecosystems and communities. Research that integrates and develops new frameworks, theory, skills, and perspective from multiple disciplines will better meet the demands of current "wicked" problems such as sustainability (UNGA, 2015).

The growing literature and practical experience of scientists involved in interdisciplinary collaboration suggests that constructive communication takes practice and patience (Van Noorden, 2015). Actively avoiding jargon, fostering empathy and respect for disciplinary norms and methods, and taking the time to evaluate what is working in the collaboration and what is not have been shown to foster interdisciplinary work. Seeking feedback early on in a project can prevent individual collaborators from going too far down a path that leads the team off course from a shared mission. Patience and persistence to forge a shared mission that is broad enough to compel collaboration even in the face of time and transaction costs are inherent in interdisciplinary research.

We sought to forge a shared mission at several scales in our work both in Belize and The Bahamas. At the full project scale, sustainable development to foster societal, economic, and environmental prosperity and integrated management of multiple coastal and ocean sectors required interdisciplinary expertise. At finer scales, each of the models for ecosystem services (e.g., fisheries, tourism, and protection from coastal hazards) we used in Belize and The Bahamas incorporates ecological, physical, societal, and economic theory and data, and were developed and validated by experts from different disciplines. Model development involved new science in several compelling ways. To estimate catch and revenue from the spiny lobster fishery in Belize, we used an age-structured model with Beverton–Holt recruitment to describe the lobster population as nine subpopulations (one per planning region) connected via immigration as lobsters move among habitats. This classic matrix model approach also incorporates the role of habitats such as mangroves, seagrass (Belize and The Bahamas), and coral (Belize) in the life history parameters and carrying capacity (Arkema et al., 2015). Surprisingly, the role of nursery and adult habitat is still often ignored in standard fisheries assessments. For tourism, we used a simple linear regression to estimate the relationships between current visitation (Wood et al., 2013) and human activities and habitats. We combined our results with data from the Belize Tourism Board and Ministry of Tourism in The Bahamas to estimate future visitation rate and tourism expenditures. This approach required new science— exploring sources of "big data" from social media sites such as Twitter and Flickr to produce information about where people choose to recreate and why (Wood et al., 2013). For storm protection, we modeled shoreline erosion, wave attenuation, and coastal exposure in the presence and absence of corals, mangroves, and seagrasses and combined these results with census data (The Bahamas) and property values (Belize) to estimate vulnerable communities and avoided damages, respectively (Guannel et al., 2015, 2016, Arkema et al., 2013, 2015). Standard coastal engineering models for estimating flooding and erosion did not previously include the ameliorating effects of habitats, but this is now changing (e.g., vegetation and reef components added to X-Beach model) due in part to the kinds of transdisciplinary research efforts described in this chapter.

In Belize, results from the service models suggest that the preferred plan will lead to greater returns from coastal protection and tourism than outcomes from scenarios oriented toward achieving either conservation or development goals alone (Fig. 16.6). The plan also is predicted to reduce impacts to coastal habitat and increase revenues from lobster fishing relative to current management (Arkema et al., 2015; CZMAI, 2016). Higher values of coastal protection (Fig. 16.7) and tourism result from the combination of both biophysical and social variables inherent in estimates of ecosystem services (NRC, 2004; Tallis

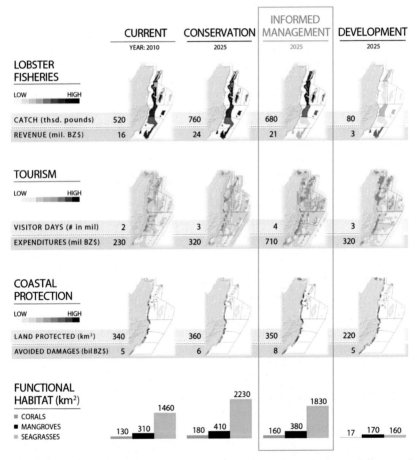

FIGURE 16.6 Biophysical and economic values for three ecosystem services and the area of habitat capable of providing services under the current situation and three future scenarios for the Integrated Coastal Zone Management Plan for Belize (see Fig. 2 in Arkema et al., 2015, Proceedings of the National Academy of Sciences).

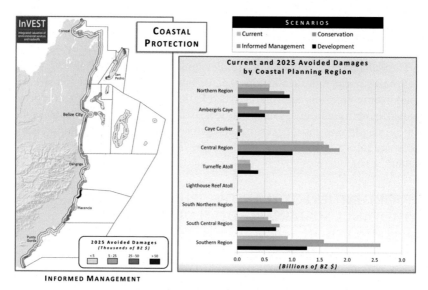

FIGURE 16.7 Annual avoided damages for the informed management scenario (2025). Bar graphs (right) show variation by planning region in avoided damages for the current situation and three future scenarios (see Fig. S11 in Arkema et al., 2015, Proceedings of the National Academy of Sciences, CZMAI, 2016).

and Polasky, 2009). Increases in the extent of activities to support economic development may lead to more cumulative impacts on coral, mangrove and seagrass, reduced nursery habitat for fishes, and reduced fisheries returns. However, even a modest increase in coastal development can lead to more land with higher property values, increases in the value of habitats for protection from storms, and more infrastructure to support tourism (Figs. 16.6 and 16.7).

Ocean conservation and management often focus on marine protected areas (MPAs) and fisheries. But focusing exclusively on these interests misses the need to balance several other sectors such as marine transportation, coastal and energy development, rising sea levels, and dredging that are central to both the challenges and opportunities in front of Caribbean countries. Including a diversity of uses in our management scenarios and articulating outcomes in terms of ecosystem services facilitated explicit consideration of multiple objectives that resource managers typically evaluate separately in decisions. Multidisciplinary science can help improve spatial planning beyond what has already been demonstrated in Belize and The Bahamas. Building upon our highly iterative and interactive modeling approach, further scenario elicitation and evaluation can reveal new ways in which coupled social–ecological systems could function to improve biodiversity, service provision, and human well-being. For example, resilience frameworks can help stakeholders envision more fundamental changes in institutions and governance that align incentives, regulation, or both, toward more sustainable outcomes (Folke et al., 2002; Berkes et al., 2003).

Another needed advance in interdisciplinary science for coastal and ocean planning is clear linkages between changes in marine ecosystems and changes in human well-being metrics that inform decisions. As exemplified in The Bahamas, first-order goals for development planning often are human well-being indicators such as jobs, health, education, and access to markets. Establishing rigorous and predictive relationships between these tangible human benefits and ecosystem services is challenging from a data availability standpoint and because of the poorly understood economic structures and informal and formal market rules in island nations (Holdschlag & Ratter, 2016).

OPPORTUNITIES FOR TRANSDISCIPLINARY SCIENCE TO ADVANCE SUSTAINABLE DEVELOPMENT AND IMPLEMENTATION

In February 2016, we traveled to Belize to celebrate the launch of the country's first ICZM Plan (CZMAI, 2016) and one of the world's first examples of a plan explicitly designed using quantitative analyses of ecosystems services, which were produced in close collaboration with policy makers and embedded in a stakeholder engagement process (http://whc.unesco.org/en/news/1455/). During the launch ceremony, the Deputy Prime Minister of Belize, Honorable Gaspar Vega, said "the Coastal Zone Management Plan is a critical component of our country's strategy for national economic development... to sustain a growing population given the effects of climate change and the central role that coastal resources play in the lives of our people and the economy." Vega's statement highlights how multisectoral ocean planning can advance sustainable economic development.

Coastal ecosystems are often sidelined in national planning processes. Yet, the vast majority of countries around the world are coastal nations. Ocean related sectors are a big part of many nations' economies and coastal and marine ecosystems support the well-being of their citizens. Belize sets a global example for how any nation can account for the role that coastal ecosystems play in the integrated management of infrastructure development, transportation, fishing, dredging, aquaculture, and tourism. Through multisectoral processes that involve information about the benefits nature provides to people such as in Belize, marine conservation biologists and practitioners have the opportunity to elevate the role of coastal and ocean ecosystems and resources into a country's national conversation and actions toward sustainable development. Such a framing broadens the constituency for ocean planning, which to some extent is overly narrowly focused on biodiversity protection and MPAs. The advantage of a sustainable development perspective is that it provides a broader set of objectives and can start with whichever are most materially relevant to society. Ecosystem services underpin many development goals (UNGA, 2015), but they do not need to be the primary objective.

One of the main reasons the Belizean government, only recently elected in fall of 2015, passed the Coastal Zone Management Plan just 4 months later was to ensure that this administration had time during its tenure to implement the plan—to make the Plan durable such that it "sticks." Implementation is a major unmet challenge not just for coastal zone management plans but for sustainable development and the outputs of many integrated planning processes. Attaching innovative financing mechanisms to spatial plans is one way to do this. For example, the IDB is using strategic loans to support sustainable tourism in the Caribbean and funding development planning in The Bahamas. Through partnerships with multilateral development banks, sustainable development plans can set the stage for future government loans and private investment.

Engaging the private sector in public–private partnerships for spatial planning is a promising, relatively new area of innovation to secure implementation and final outcomes for ecosystems and human well-being. Public–private financing for protected areas has a great potential to bolster planning, such as the $215M fund developed to secure stewardship and management for a 60 million hectare network of protected areas in the Amazon Region Protected Areas (WWF, 2015). As another example, the Chamber of Commerce is chairing the advisory committee for the national development plan in The Bahamas. The presence of business leaders reflects their incentive for spatial planning—reducing surprises and risks associated with government turnover during plan development or implementation. If business leaders can be part of a stakeholder process, designing which activities are allowed and where, the increased certainty in social license to operate into the future under a sanctioned plan is a significant advantage. From both governments' and business' perspectives, fitting infrastructure investments into a comprehensive plan can reduce the risk of "death by a 1000 projects." Integrated planning can harmonize human and ecosystem goals, while providing developers security where project and financing approvals are likely (Mandle et al., 2015).

CONCLUSION

Transdisciplinary approaches that integrate across scales, disciplines, and science–society boundaries have the potential to inform not just the planning phase but also implementation through innovative financing and connecting project-level review to approved plans. But transdisciplinary and sustainability science are not without their challenges and do not go uncontested outside transdisciplinary research communities. Several studies and the authors' practical experience note the commitment it takes for interdisciplinary collaboration to be done well. Many scientists underestimate the importance and the time it takes to develop relationships with collaborators from other disciplines, stakeholders, and policy makers. Even when researchers invest in relationships with a broad swath of actors, they are likely to face the old adage that "knowledge is power." Scientists may be perceived as taking sides, and research topics may disproportionately reflect the priorities of certain parties (Clark et al., 2016). Underrepresentation of some stakeholder groups and barriers to participation can exacerbate this problem (Lang et al., 2012). On the analysis side, vagueness and ambiguity of results can mask potential conflicts or lead to different interpretations. Furthermore, traditional academic communities are often skeptical about the reliability, validity, and methodological aspects of collaborative research. These challenges highlight the need to test transdisciplinary approaches and monitor outcomes from using ecosystem services in multisectoral coastal and ocean planning (Reyers et al., 2010; Lang et al., 2012).

Challenges of transdisciplinary research and sustainability science are very real. But a hallmark of codeveloping solutions-oriented science is to heed Voltaire's advice and to not let the perfect be the enemy of the good. More conventional research perspectives have caused some of the nation's best talent to become locked in the ivory tower. Through his interactive view of science and technology, Stokes builds a convincing case that solution-driven fundamental research will reframe the relationship between science and society. Because social and ecological systems are continually evolving and reshaping each other, efforts to develop new and useful knowledge for sustainability should be less about control and optimality and more about flexibility and adaptation. Scientists can help society see itself as a social–ecological system, emphasizing the dynamic nature, not only of the system, but also of any new understanding of how the system works.

REFERENCES

Adger, W., Hughes, T., Folke, C., et al., 2005. Social–ecological resilience to coastal disasters. Science 309, 1036–1039.

Allison, E.H., Horemans, B., 2006. Putting the principles of the sustainable livelihoods approach into fisheries development policy and practice. Marine Policy 30 (6), 757–766.

Allison, E., Perry, A.L., Badjeck, M.-C., Adger, W.N., Brown, K., Conway, D., Halis, A.S., Pilling, G.M., Reynolds, J.D., Andrew, N.L., Dulvy, N.K., 2009. Vulnerability of national economies to the impacts of climate change on fisheries. Fish and Fisheries 10, 173–196.

Arkema, K.K., Abramson, S.C., Dewsbury, B.M., 2006. Marine ecosystem-based management: from characterization to implementation. Frontiers in Ecology and the Environment 4 (10), 525–532.

Arkema, K.K., Guannel, G., Verutes, G., Wood, S.A., Guerry, A., Ruckelshaus, M., Kareiva, P., Lacayo, M., Silver, J.M., 2013. Coastal habitats shield people and property from sea-level rise and storms. Nature Climate Change 3 (10), 913–918.

Arkema, K.K., Verutes, G., Bernhardt, J.R., Clarke, C., Rosado, S., Canto, M., Wood, S.A., et al., 2014. Assessing habitat risk from human activities to inform coastal and marine spatial planning: a demonstration in Belize. Environmental Research Letters 9 (11), 114016.

Arkema, K.K., Verutes, G.M., Wood, S.A., Clarke-Samuels, C., Rosado, S., Canto, M., Rosenthal, A., et al., 2015. Embedding ecosystem services in coastal planning leads to better outcomes for people and nature. Proceedings of the National Academy of Sciences of the United States of America 112 (24), 7390–7395.

Barbier, E.B., Hacker, S.D., Kennedy, C., et al., 2011. The value of Estuarine and coastal ecosystem services. Nature Climate Ecological Monographs 81, 169–183.

Berkes, F., Colding, J., Folke, C., 2003. Navigating Social-Ecological Systems: Building Resilience for Complexity and Change. Cambridge University Press. 394 pp.

Bloomberg, M., Paulson, H., Steyer, T.F., 2014. Risky Business: The Economic Risks of Climate Change in the United States.

Brandt, P., Ernst, A., Gralla, F., Luederitz, C., Lang, D.J., Newig, J., Reinert, F., Abson, D.J., von Wehrden, H., 2013. A review of transdisciplinary research in sustainability science. Ecological Economics, Land Use 92, 1–15. http://dx.doi.org/10.1016/j.ecolecon.2013.04.008.

Brown, R.R., Deletic, A., Wong, T.H.F., 2015. Interdisciplinarity: how to catalyze collaboration. Nature News 525, 315. http://dx.doi.org/10.1038/525315a.

Bush, V., 1945. Science, the Endless Frontier: A Report to the President. [on a Program for Postwar Scientific Research]. US Government Printing Office.

Caribbean Community Climate Change Centre (CCCCC), 2009. Climate Change and the Caribbean: Regional Framework for Achieving Development Resilient to Climate Change (2009–2015). http://www.caribbeanclimate.bz/ongoing-projects/2009-2021-regional-planning-for-climate-compatible-development-in-the-region.html.

Carpenter, S., DeFries, J.R., Dietz, T., Mooney, H.A., Polasky, S., Reid, W.V., Scholes, R.J., 2006. Millenium ecosystem assessment: research needs. Science 314, 257–258.

Cash, D.W., Clark, W.C., Alcock, F., Dickson, N.M., Eckley, N., Guston, D.H., Jäger, J., Mitchell, R.B., 2003. Knowledge systems for sustainable development. Proceedings of the National Academy of Sciences of the United States of America 100 (14), 8086–8091.

Choi, B.C.K., Pak, A.W.P., 2006. Multidisciplinarity, interdisciplinarity and transdisciplinarity in health research, services, education and policy: 1. Definitions, objectives, and evidence of effectiveness. Clinical and Investigative Medicine 29, 351–364.

Clark, W.C., Tomich, T.P., Noordwijk, M., van, Guston, D., Catacutan, D., Dickson, N.M., McNie, E., 2011. Boundary work for sustainable development: natural resource management at the consultative group on international agricultural research (CGIAR). Proceedings of the National Academy of Sciences of the United States of America 113, 4615–4622. http://dx.doi.org/10.1073/pnas.0900231108.

Clark, W.C., Kerkhoff, L., van, Lebel, L., Gallopin, G.C., 2016. Crafting usable knowledge for sustainable development. Proceedings of the National Academy of Sciences of the United States of America 113, 4570–4578. http://dx.doi.org/10.1073/pnas.1601266113.

Coastal Zone Management Authority and Institute (CZMAI), 2016. Belize Integrated Coastal Zone Management Plan. (Belize City).

Cooper, E., Burke, L., Bood, N., 2009. Coastal Capital: Belize. The Economic Contribution of Belize's Coral Reefs and Mangroves. World Resources Institute, Washington.

Costanza, R., d'Arge, R., De Groot, R., Faber, S., Grasso, M., Hannon, B., Limburg, K., et al., 1997. The value of the world's ecosystem services and natural capital. Nature 387, 253–260.

Coulthard, S., Johnson, D., McGregor, J.A., 2011. Poverty, sustainability and human wellbeing: a social wellbeing approach to the global fisheries crisis. Global Environmental Change 21 (2), 453–463.

Daily, G., 1997. Nature's Services: Societal Dependence on Natural Ecosystems, fourth ed. Island Press, Washington, DC.

Day, J.W., Boesch, D.F., Clairain, E.J., Kemp, G.P., Laska, S.B., Mitsch, W.J., Orth, K., et al., 2007. Restoration of the Mississippi Delta: Lessons from Hurricanes Katrina and Rita. Science 315 (5819), 1679–1684.

Douvere, F., 2008. The importance of marine spatial planning in advancing ecosystem-based sea use management. Marine Policy 32 (5), 762–771.

Ferraro, P.J., Hanauer, M.M., Miteva, D.A., Nelson, J.L., Pattanayak, S.K., Nolte, C., Sims, K.R.E., 2015. Estimating the impacts of conservation on ecosystem services and poverty by integrating modeling and evaluation. Proceedings of the National Academy of Sciences of the United States of America 112, 7420–7425. http://dx.doi.org/10.1073/pnas.1406487112.

Fisher, B., Balmford, A., Ferraro, P.J., Glew, L., Mascia, M., Naidoo, R., Ricketts, T.H., 2013. Moving Rio forward and avoiding 10 more years with little evidence for effective conservation policy. Conservation Biology 28, 880–882.

Folke, C., 2006. Resilience: the emergence of a perspective for social–ecological systems analyses. Global Environmental Change 16, 253–267.

Folke, C., Carpenter, S., Elmqvist, T., Gunderson, L., Holling, C.S., Walker, B., 2002. Resilience and sustainable development: building adaptive capacity in a world of transformations. AMBIO: A Journal of the Human Environment 31, 437–440. http://dx.doi.org/10.1579/0044-7447-31.5.437.

Guannel, G., Ruggiero, P., Faries, J., et al., 2015. Integrated modeling framework to quantify the coastal protection services supplied by vegetation. Journal of Geophysical Research-Oceans 120, 324–345.

Guannel, G., Arkema, K., Ruggiero, P., Verutes, G., 2016. The power of three: coral reefs, seagrasses and mangroves protect coastal regions and increase their resilience. PLoS One 11, e0158094.

Guerry, A.D., Ruckelshaus, M.H., Arkema, K.K., Bernhardt, J.R., Guannel, G., Kim, C.-K., Marsik, M., et al., 2012. Modeling benefits from nature: using ecosystem services to inform coastal and marine spatial planning. International Journal of Biodiversity Science, Ecosystem Services and Management 1–15.

Guerry, A.D., Polasky, S., Lubchenco, J., Chaplin-Kramer, R., Daily Gretchen, D., 2015. Natural capital and ecosystem services informing decisions: from promise to practice. Proceedings of the National Academy of Sciences of United States of America 112 (24), 7348–7355.

Halpern, B.S., Walbridge, S., Selkoe, K.A., Kappel, C.V., Micheli, F., D'Agrosa, C., Bruno, J.F., Casey, K.S., Ebert, C., Fox, H.E., Fujita, R., Heinemann, D., Lenihan, H.S., Madin, E.M.P., Perry, M.T., Selig, E.R., Spalding, M., Steneck, R., Watson, R., 2008. A global map of human impact on marine ecosystems. Science 319, 948–952. http://dx.doi.org/10.1126/science.1149345.

He, Q., Bertness, M.D., Bruno, J.F., Li, B., Chen, G., Coverdale, T.C., Altieri, A.H., Bai, J., Sun, T., Pennings, S.C., Liu, J., Ehrlich, P.R., Cui, B., 2014. Economic development and coastal ecosystem change in China. Scientific Reports 4, 5995. http://dx.doi.org/10.1038/srep05995.

Hobday, A.J., Smith, A.D.M., Stobutzki, I.C., Bulman, C., Daley, R., Dambacher, J.M., Deng, R.A., et al., 2011. Ecological risk assessment for the effects of fishing. Fisheries Research 108 (2–3), 372–384.

Holdschlag, A., Ratter, B.M.W., 2016. Caribbean island states in a social-ecological panarchy? Complexity theory, adaptability and environmental knowledge systems. Anthropocene 13, 80–93. http://dx.doi.org/10.1016/j.ancene.2016.03.002.

IPCC, 2014. Climate change 2014: impacts, adaptation, and vulnerability. In: IPCC Working Group II Contribution to AR5, vol. II, Regional aspects, Ch. 29: Small Islands (final draft) http://ipccwg2.gov/AR5/images/uploads/WGIIAR5Chap29_FGDall.pdf.

Jasanoff, S., 2004. States of Knowledge: The Co-production of Science and the Social Order. Routledge, London.

Kates, R.W., Clark, W.C., Corell, R., Hall, J.M., Jaeger, C.C., Lowe, I., McCarthy, J.J., Schellnhuber, H.J., Bolin, B., Dickson, N.M., Faucheux, S., Gallopin, G.C., Grübler, A., Huntley, B., Jäger, J., Jodha, N.S., Kasperson, R.E., Mabogunje, A., Matson, P., Mooney, H., Moore, B., O'Riordan, T., Svedin, U., 2001. Sustainability science. Science 292, 641–642. http://dx.doi.org/10.1126/science.1059386.

Kinzig, A.P., Perrings, C., Chapin, F.S., Polasky, S., Smith, V.K., Tilman, D., Turner, B.L., 2011. Paying for ecosystem services—promise and peril. Science 334, 603–604.

Lang, D.J., Wiek, A., Bergmann, M., Stauffacher, M., Martens, P., Moll, P., Swilling, M., Thomas, C.J., 2012. Transdisciplinary research in sustainability science: practice, principles, and challenges. Sustainability Science 7, 25–43. http://dx.doi.org/10.1007/s11625-011-0149-x.

Langridge, S.M., Hartge, E.H., Clark, R., Arkema, K., Verutes, G.M., Prahler, E.E., Stoner-Duncan, S., et al., 2014. Key lessons for incorporating natural infrastructure into regional climate adaptation planning. Ocean & Coastal Management 95, 189–197.

Laurans, Y., Rankovic, A., Billé, R., Pirard, R., Mermet, L., 2013. Use of ecosystem services economic valuation for decision making: questioning a literature blindspot. Journal of Environmental Management 119, 208–219.

Ledford, H., 2015. Team science. Nature 525, 308–311.

Lee, K., 1993. Compass and Gyroscope: Integrating Science and Politics for the Environment. Island Press. 255 pp.

Levin, P.S., 2014. New conservation for the Anthropocene Ocean. Conservation Letters 7, 339–340. http://dx.doi.org/10.1111/conl.12108.

Liquete, C., Piroddi, C., Drakou, E.G., Gurney, L., Katsanevakis, S., Charef, A., Egoh, B., 2013. Current status and future prospects for the assessment of marine and coastal ecosystem services: a systematic review. PLoS One 8, e67737. http://dx.doi.org/10.1371/journal.pone.0067737.

Lubchenco, J., Sutley, N., 2010. Proposed US Policy for ocean, coast, and great Lakes stewardship. Science 328, 1485–1486. http://dx.doi.org/10.1126/science.1190041.

Mace, G.M., 2014. Whose conservation? Science 345 (6204), 1558–1560.

Mandle, L., Bryant, B.P., Ruckelshaus, M., Geneletti, D., Kiesecker, J.M., Pfaff, A., 2015. Entry points for considering ecosystem services within infrastructure planning: how to efficiently integrate conservation with development in order to aid them both. Conservation Letters. http://dx.doi.org/10.1111/conl.12201.

Masters, J., 2012. Caribbean regional fisheries mechanism. Statistics and Information Report 2010 66 pp.

McKenzie, E., Stephen, P., Patricia, T., Bernhardt, J.R., Howard, K., Rosenthal, A., 2014. Understanding the use of ecosystem service knowledge in decision making: lessons from international experiences of spatial planning. Environment and Planning C—Government and Policy 32 (2), 320–340.

McLeod, K., Leslie, H., 2009. Ecosystem-Based Management for the Oceans. Island Press, Washington, DC.

Melillo, J.M., Richmond, T.T.C., Yohe, G.W. (Eds.), 2014. Climate Change Impacts in the United States: The Third National Climate Assessment. US Global Change Research Program, p. 841. http://dx.doi.org/10.7930/J0Z31WJ2.

Menzel, S., Teng, J., 2010. Ecosystem services as a stakeholder-driven concept for conservation science. Conservation Biology 24, 907–909. http://dx.doi.org/10.1111/j.1523-1739.2009.01347.x.

Millenium Ecosystem Assessment Panel, 2005. Ecosystems and Human Well-being: Synthesis. Millenium Ecosystem Assessment Series. Island Press, Washington, DC. 3 Intergovernmental Panel on Climate Change.

Mills, M., Leon, J.X., Saunders, M.I., Bell, J., Liu, Y., O'Mara, J., Lovelock, C.E., Mumby, P.J., Phinn, S., Possingham, H.P., Tulloch, V., Mutafoglu, K., Morrison, T., Callaghan, D., Baldock, T., Klein, C., Hoegh-Guldberg, O., 2015. Reconciling development and conservation under coastal squeeze from rising sea-level. Conservation Letters. http://dx.doi.org/10.1111/conl.12213.

Naidoo, R., Weaver, L.C., Diggle, R.W., Matongo, G., Stuart-Hill, G., Thouless, C., 2016. Complementary benefits of tourism and hunting to communal conservancies in Namibia. Conservation Biology 30, 628–638. http://dx.doi.org/10.1111/cobi.12643.

National Fish and Wildlife Foundation, 2014. Hurricane Sandy Resiliency Competitive Grants Program. United States Department of Interior. www.nfwf.org/hurricanesandy.

National Research Council, 2004. Valuing Ecosystem Services: Toward Better Environmental Decision-Making. National Academies, Washington.

Ouyang, Z., Zheng, H., Xiao, Y., Polasky, S., Liu, J., Xu, W., Wang, O., Zhang, L., Xiao, Y., Rao, E., Jiang, L., Lu, F., Wang, X., Yang, G., Gong, S., Wu, B., Zeng, Y., Yang, W., Daily, G., 2016. Improvements in ecosystem services from investments in natural capital. Science 352 (6292), 1455–1459. http://dx.doi.org/10.1126/science.aaf2295.

Patrick, W.S., Spencer, P., Link, J., Cope, J., Field, J., Kobayashi, D., Lawson, P., et al., 2010. Using productivity and susceptibility indices to assess the vulnerability of United States fish stocks to overfishing. Fishery Bulletin 108 (3), 305–322.

Plaganyi, E., van Putten, I., Hutton, T., Deng, R.A., Dennis, D., Pascoe, S., Skews, T., Campbell, R.A., 2013. Integrating indigenous livelihood and lifestyle objectives in managing a natural resource. Proceedings of the National Academy of Sciences of the United States of America 110, 3639–3644. http://dx.doi.org/10.1073/pnas.1217822110.

Polasky, S., Nelson, E., Pennington, D., Johnson, K., 2011. The impact of land-use change on ecosystem services, biodiversity and returns to landowners: a case study in the State of Minnesota. Environmental and Resource Economics 48 (2), 219–242.

Posner, S.M., McKenzie, E., Ricketts, T.H., 2016. Policy impacts of ecosystem services knowledge. Proceedings of the National Academy of Sciences of the United States of America. http://dx.doi.org/10.1073/pnas.1502452113.

Reyers, B., Roux, D.J., O'farrell, P.J., 2010. Can ecosystem services lead ecology on a transdisciplinary pathway? Environmental Conservation 37, 501–511. http://dx.doi.org/10.1017/S0376892910000846.

Reyers, B., Nel, J.L., O'Farrell, P.J., Sitas, N., Nel, D.C., 2015. Navigating complexity through knowledge coproduction: mainstreaming ecosystem services into disaster risk reduction. Proceedings of the National Academy of Sciences of the United States of America 112 (24), 7362–7368.

Rosenthal, A., Verutes, G., McKenzie, E., Arkema, K.K., Bhagabati, N., Bremer, L.L., Olero, N., Vogl, A.L., 2014. Process matters: a framework for conducting decision-relevant assessments of ecosystem services International Journal of Biodiversity Science. Ecosystem Services and Management 11, 190–204.

Ruckelshaus, M., Klinger, T., Knowlton, N., DeMaster, D.P., 2008. Marine ecosystem-based management in practice: scientific and governance challenges. BioScience 58, 53–63. http://dx.doi.org/10.1641/B580110.

Ruckelshaus, M., McKenzie, E., Tallis, H., Guerry, A., Daily, G., Kareiva, P., Polasky, S., Ricketts, T., Baghabati, N., Wood, S., Bernhardt, J., 2015. Notes from the field: lessons learned from using ecosystem services to inform real-world decisions. Ecological Economics. http://dx.doi.org/10.1016/j.ecolecon.2013.07.009. Published online.

Samhouri, J.F., Levin, P.S., 2012. Linking land- and sea-based activities to risk in coastal ecosystems. Biological Conservation 145 (1), 118–129.

Schröter, M., van der Zanden, E.H., van Oudenhoven, A.P.E., Remme, R.P., Serna-Chavez, H.M., de Groot, R.S., Opdam, P., 2014. Ecosystem services as a contested concept: a synthesis of critique and counter-arguments. Conservation Letters 7, 514–523. http://dx.doi.org/10.1111/conl.12091.

Schultz, L., Folke, C., Österblom, H., Olsson, P., 2015. Adaptive governance, ecosystem management, and natural capital. Proceedings of the National Academy of Sciences of the United States of America 112 (24), 7369–7374.

Sellberg, M.M., Wilkinson, C., Peterson, G.D., 2015. Resilience assessment: a useful approach to navigate urban sustainability challenges. Ecology and Society. 20 (43). http://dx.doi.org/10.5751/es-07258-200143.

Seppelt, R., Dormann, C.F., Eppink, F.V., Lautenbach, S., Schmidt, S., 2011. A quantitative review of ecosystem service studies: approaches, shortcomings and the road ahead. Journal of Applied Ecology 48, 630–636. http://dx.doi.org/10.1111/j.1365-2664.2010.01952.x.

Small, C., Nicholls, R.J., 2003. A global analysis of human settlement in coastal zones. Journal of Coastal Research 19, 584–599.

Stokes, D.E., 1997. Pasteur's Quadrant: Basic Science and Technological Innovation. Brookings Institution Press, Washington, DC.

Tallis, H., Polasky, S., 2009. Mapping and valuing ecosystem services as an approach for conservation and natural-resource management. Annals of the New York Academy of Sciences 1162, 265–283. http://dx.doi.org/10.1111/j.1749-6632.2009.04152.x.

TEEB, 2010. The Economics of Ecosystems and Biodiversity: Mainstreaming the Economics of Nature: A Synthesis of the Approach, Conclusions and Recommendations of TEEB.

The Language of Conservation, 2013. Updated Recommendations on How to Communicate Effectively to Build Support for Conservation. Fairbank, Maslin, Maullin, Metz & Associates. Public Opinion Strategies.

UNEP and IOC-UNESCO, 2009. An Assessment of Assessments, Findings of the Group of Experts. Start-up Phase of a Regular Process for Global Reporting and Assessment of the State of the Marine Environment Including Socio-economic Aspects. ISBN: 978-92-807-2976-4.

United Nations, 2015. World Population Prospects, the 2015 Revision. Population Division, Department of Economic and Social Affairs.

United Nations General Assembly, 2015. Transforming Our World: The 2030 Agenda for Sustainable Development. Resolution adopted by the General Assembly on 25 September 2015. 70th Session, 70/1. United Nations, New York, NY.

Van Noorden, R., 2015. Interdisciplinary research by the numbers. Nature 525, 306–307.

Watts, M.E., Ball, I.R., Stewart, R.S., Klein, C.J., Wilson, K., Steinback, C., Lourival, R., Kircher, L., Possingham, H.P., 2009. Marxan with Zones: software for optimal conservation based land- and sea-use zoning. Environmental Modeling and Software 24, 1513–1521.

West, J., Brockington, D., 2006. Parks and peoples: the social impact of protected areas. Annual Review of Anthropology 35, 251–277.

White, C., Halpern, B.S., Kappel, C.V., 2012. Ecosystem service tradeoff analysis reveals the value of marine spatial planning for multiple ocean uses. Proceedings of the National Academy of Sciences of the United States of America 201114215.

Wilson, S., Sagewan-Alli, I., Calatayud, A., 2014. The ecotourism industry in the Caribbean: a value chain analysis. Inter-American Development Bank Technical Note No. IDB-TN-710, p. 51.

Wood, S.A., Guerry, A.D., Silver, J.M., Lacayo, M., 2013. Using social media to quantify nature-based tourism and recreation. Science Report 3.

Worm, B., Lenihan, H.S., 2013. Threats to marine ecosystems: overfishing and habitat degradation. In: Bertness, M., Bruno, J., Silliman, B., Stachowicz, J. (Eds.), Marine Community Ecology and Conservation. Sinauer, Sunderland, MA, pp. 449–476.

WWF (World Wildlife Fund), 2015. Project Finance for Permanence. Key Outcomes and Lessons Learned. WWF, Washington, DC. 9 pp. Available at: www.worldwildlife.org/publications/.

Yates, K.L., Schoeman, D.S., Klein, C.J., 2015. Ocean zoning for conservation, fisheries and marine renewable energy: assessing trade-offs and co-location opportunities. Journal of Environmental Management 152, 201–209. http://dx.doi.org/10.1016/j.jenvman.2015.01.045.

Chapter 17

Social–Ecological Trade-Offs in Baltic Sea Fisheries Management

Rüdiger Voss, Martin F. Quaas, Julia Hoffmann, Jörn O. Schmidt
Christian Albrechts Universität zu Kiel, Kiel, Germany

THE BALTIC SEA ECOSYSTEM

In many parts of the world, there are marine areas that are almost entirely enclosed by land, called semienclosed seas. Besides the Baltic, examples include the Mediterranean Sea, Black Sea, Chesapeake Bay, or Puget Sound. These areas are often characterized by outstanding natural attributes. At the same time, they have been the setting for intense economic, cultural, and recreational activities throughout the ages. Baltic Sea might therefore be seen as a unique case study.

The Baltic Sea is a semienclosed sea, which is suffering from multiple impacts posed by about 85 million people living in 14 countries in its catchment area extending to over 1.7 million km² (Fig. 17.1).

Human activities that impact the ecosystem occur both in the sea (pollution, maritime shipping, fisheries, tourism) and on land (e.g., airborne pollutant transfer, increased nutrient supply via riverine runoff caused by intensive agriculture). Of all the potential threats, eutrophication is one of the most important drivers affecting the Baltic Sea ecosystems and its ecosystem services, including fisheries. Although there are indications of some recent improvement, eutrophication still causes hypoxia, blooms of blue green algae, and shifts in plankton communities that have large ecological and economic consequences. The coastal fishery poses another major threat, with several coastal fisheries being currently depressed (HELCOM, 2006; Lajus et al., 2013). Shipping and associated bioinvasions also are major threats to Baltic Sea ecosystems, with several nonindigenous having pan-Baltic distribution and significant ecosystem impacts (Schaber et al., 2011; HELCOM, 2015). Like other regions, the Baltic is affected by climate change and variability, most obviously through the changed intensity and frequency of saline water inflows from the North Sea. This poses additional pressure on species living at the edge of their distribution range. Increasing pressure on coastal regions is expected in the future through multiple intensified human uses, such as the construction of new marinas, energy devices

Conservation for the Anthropocene Ocean. http://dx.doi.org/10.1016/B978-0-12-805375-1.00017-9

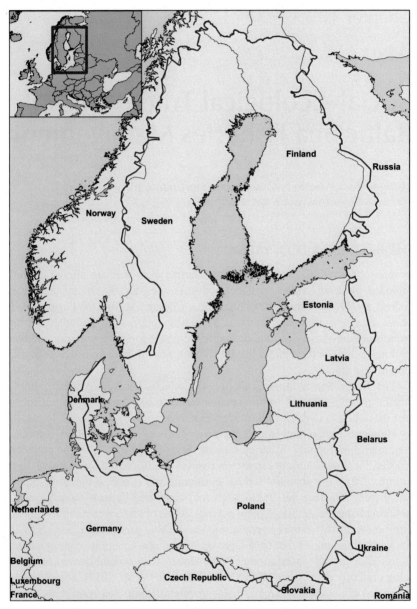

FIGURE 17.1 Baltic Sea map, including countries and catchment area.

(pipes, lines, windmills), the creation of artificial hard substrata, and tourism, or recreational activities.

The many, often conflicting, uses of the Baltic Sea force managers and scientists to examine trade-offs across social, economic, and ecological dimensions. Such trade-offs are perhaps most obvious for fisheries because there are multiple species that interact and are themselves highly dependent on habitat, water quality, and other factors that are managed and governed at multiple levels.

General Hydrography of the Baltic Sea

The topography of the Baltic is a sequence of deep basins, separated by shallow sills. There are generally high gradients in both abiotic and biotic variables due to the strong vertical stratification and the limited exchange with the North Sea. Temperature, salinity, oxygen-content, as well as prey and predator abundance vary over small horizontal and vertical scales. Oxygen renewal in the deep basins depends on irregular inflows of saline, oxygen-rich North Sea water masses the so-called Baltic inflows. These take place mainly in winter time. Accordingly, hydrographic conditions vary considerably between years depending on the magnitude of the Baltic inflow. Low inflows are characterized as a stagnation year, with hypoxia occurring within bottom waters of the deep basins. In contrast, inflow events are accompanied by lower temperatures (ca. −5°C difference in the deep Bornholm Basin), a rise in salinity, and depth of the halocline as well as greatly improved oxygen levels (up to +5 mL/L oxygen at the bottom). Abiotic conditions following a major inflow are favorable for cod recruitment, whereas mild winters benefit the sprat stock dynamics.

Ecological Interactions in the Central Baltic

The Bornholm Basin is one of the three deep basins forming the central Baltic Sea. Its ecosystem is of particular importance, as it currently forms the only available spawning ground for Eastern Baltic cod (*Gadus morhua*). Major species interactions take place, which define recruitment success with direct and indirect links to socioeconomics.

A typical, simplified annual cycle of key events in the Bornholm Basin can be described as follows (Fig. 17.2): In winter time (Fig. 17.2A), the Bornholm Basin is characterized by a permanent halocline at 50–75 m depth, which separates the less saline surface waters (salinity of 7–8) from the more saline bottom waters (salinity of 10–18) (Kullenberg and Jacobsen, 1981; Møller and Hansen, 1994). The abiotic environmental conditions (temperature, oxygen, salinity) below and within the halocline are mainly influenced by irregular saline water inflows (e.g., Matthäus and Lass, 1995).

In spring (Fig. 17.2B), a seasonal thermocline at approximately 20–30 m depth starts to form that is coupled to the onset of the spring phytoplankton bloom and subsequently leads to high zooplankton abundance. Historically, the peak spawning time of sprat (*Sprattus sprattus*) in the Baltic was in the spring. Due to their specific gravity, sprat eggs tend to accumulate in the upper portion

(A)　　　　　　　　　　　　**(B)**

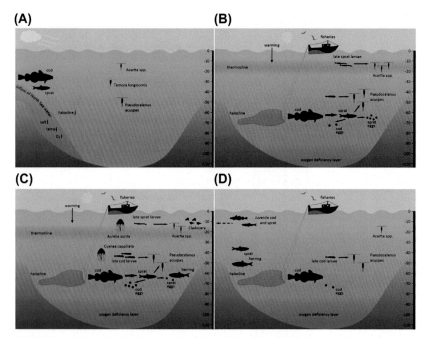

(C)　　　　　　　　　　　　**(D)**

FIGURE 17.2　Scheme of the Bornholm Basin including potential major ecological interactions, influencing fish stock dynamics: (A) winter, (B) spring, (C) summer, and (D) autumn.

of the halocline. Cod also begin its seasonal spawning in spring but exhibits peak spawning later in the early/mid-summer. Cod eggs are usually found deeper in the water column compared to sprat eggs, that is, below the permanent halocline. The region of the halocline is also the main feeding area of adult, planktivorous sprat, and thus the potential exists for adult sprat to consume both cod and sprat eggs (Köster and Möllmann, 2000).

After hatch, sprat, and cod larvae actively migrate to the upper water layers to feed (Voss, 2002), which greatly decreases the vertical overlap with feeding adults and so minimizes that potential source of predation mortality. In early spring, above-average fishing pressure can usually be observed, which is mainly the concentrations of prespawning cod.

During the summer (Fig. 17.2C), portions of the sprat population begin to migrate out of the deep basin toward their more coastal feeding grounds. Increasing abundances of spring-spawning herring (*Clupea harengus*) in the Bornholm Basin can be observed, as herring start to return from their coastal spawning grounds to feed. The thermocline gains in strength and is found in decreasingly shallower waters. The peak spawning time by cod occurs in the summer.

In autumn (Fig. 17.2D), the seasonal thermocline starts to resolve. Some cod spawning is still ongoing but large parts of all three fish populations (sprat, herring, cod) have left the deep basin toward their feeding grounds.

Cod–Herring–Sprat Fishery in Central Baltic Sea

Cod, herring, and sprat are the dominant fish species in the central Baltic Sea. The species are of high economic and ecologic importance and ecologically strongly interlinked (Fig. 17.3; Rudstam et al., 1994; Bagge et al., 1994; Kornilovs et al., 2001). Cannibalistic cod is a major predator on sprat and juvenile herring. Adult sprat and herring compete for zooplankton as food source, and prey heavily on pelagic cod and sprat eggs. Finally, adult sprat use the same main food source, that is, the copepod *Pseudocalanus acuspes*, as cod larvae. These feedbacks tend to stabilize the system, once a cod or sprat dominance has been reached, caused by unbalanced fishing.

Cod has been the most valuable commercial fish species in the Baltic Sea during the second part of the 20th century. Cod impacts humans directly through the productivity of the stock utilized by commercial and recreational fishers and indirectly by influencing the overall system productivity and food web dynamics. In the Baltic Sea, cod has traditionally been divided into two stocks: the Eastern stock found east of Bornholm Island and the Western stock (Bagge et al., 1994). Thus they are assessed and managed separately. The Eastern stock is larger in size and distribution and contributes more to the total European Union (EU) harvest.

The Baltic cod stock increased to very high levels in the late 1970s when fishing pressure was relatively low and hydrological conditions were advantageous for reproduction (Eero et al., 2011). The stock reached its peak in the 1980s (Eero et al., 2011), which was attributed to the high frequency of inflows

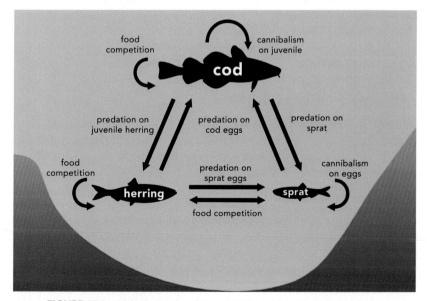

FIGURE 17.3 Major interactions between Baltic cod, herring, and sprat stocks.

from the North Sea resulting in good recruitment years. The rapid decline of the population began in the mid-1980s. Salinity and oxygen conditions were unfavorable for recruitment and fishing efforts remained high, partly due to improvements in harvest technology (MacKenzie et al., 2002; Eero et al., 2011). In addition, the waters supporting cod experienced increased eutrophication leading to hypoxia and increasing seal predation.

Spawning stock biomass of the central Baltic herring continuously decreased from 1974 to 2001. This was mainly explained by lower individual weights, a clear sign of food limitation. Total allowable catches (TACs) were reduced but still fishing mortality increased until the end of the 1990s. Since 2001, consistently low fishing mortality has led to increasing herring spawning biomass and biomass reached levels above ecological precautionary limits in 2006.

Sprat is the key prey for top predators (e.g., cod, harbor porpoise, *Phocoena phocoena*). At present, it also represents the most abundant, commercially exploited fish species in the Baltic Sea. During the previous two decades, the management of Baltic sprat has been challenged by large stock fluctuations mainly caused by highly variable recruitment success. These recruitment fluctuations are not directly coupled to sprat spawning stock biomass (Köster et al., 2003; MacKenzie and Köster, 2004) but appear to be driven by a suite of interacting environmental drivers (Voss et al., 2008).

Fluctuations in the recruitment strength of Baltic cod and sprat are also linked to larger-scale changes occurring in the Baltic Sea ecosystem. In the late 1980s, a regime shift occurred in the Baltic Sea, as shown by profound changes in the abundance and/or species composition of the fish and zooplankton assemblages (Alheit et al., 2005; Möllmann et al., 2009). This regime shift had cascading effects in the food web resulting in, among other things, decreased recruitment potential for cod (Köster et al., 2005). The sprat stock substantially benefited from the latter via decreased predation pressure. Sprat also experienced low rates of fishing mortality in the late 1980s to the early 1990s.

ECONOMICS OF THE FISHERY

The latest Annual Economic Report of the EU Fishing Fleet reports the 2013 value of landings generated by the EU Baltic Sea fleet as approximately 260 million €. In 2013, for the first time in history, herring (257,000 tons, 84 million €) and sprat (251,000 tons, 71 million €) landings were more valuable than cod landings (41,000 tons, 52 million €). The total volume landed including all species in 2013 was 586,000 tons (STECF, 2014).

Economically, the shift from a cod-dominated to a clupeid-dominated state in the late 1980s decreased the value of the catch due to the relative composition of the fish species. Increasing the cod stock and reducing sprat abundance would be economically more beneficial than the clupeid-dominated state (Döring and Egelkraut, 2008; Nieminen et al., 2012). Even so, currently sprat and herring landings increased in value by 29% and 19%, respectively, between 2012 and

2013, and cod landings decreased in value during the same period by 33%. This indicates a profound change in the economic baseline and challenges the widely accepted objective of a restoration of the cod stock.

The Eastern stock is mainly fished by Denmark, Sweden, and Poland. Profitability depends on the gear segment as well as vulnerability to the condition of the main target species (Blenckner et al., 2011). Secondary economic and social impacts include employment, retail, jobs, and income at the dockyards and work for local craftsmen. In many Baltic countries, fishermen have long fishing traditions and few job alternatives, so political decisions often imply the need to avoid the loss of jobs (Blenckner et al., 2011).

In 2013, there were 6256 vessels in the fishing fleet operating in the Baltic. These are divided into small-scale segment and large-scale segment, each with distinct economic performance. Small-scale vessels cover 88% of the boats but only account for 54% of total employment, and just 23% of revenue. The number of vessels deceased by 8% in the period 2008–2013. Overall, the performance of the large-scale fleet improved in recent years, whereas that of the small-scale sector deteriorated.

GOVERNANCE SYSTEM

The concept of the EU is an integrated Europe with common laws and regulations. This is in particular applicable to common pool resources such as fisheries. To coordinate the interests and fishing activities of countries participating in the Baltic fisheries, the International Baltic Sea Fisheries Commission was established in 1973 (Gdansk Convention, 1973). From 2005, the management of the Baltic fishery has fallen under the common fisheries policy (CFP) of the EU for all member states. The establishment of rules and laws under the CFP is a complex process, which involves political, scientific, and economic institutions (Fig. 17.4). The two most important bodies are the European Commission and the Council of the EU (Council of Ministers).

The Commission formulates draft laws and passes them on to the Council.

The Commission is supported in its work by scientific input from the International Council for the Exploration of the Sea, the Scientific, Technical and Economic Council of Fisheries of the EU, and the stakeholder input from the Baltic Sea Advisory Council (since 2004). The Council of the EU consists of 28 ministers of the EU member countries and passes all resolutions concerning the CFP.

The main instrument in CFP fisheries management is the TAC, which defines the amount of fish that is allowed to be caught from a specific stock in a year. TACs are usually set on an annual basis. The allocation of TACs between countries is based on the principle of relative stability, which states that each country receives a fixed share of the TAC (European Council, 2002; Churchill and Owen, 2010). Additional management instruments are technical specifications regarding effort and gear regulations (Churchill and Owen,

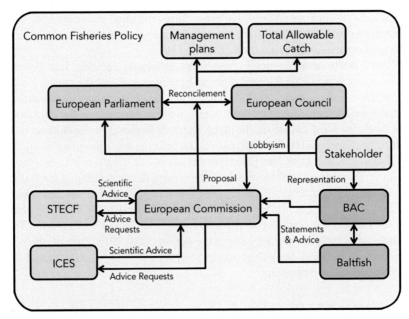

FIGURE 17.4 European common fisheries policy decision pathways.

2010), a seasonal fishing ban (Kraus et al., 2009) as well as a marine protected area to protect spawning fish (Suuronen et al., 2010).

The Baltic cod, herring, and sprat stocks used to be TAC managed. In 2007, the EU implemented a multiannual plan (MAP) for the Baltic cod stocks (European Council, 2007). In 2016, the Baltic cod MAP has been revised and a new multiannual plan has been launched that covers Baltic cod, herring, and sprat stocks. The main goals of both plans are the recovery of the stock biomass and a reduction of the fishing mortality rate by the implementation of a sustainable fishery based on maximum sustainable yield criteria (European Council, 2007; European Council, 2016.).

The CFP is the most relevant policy regarding fisheries. However, it is interrelated with other EU policies that affect the Baltic Sea environment and ecosystem, for example, the EU Maritime Strategy Framework Directive, the EU Habitat Directive, the EU Birds Directive, and the EU Blue Growth policy.

TRADE-OFFS IN BALTIC FISHERIES MANAGEMENT

The Baltic Sea could act as a practical example on how to advance fisheries management toward an approach in Europe (Fig. 17.5). To this end, we analyzed social–ecological trade-offs in a multispecies fisheries system: the trade-off between recovery of Baltic cod and the health of ecologically important forage fish stocks (herring and sprat), a challenge that we anticipate will be on the

FIGURE 17.5 Widening the scope of ecological–economy to solve imminent conservation problems and to achieve sustainable fisheries management: from a focused stock perspective to ecosystems and global markets.

near future agenda. Many of the cod stocks in the North Atlantic have suffered from overfishing and population collapse (Myers et al., 1997; Cook et al., 1997; Shelton and Lilly, 2000; Lindegren et al., 2009), with immense social and economic consequences. Moreover, decimated cod stocks have caused increases in forage fish populations (Frank et al., 2005; Casini et al., 2008). Besides being of direct commercial interest, forage fish species have an enormous indirect value as a primary food source for many marine top predators targeted by fisheries (Smith et al., 2011), as well as species of particular conservation and public concern, for example, marine mammals and birds (Cury et al., 2011).

Potential cod recovery raises two fundamental fisheries management questions involving trade-offs: (1) how much biomass and potential economic yield, provided by the high value cod stocks, needs to be sacrificed to allow for the protection of lower market value, but ecologically important, forage fish species, and (2) what are the additional costs of considering an equitable distribution of benefits between the demersal (cod) and pelagic (forage fish) fisheries sectors, given that the latter has usually (and especially in the Baltic Sea) expanded after the cod collapse?

A coupled ecological–economic optimization model framework can be used to contrast the profit-maximizing management solution for the entire multispecies fishery to two different management approaches. These focus on protecting the sprat stock for its ecological value, taking into account the secondary objectives of either profit maximization or considering equity between the demersal and pelagic fishing sectors. Finally, trade-offs exist on a regional (national) level, which arise from the principle of relative stability within the European CFP.

The combined three-species, age-structured ecological–economic model includes the predatory cod and the two forage fish species herring and sprat. The model is an extension of a single-species, age-structured fishery model by Tahvonen et al. (2013). Full detail of the model equations is given in Voss et al. (2014a,b).

Profit Maximization and the Risk of Forage Fish Stock Collapse

Simulations showed that a profit-maximizing, multispecies management strategy may indeed lead to a full recovery of the once depleted cod stock, with parent biomass reaching levels close to the historical maximum of ~700,000 tons (Fig. 17.6). The profit-maximizing solution revealed that a period of low fishing mortality (F), as presently observed, is necessary for the full recovery of the stock. Profit-maximizing multispecies harvesting would also result in a healthy and sustainable population size of herring.

In contrast to cod and herring, sprat would face the risk of stock collapse, as the equilibrium stock size would fall largely below precautionary or even limit reference points (Fig. 17.6). This outcome would be due to the higher market value of cod (compared to the forage species) that favors cod recovery and hence higher predation pressure, lower sprat biomass, and poor economic return to the forage fishing sector.

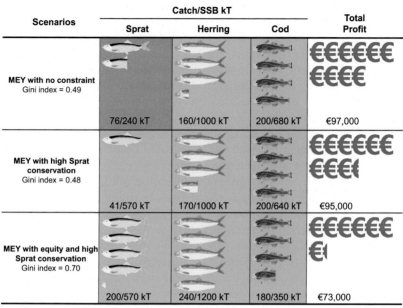

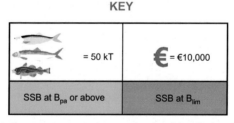

FIGURE 17.6 Baltic Sea management option overview, involving the three key target fish species sprat, herring, and cod, as well as three management scenarios.

Valuing Conservation Goals

Sprat has a key role in the Baltic Sea food web as prey for cod, marine mammals (Gårdmark et al., 2012), and birds (Österblom et al., 2006). Hence, depleting the sprat stock bears unforeseeable risks for ecosystem functioning, service provision, and protection of species with particular conservation concern. In economic terms, these are externalities that should be taken into account when designing socially reasonable policies. We evaluated the consequences of protecting the sprat stock for its ecosystem value by varying the social willingness to pay for parent biomass of sprat (the shadow price of the externality). The resulting relationships between sprat parent biomass and variables of the other two species represent efficiency frontiers, providing management options for the optimal delivery of conflicting services (Halpern et al., 2013; Polasky et al., 2008). To achieve sprat stock sizes corresponding to precautionary reference points, only a minor reduction of cod parent biomass would be necessary, that is, by 7% relative to the profit optimum of 680,000 tons (Fig. 17.6). Overall, this management strategy would cause a potential loss of profit for the combined Baltic Sea fishery amounting to ca. 2 M€, corresponding to 2.5% relative to the economically optimal management solution in the steady state.

Although this management strategy would only marginally affect cod and herring profits, the profit of the sprat fishery would collapse. Increasing sprat biomass in the steady state would need a drastic reduction of the sprat fishing mortality, causing potential sprat profit losses of 48%. The cod fishing sector would only lose 2.6% of its potential profit. The economically efficient solution to protect the sprat stock is a pronounced direct reduction of the fishing pressure on this lower market value forage fish species, in combination with a minor increase in fishing pressure directed toward its predator.

Although the pelagic herring fishery sector would actually benefit from this strategy, the sprat fishery would be marginalized, with sprat fishing license holders carrying almost the complete costs of the conservation effort. It is doubtful that such a management strategy would find acceptance by the presently expanded pelagic fishing sector, unless compensation payments are made between the different fisheries. However, a practical implementation of compensation schemes between fisheries is likely to be difficult or even infeasible.

Conservation Considering Equitable Resource Distribution

An alternative to apply an increasing value to the conservation of the sprat stock is to explore the consequences of an increasing equitable resource distribution between fishing sectors. We defined equity based on relative profits of the three interacting species using the Gini Index (Gini, 1921) and optimized the multispecies model for increasing equity levels (Fig. 17.6). Increasing equity corresponds to increasing fishing opportunities for sprat license holders and hence requires an increasingly larger sprat but a reduced cod stock. We found a slightly convex efficiency frontier (Lester et al., 2013) for this trade-off, that is, increasing equity to achieve sprat stock sizes would require a strong reduction of optimal cod stock sizes. However, these stock sizes are still above the present stock size, as well as above precautionary reference points. Overall increasing equity is positively linearly related to costs for the combined Baltic Sea fishery, which would amount to a loss of ca. 24 million € per year relative to the profit-maximizing multispecies solution.

Naturally, reduced profits of the high value cod fishery make up for most of the conservation costs inherent in the management strategy considering equity. Cod profit losses would amount to ca. 47% of the potential profit. The sprat fishing sector would achieve ca. 172% higher profits at sprat precautionary reference points, compared to the profit-maximizing multispecies solution, whereas the effect on herring profits would be negligible.

Increased equity between fishing sectors can only be achieved by a lowered predation pressure on sprat and hence a reduction of the cod stock due to a stronger fishing pressure.

Country-Specific Catch Portfolio

According to the relative stability principle, the Baltic countries hold fixed shares of cod, herring, and sprat quota. Therefore, the absolute catch values may differ between years, depending on the stock status, but not the percentage distribution of TACs to countries. Fig. 17.7 shows the country-specific catch portfolio for the year 2011. All Baltic countries are involved in all three fisheries, however, with highly variable distribution of species. The composition of each country's catch portfolio should determine its interest in (or opposition to) the plan to changing from a clupeid-dominated system back to a cod-dominated system.

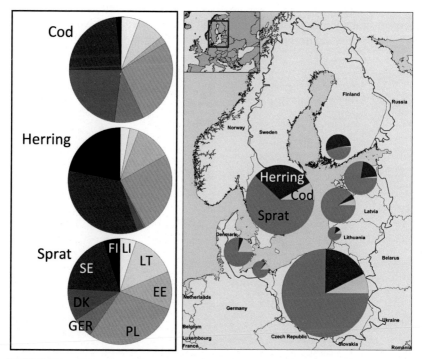

FIGURE 17.7 Country-specific catch portfolio showing each countries share of the cod, herring, and sprat quota according to the relative stability principle (left panel) and map of the Baltic Sea showing the resulting total catches in 2006 (right panel).

The country-specific economic outcome of two potential management schemes differs substantially: an open economic optimization that results in a cod-dominated system leads to overall higher profits. These are, however, mainly realized in the Western Baltic countries. Heading for (or continuing) a clupeid-dominated system will result in a more equal distribution of profits across countries (Fig. 17.8). For two countries, Finland (FI) and Estonia (EE), the return to a cod-dominated system seems economically questionable.

Ecosystem-based management (EBM) approaches to fisheries in the Baltic require model systems that account for multispecies trophic interactions and have the ability to link ecology and economy (Lindegren et al., 2009; Keller et al., 2011). Our multispecies model consequently challenges traditional single-species approaches because optimal, long-term stock sizes and profits are significantly smaller compared to species-by-species simulations.

The Baltic Sea represents a suitable case study for demonstrating the principles of trade-off evaluation in multispecies fisheries. The results confirm that triple-bottom-line management solutions are usually costly (Halpern et al., 2013). Protecting the sprat stock for its ecosystem value in an economic efficient way that disregards equity between fishing sectors would only have minor consequences for the cod stock and low costs for the overall multispecies fishery. The economically preferable approach would be to implement this management strategy together with a scheme of transfer payments that compensate the sprat fishery for forgone potential profits. Such a compensation scheme

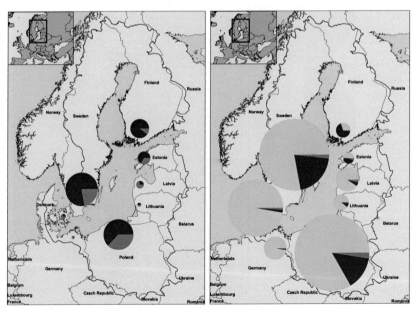

FIGURE 17.8 Country-specific yearly profits (three fisheries combined) when comparing "clupeid optimization" and "cod optimization."

could also address the potential conflicts on national level. If such a compensation scheme is not feasible, solving the emergent social conflict by achieving equity between fishing sectors would require to sacrifice a larger part of the cod stock and hence economic potential of the Baltic Sea fishery as a whole. However, a triple-bottom-line solution has the goal to maintain the Baltic Sea sprat stock at the recently determined precautionary biomass reference level while at the same time maintaining the equity level may provide a reasonable compromise, that is, a zone of "new consensus" (Hilborn, 2007) for the whole multispecies system. This management option minimizes the risk of forage fish overfishing and ensures the viability of the pelagic fishing sector. Although the cod fishing sector would lose a considerable amount of potential profit, our most equitable solution still allows for ongoing growth of the cod fishery, offering a potential win–win situation over all fishing sectors. However, it has to be noted that eventually important ecological detail is missing in the model (e.g., competition) and that critical evaluation using model ensemble approaches is warranted (Gårdmark et al., 2013). Additionally, such ensemble approaches might offer the opportunity to evaluate the effect of different fisheries management options on other parts of the ecosystem, for example, via trophic cascades.

Operationally applying ecological–economic model systems will facilitate coordinated management decisions among interacting use sectors and stakeholder involvement, which are critical components in EBM approaches, leading to increased societal values of exploited ecosystems (White et al., 2012). Through this approach, another aspect of equity, that is, participatory equity, can be addressed that increases the acceptability and hence compliance to management decisions (Halpern et al., 2013), a further footstep toward healthy fish stocks and sustainable fisheries in the world ocean.

Participatory Equity and Way Forward

Stakeholder relationships built on understanding and trust form a prerequisite to secure successful implementation of fisheries policies and regulations and to avoid unintended misguided incentives (Jentoft et al., 2009). Taking comments and feedback seriously are a prerequisite for a trusting and viable collaboration. This demands an "adaptive research agenda" approach to be able to restructure priorities, tracking the specified stakeholder priorities and needs. Furthermore, suitable, easy to understand visualization of modeling results is essential during transdisciplinary discussions. Scientific communication and dialogue should become an inherent part of "transdisciplinary" projects, with an appropriate budget allocated to support this key activity (Pascoe, 2006). During discussions, it became obvious that basic conservation goals, for example, the need for minimum stock sizes, are largely undisputed. However, it is the pathway taken toward sustainable use that is controversial. There are a number of "beyond-profit" interests in fisheries: enhanced stability and reduced uncertainty are key objectives that have been raised both from fishing industry and from conservation (i.e., environmental nongovernmental organization) representatives in the Baltic. In this context, discussions on long-term objectives and trade-off analysis are highly needed.

As the Baltic ecosystem and its associated fishery-related ecosystem, including its socioeconomic aspects, are frequently changing due to the influence of various human and naturally induced drivers (Möllmann et al., 2009), it is prudent to regularly review, revise, and adapt any multispecies multiannual management plan. Based on the Baltic experience, we assert that EBM must effectively incorporate the following four major components critical for its practical implementation: (1) diverse stakeholder involvement; (2) multispecies approaches; (3) social–ecological linkage; and (4) incorporation of environmental/ecological changes.

The success of ecosystem-based fishery management ultimately depends on the approach of implementation (Fig. 17.9). In the past, fishery management

FIGURE 17.9 The fisheries worlds of "command-and-control" and "incentive-based" management, which side will do better? *Courtesy by Daniel Freymüller.*

has relied to a large extent on technical regulations. Examples for such command-and-control measures include detailed gear prescriptions, restrictions on the days-at-sea spent fishing and vessel capacity, and minimum landing sizes. Fishery scientists, in particular, resource economists, have argued in favor of incentive-based approaches. Prominent examples include individual transferable quota systems and harvesting fees. For a system with multiple ecological and economic interactions, neither of the two approaches can ensure an ideal solution. A central future research question thus is which of the two approaches is more effective and efficient with respect to the objectives of EBM and how such a management should be deployed.

REFERENCES

Alheit, J., Möllmann, C., Dutz, J., Kornilovs, G., Loewe, P., Mohrholz, V., Wasmund, N., 2005. Synchronous ecological regime shifts in the central Baltic and the North Sea in the late 1980s. ICES Journal of Marine Science 62, 1205–1215.

Bagge, O., Thurow, F., Steffensen, E., Bray, J., 1994. The Baltic cod. Dana 10, 1–28.

Blenckner, T., Döring, R., Ebeling, M., Ho, A., Tomczak, M., Andersen, J., Kuzebski, E., Kjellstrand, J., Lees, J., Motova, A., Vatemaa, M., Vitanen, J., 2011. FishSTERN: A First Attempt at an Ecological-Economic Evaluation of Fishery Management Scenarios in the Baltic Sea Region. Naturvårdsverket, Stockholm.

Casini, M., et al., 2008. Multi-level trophic cascades in a heavily exploited open marine ecosystem. Proceedings of the Royal Society of London. Series B Biological Sciences 275, 1793–1801.

Churchill, R., Owen, D., 2010. The EC Common Fisheries Policy, first ed. Oxford University Press, Oxford.

Cook, R.M., Sinclair, A., Stefánsson, G., 1997. Potential collapse of North Sea cod stocks. Nature 385, 521–522.

Cury, P.M., et al., 2011. Global seabird response to forage fish depletion—one-third for the birds. Science 334, 1703–1706.

Döring, R., Egelkraut, T.M., 2008. Investing in natural capital as management strategy in fisheries: the case of the Baltic Sea cod fishery. Ecological Economics 64 (3), 634–642.

Eero, M., MacKenzie, B.R., Köster, F.W., Gislason, H., 2011. Multi-decadal responses of a cod (*Gadus morhua*) population to human-induced trophic changes, fishing, and climate. Ecological Applications 21 (1), 214–226.

European Council, 2002. Council Regulation (EC) No 2371/2002 on the Conservation and Sustainable Exploitation of Fisheries Resources under the Common Fisheries Policy.

European Council, 2007. Council Regulation (EC) No 1098/2007 Establishing a Multiannual Plan for the Cod Stocks in the Baltic Sea and the Fisheries Exploiting Those Stocks, Amending Regulation (EEC) No 2847/93 and Re-pealing Regulation (EC) No 779/97.

European Council, 2016. Council Regulation (EC) No 2016/1139 Establishing a Multiannual Plan for the Stocks of Cod, Herring and Sprat in the Baltic Sea and the Fisheries Exploiting Those Stocks, Amending Council Regulation (EC) No. 2187/2005 and Repealing Council Regulation (EC) No. 1098/2007.

Frank, K., Petrie, B., Choi, J., Leggett, W., 2005. Trophic cascades in a formerly cod-dominated ecosystem. Science 308, 1621–1623.

Gårdmark, A., et al., 2012. Does predation by grey seals (*Halichoerus grypus*) affect Bothnian Sea herring stock estimates? ICES Journal of Marine Science 69 (8), 1448–1456.

Gårdmark, A., et al., 2013. Biological ensemble modeling to evaluate potential futures of living marine resources. Ecological Applications 23, 742–754.

Gdansk Convention, 1973. Convention on Fishing and Conservation of the Living Resources in the Baltic Sea and the Belts.

Gini, C., 1921. Measurement of inequality and incomes. Economic Journal 31, 124–126.

Halpern, B.S., et al., 2013. Achieving the triple bottom line in the face of inherent trade-offs among social equity, economic return and conservation. Proceedings of the National Academy of Sciences of the United States of America 110 (15), 6229–6234.

HELCOM, 2015. Abundance of Coastal Fish Key Species. HELCOM Core Indicator Report. Online. Viewed 1st November 2016 http://www.helcom.fi/baltic-sea-trends/indicators/abundance-of-key-coastal-fish-species.

HELCOM, 2006. Assessment of coastal fish in the Baltic Sea. Baltic Sea Environment Proceedings No 103A.

Hilborn, R., 2007. Defining success in fisheries and conflicts in objectives. Marine Policy 31 (2), 153–158.

Jentoft, S., McCay, B.J., Wilson, D.C., 2009. Fisheries co-management: improving fisheries governance through stakeholder participation. In: Grafton, R.Q., Hilborn, R., Squires, D., Tait, M., Williams, M. (Eds.), Handbook of Marine Fisheries Conservation and Management. Oxford University Press, pp. 675–686.

Kellner, J.B., Sanchirico, J.N., Hastings, A., Mumby, P.J., 2011. Optimizing for multiple species and multiple values: tradeoffs inherent in ecosystem-based fisheries management. Conservation Letters 4 (1), 21–30.

Kornilovs, G., Sidrevics, L., Dippner, J.W., 2001. Fish and zooplankton interaction in the central Baltic Sea. ICES Journal of Marine Science 58, 579–588.

Köster, F.W., Möllmann, C., 2000. Egg cannibalism in Baltic sprat *Sprattus sprattus*. Marine Ecology: Progress Series 196, 269–277.

Köster, F.W., Möllmann, C., Neuenfeldt, S., Vinther, M., St John, M.A., Tomkiewicz, J., Voss, R., Hinrichsen, H.-H., Kraus, G., Schnack, D., 2003. Fish stock development in the central Baltic Sea (1976–2000) in relation to variability in the environment. ICES Journal of Marine Science 219, 294–306.

Köster, F.W., Möllmann, C., Hinrichsen, H.-H., Tomkiewicz, J., Wieland, K., Kraus, G., Voss, R., MacKenzie, B.R., Schnack, D., Makarchouk, A., Plikshs, M., Beyer, J.E., 2005. Baltic cod recruitment – the impact of climate and species interaction. ICES Journal of Marine Science 62, 1408–1425.

Kraus, G., Pelletier, D., Dubreuil, J., Möllmann, C., Hinrichsen, H.-H., Bas- tardie, F., Vermard, Y., Mahévas, S., 2009. A model-based evaluation of marine protected areas: the example of eastern Baltic cod (*Gadus morhua callarias* L.). ICES Journal of Marine Science 66 (1), 109–121.

Kullenberg, G., Jacobsen, T.S., 1981. The Baltic Sea: an outline of its physical oceanography. Marine Pollution Bulletin 12 (6), 183–186.

Lajus, J., Kraikovski, A., Lajus, D., 2013. Coastal fisheries in the eastern Baltic Sea (Gulf of Finland) and its basin from the 15 to the early 20th centuries. PLoS One 8 (10), e77059. http://dx.doi.org/10.1371/journal.pone.0077059.

Lester, S.E., et al., 2013. Evaluating tradeoffs among ecosystem services to inform marine spatial planning. Marine Policy 38, 80–89.

Lindegren, M., Möllmann, C., Nielsen, A., Stenseth, N.C., 2009. Preventing the collapse of the Baltic cod stock through an ecosystem-based management approach. Proceedings of the National Academy of Sciences of the United States of America 106, 14722–14727.

MacKenzie, B.R., Köster, F.W., 2004. Fish production and climate: sprat in the Baltic Sea. Ecology 85 (3), 784–794.

MacKenzie, B.R., Alheit, J., Conley, D.J., Holm, P., Kinze, C.C., 2002. Ecological hypotheses for a historical reconstruction of upper trophic level biomass in the Baltic Sea and Skagerrak. Canadian Journal of Fisheries and Aquatic Sciences 59 (1), 173–190.

Matthäus, W., Lass, H.U., 1995. The recent salt inflow into the Baltic Sea. Journal of Physical Oceanography 25, 280–286.

Møller, J.S., Hansen, I.S., 1994. Hydrographic processes and changes in the Baltic Sea. Dana 10, 87–104.

Möllmann, C., Diekmann, R., Müller-Karulis, B., Kornilovs, G., Plikshs, M., Axe, P., 2009. Reorganization of a large marine ecosystem due to atmospheric and anthropogenic pressure: a discontinuous regime shift in the Central Baltic Sea. Global Change Biology 15, 1377–1393.

Myers, R.A., Hutchings, J.A., Barrowman, N.J., 1997. Why do fish stocks collapse? The example of cod in Atlantic Canada. Ecological Applications 7, 91–106.

Nieminen, E., Lindroos, M., Heikinheimo, O., 2012. Optimal bioeconomic multispecies fisheries management: a Baltic Sea case study. Marine Resource Economics 27 (2), 115–136.

Österblom, H., Casini, M., Olsson, O., Bignert, A., 2006. Fish, seabirds and trophic cascades in the Baltic Sea. Marine Ecology: Progress Series 323, 233–238.

Pascoe, S., 2006. Economics, fisheries, and the marine environment. ICES Journal of Marine Science 63, 1–3.

Polasky, S., et al., 2008. Where to put things? Spatial land management to sustain biodiversity and economic returns. Biological Conservation 141, 1505–1524.

Rudstam, L.G., Aneer, G., Hildén, M., 1994. Top-down control in the pelagic Baltic ecosystem. Dana 10, 105–129.

Schaber, M., Haslob, H., Huwer, B., Harjes, A., Hinrichsen, H.-H., Köster, F.W., Storr-Paulsen, M., Schmidt, J.O., Voss, R., 2011. The invasive ctenophore *Mnemiopsis leidyi* in the central Baltic Sea: seasonal phenology and hydrographic influence on spatio-temporal distribution patterns. Journal of Plankton Research 33 (7), 1053–1065.

Shelton, P.A., Lilly, G.R., 2000. Interpreting the collapse of the northern cod stock from survey and catch data. Canadian Journal of Fisheries and Aquatic Sciences 57 (11), 2230–2239.

Smith, A.D.M., et al., 2011. Impacts of fishing low–trophic level species on marine ecosystems. Science 333, 1147–1150.

STECF, 2014. The 2013 Annual Economic Report on the EU Fishing Fleet. (Copenhagen).

Suuronen, P., Jounela, P., Tschernij, V., 2010. Fishermen responses on marine protected areas in the Baltic cod fishery. Marine Policy 34 (2), 237–243.

Tahvonen, O., Quaas, M.F., Schmidt, J.O., Voss, R., 2013. Effects of species interaction on optimal harvesting of an age-structured schooling fishery. Environmental and Resource Economics 54 (1), 21–39.

Voss, R., Quaas, M.F., Schmidt, J.O., Tahvonen, O., Lindegren, M., Möllmann, C., 2014a. Assessing social-ecological trade-offs to advance ecosystem-based fisheries management. PLoS One 9 (9), e107811. http://dx.doi.org/10.1371/journal.pone.0107811.

Voss, R., Quaas, M.F., Schmidt, J.O., Hoffmann, J., 2014b. Regional trade-offs from multispecies maximum sustainable yield (MMSY) management options. Marine Ecology Progress Series 498, 1–12.

Voss, R., Dickmann, M., Hinrichsen, H.-H., Floeter, J., 2008. Environmental factors influencing larval sprat *Sprattus sprattus* feeding in the Baltic Sea. Fisheries Oceanography 17 (3), 219–230.

Voss, R., 2002. Recruitment Processes in the Larval Phase: The Influence of Varying Transport on Cod and Sprat Larval Survival Ph.D. thesis. University of Kiel. 138pp.

White, C., Costello, C., Kendall, B.E., Brown, C.J., 2012. The value of coordinated management of interacting ecosystem services. Ecology Letters 15 (6), 509–519.

Chapter 18

Human Rights and the Sustainability of Fisheries

Sara G. Lewis[1], Aurora Alifano[1], Mariah Boyle[1], Marc Mangel[2]
[1]FishWise, Santa Cruz, CA, United States; [2]University of California Santa Cruz, Santa Cruz, CA, United States

INTRODUCTION

Capture fisheries and aquaculture production—like other industries—rely upon functioning environmental and social systems, and interdependencies between those systems mean human beings must be considered when addressing issues of environmental conservation. Although there can be legitimate disagreement about the level of a fished stock consistent with a well-managed fishery, the international norm is clearly that commercial industries, such as fisheries should not participate in slavery or other human rights abuses. Human rights and labor abuses within seafood supply chains have been exposed both on land and at sea. In this chapter we provide an overview of abuse aboard fishing vessels and use a simple bioeconomic model to illustrate the feedback between environmental degradation related to fishing activity and human rights. Following that, we discuss the intersection of international and national regulations of human rights and show that in the overlap there is considerable room for more policy development with respect to human trafficking, forced labor, and modern slavery[1] at sea. We close with a summary of public and private-sector initiatives that may help extend the reach of regulation beyond national boarders to reduce this worldwide problem.

---

1. The terms associated with modern slavery, forced labor, and human trafficking are likely unfamiliar to many readers of this volume, so we provide an appendix in which they are defined. In the case of fisheries, it is important to note that although human trafficking may lead to forced labor, not all forced labor is the result of human trafficking—indeed one of the greatest forced labor problems in fisheries occurs when migrants seeking better pay and working conditions are drawn into complicated debt systems that may force individuals to work for many years before receiving any income at all (Sylwester, 2014; p. 432ff).

HUMAN RIGHTS ABUSE IN SEAFOOD SUPPLY CHAINS

Until a few years ago, academics, journalists, and the media had documented relatively few instances of human trafficking or human rights abuses in the seafood sector, especially when compared to other industries (e.g., apparel, conflict minerals). In 2012, *Bloomsberg Businessweek* reported that fishermen on a South Korea-flagged ship were forced to work up to 30 h shifts in deplorable working conditions and subjected to physical and sexual abuse (Skinner, 2012). In 2014–15, investigations by *The Guardian* (Hodal et al., 2014), Associated Press (McDowell et al., 2015), and the *New York Times* (Urbina, 2015) further exposed the harsh realities of some seafood operations in mainstream media. These articles revealed evidence of human trafficking, forced labor, and other abuses occurring within some seafood operations, garnering widespread attention.

Although much of the research and media coverage of seafood-related human rights and labor abuses has focused on Thailand—where abuses have been exposed in both aquaculture and wild capture supply chains—a growing body of evidence is showing that the issues are not limited to developing countries. Human rights abuses can and do occur in developed countries with established and highly regarded fishery management systems (Kelbie, 2008; Simmons and Stringer, 2014; Stringer et al., 2011, 2016; Lawrence et al., 2015). For example, migrant fishing crew aboard South Korean vessels in New Zealand's waters were found to be victims of forced labor (Stringer et al., 2011, 2016; Skinner et al., 2012; Simmons and Stringer, 2014). Abuses have also been identified in the United Kingdom, where a legal loophole in EU transit rules enabled exploitation of Asian and African fishermen (Lawrence, 2015), as well as in Ireland (Lawrence et al., 2015) and Scotland (Kelbie, 2008), where migrant workers were recruited illegally on transport visas, charged fees for recruitment, and had their passports confiscated and their wages underpaid or unpaid.

Human rights and labor abuses in the fisheries sector can occur both on land and at sea, but some of the worst violations in the seafood industry have been reported to occur aboard fishing vessels employing migrant workers (ILO, 2013a). When cheap labor is scarce domestically, vessel owners and operators often turn to migrant workers. The use of unskilled migrant labor can reduce crew costs considerably for vessel owners who target these workers due to their willingness to accept low paying, dangerous, or temporary jobs (UNIAP, 2009; ILO, 2013a). Frequently migrant laborers find themselves without advocates or support networks, and because they are often isolated by language and cultural barriers, they are much more vulnerable to labor abuse and trafficking (ILO, 2013a). Even nationals who migrate within their country of residence can be targeted. For example, Thai men who migrate internally for work also risk being trafficked into labor exploitation on Thai fishing boats (US Department of State (USDOS), 2010).

Once aboard, crew can be contracted to a vessel or vessel owner for a period of months or years, regardless of whether they have any working knowledge of the fishing industry or desire to become fishers (EJF, 2013a). Case studies from the last decade cite examples of recruitment under false pretenses, 20 h workdays, child labor, physical and mental abuse, abandonment, and withholding of pay and identifying documents (Skinner, 2008, 2012; Surtees, 2008, 2012; Stringer et al., 2011; EJF, 2010, 2013b; Yea, 2014; ILRF, 2013; ILO, 2013a,b; USDOS, 2015). In one report, over half the victims interviewed reported seeing a fellow crew member murdered (UNIAP, 2009).

INTERRELATIONSHIP BETWEEN HUMAN RIGHTS AND ENVIRONMENTAL PROTECTION

Human rights and labor abuses such as those described earlier are clearly linked to numerous societal drivers (e.g., greed, corruption, cultural inequity, and global economic conditions, among others), but they are also rooted in environmental problems. For example, as overfishing has led to the decline of fish stocks closer to shore (FAO, 2012), vessels are traveling farther out to sea (ILO, 2013a). Fishing in more remote locations for longer periods increased fuel and operating costs for longer trips (facilitated at times by the use of transshipment vessels), and diminished catches provide ample incentive and opportunity to take advantage of the low risk of being caught when committing human rights abuses or other crimes such as illegal fishing. Economic pressures exacerbated by decreasing catch can also lead operators to cut corners with health and safety provisions aboard vessels (EJF, 2010, 2015; FAO, 2012; ILO, 2013a; Stringer et al., 2011). For example, there are reports of unhygienic working conditions, inoperable or complete lack of radio or fire safety equipment, substandard food, and poor accommodations (EJF, 2010).

The Lesson of Bioeconomics

Classical bioeconomic models (Gordon, 1954; Clark, 2010; Mangel, 2006) illustrate how the verbal arguments about human rights translate to predictions of sustainability for fisheries (cf. Brashares et al., 2014). In such models, one characterizes the biological dynamics and then economic dynamics by accounting for operational parameters (how easy it is to catch fish, q), costs c, and prices p.

The bionomic equilibrium (Gordon, 1954) corresponds to the situation in which effort is essentially uncontrolled and increases until in aggregate the rate of return from the fishery is 0. The population size at which this occurs is $N_b = c/pq$, which involves none of the biological parameters but rather the socioeconomic (c, p) and operational (q) parameters (see Mangel, 2006; p. 219).

As the right-hand side of this equation declines, for example, because costs c decrease, the left-hand side follows and may lead to a population size that is considered overfished. As the left-hand side declines, for example, because of overfishing, the right-hand side can decline in a variety of ways, but if p and q are fixed, then it will be through reducing costs c by legal (e.g., improved gear or fuel efficiency) or illegal means (e.g., by human rights and labor abuses such as forced labor, unfair wages, and long hours, or by not complying with regulations).

For a sole owner maximizing economic return, the consequence of the cost of fishing is that the economically optimal steady-state population size is often above the biologically determined maximum sustainable yield (MSY) level (Clark, 2010; Grafton et al., 2007; Mangel, 2006). According to the model, reducing costs pushes populations closer to MSY levels, but in a world of uncertain parameters and stochastic fluctuations, MSY is better as a limit than a target (Mangel et al., 2002) because it is too easy to have takes larger than MSY, leading to the decline of the stock, creating pressure to reduce costs, leading to the same kind of reinforcing cycle.

In summary, classical bioeconomics suggests that by driving down cost of fishing c, human rights and labor abuses (along with deregulation of fisheries, regulatory noncompliance, poor enforcement, uneven economic development, and other variables) can lead to environmental detriment. This conclusion illustrates the difficulty of relying on environmental policy alone to protect fishery resources from depletion.

LEGAL INSTRUMENTS ADDRESSING TRAFFICKING, FORCED LABOR, AND SLAVERY

Given the global and transboundary nature of many fisheries and seafood supply chains, international policy is an obvious place to look for human rights and labor protections for seafood workers. Between the 1920s and 2000, policy makers in the World Trade Organization (WTO), the United Nation's International Labor Organization (ILO), and the Food and Agriculture Organization adopted a range of voluntary measures, treaties, and conventions relevant to human rights and trafficking in fisheries, some of which include enforcement mechanisms (Table 18.1). However, implementation of international policy generally relies upon national-level policy adoption and enforcement. In this section, we briefly discuss how human rights in fisheries is (or is not) addressed within international conventions and national regulations, and point to some of the barriers to enforcement that hamper their effectiveness. We then discuss the jurisdictional challenges to national and international regulation of fisheries.

Fishing and International Labor Protections

International antislavery and antitrafficking laws in labor conventions date to the 1920s. In Fig. 18.1, we illustrate some of these conventions beginning

TABLE 18.1 International Conventions Relevant to Human Rights Abuses at Sea[a]

Type	Name	Provisions	Status and Ratifications
Convention—trafficking	United Nations Convention against Transnational Organized Crime (UNTOC)[b]	• Includes specific language regarding transnational trafficking and smuggling aboard fishing vessels • The UNTOC is the guardian of the protocol to prevent, suppress and punish trafficking in persons, especially women and children, and the protocol against the smuggling of migrants by land, sea, and air	Adopted—2000 Entry into force—2003 Ratifications: 187 US ratified: Yes
Convention—labor, fishing, maritime safety	International Labor Organization (ILO) work in Fishing Convention (no. 188)[c]	• Applies to all vessels engaged in commercial fishing and put responsibilities on vessel owners and skippers for ensuring crew health and safety • Crew must be old enough to work and should be provided rest, and sufficient wages, food, and medical care	Adopted—2007 Coming into force—2017 Ratifications—10 US ratified: No
Convention—labor, maritime safety	The International Labor Organization's (ILO) Maritime Labor Convention[d]	• International requirements for decent work for all seafarers • Title 5 under the convention: Outlines requirements for member state compliance and enforcement of provisions Fishing vessels exempt (see Article 2. Section 4)	Adopted—2006 Entry into force—2013 Ratifications—81 US not a party
Convention—maritime safety	(IMO) Convention for the Safety of Life at Sea (1974)[e]	• International treaty concerning safety of merchant and passenger ships • Outlines minimum standards for construction, equipment, and operation of ships Fishing vessels exempt (see Chapter 1. General provisions, Reg 3. Exceptions)	Adopted (first version)—1914 Adopted (1974 version)—1974 Entry into force—1980 Ratifications—163 US ratified: Yes

Continued

TABLE 18.1 International Conventions Relevant to Human Rights Abuses at Sea—Cont'd

Type	Name	Provisions	Status and Ratifications
Standard/ convention—fishing	(IMO) Standards of Training, Certification and Watchkeeping for Fishing vessel personnel (STCW-F Convention)[f]	• First treaty that mandates and normalizes standards of safety for crews of fishing vessels internationally through mandatory training standards • According to the IMO, it generally applies to sea-going fishing vessels 24 m in length or more	Adopted—1995 Entry into force—2012 Ratifications—20
Convention/ protocol—labor, maritime safety	(IMO) Torremolinos Convention, Torremolinos Protocol, and the Cape Town Agreement[g]	• Applies to new fishing vessels 24 m in length or more (see Chapter 1. General provisions, regulations 1. Application, in consolidated text)[h] • Cape Town Agreement allows administrations to give exemptions if viewed as unreasonable and impracticable (see Chapter 1. General provisions, regulations. 3. Exemptions, in consolidated text)[h]	Not yet entered into force US Ratified: No

[a]For the purpose of this table, we use the term "ratification" to include all parties to the treaty including signatures, ratifications, acceptances, approvals, and accessions.
[b]United Nations Office on Drugs and Crime (UNODC), 2016. Convention against Transnational Organized Crime and the Protocols Thereto. Available at: http://www.unodc.org/unodc/treaties/CTOC/.
[c]International Labor Organization (ILO), 2016. C – 188 Work in Fisheries Convention, 1997. Available at: http://www.ilo.org/dyn/normlex/en/f?p=NORMLEXPUB:12100:0::NO::P12100_ILO_CODE:C188.
[d]International Labor Organization (ILO), 2016. Compendium of Maritime Labour Instruments – second (revised) ed. Available at: http://www.ilo.org/global/publications/ilo-bookstore/order-online/books/WCMS_093523/lang--en/index.htm.
[e]International Maritime Organization (IMO), 2016. International Convention for the Safety of Life at Sea (SOLAS), 1974. Available at: http://www.imo.org/en/About/Conventions/ListOfConventions/Pages/International-Convention-for-the-Safety-of-Life-at-Sea-(SOLAS),-1974.aspx.
[f]International Maritime Organization (IMO), 2016. International Convention on Standards of Training, Certification and Watchkeeping for Fishing Vessel Personnel (STCW-F). Available at: http://www.imo.org/en/About/Conventions/ListOfConventions/Pages/International-Convention-on-Standards-of-Training,-Certification-and-Watchkeeping-for-Fishing-Vessel-Personnel-.aspx.
[g]International Maritime Organization (IMO), 2016. The Torremolinos International Convention for the Safety of Fishing Vessels. Available at: http://www.imo.org/en/About/Conventions/ListOfConventions/Pages/The-Torremolinos-International-Convention-for-the-Safety-of-Fishing-Vessels.aspx.
[h]The Torremolinos International Convention for the Safety of Fishing Vessels (Consolidated Text). A.15A, pp 183. Available at: https://www.parliament.nz/resource/en-NZ/51DBHOH_PAP68720_1/a3bce13917a069ca8b800deb29b36b4de3a4f071.

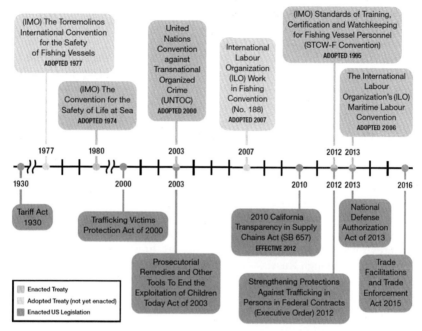

FIGURE 18.1 Timeline of national and international policy adoption relevant to trafficking and forced labor in fisheries.

with the Slavery Convention of 1926—a pre-UN convention formed through international multiparty legislation. King (2015) writes "these instruments laid the foundation for the contemporary conventions and efforts to eliminating trafficking." Unfortunately, despite the presence of numerous antitrafficking and forced labor conventions, the enforcement of their provisions aboard vessels specifically is a unique regulatory challenge.

The reach of treaties and conventions is limited by the requirement that they be ratified by the individual governments of each signatory nation. For instance, the ILO Forced Labor Convention, 1930 (No.29) has been adopted by 178 of 187 ILO members. It prohibits all forms of forced or compulsory labor and requires that ratifying states ensure that the relevant penalties imposed by law are adequate and strictly enforced. However, the United States, China, South Korea, and a handful of small Pacific island countries are among those nations that have not yet ratified the convention, leaving the crew of their fishing fleets less protected.

Another major contributor to the vulnerability of fishers at sea arises from the exclusion of fishing vessels and personnel from key provisions within maritime safety standards and international conventions and size limits aimed at large vessels tied to these requirements (Petursdottir et al., 2001; Simmons and Stringer, 2014). Fishing vessels are excluded from the majority of provisions in the Convention for Safety of Life at Sea, the International Convention on Standards of Training, Certification, and Watchkeeping for Seafarers, and the Marine Labor Convention. A notable exception that applies to all fishers and fishing vessels engaged in commercial fishing operations is the Work in Fishing Convention (No.188), which aims to ensure that fishers have decent conditions of work on board fishing vessels with regard to minimum requirements for work on board; conditions of service; accommodation and food; occupational safety and health protection; medical care and social security. The Convention may come into force 12 months after it has been ratified by 10 states, 8 of which must be coastal countries.

Human trafficking also continues to occur on a global scale, despite several international treaties aimed at combating it (USDOS, 2015). One contributing factor may be that the implementation and enforcement of treaties is largely left to the ratifying countries, yet accountability or punitive measures for countries that stop complying are limited (Gallagher, 2010). For example, although Russia—a top fisheries production and export nation (FAO, 2012)—ratified the UN Protocol to Prevent, Suppress and Punish Trafficking in Persons, Especially Women and Children, and the Protocol Against the Smuggling of Migrants by Land, Sea and Air, there are multiple reports of trafficking and forced labor practices aboard Russian fishing vessels (Surtees, 2008, 2012). In fact, due to its failure to comply with even the "minimum standards for the elimination of trafficking" (USDOS, 2013a), Russia was downgraded to the lowest possible level (Tier 3) in the USDOS (2013a) Trafficking in Persons Report and remains Tier 3 in 2016.[2] Indeed, in many cases ratifying nations have little recourse beyond political pressure or trade sanctions to ensure that treaties are enforced. These political and economic tools are primarily used to tackle the most egregious forms of abuse (e.g., slavery, human trafficking, child labor, forced labor, and workplace violence) and do not address the full spectrum of labor violation (e.g., unfair wages, worker safety, unionization, harassment). Antitrafficking policies have been useful for drawing attention and acting on criminal activities in labor recruitment and abuse but more is needed to address the underlying causes of forced labor in fisheries, including improvements in migration management and addressing the vulnerabilities of migrant workers (Marschke and Vandergeest, 2016).

---

2. The US State Department's Trafficking in Persons Report assesses how well governments are addressing and responding to the crime(s) of human trafficking. Tier three designations are assigned to countries whose governments do not fully comply with Trafficking Victims Protection Act (TVPA) standards and are not making significant efforts to do so. There are currently 27 countries and territories with a Tier three designation, including North Korea, Iran, and Saudi Arabia (USDOS, 2016).

National Policy Efforts to Address Human Rights

In many cases, national level regulations exist to protect workers from abuse but many nations still have considerable work ahead to improve protections for fishers. For instance, in response to harsh criticism Thailand has begun to overhaul its fisheries legislation and management (Ministry of Foreign Affairs of Thailand, 2015) and enact regulatory changes designed to register migrant workers and combat forced labor in the fisheries sector. With improved policies in place, the Thai government aims to implement their fisheries management plan and impose criminal and administrative sanctions for those that violate fisheries and labor-related laws. In another example, New Zealand's Fisheries (Foreign Charter Vessels and Other Matters) Amendment Act (14/60) requires that all foreign-flagged vessels operating within New Zealand's exclusive economic zone (EEZ) are reflagged to New Zealand (NZ Parliament, 2014). This change in flag means that the vessels and crew are accountable to New Zealand health, safety, and labor criminal laws. Ireland also recently overhauled its system for documenting and protecting fishing crew aboard Irish vessels from labor abuse following a Guardian report (Lawrence et al., 2015). They have introduced a system that includes provisions for clear contracts, minimum pay, and terms and conditions that are enforceable in Irish and EU law (Interdepartmental Government Task Force, 2015).

THE LIMITED REACH OF NATIONAL AND INTERNATIONAL ENFORCEMENT

Unfortunately, enforcing regulations in fisheries—both national or multinational—is not generally straightforward. Responsibility for prosecuting a given violation could require action by several local and national regulatory bodies within both the country of harvest and the vessel's flag nation. It can be challenging to enforce domestic or international laws on another country's vessels due in part to customary international law pertaining to the limits of national regulatory domains. Under the UN Convention on the Law of the Sea (UNCLOS), countries are typically unable to enforce many regulations or treaties on foreign flagged vessels outside of their 12 nautical mile territorial sea, and they have only limited power to police vessels operating within their EEZ (12–200 nautical miles). The broad-scale ratification of UNCLOS—167 nations including the EU as of 2017—means that even a state that is not party to the convention (e.g., the United States) must recognize UNCLOS as customary international law. The limits of domain are particularly problematic for monitoring distant water fleets fishing on the high seas, outside the customary jurisdiction of nation states. In these fisheries, vessels are only held to the laws of their flag nation and those of applicable regional fisheries management organizations (RFMOs).

Another enforcement challenge arises when countries attempt to monitor vessels fishing under their flag in foreign waters. Fishing vessels can often go for long periods without coming into port or reentering their flag country, and monitoring a global fishing fleet can be very resource intensive, and the flags of countries that do not properly monitor their fleets can be exploited by those engaging in illegal and unethical labor or fishing practices. So-called "flags of convenience," issued by poorly monitored flag states, shift the enforcement burden over to the harvest and port states. For example, though New Zealand is a country with some of the best managed fisheries in the world from an ecological standpoint (Ministry of Primary Industries, 2016), until recently their domestic agencies failed to recognize the deception, exploitation, and coercion occurring aboard some foreign charter vessels in New Zealand waters (Stringer et al., 2011, 2016; Simmons and Stringer, 2014).

PROMISING AVENUES FOR CHANGE

Given the limited reach of most national regulations and treaties relevant to human rights and trafficking, some of the most exciting and creative regulations for protecting workers at sea have placed greater reporting and enforcement responsibilities on private sector businesses. For example, the California Transparency in Supply Chains Act (2010) requires large retailers and manufacturers doing business in the state of California to disclose on their web sites their efforts to eradicate slavery and human trafficking from their direct supply chains of goods offered for sale. It also requires disclosure concerning product supply chains, supplier audits and certifications, and internal accountability (USDOL, 2016).

The UK Modern Slavery Act (2015) is even more stringent than the California Transparency in Supply Chains Act. It requires that as of March 31, 2016, commercial organizations above a threshold size must prepare a slavery and human trafficking statement each year and indicate steps taken during the applicable year to ensure that human trafficking is not occurring in the supply chain or business. Topics requiring disclosure include the organizational structure, business model and supply chain relationship, applicable policies, due diligence and auditing processes, human trafficking risks, and steps taken to assess and mitigate risk, compliance effectiveness, and training.

Transparency regulations have numerous advantages for addressing human rights in fisheries. First, they take advantage of the transnational nature of many seafood companies—harvesting from multiple EEZs or across RFMOs on the high seas. Making these companies become agents of labor change could help regulators get around the limitations to national regulatory domain by allowing the countries buying the seafood products to influence fishing activities occurring beyond their boarders and outside of their flagged fleets. Second, by not being overly prescriptive about how companies approach the task of monitoring and addressing human rights and trafficking risks in their specific supply chains, this kind of regulation could provide companies with the flexibility to develop their own tailored and innovative approaches. Policy making can be very slow—particularly at the international level where it is hindered by the need for national level adoption and ratification of treaties—whereas private businesses can implement changes much more rapidly. Third, improving transparency allows nongovernment individuals to fact check information and encourage laggards to improve.

Unfortunately, to date many of these efforts still fall short regarding enforcement. For example, the California Transparency in Supply Chains Act has only resulted in one court opinion since its passage (Barber v. Nestle USA), where it was ruled that companies do not need to actively improve labor practices, only to disclose their company policies (US District Court, 2015). In addition, from 1932 until 2016 there were only 40 enforcement actions were taken under the United States; Tariff Act of 1930 (Bajaj, 2015). Even so, such requirements are promising options for expanding the reach of domestic policy far beyond national borders and are beginning to address the jurisdictional challenge facing other types of country-based public regulation.

The Role of Industry

Although the primary duty to protect human rights remains with national governments, companies have an internationally recognized responsibility to respect human rights in their operations—as was underscored by the United Nations Human Rights Council in 2011 in their "Guiding Principles on Business and Human Rights" (UN, 2011). Seafood companies are increasingly incorporating human rights in sustainability initiatives into their business plans, in part due to the benefits of mitigating the legal and reputational risks associated with human rights violations or illegality. An increasing number of companies are recognizing the need for an expanded vision of sustainable seafood that embraces social, environmental, and economic considerations and hoping to meet consumer demands while gaining positive press and visibility for their brand and product (Boyle, 2014; FishWise, 2016). Many are voluntarily investing in traceability and monitoring technology (e.g., Bumble Bee), committing to respect human rights (e.g., Nestlé), and working in partnership with governments, international organizations, and civil society initiatives to promote sustainable seafood (Safeway and Fair Trade Seafood). Still others struggle with new concerns about monitoring and enforcing working conditions and labor standards throughout the complex supply chains of multinational companies around the world. Clearly, opportunities exist for regulators and private industry to work together to address human rights abuses as neither can solve the problem alone. For example, Issara Institute, a public–private–social partnership, brings together a wide range of global brands, retailers, nongovernmental organizations, academics, and technical experts to investigate and resolve labor issues in export-oriented seafood supply chains and progressive companies have joined the Seafood Task Force to precompetitively address issues of collective concern and communicate with local government. Whether through regulations like the UK Modern Slavery Act and the Trade Facilitation and Trade Enforcement Act of 2015 (H.R. 644) or through voluntary measures, industry initiatives must incorporate greater transparency and traceability within seafood supply chains, genuine worker feedback and representation, and remedies for victims.

DISCUSSION AND CONCLUSION

There is progress, represented by the adoption of policies, the rise of voluntary initiatives, and the expansion of legal and voluntary frameworks steering the international community toward social responsibility. However, even with the growing number of government and industry initiatives, workers remain vulnerable, documentation of significant concerns continue, and additional mechanisms for accountability are still urgently needed. In addition to working with businesses, national and international policy makers need to work on mainstreaming human rights into environmental and trade considerations. Dalal-Clayton and Bass (2009, p. 11), writing in the context of development, describe environmental mainstreaming as the "informed inclusion of relevant environmental concerns into the decisions of institutions that drive national, local and sectorial development policy, rules, plans, investment and action." In the long-term, management of human rights abuses in seafood will benefit from environmental mainstreaming in human rights law and treaties and human rights mainstreaming in environmental and trade law and treaties (Sylwester, 2014).

Given the international nature of fisheries supply chains, trade law could become another crucial tool for combatting human trafficking and labor abuse. The WTO has focused on promoting free trade for many years, but in almost five decades there was barely a reference to the notion of human rights in the General Agreement on Tariffs and Trade (GATT, 1947) system. Although the WTO has recently collaborated on several documents relevant to human rights with UN organizations, each body has a unique mandate that can lead them to prioritize issues differently and to prefer different policy approaches (Bartels, 2009). Indeed, human rights and labor protections may conflict with trade liberalization policies (Bartels, 2009, p. 593). Human rights institutions (e.g., the ILO and other UN agencies) have recognized that trade liberalization may lead to human rights violations (Bartels, 2009), and in 2002 and 2003 the UN Commission on Human Rights produced two reports (UN, 2002, 2003) on globalization and human rights. Reflecting on these reports, Bartels (2009) notes that trade liberalization and human rights—much like environmental policy and human rights—share the common goal of bettering the human condition by improving understanding between nations and improving human welfare. Thus although the WTO is a less obvious source of human rights policy than the ILO, human rights protections may yet emerge through trade liberalization.

There are also opportunities to link international environmental law with the laws of human and labor rights and human trafficking. Historically, if humans were considered at all in international and domestic environmental law the focus was solely on protections for health, children, standard of living, cultural dignity, safety, and the pursuit of social and economic development (Birnie et al., 2009, pp. 271–272; Sands et al., 2012, p. 780), and not explicitly on ensuring free labor or preventing trafficking. For many years some have claimed that degraded environments could be seen as a violation of both individual and collective human rights (Birnie et al., 2009, p. 271, also see Fitzmaurice, 2010, p. 623ff). Our bioeconomic model also implies that human rights violations and their affiliated drivers could lead to environmental degradation, and thus a purely environmental approach to fishery policy—one that does not account for human rights protection—may not be enough to ensure the sustainability of fishery resources.

There is much to be done and indeed not a minute to be lost.

APPENDIX: A GLOSSARY OF HUMAN RIGHTS TERMS

Bonded labor or Debt bondage The use of a bond, debt, or other threats of financial harm as a form of coercion for the purpose of forced labor or services or practices similar to slavery or servitude. Some workers inherit debt; others fall victim to traffickers or recruiters who unlawfully exploit an initial debt assumed as a term of employment (USDOS, 2014).

Child labor The International Labor Organization's (ILO) Website defines "child labor" as "work that deprives children of their childhood, their potential and their dignity, and that is harmful to physical and mental development." This includes work that is "mentally, physically, socially or morally dangerous and harmful to children; and interferes with their schooling by depriving them the opportunity to attend school." The ILO's Worst Forms of Child Labor Convention (No. 182) considers a "child" to be any person under the age of 18 (ILO, 1999b).

Forced labor Forced labor, sometimes also referred to as labor trafficking, encompasses the range of activities—recruiting, harboring, transporting, providing, or obtaining—involved when a person uses force or physical threats, psychological coercion, abuse of the legal process, deception, or other coercive means to compel someone to work. Once a person's labor is exploited by such means, the person's prior consent to work for an employer is legally irrelevant: the employer is a trafficker and the employee a trafficking victim (USDOS, 2016).

Human rights The rights people are entitled to simply because they are human beings, irrespective of their citizenship, nationality, race, ethnicity, language, gender, etc. This term refers to the UN Universal Declaration of Human Rights which lists 30 articles defining those rights, including that "all humans are born free and equal…have a right to life, liberty and security of person…shall not be held in slavery or servitude…everyone has a right to leave any country…everyone has the right to work, to free choice of employment, to just and favourable conditions of work and to protection against unemployment" (UN, 1948).

Human trafficking (trafficking in persons) The act of recruiting, harboring, transporting, providing or obtaining a person for compelled labor or commercial sex through the use of force, fraud, or coercion (USDOS, 2014). Trafficking victims can include individuals born into servitude, exploited in their hometown, or smuggled to the exploitative situation as well as individuals who previously agreed to work for a trafficker or participated in a crime as a result of being trafficked (USDOS, 2013b). At the core of this issue is the traffickers' intention to exploit or enslave another human being, and the coercive, underhanded practices they engage in to do so (USDOS, 2013b). The international definition set forth by the United Nations (UN) Office on Drugs and Crime (ODC) defines Trafficking in Persons as "the recruitment, transportation, transfer, harboring or receipt of persons, by means of the threat or use of force or other forms of coercion, of abduction, of fraud, of deception, of the abuse of power or of a position of vulnerability or of the giving or receiving of payments or benefits to achieve the consent of a person having control over another person, for the purpose of exploitation" (UNODC, 2013).

Labor rights Labor rights refer to a broader category of issues than trafficking or modern slavery. The International Labor Organization's (ILO) "Declaration of the Fundamental Principles and Rights at Work" places these rights into core standards: freedom of association, right to collective bargaining, prohibition of forced labor, elimination of the worst forms of child labor, and nondiscrimination in employment (ILO, 1999a,b). ILO has adopted 184 Conventions that establish standards for a range of workplace issues including (but not limited to) Weekly Rest, Forced Labor, Hours of Work, Minimum Wage, Safety and Health, Rights of Rural Workers, Migrant Labor Protections, and Workers' Compensation.

Modern slavery This is a general term often used when referring to holding a person in compelled service, including trafficking, forced labor, involuntary servitude, and bonded labor (USDOS, 2013b).

Smuggling of migrants The United Nations Convention against Transnational Organized Crime defines "smuggling of migrants" as "the procurement, in order to obtain, directly or indirectly, a financial or other material benefit, of the illegal entry of a person into a State Party of which the person is not a national or a permanent resident" (UN, 2001).

UN Guiding Principles on Business and Human Rights (Ruggie Principles) The United Nations Human Rights Council endorsed a set of Guiding Principles for Business and Human Rights designed to provide a global standard for preventing and addressing the risk of adverse impacts on human rights linked to business activity. They set out, in three pillars, principles concerning the State duty to protect human rights, the corporate responsibility to respect human rights, and access to remedy for victims of human rights abuse. The "corporate responsibility to respect" exists independently of States' abilities or willingness to fulfill their own human rights obligations. The Guiding Principles require that companies have a policy commitment to respect human rights, and proactively take steps to prevent, mitigate and, where appropriate, remediate, their adverse human rights impacts. These Guiding Principles apply to all States and to all business enterprises, both transnational and others, regardless of their size, sector, location, ownership, and structure.

ACKNOWLEDGMENTS

We thank Christina Stringer and Melissa Marschke for their thoughtful feedback and two anonymous reviewers and Phil Levin for comments on a previous version of the manuscript.

REFERENCES

Bajaj, V., 2015. A Much-Needed Rule on Force Labor. New York Times. May 15, 2015 http://takingnote.blogs.nytimes.com/2015/05/15/a-much-needed-rule-on-forced-labor/.

Bartels, L., 2009. Trade and human rights. In: Bethlehem, D. (Ed.), The Oxford Handbook of International Trade Law. Oxford University Press, Oxford, UK, pp. 571–596.

Birnie, P., Boyle, A., Redgwell, C., 2009. International Law and the Environment, third ed. Oxford University Press, Oxford.

Boyle, M., 2014. Trafficked II: An Updated Summary of Human Rights Abuses in the Seafood Industry. FishWise. Available at: http://fishwise.org/images/pdfs/Trafficked_II_FishWise_2014.pdf.

Brashares, J., Abrahms, B., Fiorella, K.J., Golden, C.D., Hojnowski, C.E., Marsh, R.A., McCauley, D.J., Nuñez, T.A., Seto, K., Withey, L., 2014. Wildlife decline and social conflict. Science 345, 376–378.

Clark, C.W., 2010. Mathematical Bioeconomics, third ed. Wiley, New York.

Dalal-Clayton, B., Bass, S., 2009. The Challenges of Environmental Mainstreaming: Experience of Integrating Environment into Development Institutions and Decisions. Environmental Governance No. 3. International Institute for Environment and Development, London. Available at: http://pubs.iied.org/pdfs/17504IIED.pdf.

Environmental Justice Foundation (EJF), 2010. All at Sea: The Abuse of Human Rights Aboard Illegal Fishing Vessels. Available at: http://ejfoundation.org/sites/default/files/public/media/report-all%20at%20sea_0.pdf.

Environmental Justice Foundation (EJF), 2013a. Sold to the Sea: Human Trafficking in Thailand. Available at: http://ejfoundation.org/sites/default/files/public/Sold_to_the_Sea_report_lo-res-v2.pdf.

Environmental Justice Foundation (EJF), 2013b. The Hidden Cost: Human Rights Abuses in Thailand's Shrimp Industry. Available at: http://ejfoundation.org/sites/default/files/public/shrimp_report_v44_lower_resolution.pdf.

Environmental Justice Foundation (EJF), 2015. Pirates and Slaves: How Overfishing in Thailand Fuels Human Trafficking and the Plundering of Our Oceans. Environmental Justice Foundation, London. Available at: http://ejfoundation.org/report/pirates-and-slaves-how-overfishing-thailand-fuels-human-trafficking-and-plundering-our-oceans.

Fitzmaurice, M., 2010. Environmental degradation. In: Moeckli, D., Shah, S., Sivakumaran, S. (Eds.), International Human Rights Law. Oxford University Press, Oxford, pp. 622–642.

FishWise, 2016. Social Responsibility in the Global Seafood Industry: Background and Resources. Available at: http://www.fishwise.org/traceability/Social_Responsibility_White_Paper.

Food and Agriculture Organization of the United Nations (FAO), 2012. State of the World Fisheries and Aquaculture. FAO Fisheries & Aquaculture Department. Available at: http://www.fao.org/docrep/016/i2727e/i2727e00.htm.

Gallagher, A.T., 2010. The International Law of Human Trafficking. Cambridge University Press, Cambridge.

GATT (General Agreement on Trade and Tariffs), 1947. Available at: https://www.wto.org/english/docs_e/legal_e/gatt47_e.pdf.

Gordon, H.S., 1954. The economic theory of a common property resource: the fishery. Journal of Political Economy 62, 124–142.

Grafton, R.Q., Kompas, T., Hilborn, R.W., 2007. Economics of overexploitation revisited. Science 318, 1601.

Hodal, K., Kelly, C., Lawrence, F., 2014. Revealed: Asian Slave Labour Producing Prawns for Supermarkets in US, UK. The Guardian. Available at: https://www.theguardian.com/global-development/2014/jun/10/supermarket-prawns-thailand-produced-slave-labour.

H.R. 644 — 114th Congress: Trade Facilitation and Trade Enforcement Act of 2015. Available at: www.GovTrack.us. 2015. February 23, 2017. https://www.govtrack.us/congress/bills/114/hr644.

Interdepartmental Government Task Force, 2015. Report of the Government Task Force on Non-EEA Workers in the Irish Fishing Fleet. Government of Ireland. Available at: https://www.agriculture.gov.ie/media/migration/publications/2015/TaskForceReport141215.pdf.

International Labor Rights Forum (ILRF) and Warehouse Workers United (WWU), 2013. The Walmart Effect: Child and Worker Rights Violations at Narong Seafood, Thailand's Model Shrimp Processing Factory. Briefing Paper. Available at: http://www.laborrights.org/sites/default/files/publications/The_Walmart_Effect_-_Narong_Seafood.pdf.

International Labour Organisation (ILO), December 1999a. Tripartite Meeting on Safety and Health in the Fishing Industry. Geneva, pp. 13–17.

International Labour Organisation (ILO), 1999b. Worst Forms of Child Labour Convention (No. 182). Available at: http://www.ilo.org/dyn/normlex/en/f?p=NORMLEXPUB:12100:0::NO:12100:P12100_ILO_CODE: C182.

International Labour Organisation (ILO), 2013a. Caught at Sea: Forced Labour and Trafficking in Fisheries. Available at: http://www.ilo.org/global/topics/forced-labour/publications/WCMS_214472/lang–en/index.htm.

International Labour Organisation (ILO), 2013b. Employment Practices and Working Conditions in Thailand's Fishing Sector/International Labour Organization Office Southeast Asia Regional Office. ILO, Bangkok. http://www.ilo.org/wcmsp5/groups/public/—asia/—ro-bangkok/documents/publication/wcms_220596.pdf.

Kelbie, P., December 13, 2008. Scottish Fishermen Accused of Exploiting Migrant Boat Crew. The Guardian. https://www.theguardian.com/uk/2008/dec/14/immigration-fishing-scotland-filipinos.

King, L., 2015. International Law and Human Trafficking. Topical Research Digest: Human Rights and Human Trafficking. Available at: https://scholar.google.com/scholar?hl=en&q=King+L.%2C+International+Law+and+Human+Trafficking.+Topical+Research+Digest%3A+Human+Rights+and+Human+Trafficking&btnG=&as_sdt=1%2C5&as_sdtp= http://www.du.edu/korbel/hrhw/researchdigest/trafficking/InternationalLaw.pdf.

Lawrence, F., August 17, 2015. The Exploitation of Migrants Has Become Our Way of Life. The Guardian. Available at: https://www.theguardian.com/commentisfree/2015/aug/17/exploitation-migrants-way-of-life-immigration-business-model.

Lawrence, F., McSweeney, E., Kelly, A., Heywood, M., Susman, D., Kelly, D., Domokos, J., November 2, 2015. Revealed: Trafficked Migrant Workers Abused in Irish Fishing Industry. The Guardian. http://www.theguardian.com/global-development/2015/nov/02/revealed-trafficked-migrant-workers-abused-in-irish-fishing-industry.

Mangel, M., 2006. The Theoretical Biologist's Toolbox. Cambridge University Press, Cambridge.

Mangel, M., Marinovic, B., Pomeroy, C., Croll, D., 2002. Requiem for Ricker: unpacking MSY. Bulletin of Marine Science 70, 763–781.

Marschke, M., Vandergeest, P., 2016. Slavery scandals: unpacking labour challenges and policy responses within the off-shore fisheries sector. Marine Policy 68, 39–46.

McDowell, R., Mason, M., Mendoza, M., 2015. AP Investigation: Are Slaves Catching the Fish You Buy? The Associated Press. Available at: https://www.yahoo.com/news/ap-investigation-slaves-catching-fish-buy-012742683.html?ref=gs.

Ministry of Foreign Affairs of Thailand, 2015. Thai Government Approves New Fisheries Legislation and Major Plans to Combat IUU Fishing and Trafficking in Persons in Fisheries. Available at: http://www.thaiembassyuk.org.uk/?q=node/573.

Ministry of Foreign Affairs of Thailand, 2016. Thailand's Reform of Fishing License Regime. Department of European Affairs. Available at: http://www.thaiembassy.org/warsaw/en/information/64887-Thailand's-Reform-of-Fishing-License-Regime.html.

Modern Slavery Act, 2015. United Kingdom Parliament. Available at: http://www.legislation.gov.uk/ukpga/2015/30/notes/contents.

New Zealand (NZ) Parliament, 2014. Fisheries (Foreign Charter Vessels and Other Matters) Ammendment Bill 2012. Bills Digest No. 2070. Available at: https://www.parliament.nz/en/pb/bills-and-laws/bills-proposed-laws/document/00DBHOH_BILL11820_1/fisheries-foreign-charter-vessels-and-other-matters-amendment#.

Petursdottir, G., Hannibalsson, O., Turner, J.M., 2001. Safety at Sea as an Integral Part of Fisheries Management. FAO Fisheries Circular. No. 966 FAO, Rome. Available at: http://www.fao.org/docrep/003/x9656e/x9656e00.htm.

Sands, P., Peel, J., Fabra, A., MacKenzie, R., 2012. Principles of International Environmental Law. Cambridge University Press, Cambridge.

Simmons, G., Stringer, C., 2014. New Zealand's fisheries management system: forced labour an ignored or overlooked dimension? Marine Policy 50, 74–80.

Skinner, E.B., 2008. A Crime So Monstrous. Free Press, New York. Quoted in Osgood, S., 2012. Slaves in the food chain: when compliance isn't enough. Institute for Human Rights and Business. Available at: http://www.ihrb.org/commentary/guest/slaves-in-the-food-chain-when-compliance-isnt-enough.html.

Skinner, E.B., February 23, 2012. The Fishing Industry's Cruelest Catch. Bloomberg Businessweek. http://www.bloomberg.com/news/articles/2012-02-23/the-fishing-industrys-cruelest-catch.

Stringer, C., Simmons, G., Coulston, D., 2011. Not in New Zealand's Waters, Surely? Labour and Human Rights Abuses Aboard Foreign Fishing Vessels. Working Paper 11–01 New Zealand Asia Institute, University of Auckland. https://researchspace.auckland.ac.nz/handle/2292/22016.

Stringer, C., Whittaker, D.H., Simmons, G., 2016. New Zealand's turbulent waters: the use of forced labour in the fishing industry. Global Networks 16, 3–24.

Surtees, R., 2008. Trafficking of Men – A Trend Less Considered: The Case of Belarus and Ukraine. International Organization for Migration. Available at: http://publications.iom.int/bookstore/free/MRS_36.pdf.

Surtees, R., 2012. Trafficked at Sea. The Exploitation of Ukrainian Seafarers and Fishers. International Organization for Migration. Available at: http://publications.iom.int/bookstore/free/Trafficked_at_sea_web.pdf. Quoted in ILO, 2013. Caught at sea: Forced labour and trafficking in fisheries. Available at: http://www.ilo.org/wcmsp5/groups/public/--ed_norm/--declaration/documents/newsitem/wcms_214472.pdf.

Sylwester, J., 2014. Fishers of men: labor trafficking and environmental degradation in the Thai fishing industry. Pacific Rim Law & Policy Journal 23, 423–459.

United Nations Inter-Agency Project on Human Trafficking (UNIAP), 2009. Exploitation of Cambodian Men at Sea: Facts about Trafficking of Cambodian Men onto Thai Fishing Boats. SIREN Case Analysis. Available at: http://www.ilo.org/wcmsp5/groups/public/--ed_norm/--declaration/documents/publication/wcms_143251.pdf.

United Nations Office of Drugs and Crime (UNODC), 2013. Human Trafficking. Available at: http://www.unodc.org/unodc/en/human-trafficking/what-is-human-trafficking.html.

United Nations (UN), 1948. Universal Declaration of Human Rights. Available at: http://www. un.org/en/documents/udhr/.

United Nations (UN), 2001. Protocol against the Illicit Manufacturing of and Trafficking in Firearms, Their Parts and Components and Ammunition, Supplementing the United Nations Convention against Transnational Organized Crime. General Assembly Resolution 55/225. Available at: http://www.unodc.org/documents/treaties/UNTOC/Publications/A-RES%2055-255/55r255e.pdf.

United Nations (UN), 2002. Globalization and Its Impact on the Full Enjoyment of Human Rights. Report of the High Commissioner on Human Rights. GE.02–10108 (E).

United Nations (UN), 2003. Human Rights, Trade and Investment. Report of the High Commissioner for Human Rights. GE.03-14847 (E) 310703.

United Nations (UN), 2011. Guiding Principles on Business and Human Rights. Implementing the United Nations 'Protect, Respect and Remedy Framework'. United Nations, New York and Geneva HR/PUB/11/04.

Urbina, I., 2015. The Outlaw Ocean. The New York Times. July 25, 2015–December 28, 2015. Available at: http://www.nytimes.com/interactive/2015/07/24/world/the-outlaw-ocean.html.

US Department of Labor (USDOL), 2016. California Transparency in Supply Chains Act. Available at: https://www.dol.gov/ilab/child-forced-labor/California-Transparency-in-Supply-Chains-Act.htm.

US Department of State (USDOS), 2010. Trafficking in Persons Report 2010. Washington, DC. Available at: https://www.state.gov/j/tip/rls/tiprpt/2010/.

US Department of State (USDOS), 2013a. Trafficking in Persons Report 2013. Washington, DC. Available at: https://www.state.gov/j/tip/rls/tiprpt/2013/.

US Department of State (USDOS), 2013b. What Is Modern Slavery? Available at: http://www.state. gov/j/tip/what/.

US Department of State (USDOS), 2014. What Is Trafficking in Persons? Available at: https:// www.state.gov/documents/organization/233944.pdf.

US Department of State (USDOS), 2015. Trafficking in Persons Report 2015. Washington, DC. Available at: https://www.state.gov/j/tip/rls/tiprpt/2015/.

US Department of State (USDOS), 2016. Trafficking in Persons Report 2016. Washington, DC. Available at: https://www.state.gov/j/tip/rls/tiprpt/2016/.

US District Court, December 9, 2015. Barber v. Nestle USA. Available at: http://www.csrandthelaw. com/wp-content/uploads/sites/2/2016/01/Nestle-dismissal.pdf.

Yea, S., 2014. Trafficking on the high seas: the exploitation of migrant fishermen in SouthEast Asia's long Haul fishing industry. In: Wilhelm Hofmeister, W., Rueppel, P. (Eds.), Trafficking in Human Beings: Learning from Asian and European Experiences. Konrad-Adenauer-Stiftung and European Union, Singapore, p. 8596.

Section IV

Looking Forward

Chapter 19

Implications of a Changing Climate for Food Sovereignty in Coastal British Columbia

Terre Satterfield, Leslie Robertson, Nathan Vadeboncoeur, Anton Pitts
University of British Columbia, Vancouver, BC, Canada

INTRODUCTION

Approximately 4 years ago, we stood around a large table stacked with territory maps in the 'Namgis First Nation band office, discussing the consequences of climate change. A pithy set of questions was posed to us as researchers. It went something like this: We want to be food sovereign. We want to eat traditional foods; that's what we think about when we think about climate change. How could it be done? What would it take?

In recent years, scholars have considered the implications of climate-driven ecosystem change for communities highly dependent on marine resources (Turner and Clifton, 2009). The social consequences are both acute (e.g., relocation of peoples living at or below sea level, see Douglas et al., 2008) and chronic (e.g., compromised availability of marine and "country" foods, see Lipset, 2013). And yet, at the very moment that communities are coming to regard the exercise of rights to fish and hunt as a key means for practicing sovereignty, there exists a dual burden: limited access to food (both physical and political), alongside the daunting effects of climate change on key food species.

This chapter explores both the material and political implications of climate change for indigenous food security and sovereignty. Our focus is a collaborative project with the 'Namgis First Nation, one of several Kwakwaka'wakw nations. Their territorial home at the northern end of Vancouver Island, British Columbia, encompasses 2623 square kilometers of terrestrial and marine territory from the Nimpkish River watershed to the Broughton Archipelago, and is inclusive of interisland waters and fjords along the central coast of BC. Inspired by their hope of becoming food sovereign, we investigated four central questions: First we ask what food security and sovereignty is, theoretically, and how indigenous scholars among others have redefined that meaning

Conservation for the Anthropocene Ocean. http://dx.doi.org/10.1016/B978-0-12-805375-1.00019-2
399

to better represent the needs of local food economies long effected by colonialism. We then consider the dense puzzle posed by the 'Namgis Nation's desire to feed a local population of approximately 1000 residents. That is, how might a diet that is largely "traditional" be constructed anew when food knowledge is available and many historically important foods are still eaten, but where uncertainty about the required food resources is also high? We then ask which foods are also fundamental to feasting and potlatching, practices central to 'Namgis identity and resilience in the shadow of colonial pasts, and ongoing challenges. Lastly, we investigate how the dual burdens mentioned above—the politics of access to food and the consequences of environmental change—enable or disable a food secure and sovereign future, identify concrete policy needs under conditions of environmental change and ongoing colonial constraints.

DEFINING FOOD SECURITY AND SOVEREIGNTY

In the epoch of the Anthropocene and its signature—global environmental change—scholars have begun to consider the rise of "community economies" that involve new ethical arrangements and political strategies linked to their local ecologies (Gibson-Graham and Roelvink, 2012, p. 323). Scholarship concerned with different social and physical vulnerabilities experienced by communities has also recently placed food security and sovereignty at the center of understanding the social–ecological resilience of local economies and the household needs they seek to meet (Pinstrup-Andersen, 2009). In its most basic form, food security involves the ability to produce, procure, or purchase food sufficient to meet the nutritional needs of an individual or a household on an ongoing basis, such that hunger, malnutrition, and starvation are nonexistent. Measures of security vary, but most of them seek to define minimum dietary needs, and to reduce or eliminate periods of insecurity (FAO et al., 2012). Food sovereignty focuses on the ability and right to access, control, and produce what is needed for a healthy—and culturally and socially relevant—diet. Attaining sufficient food is distinctly different than procuring or producing sufficient food while also keeping intact local meanings and governance of the overall food system. In particular, food sovereignty prioritizes "those who produce, distribute and consume food at the heart of food systems" and its policies (Declaration of Nyéléni, 2007). Food sovereignty also focuses on the quality of local governance; land tenure and rights; access to credit; the social and policy support necessary for cultivating postindustrial agricultural systems; and the recognition of traditional and local knowledges that accompany food systems (Reardon and Pérez, 2010).

Indigenous food sovereignty has concurrently evolved as its own school of thought (Manson, 2015). It emphasizes much of the above, as well as the indigenous food practices necessary for continuity of identity; recognizing obligations of reciprocity to the nonhuman world that provides food in the first place, and harvesting as a cultural practice and enactment of communal and spiritual relationships (Coulthard, 2010). The indigenous food sovereignty literature also departs from conventional food sovereignty scholarship by focusing on the ways in which colonial dispossession has and continues to constrain the ability of Indigenous people to sustain and renew food regimes (Grey and Patel, 2015), and the philosophical principles and the cultural practices that accompany these (Atleo, 2012).

The geographically specific histories of dispossession of lands and marine resources distinguish indigenous food sovereignty (Grey and Patel, 2015) as often producing sustained "food injustice" (Whyte, 2017). Colonial dispossession has included explicit and often brutal policies aimed at restructuring indigenous food regimes (Daschuk, 2013; Harris, 2001, 2009). In British Columbia, as elsewhere, colonial policies did so by eroding territorial jurisdiction and absorbing indigenous people into cash economies (Knight, 1996; Coulthard, 2007). Food sovereignty was first compromised for 'Namgis through Indian Act legislation (1884–1951) and the rapid criminalization of harvesting activities that followed. People were jailed and fined for participating in feasting and potlatching—ceremonials at the center of indigenous systems of governance and law, the management of resources, recordkeeping, and economic relations. Subsequent provincial legislation banned customary technologies (e.g., weirs and spears), enforced discriminatory licensing (in fishing, hunting, and access to timber), and limited harvesting activities to subsistence only (Harris, 2009). Colonial interventions also forbade access to harvesting sites (thus, to food species); and criminalized transactions such as trade. Since the 1880s, the government controlled the aboriginal food fishery, and issued permits "to allow Indians to catch fish for the purpose of providing food for themselves and their families but for no other purpose" (Harris, 2009, p. 112). This intervention made way for an expanding commercial fishery in BC waters and excluded indigenous fishers. In the late 19th century, the Canadian Fisheries Act further separated commercial and food fisheries, thus requiring special licenses to sell or barter fish (Weinstein, 2007, p. 11). By 1894 a more elaborated insult was the requirement that aboriginal fishers had to acquire permission from the Inspector of Fisheries to fish for food whereas nonaboriginal fishers were able to purchase a license for one dollar and did not require permission (Schreiber, 2008, p. 89).

When 'Namgis chiefs testified at the McKenna-McBride Royal Commission (MMRC) in 'Yalis (Alert Bay) in 1913, aboriginal fishers were still forbidden from using traps, weirs, and spears; they had little access to cannery licenses and no access to "independent licenses" (UBCIC, p. 142). In response to a growing land claims movement, Canada amended the Indian Act in 1927 making it illegal for First Nations to hire legal experts to pursue aboriginal rights and land title. With the expansion of the Pacific salmon industry and its dependence on indigenous labor for fishing and canning, First Nations people were propelled into the market food system, and consequently became dependent on cannery stores for food

(UBCIC, p. 69). Following these significant colonial acts, a more contemporary struggle for food sovereignty and food security played out in Canada through a series of legal cases. Beginning with Sparrow v Queen (1990), the aboriginal right to fish for food, social, and ceremonial purposes was upheld, and the possibility of fishing rights extending to commercial sale has also been recognized (Ahousaht First Nation v. Canada, 2008). Ultimately, indigenous food sovereignty includes, but is much more than, control over local food systems. It is embedded in a long history of policies and political struggle, as well as the social, ceremonial, and cultural practices to which these histories are linked (Elliott et al., 2012).

Food, then, is much more than nutrition. For First Nations, it is also wealth (Brown, 1993), resilience (Alfred, 2011), and a vital expression of continuance and identity. Indigenous food practices are a powerful medium for knowledge transmission across generations, which occur at harvesting sites and during food production, often in linguistic and metaphorical languages that are a key to processes of cultural resurgence (Mihesuah, 2003; Satterfield et al., 2013). And food is a central currency of ceremonial practice and cultural "business" on the NW coast wherein feasting and potlatching are the primary institutions of cultural reproduction through naming, gifting, and the recognition of chiefly responsibilities that occurs therein (Robertson and Kwagu'ł Gixsam, 2012). For the 'Namgis, the very concept of a "traditional diet" cannot be decoupled from aboriginal rights in general or from particular knowledge about the cultural and political meanings associated with the harvesting, production, and distribution of foods that originate in their unceded territories.[1]

BECOMING FOOD SECURE: WHAT MIGHT IT TAKE?

Food sovereignty remains pertinent today; is evident in many aspects of daily social, cultural, and physical life; and counters the colonial pasts that have long devastated communities (McMichael, 2014). The 'Namgis Nation engages in myriad food related climate–adaptation activities (e.g., through involvement in Adapting to Uncertain Futures workshops) and in the revitalization of traditional foods more broadly (e.g., as seen in the documentary film "My Big Fat Diet," Bisell, 2008). The research challenge posed to us is therefore both prosaic and historically profound: What would it take to become food sovereign and to eat a traditional diet? Is that possible given a long history of restrictions and negative impacts of climate change? Could daily "traditional" diets be fulfilled, moreover, what is the potential for procuring and producing ceremonial foods required to feed the thousands who attend feasts and potlatches every year (explained below)?

---

1. Unceded territory refers to lands and marine areas where First Nation's title was not surrendered nor acquired through treaty. A modern treaty process began in British Columbia in 1992 (BCTC http://www.bctreaty.net). In 2014, the Supreme Court decision in Tsilhqot' in Nation v. British Columbia granted aboriginal title in Canada for the first time.

APPROXIMATING A CONTEMPORARY VERSION OF A TRADITIONAL DIET

Any estimate of a traditional diet is by definition uncertain, and no gold standard yet exists for ascertaining such diets, despite having traditional foods experts in most communities. Nor does a perfect historical baseline of food volume and historical intake exist. But, we argue, a reasonable contemporary expression is possible. Weinstein and Morrell (1994) give the most comprehensive estimates available for the 'Namgis First Nation. And while "need is not a number," abandoning efforts to estimate diets is intellectually and socially insufficient (Weinstein and Morrell, 1994, pp. 39–41). Facing similar difficulties in establishing system "baselines" that have plagued restoration ecologists (Lennon, 2015), we nevertheless estimate a food secure diet for a particular community to avoid nihilistic pitfalls and instead ignite efforts to operationalize better food futures (see also Lavallée-Picard, 2016).

Our contemporary reanimation of a 'Namgis-specific traditional diet is motivated by four goals. First, to discern as fully as possible, from the oral historical and archaeological record, what was eaten in the past (Boas, 1921; Pasco et al., 1998; Turner, 1995). Second, reduce the total possible foods to those still eaten and currently available, harvested, and consumed. Third, use an existing food guide, in our case the Canada food guide, to estimate diet composition and caloric need (estimated at 2000 calories per person) (Health Canada, 2012). Finally, modify the guide-advised diet to reflect traditional diet features, most notably a diet higher in protein as consistent with evidence about the historical food consumption in the region (e.g., globally, seafoods consumed by coastal peoples are estimated to be 15% higher than nonindigenous country food diets, see Cisneros-Montemayor et al., 2016).

Toward this end, we generated a list of 120 culturally relevant foods, and then reduced this to foods widely available and desired by many, based on information gathered in interviews conducted by 'Namgis historian Diane Jacobson, and discussions with employees of the 'Namgis natural resource office. Then, using the Canadian food guide's recommendation of 2000 calories per person per day, we found macronutrient compositions for 27 central foods (e.g., maƚik/sockeye) or food groups (e.g., tak'astan/seaweeds). For some foods, nutritional information was readily available from composition databases[2]. When nutritional information was not available, we approximated from a close representative food item. For instance, we assumed that various roots consumed as part of a historical diet (xatam/Tslōsaxa hōqlwalē/dłaxsam/ƚaxwsus) would roughly have similar macronutrient content as potato, providing the closest available surrogate measure. Finer macronutrient detail was used for meat and seafood sources, while plant foods were grouped as there is a higher diversity of plant foods and breaking these down to a species level

---

2. http://nutritiondata.self.com/.

would unnecessarily complicate our calculations. Available nutrient databases contain a lot of "country" (wild) meats and berries, but they do not have data on plant foods such as roots, seeds, and other starchy crops.

To construct a reasonable macronutrient content for a diet high in protein, we consulted several streams of evidence. These include records of fish consumption rates in nearby US Native American communities (e.g., roughly 148 grams per day currently, with double and more for traditional diets, see Donatuto and Harper, 2008). In addition, Weinstein and Morrell (1994) found that 25%–45% of calories were from salmon alone, with an additional 10% of caloric intake from terrestrial resources. We held our estimate for salmon to the lower level of 25% due to declining runs in the region (Schindler et al., 2003), with another 5% each made up of pink and chum salmon. These estimates were calibrated with additional evidence for precontact diets in the region (Chisholm et al., 1982), which suggests 90% of proteins from fish and seafoods based on the evidence from multiple sites between the Nass River (on the north coast of BC) and Crescent Beach (near the Washington state/British Columbia border). Corroboration also stems from Ströhle and Hahn (2011) who looked at the carbohydrate intake of peoples considered to be contemporary "hunter-gatherers" (i.e., those who did not need to maintain an agricultural food economy because marine foods were so fully available within their territory). Their eco-cultural classification places the 'Namgis as either "northern coniferous people" for whom carbohydrates were found to range from <15% of the diet, or "temperate forest (mostly mountain) people" whose carbohydrate intake ranged between 23% and 26%. We again chose the more conservative proportion of 24%. In summary, we selected a traditional diet comparatively high in protein (but not as high as might be justifiable) and low in carbohydrates.

FROM HISTORICAL INSIGHT TO A POSSIBLE CONTEMPORARY-TRADITIONAL DIET

Our estimated 'Namgis traditional diet is shown in Table 19.1 below and means to serve an on-reserve population of roughly 1000 people (n=971 as of govcan https://www.aadnc-aandc.gc.ca/eng/1357840941761/1360160050632). Total yearly needs were calculated in units that made sense for harvesting (e.g., one does not harvest half an elk and so on).

Calibrating these to an annual per person diet suggests a possible diet of roughly 24 sockeye, 30 crabs, 15 bags of clams, 80 kilograms (kg) of berries and fruits, 12 waterfowl, 16 kg of elk, and roughly 60 kg of seaweeds before drying. In this historically proximate combination, about 48% of all calories are provided by different fish species, especially sockeye/maɫik with 30% of calories. A further 20% (approximately) of calories are provided by meats and shellfish, with a remaining 27% provided by nuts, fruits and berries, vegetables, seaweeds, roots, and grains, and a final 10% provided by fats and oils (e.g., t'ɫi'na/eulachon grease) and sugars/

TABLE 19.1 Yearly Food Need by Kilogram for a Traditional 'Nam̲gis Diet for 1000 People per Year

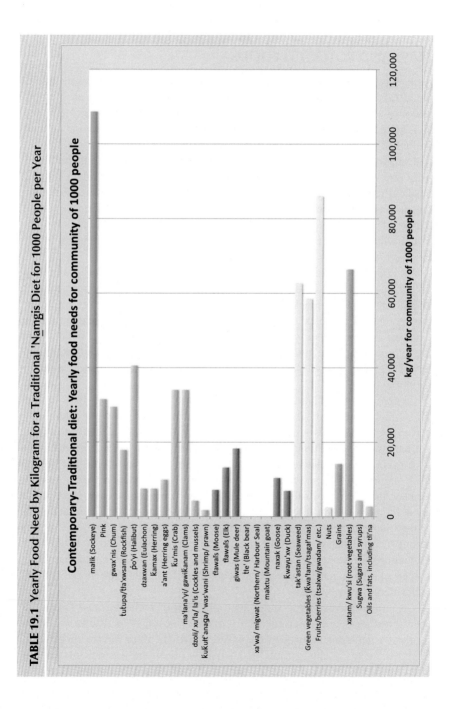

Contemporary-Traditional diet: Yearly food needs for community of 1000 people

kg/year for community of 1000 people

syrups. The high proportion of total calories from fish, meats, and seafood is not surprising as these are highly caloric foods. A very large volume of plant foods is still required (despite their lesser caloric load), at roughly 230 kilos per person per year.

FEAST AND POTLATCH FOODS

Archaeological evidence shows habitation in Kwak'wala-speaking territories for at least 8000 years. The sheer abundance of preservable food enabled the return to winter villages for several months each year for intensive ceremonial and artistic activities (Cranmer-Webster, 2006). To this day, traditional foods are linked to ancestral rights to particular sites; and histories and memories are tied to ancestors' taste preferences, to social status, and reputations for generosity and wealth. Through feasting, traditional name bearers conduct cultural business in front of guests who are also given gifts in order to acknowledge the business they have witnessed. Those who attend are served foods (usually) harvested by family members and others who possess rights to access sites associated with family/clan histories; the foods are prepared by extended family members who usually also serve them at the feast or potlatch.

In 'Yalis, the ceremonial big house or gok'wamdzi is where feasting and potlatching occurs. It holds up to 1000 people who are generally served lunch and dinner as well as snacks for each day of the feast or potlatch. Writing about the ceremonial complex that surrounds potlatching, 'Namgis scholar Gloria Cranmer-Webster (1991, p. 229) notes, "The reasons for giving a potlatch are the same as they were in the past—naming children, mourning the dead, transferring rights and privileges, and, less frequently, marriages or the raising of memorial poles...".

In the case of the 'Namgis Nation, feast and potlatching foods have powerful symbolic and political properties and are necessarily served at ceremonial occasions. Such foods are often regarded as the "wealth" of the 'na'mi'ma, are harvested in traditional territories or acquired through trade or purchase by hosting Chiefs, relatives, and supporters. Family groups contribute meats, fresh and frozen fish, and processed foods (e.g., tlina (eulachon grease), jarred, canned, and or smoked salmon, eulachon, clams, berry jams, etc.). Knowledge about food production constitutes a kind of cultural property held within family groups and transmitted among generations (see Thom, 2003 for a discussion among Coast Salish peoples). Feast and potlatch foods can also be understood as "cultural keystones" (Garibaldi and Turner, 2004), which single out those species so fully salient to key aspects of cultural life that they are literally linked to the identity of a people.

To further understand 'Namgis food practices of potlatches and feasts, we codesigned an interview protocol to provide some sense of prioritized foods for these events. Interviewees were asked to list: (1) all of the traditional seafood, wild meats, grains, grasses, and berries or other land-based plants that are often eaten at feasts and potlatches; and (2) of those, essential foods for feasting

TABLE 19.2 Preferred Feasting Foods—2013 as Example Only

Food Source	Total Consumption During 3 Potlatches and 6 Feasts
Sockeye salmon (małik)	3265 fish
Pink salmon (hanu'n)	553 fish
łak'astan seaweed	81 Gal
p̓o'yi halibut	2444 lb
a'ant herring roe	123 Gal
ła'xwsam cod (rockfish)	2444 lb
dzaxwan eulachon	15,550 pieces
'was'wani prawns	110,500 pieces
gawiḵanam clams	96 sacks 1 sack=1 Gal bucket cleaned.
dzoli cockles	10 5/8 sacks and 3 pint jars
ḵu'mis crabs	3525 pieces served in halves, legs separate
tłi'na eulachon grease	11 Gal
ławal's elk	4525 lb
ławal's moose	4525 lb

and potlatching (i.e., foods "you cannot imagine doing without"). Interviews about ceremonial foods were conducted by Diane Jacobsen and other 'Namgis researchers who spoke with food specialists—harvesters, producers, and caterers.

The demand for ceremonial foods varies considerably for a single year. The second part of the interview thus included questions about the approximate volume of each food species served at feasts and potlatches held at the Big House in 2013. Table 19.2 provides these estimates, which involved feeding 6500 people over the course of 13 days of ceremonial events that year, as an example only. The list of preferred ceremonial foods indicates likely cultural keystone foods.

ON THE POLITICS OF ACCESS AND THE SPECTER OF ENVIRONMENTAL CHANGE

By any measure, the annual procurement of marine resources that could provide the 'Namgis First Nation with a secure traditional diet is substantial, and must include additional considerations such as the availability of boats and trucks for harvesting, available labor, and necessary storage and processing capacities (Satterfield et al., 2015). Regardless, food security in dietary terms should be understood as discrete from sovereignty where the latter can be thought of as a more complete capacity to designate and live a 'Namgis-specific food regime. This includes potlatching and feasting as important components of indigenous sovereignty as defined above, and can be thought of as "feeding the spirit" as different from "feeding the body" (Donatuto et al., 2011). But the dilemma surrounding the capacity to be both food secure and food sovereign is also a matter of policy and politics allowing for unrestricted access and capacity—to fish, to hunt, to be active on the land and marine-scape, "in territory" or "exercising one's treaty rights" (Williams and Hardison, 2013).

On paper Canadian law provides for the aboriginal right to "food fish" (albeit a law that was highly restrictive in the first place as detailed above), and was given constitutional protection under s. 35(1) in R v. Sparrow, the Supreme Court of Canada. That law states that the right to fish for food preexisted the enactment of the British Columbia Fisheries Act, which sought to significantly limit this right. Further, the right is meant to supersede other fishing, both recreational and commercial. But many factors are at odds with this most fundamental right to fish and to food more broadly. The decline in fishing licenses (bought up by corporate fishers) has, for example, resulted in the reduction of boats available for fishing and has been a blow to local household economies (Burke, 2010). Exceptionally poor fishing years, such as 2009 with almost no food fish from sockeye runs, have delivered additional blows. Most recently, the 2016 Fraser River run was characterized as the worst year in more than a century (Hoekstra, 2016).

Allocations of all foods are thus well below the volumes necessary for a traditional diet procured locally. Records provided for 2006–2011 from the 'Namgis band office indicate, for example, that food harvests and allocations sit somewhere between less than 1% for shellfish and mollusks through to 30% of estimated need for sockeye (see Table 19.3 below). Sockeye, for example, the food most sought by all 'Namgis, is well below our estimate of just under 24,000 fish annually, with procurement averaging about 7000 fish between 2006 and 2011, exempting the outlier year 2010. Year on year fluctuations are normal and reflected in the provided averages. Averages across years are more viable as storage including freezing, canning, smoking, and drying are widely practiced. Even a modicum of proteins from elk (the most prized country/wild meat) is currently out of reach given the minimal number of hunting tags allotted to 'Namgis, and the poor status of herds at the north end of Vancouver Island and beyond. Like many First Nations, the 'Namgis live in a "food desert" where fresh food retail is scarce, coupled with a high rate of unemployment, procuring food through the market presents challenges. For some, local sources of protein may be one of the few ways to prevent hunger and malnutrition during winter.

TABLE 19.3 Food Availability and Expectations of Climate Effects

	Estimated Need	Estimated Availability	% Need Under-Served	Estimated Change	Direction of Change
Sockeye	~24000 fish	~6900 (exclusive of 2010)	~70%	10.0%–19%	Negative
Pink	~6900 fish	1740	75%	10.0%–19%	Negative
Chum	~3600 fish	274	92%	10.0%–19%	Negative
Halibut	~4500 fish	310	93%	0%–10%	Likely positive
Herring	~7500 kg	n/a		28.1%–49.5%	Negative
Rockfish	~1900 fish	578	70%	Estimate unknown	Neutral
Crab	~29,900 whole crabs	112	<1%	No change	Neutral
Pink shrimps	190 bags	N/A		Stable or upward, estimate unknown	Neutral–positive
Clams	~15,000 bags or 300,000 lb	28.5 bags or 570 lb OR 62 bags or 570 kg	Less than 1% in either case	Estimate unknown	Negative

For salmonids, the northern range is lower at 3.2%–8.2%; but in the 'Namgis case, a large portion of catch comes from the Fraser River run that travels over the north end of Vancouver Island and through Johnstone Strait and whose decline is estimated at 17.1%–29.2%. Thus a very rough proxy is this average of the high and low estimates. Assumption is 55 clams per kilogram.

TRADITIONAL DIETS AND ENVIRONMENTAL CHANGE

Regional environmental changes further erode the current (and already subop-timal) availability of foods that might enable a secure (and possibly sovereign) traditional diet (Helmuth et al., 2014). On the north coast of BC, Turner and Clifton (2009) have observed, for example, that newer, faster environmental changes are evident. Among the observed changes are species declines, the appearance of novel species, extreme weather, changing tides and currents, and altered harvest windows (in one example, the window for seaweed harvest shifted by as much as a month) (p. 184). The challenge of procuring available food is profound, even when considering only the needs of one First Nation at the north end of Vancouver Island—a region historically rich in marine and ter-restrial foods.

Climate change effects are compounded by other pressures (notably, over-fishing) (Hughes, 2000; McCarty, 2001; Coristine and Kerr, 2011; Okey et al., 2014; Cheung et al., 2013). Changes in species distributions and ecosystem composition, in particular, will affect food systems and the cultural and social worlds that rely on them. The degree of impact will vary regionally and will depend on the nature and magnitude of local climate changes, the influence of nonclimate stressors, and the vulnerability and adaptive responses of eco-systems, individual species, and populations (Okey et al., 2014). While the climate of coastal BC is highly variable at the scale of years and decades owing to the strong influence of the El Niño/Southern Oscillation (ENSO) and the Pacific Decadal Oscillation (PDO), there is a long-term warming trend (see Vadeboncoeur, 2016 for a summary). These highly variable cycles are the prin-cipal influence on the climate of the region and predicted to be a dominant driver of marine-ecosystem change over the next several decades (Overland and Wang, 2007), with effects on water and air temperatures, timing, and other atmospheric variables (Fleming and Whitfield, 2010).

Just how these impacts may affect the species of special concern for the 'Namgis Nation is largely unstudied, requires a broader engagement with the human dimensions of climate change, and the diverse human values behind future cultural–ecological systems. But, at the very least, if we focus on highly prized feasting and potlatching foods, several of which also comprise a signifi-cant portion of a defensible contemporary-traditional diet, a few observations regarding climate impacts are possible. This is particularly so as warming is likely to influence the distribution of marine species and have considerable eco-logical implications (Brierley and Kingsford, 2009; Cheung et al., 2010, 2013; Blanchard et al., 2012). More locally, a coast-wide decline in the abundance of species key to First Nations is expected to involve on average a 4.5%–10.7% decline between 2000 and 2050; with relatively greater declines further south (Weatherdon et al., 2016).

Overall, the forecast for five of the nine key marine ceremonial foods (sock-eye, pink and chum salmon, herring, and mollusks) is negative. Projections are neutral for crab and rockfish; and neutral-to-positive for pink shrimp. In 'Namgis territory, neutral or positive shifts in catch potential for white sturgeon, kelp greenling, and two species of perch are also anticipated, but are not part of a generally accepted traditional diet. Positive changes to halibut and some other species of rockfish might be expected to make up for a small portion of the loss of other foods. A summary of these expectations is provided in Table 19.3 below, with specific arguments and empirical discussion in the text that follows.

SALMONIDS, GROUND FISH, ROCKFISH, AND HERRING

The forecast for salmonids is, above all else, of highest concern with most research anticipating a poleward/northern movement in response to warmer ocean and river waters (Hinch and Martins, 2011; Irvine and Crawford, 2012; Peterman et al., 2012). The impacts of warmer water on Fraser River sockeye may include reduced upriver migration, lower survival rates of salmon during the freshwater stages of their life cycle (Welch et al., 1998; Irvine and Fukuwaka, 2011; Rogers and Schindler, 2011), and prespawning mortality (Rand et al., 2006; Hinch and Martens, 2011). These effects will be most pronounced in the southern BC, near the southern limit of salmonid range. But such effects could also impact salmon populations in 'Namgis territory as the climate continues to warm and because much 'Namgis fishing occurs in that portion of the renowned Fraser River run that travels over the northern end of Vancouver and through Johnstone Strait. Natural variability has affected salmon throughout the history, but more extreme seasonal changes, particularly those that can affect river temperature in autumn, are expected along the BC coast. Collectively, salmon (*Oncorhynchus* spp.) are projected to exhibit cumulative declines in catch potential of 17.1%–29.2% within BC's marine environment (Weatherdon et al., 2016), with declines for the 'Namgis said to be somewhat less severe when excluding the Fraser River portion of the 'Namgis fishery (p. 7).[3]

The other species of great concern is Pacific herring, including two primary sources: the north coast stock (Haida Gwaii, Prince Rupert, and Central Coast), and south coast stock (Strait of Georgia and west coast of Vancouver Island). Together, these comprise 17 distinct populations (Hay et al., 2008; Moody and Pitcher, 2010). Herring group in both large migratory and small, local, populations along the west coast and although they are relatively resilient to environmental change, strong climate impacts in nearshore areas could affect local populations (ibid). For example, sea-level rise can affect the recruitment and abundance of herring by altering nearshore habitats important for spawning,

---

3. An important basis for uncertainty is the fact that specific estimates (e.g., Weatherdon et al., 2016) are derived using global climate models. Much less easy to anticipate is the presence and impact of seasonal variation, especially extreme variation as was the case for 2009 and appears to be the case as well for 2016.

while temperature, salinity, and ocean circulation can affect survival of eggs and larvae (Beamish et al., 2009). The relatively large number of small, local herring populations may provide some resistance to climate change if different populations are subject to different stressors, or capable of adapting to different conditions.

Temperature changes could have a considerable effect on Pacific herring if it impacts growth rate/body size and maturity, mortality, and fecundity (Rose et al., 2001, 2008). The estimates provided by Weatherdon et al. (2016) are particularly foreboding as they predict relatively severe declines in catch potential ranging between 28.1% and 49.2% across coastal BC (p. 6). The herring fishery also provides herring, roe, spawn-on-kelp, and bait, and important potlatching and feasting foods.

Eulachon/t'lina is perhaps the most prized noncommercial food in the traditional diet of coastal First Nations (Moody and Pitcher, 2010), but are not included here because they are often gifted and traded rather than harvested in 'Namgis territory—though many do have rights to render the prized 't'lina (grease) at Dzawadi (Knight Inlet) due to marriage and decent relations among A'wa'etlala people. Gifting and trade are vibrant activities across coastal communities that sometimes ensure the provision of foods on failing years or in times when a resource is no longer accessible in home territories.

Only halibut populations appear to be at an all-time high (Beamish et al., 2009), and climate change may benefit this species overall (Okey et al., 2014). For example, the growth of juvenile halibut increases with temperature (Hurst et al., 2005). Recruitment is established within the first year of life, and influenced heavily by climate and weather (Clark and Hare, 2002). Many of the halibut caught in BC waters hatch in Alaska and migrate south, or off the continental shelf and drift landwards. Others have argued, however, that estimates fail to properly account for the negative effects of ocean acidification and thus underestimate impacts (Weatherdon, p. 12).

A somewhat similar set of expectations holds for red snapper (*Sebastes ruberrimus*), also known as yellow-eye rockfish. These populations are relatively stationary with some fish living most of their lives within a few square kilometers (Black et al., 2008). The species is very long-lived (up to 120 years) and its reproductive success is therefore relatively resilient to short-term environmental change (Berkeley et al., 2004). However, growth is sensitive to climate oscillations as it may be affected by long- and medium-term climate cycles (Black et al., 2008). Red snapper populations have been overfished and until populations recover, their adaptive capacity will likely be reduced from historical levels. But the relatively stationary regional populations suggest that only direct fishing impacts on the stocks accessed by the 'Namgis will be a concern to their food security.

Among common ground fish, only Pacific Ocean perch (*Sebastes alutus*), remain a concern, especially given large catch rates in years prior. The species is sensitive to climate variability and the current population is supported by the 1977–1988 cohorts, which benefited from favorable ocean conditions (Beamish et al., 2009). Maintaining sufficient adult biomass of Pacific Ocean perch can allow this species to take advantage of favorable conditions when they reemerge, and thus there is hope for future stocks (Schnute et al., 2001).

MOLLUSKS AND SHELLFISH

Dungeness crab (*Cancer magister*) is a highly prized marine and ceremonial food, but is also vulnerable to a range of stressors including fishing, pollution, and estuarine and coastal modifications (Okey et al., 2014). Climate change may also impact this species by changing habitat characteristics. These changes include increased water temperature that leads to earlier hatching (Park and Shirley, 2008) and competition and predation from more southerly species such as the non-native European green crab (Pellegrin et al., 2007). Sea level rise–driven changes can affect this species, but are of lesser concern in 'Namgis territory, given the relatively slow rate of sea-level change in that region.

Mollusks, on the other hand, are most likely threatened by increased ocean acidification and temperature, and sea-level rise will restrict human access to intertidal populations.

Ocean acidification is the most significant among these threats to shellfish, and is expected to affect the reproductive success of oysters (Kurihara et al., 2007), mussels (Melzner et al., 2011), clams (Ries et al., 2009), and sea urchins (Reuter et al., 2011). The impacts of acidification will be particularly severe for coastal BC because the North Pacific has one of the lowest carbonate saturation states on the earth, and is already very close to the level considered corrosive (Feely et al., 2004; Harley et al., 2006). It is also possible that because the BC coast is highly variable (Nemcek et al., 2008), carbonate populations may remain in relatively good health at sites where increases in acidity are slight. Regardless, long-term assessments of acidity in coastal BC are not currently possible due to data limitations (Ianson, 2013).

Among invertebrates prized by 'Namgis, only pink shrimp (*Pandalus jordani*) appear stable or more abundant, particularly as populations move northwards into part of the range of the endemic Northern shrimp (*Pandalus borealius*). This appears to be a result of warmer water (Okey et al., 2014), which favorably influences the spring phytoplankton bloom (Koeller et al., 2009). If this is the case, this species should remain in the region as water continues to warm and the southern range of species expands northward.

POPULATIONS NOT PREVIOUSLY ABUNDANT OR PRESENT

Increases in previously rare species are also a possibility. Indeed, one of the most easily observable impacts of climate change will be the general northward movement of marine species (Brodeur et al., 2003; Harding et al., 2011). More frequent occurrences of species such as the Humboldt squid (*Dosidicus gigas*) (Cosgrove, 2005) and increased biomass of "California" sardine populations (*Sardinops sagax*) (Ishimura et al., 2013) are two notable examples of species that are beginning to expand into BC waters. These temporary changes could become permanent over the next several decades as climate change becomes a dominant influence on marine conditions in the region (Overland and Wang, 2007). In particular, the ranges of 28 pelagic fish species in coastal BC are expected to move northward by approximately 30 kilometers per decade as a result of climate change (Cheung et al., 2015).

In total, the future of marine foods on which both food security and sovereignty depend are likely compromised with only a few species expected to remain stable or increase in abundance. The introduction of new species previously outside the range of harvested foods is also likely. The significant blow to salmon and herring stocks is already widely felt across coastal First Nations, and will be difficult to counter or mitigate in the coming years. Sockeye in particular (and to a lesser extent pinks and chum salmon) is the most culturally significant food for the 'Namgis nation. It represents a huge portion of their diet historically, and is a cultural keystone as a species is linked to (and indeed a medium for) all culturally important practices or transactions. They are also a species high in caloric load and thus are not easily replaced by an increase in groundfish already consumed, notably halibut and species of rockfish, which have some potential to increase in abundance. Similarly, Dungeness crab stocks may remain stable, but face demand pressures from international markets and high-end restaurants. Rockfish, mollusks, and shellfish (especially pink shrimp) thus face pressure to make up the difference. With herring also likely compromised, the question of whether halibut and more resilient species of rockfish might make up the difference or a significant portion thereof is the key, as is the very notion of what is, in any way, substitutable.

DISCUSSION: ON THE MEANING OF FOOD—MATERIAL AND INTANGIBLE

Aspirations toward greater food sovereignty and renewed respect for traditional diets is painfully incongruous with the impacts of climate change. Human activity in watersheds, including industrial and residential developments and spills affecting water quality in rivers, impact marine foods (Lohse et al., 2008). Commercial fishing itself presents an additional threat to all species by reducing the age, size, and biodiversity of stocks (Ainsworth et al., 2011).

Procuring basic nutrition is a fundamental right and whether that is met currently, let alone in the future, is questionable at best. Most somber is the reality that for the First Nations in Canada, possibilities for food sovereignty are overdetermined in a colonial context within which the Indian Act remains a major piece of legislation. Hence, our posited traditional diet is currently breached by the conditions of access, however much the law might suggest prioritizing "food fish."

Any consideration of culturally important foods must also consider the intangible dimensions. By intangible, we refer to the meaning and identity-linked social practices that comprise many indigenous food systems. For example, *traditional knowledge or skill* is the knowledge borne of practice, including maintaining and perpetuating collective habits and ways of being that are learned and transmitted by hunting, fishing and harvesting, gathering, processing, and storing of foods. To say that food is knowledge and knowledge is food is to say that people continually go to harvesting sites to prepare for feasting, to harvest, and to process foods (cleaning, drying, smoking, canning), often within multigenerational family and clan groups. They acquire skills and understanding about cultural lifeways, ongoing activities through which people develop, and nurture relationships across generations, which are the key to continuity and identity.

Potlatching and feasting are important cases in point. Foods central to these are the material currency of enduring social relationships. Ceremonial events are sites where chiefs and their clan groups pass on names, fulfill responsibilities to other clan groups, mark important rites of passage, and affirm their relationships to their territories through inviting the community to witness the assignation of rights and responsibilities and status changes (Robertson and Kwagu'ł Gixsam, 2012). Potlatches are also public legal occasions where clan groups acknowledge grieving of those passed, or transfer names and their responsibilities in the event of a birth, a death, or an adoption. They are where crests, dances, songs, and accompanying regalia (e.g., blankets, coppers, and masks) express and confirm peoples' histories in their territories.

South of 'Namgis territory, in coastal Washington State, Quinault people speak of "clam hunger" in the aftermath of an algae bloom attributed to climate change, and in reference to the very culturally important razor clam. The analogy is fitting here:

> *Many Quinault people speak of "clam hunger," a physical, emotional, and spiritual craving for a food that connects them to their native landscape, their ancestors, and their very existence as a people. Clam hunger can even drive people to eat this food when scientists and resource managers tell them that toxins render it unsafe.*
>
> DeWeerdt (2016)

In the end, loss of species and the intangible aspects of their meaning is not only a shock to the ecological health of 'Namgis territory, but equally a shock to cultural well-being. Such losses are often invisible in cost–benefit calculations, but may well constitute an even greater cost than more conventionally recognized economic losses (Turner et al., 2008).

Ultimately fisheries management in BC, Alaska, and off the continental shelf will play an important role in the future and in the face of multiple pressures, climate change in particular. But managing for the deeper and more enduring consequences for food sovereignty is another matter altogether, be that cultural resilience, identity, or rights at risk. Each of these consequences warrant serious attention which might be mitigated and if so, how. Communities are already grappling with these challenges on a near-daily basis (Berkes and Jolly, 2002); academic practitioners have only just begun.

ACKNOWLEDGMENTS

We are grateful for the hospitality and spirit of people in 'Yalis and the staff of the 'Namgis band office, and insights provided by Shannon Hagerman, University of British Columbia. Funding provided by the Solutions Initiative, Peter Wall Institute for Advanced Studies at the UBC was critical to every stage of this work.

REFERENCES

Ahousaht First Nation v. Canada (Fisheries and Oceans), 2008 FCA 212.

Ainsworth, C.H., Samhouri, J.F., Busch, D.S., Cheung, W.W.L., Dunne, J., Okey, T.A., 2011. Potential impacts of climate change on Northeast Pacific marine foodwebs and fisheries. ICES Journal of Marine Science 68 (6), 1217–1229.

Alfred, D., 2011. My Life With the Salmon. Theytus Books.

Atleo, R., 2012. Principles of Tsawalk: An Indigenous Approach to Global Crisis. UBC Press, Vancouver.

Beamish, R.J., King, J.R., McFarlane, G.A., 2009. Impacts of climate and climate change on the key species in the fisheries in the North Pacific. In: Beamish, R.J. (Ed.), PICES Scientific Report No. 35 PICES Working Group on Climate Change. Sidney, B.C. North Pacific Marine Science Organization (PICES), Secretariat, pp. 14–55.

Berkeley, S.A., Chapman, C., Sogard, S.M., 2004. Maternal age as a determinant of larval growth and survival in a marine fish, *Sebastes melanops*. Ecology 85 (5), 1258–1264.

Berkes, F., Jolly, D., 2002. Adapting to climate change: social-ecological resilience in a Canadian western Arctic community. Conservation Ecology 5 (2).

Bisell, M., 2008. My Big Fat Diet. Barebones Production Ltd. and CBC Newsworld.

Black, B.A., Boehlert, G.W., Yoklavich, M.M., 2008. Establishing climate–growth relationships for yelloweye rockfish (*Sebastes ruberrimus*) in the northeast Pacific using a dendrochronological approach. Fisheries Oceanography 17 (5), 368–379.

Blanchard, J.L., Jennings, S., Holmes, R., Harle, J., Merino, G., Allen, J.I., et al., 2012. Potential consequences of climate change for primary production and fish production in large marine ecosystems. Philosophical Transactions of the Royal Society of London. Series B, Biological Sciences 367 (1605), 2979–2989.

Boas, F., 1921. Ethnology of the Kwakiutl. Thirty-Fifth Annual Report. Bureau of American Ethnology, Washington, DC.

Brierley, A.S., Kingsford, M.J., 2009. Impacts of climate change on marine organisms and ecosystems. Current Biology 19, R602–R614.

Brodeur, R.D., Pearcy, W.G., Ralston, S., 2003. Abundance and distribution patterns of Nekton and Micronekton in the northern California current transition zone. Journal of Oceanography 59 (4), 515–535.

Brown, P., 1993. Cannery Days: A Chapter Is the Lives of Heiltsuk. Master's Thesis. University of British Columbia.

Burke, C.L., 2010. When the Fishing's Gone. MA Thesis University of British Columbia. https://open.library.ubc.ca/cIRcle/collections/ubctheses/24/items/1.0069876.

Cheung, W.W.L., Lam, V.W.Y., Sarmiento, J.L., Kearney, K., Watson, R., Zeller, D., Pauly, D., 2010. Large-scale redistribution of maximum fisheries catch potential in the global ocean under climate change. Global Change Biology 16, 24–35.

Cheung, W.W., Watson, R., Pauly, D., 2013. Signature of ocean warming in global fisheries catch. Nature 497 (7449), 365–368.

Cheung, W.W.L., Brodeur, R.D., Okey, T.A., Pauly, D., 2015. Projecting future changes in distributions of pelagic fish species of Northeast Pacific shelf seas. Progress in Oceanography 130, 19–31.

Chisholm, B.S., Erle Nelson, D., Schwarcz, H.P., 1982. Stable-carbon isotope ratios as a measure of marine versus terrestrial protein in ancient diets. Science 216 (4550), 1131–1132.

Cisneros-Montemayor, Andrés, M., et al., 2016. A global estimate of seafood consumption by coastal indigenous peoples. PLoS One 11, 12.

Clark, W.G., Hare, S.R., 2002. Effects of climate and stock size on recruitment and growth of Pacific halibut. North American Journal of Fisheries Management 22 (3), 852–862.

Coristine, L.E., Kerr, J.T., 2011. Habitat loss, climate change, and emerging conservation challenges in Canada. Canadian Journal of Zoology 89 (5), 435–451.

Cosgrove, J.A., 2005. The First Specimens of Humboldt Squid in British Columbia. PICES Press.

Coulthard, G.S., 2007. Subjects of empire: indigenous peoples and the "politics of recognition" in Canada. Contemporary Political Theory 6 (4), 437–460.

Coulthard, G.S., 2010. Place against empire: understanding indigenous anti-colonialism. Affinities: A Journal of Radical Theory, Culture, and Action, vol. 4.

Cranmer-Webster, G., 1991. The contemporary potlatch. In: Jonaitis, A. (Ed.), Chiefly Feasts: The Enduring Kwakiutl Potlatch. Douglas and McIntyre, Vancouver.

Cranmer-Webster, G., 2006. Kwakwaka'wakw, Kwakiutl. The Canadian Encyclopedia.

Daschuk, J.W., 2013. Clearing the Plains: Disease, Politics of Starvation, and the Loss of Aboriginal Life. University of Regina Press.

Declaration of Nyéléni. February 27, 2007. Forum for Food Sovereignty in Sélingué, Mali. https://nyeleni.org/spip.php?article290.

DeWeerdt, S., 2016. Clam Hunger. Puget Sound Institute. Encyclopedia of Puget Sound, August 31st https://www.eopugetsound.org/magazine/clam-hunger.

Donatuto, J., Harper, B., 2008. Issues in Evaluating fish consumption rates for native American Tribes. Risk Analysis 28 (6), 1497–1506.

Donatuto, J.L., Satterfield, T., Gregory, R., 2011. Poisoning the body to nourish the soul: prioritising health risks and impacts in a Native American community. Health, Risk & Society 13 (2), 103–127.

Douglas, I., Alam, K., Maghedna, M., Mcdonnell, Y., Mclean, L., Campbell, J., 2008. Unjust waters: climate change, flooding and the urban poor in Africa. Environment and Urbanization 20, 1876205.

Elliott, B., Jayatilaka, D., Brown, C., Varley, L., Corbett, K.K., 2012. "We are not being heard": aboriginal perspectives on traditional foods access and food security. Journal of Environmental and Public Health 2012 (6), 130945–130949.

FAO, WFP, IFAD, 2012. The State of Food Insecurity in the World 2012 (PDF). (Rome).

Feely, R.A., Sabine, C.L., Lee, K., Berelson, W., Kleypas, J., Fabry, V.J., Millero, F.J., 2004. Impact of anthropogenic CO_2 on the $CaCO_3$ system in the oceans. Science 305 (5682), 362–366.

Fleming, S.W., Whitfield, P.H., 2010. Spatiotemporal mapping of ENSO and PDO surface meteorological signals in British Columbia, Yukon and Southeast Alaska. Atmosphere-Ocean 48 (2), 122–131.

Garibaldi, A., Turner, N., 2004. Cultural keystone species: implications for ecological conservation and restoration. Ecology and Society 9, 3.

Gibson-Graham, J.K., Roelvink, G., 2012. An economic ethics for the Anthropocene. In: The Point Is to Change it. John Wiley & Sons, Ltd, Chichester, UK, pp. 320–346.

Grey, S., Patel, R., 2015. Food sovereignty as decolonization: some contributions from Indigenous movements to food system and development politics. Agriculture and Human Values 32 (3), 431–444.

Harding, J.A., Ammann, A.J., MacFarlane, R.B., 2011. Regional and seasonal patterns of epipelagic fish assemblages from the central California Current. Fishery Bulletin 109 (3), 261–281.

Harley, C.D.G., Randall Hughes, A., Hultgren, K.M., Miner, B.G., Sorte, C.J.B., Thornber, C.S., et al., 2006. The impacts of climate change in coastal marine systems. Ecology Letters. 9 (2), 228–241. http://doi.org/10.1111/j.1461-0248.2005.00871.x.

Harris, D., 2001. Fish, Law and Colonialism. University Toronto Press, Toronto.

Harris, D., 2009. Landing Native Fisheries: Indian Reserves & Fishing Rights in British Columbia, 1849–1925. University of British Columbia, Vancouver.

Hay, D.E., Rose, K.A., Schweigert, J., Megrey, B.A., 2008. Geographic variation in North Pacific herring populations: Pan-Pacific comparisons and implications for climate change impacts. Progress in Oceanography 77 (2–3), 233–240.

Health Canada, 2012. Eating Well With Canada's Food Guide. Health Canada, Ottawa, Ontario.

Helmuth, B., Russell, B.D., Connell, S.D., Dong, Y., Harley, C.D., Lima, F.P., et al., 2014. Beyond long-term averages: making biological sense of a rapidly changing world. Climate. 1 (1), 1015. http://doi.org/10.1186/s40665-014-0006-0.

Hinch, S.G., Martins, E.G., 2011. A review of potential climate change effects on survival of Fraser River sockeye salmon and an analysis of Interannual trends in En Route loss and pre-spawn mortality. Cohen Commission Technical Report.

Hoekstra, G., August 8, 2016. Grim Fraser River salmon runs even worse than forecast. The Vancouver Sun. http://vancouversun.com/business/local-business/grim-fraser-river-salmon-runs-even-worse-than-forecast.

Hughes, L., 2000. Biological consequences of global warming: is the signal already apparent? Trends in Ecology & Evolution 15 (2), 56–61.

Hurst, T.P., Spencer, M.L., Sogard, S.M., Stoner, A.W., 2005. Compensatory growth, energy storage and behavior of juvenile Pacific halibut *Hippoglossus stenolepis* following thermally induced growth reduction. Marine Ecology Progress Series 293, 233–240.

Ianson, D., 2013. The increase in carbon along the Canadian Pacific coast. In: Christian, J.R., Foreman, M.G.G. (Eds.), Climate Trends and Projections for the Pacific Large Area Basin. Canadian Technical Report of Fisheries and Aquatic Sciences. , 3032, pp. 57–66. http://publications.gc.ca/collections/collection_2014/mpo-dfo/Fs97-6-3032-eng.pdf.

Irvine, J.E., Crawford, W.R., 2012. State of the ocean report for the Pacific North Coast Integrated Management Area (PNCIMA). Canadian Manuscripts Reports of Fisheries and Aquatic Sciences 2971 xii±561.

Irvine, J.R., Fukuwaka, M.A., 2011. Pacific salmon abundance trends and climate change. Ices Journal of Marine Science 68 (6), 1122–1130.

Ishimura, G., Herrick, S., Sumaila, U.R., 2013. Stability of cooperative management of the Pacific sardine fishery under climate variability. Marine Policy 39, 333–340.

Knight, R., 1996. Indians at Work: An Informal History of Native Labour in British Columbia 1858–1930. New Star, Vancouver.

Koeller, P., Fuentes-Yaco, C., Platt, T., Sathyendranath, S., Richards, A., Ouellet, P., et al., 2009. Basin-scale coherence in phenology of shrimps and phytoplankton in the North Atlantic Ocean. Science 324 (5928), 791–793.

Kurihara, H., Kato, S., Ishimatsu, A., 2007. Effects of increased seawater pCO_2 on early development of the oyster *Crassostrea gigas*. Aquatic Biology 1, 91–98.

Lavallée-Picard, V., 2016. Planning for food sovereignty in Canada? A comparative case study of two rural communities. Canadian Food Studies/La Revue Canadienne Des Études Sur L'alimentation 3 (1), 71–95.

Lennon, M., 2015. Nature conservation in the Anthropocene: preservation, restoration and the challenge of novel ecosystems. Planning Theory & Practice 16 (2), 285–290.

Lipset, D., 2013. The new state of nature: rising sea-levels, climate Justice, and community-based adaptation in Papua New Guinea (2003–2011). Conservation and Society 11 (2), 144–215.

Lohse, K.A., Newburn, D.A., Opperman, J.J., Merenlender, A.M., 2008. Forecasting relative impacts of land use on anadromous fish habitat to guide conservation planning. Ecological Applications 18 (2), 467–482.

Manson, J., January 6, 2015. Relational Nations: Trading and Sharing Ethos for Indigenous Food Sovereignty on Vancouver Island (MA thesis). University of British Columbia.

McCarty, J.P., 2001. Ecological consequences of recent climate change. Conservation Biology 15 (2), 320–331.

McMichael, P., 2014. Historicizing food sovereignty. The Journal of Peasant Studies 41.6, 933–957.

Melzner, F., Stange, P., Trübenbach, K., Thomsen, J., Casties, I., Panknin, U., Gorb, S.N., Gutowska, M.A., 2011. Food supply and seawater pCO_2 impact calcification and internal shell dissolution in the blue mussel *Mytilus edulis*. PLoS One 6 (9), e24223. http://dx.doi.org/10.1371/journal.pone.0024223.

Mihesuah, D., 2003. Decolonizing our diets by recovering out ancestors gardens. American Indian Quarterly 27 (3/4), 807–839.

Moody, M., Pitcher, T., 2010. Eulachon (*Thaleichthys Pacificus*): Past and Present. Fisheries Centre Research Reports, vol. 18. The University of British Columbia, pp. 1198–6727.

Nemcek, N., Ianson, D., Tortell, P.D., 2008. A high-resolution survey of DMS, CO_2, and O_2/Ar distributions in productive coastal waters. Global Biogeochemical Cycles 22 (2).

Okey, T.A., Alidina, H.M., Lo, V., Jessen, S., 2014. Effects of climate change on Canada's Pacific marine ecosystems: a summary of scientific knowledge. Rev Fish Biol Fisheries 1–41.

Overland, J.E., Wang, M., 2007. Future climate of the North Pacific ocean. Eos, Transactions American Geophysical Union 88 (16), 178–182.

Park, W., Shirley, T.C., 2008. Variations of abundance and hatch timing of dungeness crab larvae in Southeastern Alaska: implications for climate effect. Animal Cells and Systems 12 (4), 287–295.

Pasco, J., Compton, B.D., Hunt, L., 1998. The Living World: Plants and Animals of the Kwakwaka'wakw. U'mista Cultural Society, Alert Bay, British Columbia, Canada.

Pellegrin, N., Boutillier, J., Lauzier, R., Verrin, S., Johannessen, D., 2007. Appendix F: invertebrates. In: Lucas, B.G., Verrin, S., Brown, R. (Eds.), Ecosystem Overview: Pacific North Coast Integrated Management Area (PNCIMA)Can. Tech. Rep. Fish. Aquat. Sci, vol. 2667 iii±37.

Peterman, R.M., Dorner, B., Rosenfeld, J.S., 2012. A widespread decrease in productivity of sockeye salmon (*Oncorhynchus nerka*) populations in western North America. Canadian Journal of Fisheries and Aquatic Sciences 69 (8), 1255–1260.

Pinstrup-Andersen, P., 2009. Food security: definition and measurement. Food Security 1 (1), 5–7.

R. *v.* Sparrow, [1990] 1 S.C.R. 1075.

Rand, P.S., Hinch, S.G., Morrison, J., Foreman, M., MacNutt, M.J., Macdonald, J.S., et al., 2006. Effects of river discharge, temperature, and future climates on energetics and mortality of adult migrating Fraser River sockeye salmon. Transactions of the American Fisheries Society 135 (3), 655–667.

Reardon, J.A.S., Pérez, R.A., 2010. Agroecology and the development of indicators of food sovereignty in Cuban food systems. Journal of Sustainable Agriculture 34 (8), 907–922.

Reuter, K.E., Lotterhos, K.E., Crim, R.N., Thompson, C.A., Harley, C.D.G., 2011. Elevated pCO_2 increases sperm limitation and risk of polyspermy in the red sea urchin *Strongylocentrotus franciscanus*. Global Change Biology 17 (7), 2512.

Ries, J.B., Cohen, A.L., McCorkle, D.C., 2009. Marine calcifiers exhibit mixed responses to CO_2-induced ocean acidification. Geology 37 (12), 1131–1134.

Robertson, L., Kwagu'ł Gixsam, 2012. Standing up With Ga'axsta'las: Jane Constance Cook and the Politics of Memory, Church, and Custom. UBC Press.

Rogers, L.A., Schindler, D.E., 2011. Scale and the detection of climatic influences on the productivity of salmon populations. Global Change Biology. 17 (8), 2546–2558. http://doi.org/10.1111/j.1365-2486.2011.02415.x.

Rose, K.A., Cowan, J.H.J., Winemiller, K.O., Myers, R.A., Hilborn, R., 2001. Compensatory density dependence in fish populations: importance, controversy, understanding and prognosis. Fish and Fisheries 2 (4), 293–327.

Rose, K.A., Megery, B.A., Hay, D., Werner, F., Schweigert, J., 2008. Climate regime effects on Pacific herring growth using coupled nutrient-phytoplankton–zooplankton and bioenergetics models. Transactions of the American Fisheries Society 137, 278–297.

Satterfield, T., et al., 2013. Culture, intangibles and metrics in environmental management. Journal of Environmental Management 117, 103–114.

Satterfield, T., Robertson, L., Pitts, A., Jacobson, D., 2015. Reasserting 'Namgis Food Sovereignty aka Three Boats and Pick up Truck. https://open.library.ubc.ca/cIRcle/collections/facultyresearchandpublications/52383/items/1.0307423.

Schindler, D.E., Scheuerell, M.D., Moore, J.W., Gende, S.M., Francis, T.B., Palen, W.J., 2003. Pacific salmon and the ecology of coastal ecosystems. Frontiers in Ecology and the Environment 1 (1), 31–37.

Schreiber, D., 2008. "A liberal and paternal spirit": Indian agents and native fisheries in Canada. Ethnohistory 55 (1), 87–118.

Schnute, J.T., Haigh, R., Krishka, B.A., Starr, P., 2001. Pacific ocean perch assessment for the west coast of Canada in 2001. Canadian Science Advisory Secretariat Research Document 2002/138 96.

Ströhle, A., Hahn, A., 2011. Diets of modern hunter-gatherers vary substantially in their carbohydrate content depending on eco-environments: results from an ethnographic analysis. Nutrition Research 31 (6), 429–435.

Thom, B., 2003. Intangible property within coast salish first Nations communities, British Columbia. In: Presented at the WIPO North American Workshop on Intellectual Property and Traditional Knowledge, Ottawa, September 9, 2003 Available: http://www.hulquminum.bc.ca/pubs/paper-wipo.pdf?lbisphpreq=1http://www.pac.dfompo.gc.ca/sci/psarc/OSRs/Ocean_SSR_e.htm.

Turner, N.J., Clifton, H., 2009. "It's so different today": climate change and indigenous lifeways in British Columbia, Canada. Global Environmental Change 19 (2), 180–190.

Turner, N.J., Gregory, R., Brooks, C., Failing, L., Satterfield, T., 2008. From invisibility to transparency: identifying the implications. Ecology and Society 13 (2).

Turner, N., 1995. Plant Foods of Coastal First Peoples. UBC Press.

UBCIC Union of British Columbia Indian Chiefs n.d. Our homes are bleeding: McKenna-McBride royal commission, "Kwawkewlth Agency Testimony," 1 June 1914, 109. http://ubcic.bc.ca/Resources/ourhomesare/testimonies.htm.

Vadeboncoeur, N., 2016. Perspectives on Canada's west coast region. In: Lemmen, D.S., Clarke, C.M., Warren, F. (Eds.), Canada's Coasts in a Changing Climate: Understanding Impacts and Adaptation. Government of Canada, Ottawa, ON.

Weatherdon, L.V., Ota, Y., Jones, M.C., Close, D.A., Cheung, W.W.L., 2016. Projected scenarios for coastal first Nations' fisheries catch potential under climate change: management challenges and opportunities. PLoS One 11 (1).

Weinstein, M., Morrell, M., 1994. Need Is Not a Number: Report of the Kwakiutl Marine Food Fisheries Reconnaissance Survey. BC: Kwakiutl Territorial Fisheries Commission, Alert Bay.

Weinstein, M. (2007) Teaching each other: the Canadian struggle over aboriginal rights to marine fish and fisheries. A Chapter Prepared for the Norwegian Whitepaper on Global Aboriginal Rights to Fish and Fisheries Relevant to the Implementation of Sámi Aboriginal Fish and Fisheries Rights within Norway. (The Norwegian Commission: Kystfiskeutvalget for Finnmark.)

Welch, D.W., Ishida, Y., Nagasawa, K., 1998. Thermal limits and ocean migrations of sockeye salmon (*Oncorhynchus nerka*): long-term consequences of global warming. Canadian Journal of Fisheries and Aquatic Sciences 55 (4), 937–948.

Whyte, K.P., February 28, 2017. Food sovereignty, Justice and indigenous peoples: an essay on settler colonialism and collective continuance. In: Barnhill, A., Doggett, T., Egan, A. (Eds.), Oxford Handbook on Food Ethics. Oxford University Press (forthcoming). Available at: https://ssrn.com/abstract=2925285.

Williams, T., Hardison, P., 2013. Culture, law, risk and governance: contexts of traditional knowledge in climate change adaptation. Climatic Change 120 (3), 531–544.

FURTHER READING

Orsi, J.A., Harding, J.A., Pool, S.S., 2007. Epipelagic fish assemblages associated with juvenile Pacific salmon in neritic waters of the California Current and the Alaska Current. American Fisheries.

Suttles, W., 1991. Streams of property, Armor of wealth: the traditional Kwakiutl Potlatch. In: Jonaitis, A. (Ed.), Chiefly Feasts. Douglas and McIntyre, Vancouver, pp. 71–134.

Chapter 20

The Future of Modeling to Support Conservation Decisions in the Anthropocene Ocean

Éva E. Plagányi[1], Elizabeth A. Fulton[2,3]

[1]CSIRO Oceans & Atmosphere, St Lucia, QLD, Australia; [2]CSIRO Oceans & Atmosphere, Hobart, TAS, Australia; [3]Centre for Marine Socioecology, Hobart, TAS, Australia

INTRODUCTION

Evidence of the use of terrestrial and marine resources is equally old, extending back at least 40,000–50,000 years (O'Connor et al., 2011). However, the oceans (especially the deep ocean) were not as readily accessible until the mid- to late 20th century, with few deep water resources extensively exploited before then (with whales being a notable exception). As such, the extent of anthropogenic uses of the ocean has grown much more slowly and largely been constrained to coasts and near shore waters. Until very recently, the average vision of the use of the oceans centered on fisheries, shipping, and military strategy (with oil exploration added in the last century), but this is now changing rapidly.

Terrestrially, broad-scale industrialization of landscapes has reshaped the world's land surfaces, especially in the last 250 years (since the dawn of the industrial revolution) and particularly post-World War II and the period known as the "Great Acceleration." Now, as accessible terrestrial resources have dwindled, more of the activities we conceive of as terrestrial are transitioning to marine industries, with many indicators suggesting that the oceans are on the cusp of their own industrial transformation (Fig. 20.1). Within the next two decades, it is likely that marine farming (aquaculture and mariculture) will provide more for human consumption than wild capture fisheries (Msangi et al., 2013). Thus the use of oceans for hunting and farming will be one small part of a mosaic that includes mining, waste disposal, energy generation, transport, leisure (recreation and tourism), pharmaceutical production, conservation areas, and perhaps even geo-engineering. In addition, land reclamation and sea steading (creation of islands or floating settlements) are likely to become far more common, particularly as sea-level rise impacts low-lying nations (e.g., Holland has already opened some of its first floating neighborhoods).

Conservation for the Anthropocene Ocean. http://dx.doi.org/10.1016/B978-0-12-805375-1.00020-9

423

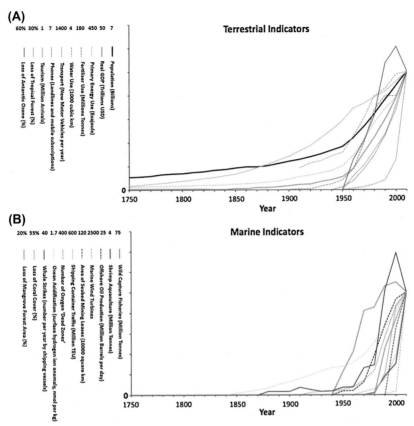

FIGURE 20.1 Time series of indicators of (A) terrestrial and (B) marine uses and indicators, showing the level of industrialization of global landscapes and seascapes. Note that the oceans appear to be on the cusp (or in the midst) of the kind of broad-scale and diverse industrialization that began on land 50–200 years ago. *Data from McCauley, D.J., Pinsky, M.L., Palumbi, S.R., Estes, J.A., Joyce, F.H., Warner, R.R., 2015. Marine defaunation: animal loss in the global ocean. Science 347, 1255641; Steffen, W., Broadgate, W., Deutsch, L., Gaffney, O., Ludwig, C., 2015. The trajectory of the anthropocene: the great acceleration. The Anthropocene Review 2, 81–98.*

Just as on land, conservation issues need to be considered as part of planning and managing ocean uses. Decision making pertaining to conservation issues is already being challenged by ever increasing complexity, as the number and types of ocean users' increase, in synchrony with growing interdependencies between the ecosystem and human well-being (and recognition of those interdependencies). This challenge will only increase as anthropogenic ocean uses become more diverse and intense and also as climate change and ocean acidification influence ocean conditions to an extent not seen in many millions of years (Fig. 20.2; Turley et al., 2006). Acclimatization and evolution may moderate or exacerbate the intensity of effects on marine fauna and flora, and the

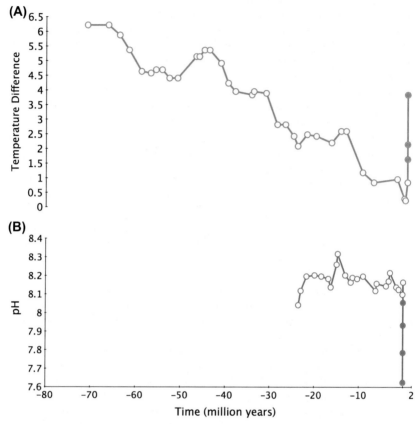

FIGURE 20.2 Time series of observed and projected global (A) average temperature and (B) ocean pH values. The open circles are from various data sources, solid circles are model-based projections. *Redrawn from Turley, C., Blackford, J., Widdicombe, S., Lowe, D., Nightingale, P., Rees, A., 2006. Reviewing the impact of increased atmospheric CO$_2$ on oceanic pH and the marine ecosystem. Avoiding Dangerous Climate Change 8, 65–70.*

rapidity of change will stress marine ecosystems and those charged with their management and sustainability. To be successful in this context, conservation decisions will need to directly address the role of humans as one of the great shapers of modern seascapes and coastal landscapes, on par with geological and climatological forces (Steffen et al., 2015).

Research and management to achieve conservation targets will need to (already needs to) synthesize information across a broad range of disciplines, deal with nonstationary complexity, where conditions and interactions change rapidly and are dependent on historical or recent decisions, and where there is a high degree of uncertainty (both around the understanding of novel ecosystems forming in the changing conditions, but also about future potential trajectories). Modeling is a valuable tool for synthesizing and analyzing this complexity, as well as simulating innumerable alternative scenarios that would be impossible to test empirically. Indeed, models themselves have not stayed static but scientists are constantly advancing models to equip them with the structures and capabilities needed to support conservation decisions in the Anthropocene Ocean (Fig. 20.3).

In the past, individual disciplines often developed in relative isolation with resulting divergence in perspective on related topics. For example, fisheries and conservation have contrasting views on the implications of changes in population biomass—to a classical fisheries scientist, achieving target biomass levels (i.e., the point of maximum sustainable yield from a fish population) required a 50% (or greater) reduction in biomass, whereas a conservation ecologist would find such a decline deeply troubling. As understanding of system dynamics has grown, and as modern management approaches have reflected this understanding [e.g., fisheries management approaches are often obligated to account for "ecosystem considerations" and to achieve an ecosystem approach to fisheries (EAF), e.g., Garcia, 2003], the need to move to interdisciplinary approaches has been realized, with the coincident need to reconcile and merge the different disciplines. Initially, this linkage was achieved via the inclusion of weak (often unidirectional) linkages between models developed by biologists, oceanographers, economists, and social scientists, with the focus restricted to improving

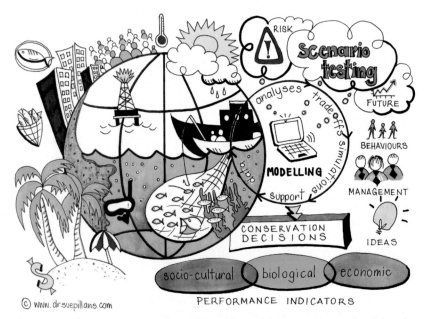

FIGURE 20.3 Graphic showing the complex interdependencies that characterize the modern Anthropocene Ocean, and the role of models in synthesizing and analyzing this complexity, as well as testing alternative scenarios within the safety of cyber space. *Graphic by Dr Sue Pillans, Copyright © 2016 www.drsuepillans.com.*

the depth and sophistication of these discipline-specific descriptions of the ocean's biology, functioning, climate, and human use (Clark, 1980; Fasham et al., 1990; Jennings et al., 2001; Travers et al., 2007). More recently, there is increased focus on broadening and connecting these components in a far more integrated way. In this chapter, we present examples of current advances in developing more holistic models and forecast future trajectories describing model development pathways.

ASPECTS OF THE FUTURE OF MODELING

Marine population models have a long history (Schaefer, 1954) of development and implementation as tools for assessing the current status and trends of populations (Quinn and Deriso, 1999). In this way, the models have played a critical role in supporting decision making pertaining to both fisheries management and conservation of species and ecosystems. Single-species assessment models remain useful as tools for providing tactical management advice based on robust predictions that have been validated using historical data. However, the scope of these models can be overly narrow and insufficient as a base for conservation-management recommendations, when considering that other drivers such as climate change and globalization are rapidly influencing ocean dynamics, and there are no historical analogs for these processes (or in some cases the novel communities they are producing). In addition, accurate representation of a species' population dynamics in isolation has limited utility in securing a desired conservation (and fisheries) outcome if there are important environmental drivers, ecological interactions, or sociopolitical factors that need to be simultaneously considered (Raemaekers et al., 2011).

A good deal of the modeling effort of biophysical modelers has focused on better integrating physical processes, biogeochemical processes, and ecosystem dynamics (particularly food web and habitat-related interactions; Fulton and Link, 2014). Such work is ongoing with new processes, such as acclimatization and evolution being added to biological and ecosystem models (De Roos et al., 2006; Zhang et al., 2015) to better understand future biological adaptation capacities. This work builds off research into the evolutionary adaptation of animals to key variables such as temperature (e.g. Clarke, 2003; Russo et al., 2010) and the interaction of evolution and ecological processes (Hairston et al., 2005). This information is being increasingly incorporated into models to reduce the uncertainty associated with projections incorporating future changes in survival, distribution, and productivity of species (Pörtner and Gutt, 2016). Indeed, a broad spectrum of novel approaches is constantly evolving in ecology, with the potential to dramatically influence and improve future modeling efforts. These include refinements of alternative ways to consider growth and change at an individual level and the implications for communities (e.g., dynamic energy budget modeling; Nisbet et al., 2012), but also fundamentally different ways of considering system structure and its representation (e.g., at the global scale contrast the size-based models of Blanchard et al. (2012) with the agent-based approaches of Harfoot et al. (2014)). In addition, completely new modeling approaches are being applied to modeling marine and other systems. For example, Sugihara et al. (2012) highlight the importance of nonlinear system behavior and are developing methods based on nonlinear state space reconstruction for identifying causality in complex systems. Similarly, network theory (Gao et al., 2016; Yu et al., 2016) is being used to explore the dynamic nature of interconnected systems, such as food webs, and how these change under pressure. Another research area that has been identified is the need for a new generation of models for representing disease threats affecting ocean systems, noting that the knowledge needed to transfer terrestrial-based epidemiological theory to marine systems is still in its infancy (McCallum et al., 2004).

However, this purely biophysically oriented work is insufficient in isolation, and further integration, with sciences dealing with human behavior and processes, including economics, psychology, anthropology, and other social sciences, is important, as there is increasing recognition that social factors are critical in natural resource management. Methods that incorporate so-called "social–ecological" considerations can enhance natural resource management decision making (e.g., Österblom et al., 2013; van Putten et al., 2013). Indeed, it is well recognized that quantifying ecosystem services and analyzing trade-offs, such as between resource status, economic gains, and social outcomes, can improve the effectiveness, efficiency, and defensibility of natural resource decisions (Nelson et al., 2009). Policy makers have also recognized this, with the requirement for "triple-bottom-line" (environmental, economic, and social) outcomes and the inclusion of a wide range of social objectives, alongside economic and ecological criteria, in fisheries management (Pascoe et al., 2014; Wadsworth et al., 2014).

Recently, much attention has focused on methods for incorporating social considerations into social–ecological models (Ostrom, 2009) and the field is a rapidly growing area of research. One example is the work of Hicks et al. (2016), which showed that the human activities that put pressure on marine ecosystems and can cause those systems to shift in state can also act as early warning signals for those shifts. This builds on a large body of work that shows common dynamic signatures and patterns of behavior of many complex systems from financial markets to psychological health and ecosystems (Scheffer et al., 2012). This interest in complex systems and the desire to explicitly include social aspects in integrated models is driven by direct curiosity around system understanding but also because tools are needed to assist management. For instance, tools are needed to identify the potential consequences of trade-offs among fisheries management and conservation objectives (Christensen and Walters, 2004; Gaichas et al., 2012; Levin et al., 2013; Link, 2010). The use of appropriate models and management strategy evaluation (MSE) (Smith et al., 2007) are widely accepted as integral components of advancing modern fisheries (and conservation) science and achieving an EAF approach (Levin et al., 2009; Link, 2010). This is also considered best practice by the Food and Agriculture Organization of the United Nations (FAO, 2008).

Models and MSE come in many forms, from qualitative to quantitative, and are useful for particular aspects of management decision support. Tactical models are intended to inform specific operational management decisions, such as recommending the total catch (either single-species or for a multispecies complex, or the catch of a target species that takes into account impacts on other species in the ecosystem). Strategic models are focused on a broad-scale assessment of direction and change in the ecosystem. Their uses include improving the understanding of the impact of different management alternatives on the structure and functioning of an ecosystem, as well as the social and economic consequences.

Both kinds of models are needed, both for managing the oceans now and for thinking about what the future will bring. The Anthropocene Ocean will be a place where past observations may provide only limited insight, as shifting species distributions will see species combinations and ecosystem structures previously unobserved (Hobday and Pecl, 2014). Similarly, human uses will interact with the ecosystem in ways we may not yet be able to anticipate. Experience from the terrestrial sphere will provide some information but the highly interconnected and fluid nature of marine systems will undoubtedly lead to some novel outcomes.

An uncertain future without precedents does not mean modeling is a futile exercise, models can still provide significant insights and highlight possibilities: synthesizing information; identifying critical information gaps; and allowing an exploration of existing understanding and alternatives within a consistent framework and under conditions that may be hard to observe in reality, or have yet to have ever been experienced (Fulton et al., 2015). Used in this way models can allow managers (as well as users and society) to understand the potential outcomes and costs associated with alternative management options (including the counterfactual of no action). In this way, they can gain insight into strengths and weaknesses of the management approaches. They facilitate comprehending how the various options function under uncertainty or directional drivers of change. This allows managers to anticipate future potential issues and crises, providing for proactive preparation rather than only reactive responses (Fulton et al., 2015).

A wide variety of ecological simulation models exist and have been reviewed elsewhere (Hollowed et al., 2000, 2009; Plagányi, 2007), ranging from more tactical single-species models coupled to physical oceanographic models (Hobday et al., 2011), through models that focus on a key subset of the ecosystem only, such as MICE (Models of Intermediate Complexity for Ecosystem assessments) (Plagányi et al., 2014b), to more strategic whole ecosystem or end-to-end models such as Atlantis (Fulton et al., 2005, 2014; Rose et al., 2010). Here, we provide three examples of how socioecological frameworks and two-way dynamic feedbacks can be incorporated into marine conservation models and decision support. Although we cannot anticipate the myriad ways models will develop (or need to be developed) in the future to deal with currently unthought-of conservation challenges, these examples provide some immediate practical examples on how to begin extending marine models to explore the relationships among human activities, natural perturbations, and risks and status of human biophysical dimensions of the ecosystem.

Extending Existing Bioeconomic Models into the Social Domain

Broadening of societal objectives for its shared resources has seen a move away from a focus on economic returns in isolation from other socioecological aspects. An excellent example of this is in the Torres Strait tropical rock lobster (TRL), *Panulirus ornatus* fishery, which is the most important commercial fishery to Torres Strait Islanders (in the waters immediately north of the northern tip of Queensland, Australia). The fishery is managed by the Protected Zone Joint Authority, comprising representatives from the Australian and Queensland governments under Article 22 of the Torres Strait Treaty (February 1985) between Australia and Papua New Guinea. The Treaty refers not only to the need for optimal utilization of the resource, but also to the need for maximizing the opportunities for participation by the traditional inhabitants of both countries. This poses complex new challenges in terms of using modeling to support conservation decision making. Although bioeconomic modeling has been a recognized modeling approach for more than 30 years (Charles, 1983), biological and social models have seldom been coupled; in part because of the discrepancy between using quantitative data pertaining to the resource as opposed to more qualitative sociocultural information.

Plagányi et al. (2013b) extended a more traditional biological assessment framework to include sociocultural considerations as well as economic information for different sectors of the fishery (Fig. 20.4). Representation of the social dimension was underpinned by extensive stakeholder consultation that included a series of dedicated workshops and individual interviews. To simulate the outcomes of different management systems, it was necessary to quantify how participation by each fisher subgroup would change. To estimate participation rate changes for the Traditional Inhabitant Boat license (TIB) sector, which depends on social as well as economic drivers, a Bayesian Network (BN) analysis was used (van Putten et al., 2013). BNs use conditional probability tables to represent dependence between variables as a basis for evaluating the likelihood outcomes of parameters of interest. The TIB fleet was divided into subfleets based on a typology of activity and alternative licensing arrangements (as well as technical and economic factors) and this facilitated prediction of the changes in participation under alternative scenarios. The coupled model calculated profit per sector/subfleet but also incorporated a production frontier analysis to derive estimates of the marginal value product for each sector (Pascoe et al., 2013a). Data envelopment analysis (DEA) (Pascoe et al., 2013b) was used to estimate which nonindigenous vessels might exit the fishery with lower quota levels. DEA is an economics linear programming method used for capacity

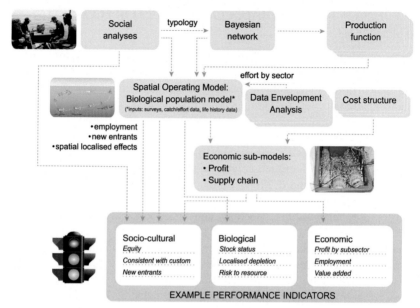

FIGURE 20.4   Schematic of the content of the Plagányi et al. (2013b) model of the Torres Strait tropical rock lobster fishery.

assessment, based on estimation of technical efficiency and capacity utilization. The model included estimates of owner operator returns to labor for the islander traditional boat holder license operators. Finally, the framework included the supply chain, which enabled prediction of an additional performance indicator, namely value added (Fig. 20.4).

Briefly, the model illustrated significant conflicts and trade-offs in traditional communities such as the Torres Strait Islander fishers, where value was found to depend also on principles of equity, community coherence, resemblance to sea country (relating to the authority held and responsibilities of particular groups of Traditional Owners to particular areas of sea), and Island custom (Plagányi et al., 2013b). For example, there are complex trade-offs between economic indicators (such as profit) and social indicators (such as lifestyle preferences and community coherence). The study highlighted two key social indicators, namely equity and a sense of self-determination, both of which correlate strongly with the level of buy-in to a management approach, the lack of which is one of the fundamental causes of failure of management systems globally (Defeo et al., 2014; Raemaekers et al., 2011). These MSE outputs underscored the complexity of trade-offs between social and economic considerations for islander (indigenous) and nonislander participants, and hence is one way forward in terms of quantifying and making explicit the trade-offs in the impact of alternative management strategies on the "triple-bottom-line" sustainability objectives (Plagányi et al., 2013b).

MODELS OF INTERMEDIATE COMPLEXITY FOR ECOSYSTEM ASSESSMENTS

MICE models bring a system's perspective to the modeling of marine ecosystems, but maintain models focused enough to allow for the weight and power of statistical methods to be brought to bear. These models are fitted to all available data using the same methods as applied in stock assessment models, and hence provide rigorous multispecies assessments together with quantification of the associated uncertainty (Plagányi et al., 2014b). These models are focused on the interaction of a small number of key processes (environmental, biological, economic, and social) and are constructed to focus on targeted conservation and management questions. Model development, therefore, starts with consideration of the focus species or species complex, and includes linkages such as predator–prey relationships, which are considered key and account for a substantial proportion of the mortality on a species. Similarly key environmental, social, or economic relationships could be included. These models are typically restricted to 10 or fewer components given there are usually too few data available to represent more species in the system. The approach focuses on finding the sweet spot where uncertainty is reduced (by including key drivers considered as externalities in past single species models) and the utility of these models to management is maximized (Collie et al., 2014).

There are a growing number of MICE models being developed (Angelini et al., 2016; Blamey et al., 2013; Morello et al., 2014; Punt et al., 2016). The South African abalone *Haliotis midae* fishery provides a compelling example of the need to account for ecosystem considerations, including the role of humans in fisheries management. Although fairly sophisticated stock assessment approaches are used in managing the fishery (Plagányi et al., 2011a), the complex social, political, and economic drivers influencing fisher behavior needed to be understood to develop more holistic solutions for management. Raemaekers et al. (2011) propose that rampant illegal fishing could have been prevented or reduced in the early stages through a more integrated governance approach that accommodated traditional fishers in a legal fishing rights framework. In addition, changes in environmental conditions during the early 1990s led to an eastward shift in the West Coast rock lobster into the area that comprised the heart of the commercial abalone fishery. The rock lobsters caused a decline in urchins, *Parechinus angulosus* (Mayfield and Branch, 2000). Juvenile abalone have a close association with the urchin, whereby they take shelter beneath urchin spines receiving both nourishment and protection, and hence this resulted in further depletion of abalone in the areas where lobsters occurred.

To facilitate understanding of the complex interactions and to test management alternatives, a MICE was developed and included the key groups only, namely abalone, urchins, lobsters, and fish (Blamey et al., 2013). The model was useful in providing insight on the performance of alternative management approaches (such as ensuring that fish stocks are sustainably harvested) to avoid future ecosystem shifts. This was because under the observed overfishing scenario; the model predicted that lobsters invaded the range of abalone, depleted the urchins, changed the benthos, and crashed the abalone population. In contrast, under the sustainable fishing scenario, although lobsters invaded the range of abalone, they were kept in check by higher fish density and predation and hence the system was more resilient to changes (Blamey et al., 2014).

Although MICE originally focused on biophysical aspects of ecosystems, the approach has been successfully extended to include economic aspects (Plagányi et al., 2011b, 2014b) and is now being extended to social aspects; that is, encompassing the principles outlined by Österblom's social–ecological systems (SES) framework (Österblom et al., 2013), but focusing on key linkages only and representing these in as simple a manner as possible while still capturing the important dynamics. One such approach currently being trialed involves quantifying the social science concept of sense of place. Sense of place is a dynamic concept as people develop an attachment, or aversion, to a place over time (Zia et al., 2014). Changes in sense of place influence movement rates and utilization pressure by humans on natural environments, and this, in turn, alters the magnitude and types of pressures exerted on local natural resources.

MICE models are thus being developed that include a Sense of Place Index (SoPI) to dynamically link the two-way feedbacks between ecological systems and socioeconomic systems (Van Putten et al., 2017). Used as a model variable, SoPI allows for the quantitative integration of environmental psychology into socioecological models (Van Putten et al., 2017). As an example of two-way dynamic feedbacks between ecological and human systems, consider the impact of a development scenario that increases jobs and migration to an area. But this in turn decreases water quality which in turn results in fewer fish and hence less tourists and users, and this again influences fish abundance so that the net reduction in fish abundance is influenced by feedbacks due to human perceptions of the environment. Explicitly accounting for SoPI feedback in marine SES models can better inform local planners and decision makers to potential over- or underprediction of flow-on effects of tourist numbers or residents. At the same time, using SoPI to dynamically model changes in human sociodemographics could improve predictions of future direct environmental impacts such as fishing, as well as indirect impacts such as water quality due to development.

Integrated and Multiple Use Models

Much broader scale models are required to consider the multitude of connections and feedbacks active in multiple use areas; where multiple industries are operating, each with its own effects on the underlying ecosystem and surrounding community, and each with its own objectives. Such models are much broader in scope (incorporating many more sectors than MICE) and also much more uncertain. This uncertainty stems not only from the sheer scale of the models (and the need to parameterize them with incomplete observations), but also from fundamental uncertainty around system function and how that will evolve into the future.

Marine Integrated Assessment Models, as large multisector system models may be known (particularly at the global scale), or whole-of-system models (the term used for more regional models that span all socioecological aspects) are still few in number but the approach is under active development and use. The simplest representation of human action in these models is through the forced evolution of major environmental or economic drivers, with the shape of those drivers representing the outcomes of scenario narratives (i.e., storylines around potential changes; O'Neill et al., 2013; Van Vuuren et al., 2011). More dynamically, instead of using forced time series, system models have been used to create impact functions for use in ecosystem models focusing on fisheries and their futures under climate change in Australia (Fulton and Gorton, 2014). Just as functional responses represent how feeding pressure shifts with altered predator and prey biomass, so do impact functions dictate how the level of pressure on an ecosystem variable (or variables) shifts with changing human activity levels. Impact functions were used because the ecosystem models themselves were quite complex (dynamically representing the physical drivers, biogeochemical cycles, food web and habitats, fisheries sectors, and fisheries and conservation management processes associated with 3.7 million km^2 of Australia's southeastern marine estate) and so the additional computational costs of dynamically coupling to a multi-industry systems model were prohibitive. The impact function was not static but was a simplified emulator of the output of a more complex model that explicitly represented the activities (stocks and flows) of coastal industries (including agriculture and urban centers) and their impacts on river flow, sediments, pollutants, and salinity levels. The simple economics-based representation of coastal industries also allowed for shifting demography, effects on available labor sources and demands on services, innovation and technological uptake (changing costs, prices, and efficiencies), and shifts in infrastructure (capacity, costs, and maintenance). By conditioning the emulator on a systems model that had explicitly incorporated dynamic social and economic interactions, the final impact function successfully represented the way in which cumulative impacts of the development of marine industries (other than fishing) and coastal population centers could influence the sustainability of fisheries over the next 50 or more years, without incurring a significant computational penalty.

Models with shorter projection periods, or smaller spatial extents, can allow for a more even and explicit handling of the different sectors without resorting to emulators and impact functions. One example is the in vitro modeling framework, which has been applied in northwestern Australia to explore the dynamics of many aspects of coastal and marine systems, including: environmental drivers, ecological ecosystem components, fine-scale human decision models of local communities and industry operators (across multiple industries), regional economics, management, monitoring, and assessment (Fulton et al., 2011; Gray et al., 2006). This model employs a modular hybrid approach likely to become more widely used when modeling socioecological systems in the future. This approach allows for alternative representations for each part of the system, representing the different processes using a mix of analytical equation-based formulations and algorithmic individual-based (agent-based) behavioral models, employing empirically derived behavioral rules (or decision trees).

The analytical formulations were often employed for system-scale economic drivers or for physical, chemical, and lower trophic level processes; whereas the agent-based representations were often employed for rare species, top-level predators, or human activities occurring at the finest scales. Each submodel was executed at the temporal and spatial scales appropriate for that set of processes. This hybrid model structure allows for a much broader system scope to be captured than would be possible if all had to be done to the same resolution and in the same model style. The final model allowed for the exploration of sustainable multiple use management questions and the realization by managers across multiple jurisdictions that (1) there are complex relationships between development and environmental status in the region; and (2) that there is the need for system-level management and decision making across sectors (e.g., growing industrial development may meet local employment needs, but the additional recreational pressure applied by the growing population may be sufficient, especially in combination with climate-induced coral bleaching, to cause a collapse in some key target reef fish species).

MODELS AS RISK MANAGEMENT TOOLS

Management of risk, especially cumulative risk, is already a central concern for management agencies in places such as Europe, North America, and Australia. Spatial planning has been used in the short term to deal with the increasingly crowded nature of places such as the North Sea, but as development on land has already shown us, this is not a long-term solution in isolation (McCauley et al., 2015). Moreover, it is projected that more than a billion people will live along coastlines by 2050 (Neumann et al., 2015). The stresses being faced in busy trade ports and the fuzzy nature of coastlines in Asia (where human use of the coastal margin means it can be unclear where the land ends and the water begins) provide early warnings that the future industrialization of the oceans will mean that marine risk management will be in spiraling demand.

MSE approaches can serve as formal risk assessment methods, given their focus on the identification and modeling of uncertainties as well as in balancing different representations of resource dynamics (Sainsbury et al., 2000). This includes consideration of the implications, for both the resource and its stakeholders, of alternative combinations of monitoring data, analytical procedures, and decision rules (Rademeyer et al., 2007; Sainsbury et al., 2000;

Smith et al., 2007). By identifying and evaluating trade-offs in performance across a range of management objectives, it provides indicators on whether different objectives can be reconciled and whether the outcomes are robust to inherent uncertainties in the inputs and assumptions on which decisions are based (Cooke, 1999).

Plagányi et al. (2013a) used MSE to integrate across biological and climate uncertainties, and test the performance and risks (biological, multispecies, economic) of alternative management strategies applied to the Torres Strait bêche-de-mer (sea cucumber) fishery. A Reference Set (Rademeyer et al., 2007) of alternative model parameterizations was used to collectively capture some of the key biological uncertainties (e.g., alternative natural mortality estimates and steepness of the stock–recruitment relationship), as well as uncertainty of the likelihood and severity of climate-change effects; the former by using high-risk scenarios only versus assuming both high- and medium-risk scenarios occur and the latter by accounting for a doubling of the severity of each postulated effect. In this way, Plagányi et al. (2013a) simultaneously integrated across a range of biological and climate-impact uncertainties, and thereby tested a range of alternative harvest strategies (including differences with respect to investment in monitoring) to evaluate performance under changing climate.

Plagányi et al. (2013a) found that *status quo* management would result in half the species falling below target levels, moderate risks of overall and local depletion, and significant changes in species composition. The three strategies with minimal investment in monitoring (spatial rotation, closed areas, multispecies composition) were all successful in reducing these risks, but with fairly substantial decreases in the average profit. Higher profits (for the same risk levels) could only be achieved with strategies that included monitoring and hence adaptive management. Spatial management approaches based on adaptive feedback performed best overall. Their study provides a demonstration of use of MSE to test the performance and adaptability of alternative harvest strategies in meeting fishery objectives, such as ensuring, low risk of depletion (overall and local), high probability of good catch and profit, low risk of changing the multispecies community composition, and a high probability of managing through climate variability and change.

Models to Inform Strategic Planning to Improve Future Conservation Outcomes

One of the biggest challenges for the future involves supporting the resilience of ecosystems and managing human activities to ensure that they do not trigger abrupt and irreversible shifts as a result of tipping points being transgressed (Hughes et al., 2013). Abrupt changes in ecosystems can have major consequences and be impossible to reverse, but there is currently limited biological understanding available, which makes it difficult to model these threshold responses (Bakun and Weeks, 2006; Hughes, 1994; Scheffer et al., 2001). It is generally accepted that resilience can be enhanced, for example, by conserving biodiversity and avoiding excessive depletion of individual, and particularly key, species (Folke et al., 2004; Hughes, 1994; Hughes et al., 2007). Models are a useful tool for understanding underlying drivers as well as system responses and can simulate the efficacy of alternative management actions to maintain or enhance ecosystem resilience. As an example, Blamey et al. (2014) used a MICE model of a kelp–forest ecosystem along the southwest coast of South Africa, to show that a regime shift might have been avoided in this system had linefish not been overfished.

Resource managers would benefit from having early warning signals to identify approaching tipping points, and need methods to distinguish early warning behaviors from noise (Carpenter and Brock, 2006; Boettiger and Hastings, 2012; Boettiger et al., 2013). Rising variance or standard deviation associated with time-series observations of ecosystem state variables, such as biomass, catches, or chemical concentrations, has been identified as a leading statistical early warning signal of regime shifts and may be present in a variety of systems, including ecological, physical, social, and financial (Carpenter et al., 2008; Scheffer et al., 2009). Plagányi et al. (2014a) used modeling and empirical observations to show that increasing variance in population monitoring data from three marine systems (a kelp forest, coral reef, and open ocean ecosystem) may be a simple and useful diagnostic to signal a forthcoming abrupt decline. In this way, models can be used to strategically plan for the changes ahead, or inform on ways to mediate or avoid undesirable ecosystem states. The growing adoption of adaptive management principles facilitates the ability to use feedback from monitoring to adjust management in line with agreed goals and targets.

Over the past decade, the divide between the science required by fisheries and conservation has dwindled, the same science and the same tools can serve both communities (Jennings et al., 2014). Indeed, the objectives are also aligning with components of conservation clear in ecosystem-based approaches to fisheries management and sustainable provisioning of society being posited as a driver for marine conservation by at least some in the conservation community (Sharpless and Evans, 2013). Although conservation planning and management has not embraced the use of models as fundamentally as fisheries management has (Bunnefeld et al., 2011; Nicholson et al., 2012), this does not mean that models have nothing to offer or that any distrust of modeling methods is permanent. It took time for the approaches to be accepted in fisheries but models are now one of many information sources regularly used by fisheries management when making management decisions on both annual and longer-term scales (e.g., Dichmont et al., 2006; Fulton et al., 2014; Klaer et al., 2012; Plagányi et al., 2015).

Modeling in support of strategic planning is a tool with broad potential benefit to fisheries and conservation alike. An example of this is the work by Fulton and Gorton (2014), mentioned earlier, where a whole-of-system model was used to consider the future of fisheries (and conservation) in southeastern Australia. The core finding of this work was that sectoral management in isolation was rapidly overwhelmed by cumulative impacts and that only integrated management (across sectors and dynamically accounting for climate influences) successfully met the many objectives that society has expressed for the region, around food security, economic returns, employment, and conservation. The work highlighted the potential for climate and shifting human uses to complicate cross-jurisdictional arrangements and to undermine the effectiveness of existing conservation measures. This need for cross-sector (e.g., on and off reserve management) has been highlighted by other modeling studies (Brown and Mumby, 2014; Dichmont et al., 2013; Savina et al., 2013), but also by experience derived from observing how the mismatch of natural and management scales on land has undermined terrestrial conservation efforts (McCauley et al., 2015).

CONCLUSIONS

The Anthropocene has arrived much later in the oceans than on land. This has provided managers and scientists focusing on the marine realm with an unprecedented opportunity. Not only are there 50–250 years of experience of what industrialization has done on land to act as a database to inform potential marine trajectories, but the sciences—and in particular the models—available today far outstrip those available as the early ecologists and sectoral managers struggled with the consequences of the reshaping of terrestrial landscapes during periods such as the Great Transition. It is a particularly intriguing qualitative counterfactual to ponder how different our world might now be if the scientists of 50–100 years ago had realized that they were living through a period of rapid transition and if they had had access to integrated modeling tools for improving understanding, exploring the options, highlighting undesirable states, and pathways to avoid them; tools for understanding how history shapes the future; and tools for finding new ways of managing things that were not only robust to change but actively acknowledged and worked with it (such as dynamic ocean management, where the boundaries of management zoning can shift in near real time with critical ocean properties; Hobday and Hartog, 2014).

Uncertainty remains a significant issue bedeviling model utility. Better integration of different disciplines is constantly advancing the frontiers of what is possible in models. Although considerable further work is needed to validate and improve models, significant effort is being dedicated to data collection, laboratory experimentation, and theoretical advances to better ground the complex modeling tools. Improvements in the use of uncertainty handling methods, data-model fusion (Zobitz et al., 2011) and skill assessments (Olsen et al., 2016; Stow et al., 2009) are also advancing the robustness and reliability of the models. Moreover, the wise use of models and modeling approaches, such as ensembles, already facilitates the wise use of models, cognizant of their uncertainty, strengths, and weaknesses. As highlighted in our examples, models of different complexity and rigor will be needed to address a range of research and management questions, from those providing more precise short-term predictions to those intended for use to broadly identify longer-term plausible outcome ranges.

The Anthropocene has brought great change to land and will do so to the sea but it presents as many opportunities as challenges. Modeling provides a means for initiating and supporting discussions around potential future threats and responses and of identifying trade-offs in meeting multiple and potentially conflicting objectives. Armed with such tools, the conservation managers of the Anthropocene Ocean are much more likely to stay safely away from tipping points (from marine planetary boundaries; Rockström et al., 2009) both in the biophysical and sociopolitical world that may jeopardize the health and resilience of the world's oceans.

REFERENCES

Angelini, S., Hillary, R., Morello, E.B., Plagányi, É.E., Martinelli, M., Manfredi, C., Isajlović, I., Santojanni, A., 2016. An ecosystem model of intermediate complexity to test management options for fisheries: a case study. Ecological Modelling 319, 218–232.

Bakun, A., Weeks, S.J., 2006. Adverse feedback sequences in exploited marine systems: are deliberate interruptive actions warranted? Fish and Fisheries 7, 316–333.

Blamey, L.K., Plagányi, É.E., Branch, G.M., 2013. Modeling a regime shift in a kelp forest ecosystem caused by a lobster range expansion. Bulletin of Marine Science 89, 347–375.

Blamey, L.K., Plagányi, É.E., Branch, G.M., 2014. Was overfishing of predatory fish responsible for a lobster-induced regime shift in the Benguela? Ecological Modelling 273, 140–150.

Blanchard, J.L., Jennings, S., Holmes, R., Harle, J., Merino, G., Allen, J.I., Holt, J., Dulvy, N.K., Barange, M., 2012. Potential consequences of climate change for primary production and fish production in large marine ecosystems. Philosophical Transactions of the Royal Society of London B: Biological Sciences 367, 2979–2989.

Boettiger, C., Hastings, A., 2012. Early warning signals and the prosecutor's fallacy. Proceedings of the Royal Society of London B: Biological Sciences. http://dx.doi.org/10.1098/rspb. 2012.2085.

Boettiger, C., Ross, N., Hastings, A., 2013. Early warning signals: the charted and uncharted territories. Theoretical Ecology 6 (3), 255–264.

Brown, C.J., Mumby, P.J., 2014. Trade-offs between fisheries and the conservation of ecosystem function are defined by management strategy. Frontiers in Ecology and the Environment 12, 324–329.

Bunnefeld, N., Hoshino, E., Milner-Gulland, E.J., 2011. Management strategy evaluation: a powerful tool for conservation? Trends in Ecology & Evolution 26, 441–447.

Carpenter, S.R., Brock, W.A., 2006. Rising variance: a leading indicator of ecological transition. Ecology Letters 9 (3), 311–318.

Carpenter, S.R., Brock, W.A., Cole, J.J., Kitchell, J.F., Pace, M.L., 2008. Leading indicators of trophic cascades. Ecology Letters 11 (2), 128–138.

Charles, A.T., 1983. Optimal fisheries investment: comparative dynamics for a deterministic seasonal fishery. Canadian Journal of Fisheries and Aquatic Sciences 40, 2069–2079.

Christensen, V., Walters, C.J., 2004. Ecopath with Ecosim: methods, capabilities and limitations. Ecological Modelling 172, 109–139.

Clark, C.W., 1980. Towards a predictive model for the economic regulation of commercial fisheries. Canadian Journal of Fisheries and Aquatic Sciences 37, 1111–1129.

Clarke, A., 2003. Costs and consequences of evolutionary temperature adaptation. Trends in Ecology & Evolution 18, 573–581.

Collie, J.S., Botsford, L.W., Hastings, A., Kaplan, I.C., Largier, J.L., Livingston, P.A., Plagányi, É., Rose, K.A., Wells, B.K., Werner, F.E., 2014. Ecosystem models for fisheries management: finding the sweet spot. Fish and Fisheries 17, 101–125.

Cooke, J.G., 1999. Improvement of fishery-management advice through simulation testing of harvest algorithms. ICES Journal of Marine Science 56, 797–810.

De Roos, A.M., Boukal, D.S., Persson, L., 2006. Evolutionary regime shifts in age and size at maturation of exploited fish stocks. Proceedings of the Royal Society of London B: Biological Sciences 273, 1873–1880.

Defeo, O., Castrejón, M., Pérez-Castañeda, R., Castilla, J.C., Gutiérrez, N.L., Essington, T.E., Folke, C., 2014. Co-management in Latin American small-scale shellfisheries: assessment from long-term case studies. Fish and Fisheries 17, 176–192.

Dichmont, C.M., Deng, A.R., Punt, A.E., Venables, W., Haddon, M., 2006. Management strategies for short-lived species: the case of Australia's Northern Prawn Fishery: 1. Accounting for multiple species, spatial structure and implementation uncertainty when evaluating risk. Fisheries Research 82, 204–220.

Dichmont, C.M., Ellis, N., Bustamante, R.H., Deng, R., Tickell, S., Pascual, R., Lozano-Montes, H., Griffiths, S., 2013. Evaluating marine spatial closures with conflicting fisheries and conservation objectives. Journal of Applied Ecology 50, 1060–1070.

FAO, 2008. Best practices in ecosystem modelling for informing an ecosystem approach to fisheries. FAO Fisheries Technical Guidelines for Responsible Fisheries 4 (Suppl. 2) Add. 1, 78.

Fasham, M., Ducklow, H., McKelvie, S., 1990. A nitrogen-based model of plankton dynamics in the oceanic mixed layer. Journal of Marine Research 48, 591–639.

Folke, C., Carpenter, S., Walker, B., Scheffer, M., Elmqvist, T., Gunderson, L., Holling, C.S., 2004. Regime shifts, resilience, and biodiversity in ecosystem management. Annual Review of Ecology, Evolution, and Systematics 35, 557–581.

Fulton, E., Gorton, R., 2014. Adaptive Futures for SE Australian Fisheries and Aquaculture: Climate Adaptation Simulations. CSIRO, Australia, Hobart.

Fulton, E.A., Bax, N.J., Bustamante, R.H., Dambacher, J.M., Dichmont, C., Dunstan, P.K., Hayes, K.R., Hobday, A.J., Pitcher, R., Plagányi, É.E., 2015. Modelling marine protected areas: insights and hurdles. Philosophical Transactions of the Royal Society of London. Series B, Biological Sciences 370 20140278.

Fulton, E.A., Gray, R., Sporcic, M., Scott, R., Little, R., Hepburn, M., Gorton, B., Hatfield, B., Fuller, M., Jones, T., De la Mare, W., Boschetti, F., Chapman, K., Dzidic, P., Syme, G., Dambacher, J., McDonald, D., 2011. Ningaloo Collaboration Cluster: Adaptive Futures for Ningaloo, Ningaloo Collaboration Cluster Final Report No. 5.3, Australia. .

Fulton, E.A., Link, J.S., 2014. Modeling approaches for marine ecosystem-based management. In: Fogarty, M.J., McCarthy, J.J. (Eds.), Marine Ecosystem-Based Management. The Sea. Harvard University Press.

Fulton, E.A., Smith, A.D., Punt, A.E., 2005. Which ecological indicators can robustly detect effects of fishing? ICES Journal of Marine Science 62, 540–551.

Fulton, E.A., Smith, A.D., Smith, D.C., Johnson, P., 2014. An integrated approach is needed for ecosystem based fisheries management: insights from ecosystem-level management strategy evaluation. PLoS One 9, e84242.

Gaichas, S., Gamble, R., Fogarty, M., Benoît, H., Essington, T., Fu, C., Koen-Alonso, M., Link, J., 2012. Assembly rules for aggregate-species production models: simulations in support of management strategy evaluation. Marine Ecology Progress Series 459, 275–292.

Gao, J., Barzel, B., Barabási, A.-L., 2016. Universal resilience patterns in complex networks. Nature 530, 307–312.

Garcia, S.M., 2003. The Ecosystem Approach to Fisheries: Issues, Terminology, Principles, Institutional Foundations, Implementation and Outlook, vol. 443. Food & Agriculture Organisation, Rome.

Gray, R., Fulton, E., Little, L., Scott, R., 2006. Operating Model Specification Within an Agent Based Framework. North West Shelf Joint Environmental Management Study Technical Report. CSIRO, Hobart. 127.

Hairston, N.G., Ellner, S.P., Geber, M.A., Yoshida, T., Fox, J.A., 2005. Rapid evolution and the convergence of ecological and evolutionary time. Ecology Letters 8, 1114–1127.

Harfoot, M.B., Newbold, T., Tittensor, D.P., Emmott, S., Hutton, J., Lyutsarev, V., Smith, M.J., Scharlemann, J.P., Purves, D.W., 2014. Emergent global patterns of ecosystem structure and function from a mechanistic general ecosystem model. PLoS Biology 12, e1001841.

Hicks, C.C., Crowder, L.B., Graham, N.A., Kittinger, J.N., Cornu, E.L., 2016. Social drivers fore-warn of marine regime shifts. Frontiers in Ecology and the Environment 14, 252–260.

Hobday, A.J., Hartog, J.R., 2014. Derived ocean features for dynamic ocean management. Oceanography 27, 134–145.

Hobday, A.J., Hartog, J.R., Spillman, C.M., Alves, O., Hilborn, R., 2011. Seasonal forecasting of tuna habitat for dynamic spatial management. Canadian Journal of Fisheries and Aquatic Sciences 68, 898–911.

Hobday, A.J., Pecl, G.T., 2014. Identification of global marine hotspots: sentinels for change and vanguards for adaptation action. Reviews in Fish Biology and Fisheries 24, 415–425.

Hollowed, A.B., Bax, N., Beamish, R., Collie, J., Fogarty, M., Livingston, P., Pope, J., Rice, J.C., 2000. Are multispecies models an improvement on single-species models for measuring fishing impacts on marine ecosystems? ICES Journal of Marine Science 57, 707–719.

Hollowed, A.B., Bond, N.A., Wilderbuer, T.K., Stockhausen, W.T., A'mar, Z.T., Beamish, R.J., Overland, J.E., Schirripa, M.J., 2009. A framework for modelling fish and shellfish responses to future climate change. ICES Journal of Marine Science 66, 1584–1594.

Hughes, T.P., 1994. Catastrophes, phase-shifts, and large-scale degradation of a Caribbean coral-reef. Science 265, 1547–1551.

Hughes, T.P., Carpenter, S., Rockström, J., Scheffer, M., Walker, B., 2013. Multiscale regime shifts and planetary boundaries. Trends in Ecology & Evolution 28, 389–395.

Hughes, T.P., Rodrigues, M.J., Bellwood, D.R., Ceccarelli, D., Hoegh-Guldberg, O., McCook, L., Moltschaniwskyj, N., Pratchett, M.S., Steneck, R.S., Willis, B., 2007. Phase shifts, herbivory, and the resilience of coral reefs to climate change. Current Biology 17, 360–365.

Jennings, S., Kaiser, M., Reynolds, J., 2001. Marine Fisheries Ecology. Blackwell, Malden, Oxford, Carlton.

Jennings, S., Smith, A.D., Fulton, E.A., Smith, D.C., 2014. The ecosystem approach to fisheries: management at the dynamic interface between biodiversity conservation and sustainable use. Annals of the New York Academy of Sciences 1322 (1), 48–60.

Klaer, N.L., Wayte, S.E., Fay, G., 2012. An evaluation of the performance of a harvest strategy that uses an average-length-based assessment method. Fisheries Research 134, 42–51.

Levin, P.S., Fogarty, M.J., Murawski, S.A., Fluharty, D., 2009. Integrated ecosystem assessments: developing the scientific basis for ecosystem-based management of the ocean. PLoS Biology 7, e1000014.

Levin, P.S., Kelble, C.R., Shuford, R.L., Ainsworth, C., deReynier, Y., Dunsmore, R., Fogarty, M.J., Holsman, K., Howell, E.A., Monaco, M.E., Oakes, S.A., Werner, F., 2013. Guidance for implementation of integrated ecosystem assessments: a US perspective. ICES Journal of Marine Science. http://dx.doi.org/10.1093/icesjms/fst112.

Link, J., 2010. Ecosystem-Based Fisheries Management: Confronting Tradeoffs. Cambridge University Press.

Mayfield, S., Branch, G.M., 2000. Interrelations among rock lobsters, sea urchins, and juvenile abalone: implications for community management. Canadian Journal of Fisheries and Aquatic Sciences 57, 2175–2185.

McCallum, H.I., Kuris, A., Harvell, C.D., Lafferty, K.D., Smith, G.W., Porter, J., 2004. Does terrestrial epidemiology apply to marine systems? Trends in Ecology & Evolution 19, 585–591.

McCauley, D.J., Pinsky, M.L., Palumbi, S.R., Estes, J.A., Joyce, F.H., Warner, R.R., 2015. Marine defaunation: animal loss in the global ocean. Science 347, 1255641.

Morello, E.B., Plagányi, É.E., Babcock, R.C., Sweatman, H., Hillary, R., Punt, A.E., 2014. Model to manage and reduce crown-of-thorns starfish outbreaks. Marine Ecology Progress Series 512, 167–183.

Msangi, S., Kobayashi, M., Batka, M., Vannuccini, S., Dey, M., Anderson, J., 2013. Fish to 2030: Prospects for Fisheries and Aquaculture. (World Bank Report).

Nelson, E., Mendoza, G., Regetz, J., Polasky, S., Tallis, H., Cameron, D.R., Chan, K.M.A., Daily, G.C., Goldstein, J., Kareiva, P.M., Lonsdorf, E., Naidoo, R., Ricketts, T.H., Shaw, M.R., 2009. Modeling multiple ecosystem services, biodiversity conservation, commodity production, and tradeoffs at landscape scales. Frontiers in Ecology and the Environment 7, 4–11.

Neumann, B., Vafeidis, A.T., Zimmermann, J., Nicholls, R.J., 2015. Future coastal population growth and exposure to sea-level rise and coastal flooding-a global assessment. PLoS One 10, e0118571.

Nicholson, E., Collen, B., Barausse, A., Blanchard, J.L., Costelloe, B.T., Sullivan, K.M., Underwood, F.M., Burn, R.W., Fritz, S., Jones, J.P., 2012. Making robust policy decisions using global biodiversity indicators. PLoS One 7, e41128.

Nisbet, R.M., Jusup, M., Klanjscek, T., Pecquerie, L., 2012. Integrating dynamic energy budget (DEB) theory with traditional bioenergetic models. Journal of Experimental Biology 215, 892–902.

O'Connor, S., Ono, R., Clarkson, C., 2011. Pelagic fishing at 42,000 years before the present and the maritime skills of modern humans. Science 334, 1117–1121.

O'Neill, B.C., Kriegler, E., Riahi, K., Ebi, K.L., Hallegatte, S., Carter, T.R., Mathur, R., Vuuren, D.P., 2013. A new scenario framework for climate change research: the concept of shared socioeconomic pathways. Climatic Change 122, 387–400.

Olsen, E., Fay, G., Gaichas, S., Gamble, R., Lucey, S., Link, J.S., 2016. Ecosystem model skill assessment. Yes we can!. PLoS One 11, e0146467.

Österblom, H., Merrie, A., Metian, M., Boonstra, W.J., Blenckner, T., Watson, J.R., Rykaczewski, R.R., Ota, Y., Sarmiento, J.L., Christensen, V., Schlüter, M., Birnbaum, S., Gustafsson, B.G., Humborg, C., Mörth, C.-M., Müller-Karulis, B., Tomczak, M.T., Troell, M., Folke, C., 2013. Modeling social—ecological scenarios in marine systems. BioScience 63, 735–744.

Ostrom, E., 2009. A general framework for analyzing sustainability of social-ecological systems. Science 325, 419–422.

Pascoe, S., Brooks, K., Cannard, T., Dichmont, C.M., Jebreen, E., Schirmer, J., Triantafillos, L., 2014. Social objectives of fisheries management: what are managers' priorities? Ocean & Coastal Management 98, 1–10.

Pascoe, S., Hutton, T., van Putten, I., Dennis, D., Plagányi, E.E., Deng, R., 2013a. Implications of quota reallocation in the Torres Strait tropical rock lobster fishery. Canadian Journal of Agricultural Economics 61, 335–352.

Pascoe, S., Hutton, T., van Putten, I., Dennis, D., Skewes, T., Plagányi, É., Deng, R., 2013b. DEA-based predictors for estimating fleet size changes when modelling the introduction of rights-based management. European Journal of Operational Research 230, 681–687.

Plagányi, É., 2007. Models for an Ecosystem Approach to Fisheries. FAO Fisheries Technical Paper 477.

Plagányi, É., Butterworth, D., Burgener, M., 2011a. Illegal and unreported fishing on abalone—quantifying the extent using a fully integrated assessment model. Fisheries Research 107, 221–232.

Plagányi, É.E., Ellis, N., Blamey, L.K., Morello, E.B., Norman-Lopez, A., Robinson, W., Sporcic, M., Sweatman, H., 2014a. Ecosystem modelling provides clues to understanding ecological tipping points. Marine Ecology Progress Series 512, 99.

Plagányi, É.E., Punt, A.E., Hillary, R., Morello, E.B., Thébaud, O., Hutton, T., Pillans, R.D., Thorson, J.T., Fulton, E.A., Smith, A.D., 2014b. Multispecies fisheries management and conservation: tactical applications using models of intermediate complexity. Fish and Fisheries 15, 1–22.

Plagányi, É.E., Skewes, T., Murphy, N., Pascual, R., Fischer, M., 2015. Crop rotations in the sea: increasing returns and reducing risk of collapse in sea cucumber fisheries. Proceedings of the National Academy of Sciences 112, 6760–6765.

Plagányi, É.E., Skewes, T.D., Dowling, N.A., Haddon, M., 2013a. Risk management tools for sustainable fisheries management under changing climate: a sea cucumber example. Climatic Change 119, 181–197.

Plagányi, É.E., van Putten, I., Hutton, T., Deng, R.A., Dennis, D., Pascoe, S., Skewes, T., Campbell, R.A., 2013b. Integrating indigenous livelihood and lifestyle objectives in managing a natural resource. Proceedings of the National Academy of Sciences 110, 3639–3644.

Plagányi, É.E., Weeks, S.J., Skewes, T.D., Gibbs, M.T., Poloczanska, E.S., Norman-López, A., Blamey, L.K., Soares, M., Robinson, W.M., 2011b. Assessing the adequacy of current fisheries management under changing climate: a southern synopsis. ICES Journal of Marine Science 68, 1305–1317.

Pörtner, H.O., Gutt, J., 2016. Impacts of climate variability and change on (marine) animals: physiological underpinnings and evolutionary consequences. Integrative and Comparative Biology 56, 31–44.

Punt, A.E., MacCall, A.D., Essington, T.E., Francis, T.B., Hurtado-Ferro, F., Johnson, K.F., Kaplan, I.C., Koehn, L.E., Levin, P.S., Sydeman, W.J., 2016. Exploring the implications of the harvest control rule for Pacific sardine, accounting for predator dynamics: a MICE model. Ecological Modelling 337, 79–95.

Quinn, T.J., Deriso, R.B., 1999. Quantitative Fish Dynamics. Oxford University Press.

Rademeyer, R.A., Plagányi, E.E., Butterworth, D.S., 2007. Tips and tricks in designing management procedures. Ices Journal of Marine Science 64, 618–625.

Raemaekers, S., Hauck, M., Bürgener, M., Mackenzie, A., Maharaj, G., Plagányi, É.E., Britz, P.J., 2011. Review of the causes of the rise of the illegal South African abalone fishery and consequent closure of the rights-based fishery. Ocean & Coastal Management 54, 433–445.

Rockström, J., Steffen, W.L., Noone, K., Persson, Å., Chapin III, F.S., Lambin, E., Lenton, T.M., Scheffer, M., Folke, C., Schellnhuber, H.J., 2009. Planetary boundaries: exploring the safe operating space for humanity. Ecology and Society 14 (2), 32.

Rose, K.A., Allen, J.I., Artioli, Y., Barange, M., Blackford, J., Carlotti, F., Cropp, R., Daewel, U., Edwards, K., Flynn, K., Hill, S.L., HilleRisLambers, R., Huse, G., Mackinson, S., Megrey, B., Moll, A., Rivkin, R., Salihoglu, B., Schrum, C., Shannon, L., Shin, Y.-J., Smith, S.L., Smith, C., Solidoro, C., St John, M., Zhou, M., 2010. End-to-end models for the analysis of marine ecosystems: challenges, issues, and next steps. Marine and Coastal Fisheries 2, 115–130.

Russo, R., Riccio, A., di Prisco, G., Verde, C., Giordano, D., 2010. Molecular adaptations in Antarctic fish and bacteria. Polar Science 4, 245–256.

Sainsbury, K.J., Punt, A.E., Smith, A.D.M., 2000. Design of operational management strategies for achieving fishery ecosystem objectives. ICES Journal of Marine Science 57, 731.

Savina, M., Condie, S.A., Fulton, E.A., 2013. The role of pre-existing disturbances in the effect of marine reserves on coastal ecosystems: a modelling approach. PLoS One 8, e61207.

Schaefer, M.B., 1954. Some aspects of the dynamics of populations important to the management of the commercial marine fisheries. Inter-American Tropical Tuna Commission Bulletin 1, 23–56.

Scheffer, M., Bascompte, J., Brock, W.A., Brovkin, V., Carpenter, S.R., Dakos, V., Held, H., Van Nes, E.H., Rietkerk, M., Sugihara, G., 2009. Early-warning signals for critical transitions. Nature 461 (7260), 53–59.

Scheffer, M., Carpenter, S., Foley, J.A., Folke, C., Walker, B., 2001. Catastrophic shifts in ecosystems. Nature 413, 591–596.

Scheffer, M., Carpenter, S.R., Lenton, T.M., Bascompte, J., Brock, W., Dakos, V., Van De Koppel, J., Van De Leemput, I.A., Levin, S.A., Van Nes, E.H., 2012. Anticipating critical transitions. Science 338, 344–348.

Sharpless, A., Evans, S., 2013. The Perfect Protein: The Fish Lover's Guide to Saving the Oceans and Feeding the World. Rodale, New York.

Smith, A.D.M., Fulton, E.J., Hobday, A.J., Smith, D.C., Shoulder, P., 2007. Scientific tools to support the practical implementation of ecosystem-based fisheries management. ICES Journal of Marine Science 64, 633–639.

Steffen, W., Broadgate, W., Deutsch, L., Gaffney, O., Ludwig, C., 2015. The trajectory of the anthropocene: the great acceleration. The Anthropocene Review 2, 81–98.

Stow, C.A., Jolliff, J., McGillicuddy, D.J., Doney, S.C., Allen, J.I., Friedrichs, M.A., Rose, K.A., Wallhead, P., 2009. Skill assessment for coupled biological/physical models of marine systems. Journal of Marine Systems 76, 4–15.

Sugihara, G., May, R., Ye, H., Hsieh, C.-H., Deyle, E., Fogarty, M., Munch, S., 2012. Detecting causality in complex ecosystems. Science 338, 496–500.

Travers, M., Shin, Y.-J., Jennings, S., Cury, P., 2007. Towards end-to-end models for investigating the effects of climate and fishing in marine ecosystems. Progress in Oceanography 75, 751–770.

Turley, C., Blackford, J., Widdicombe, S., Lowe, D., Nightingale, P., Rees, A., 2006. Reviewing the impact of increased atmospheric CO_2 on oceanic pH and the marine ecosystem. Avoiding Dangerous Climate Change 8, 65–70.

van Putten, I., Lalancette, A., Bayliss, P., Dennis, D., Hutton, T., Norman-López, A., Pascoe, S., Plagányi, E., Skewes, T., 2013. A Bayesian model of factors influencing indigenous participation in the Torres Strait tropical rocklobster fishery. Marine Policy 37, 96–105.

Van Putten, E.I., Plaganyi, E.E., Richards, S.A., Fulton, E., Fleming, A., Punt, A., 2017. A Framework for Including Psychological Factors into Socio-Ecological Models Relevant to Natural Resource Management (in preparation).

Van Vuuren, D.P., Edmonds, J., Kainuma, M., Riahi, K., Thomson, A., Hibbard, K., Hurtt, G.C., Kram, T., Krey, V., Lamarque, J.-F., 2011. The representative concentration pathways: an overview. Climatic Change 109, 5–31.

Wadsworth, R.M., Criddle, K., Kruse, G.H., 2014. Incorporating stakeholder input into marine research priorities for the Aleutian Islands. Ocean & Coastal Management 98, 11–19.

Yu, Y., Xiao, G., Zhou, J., Wang, Y., Wang, Z., Kurths, J., Schellnhuber, H.J., 2016. System crash as dynamics of complex networks. Proceedings of the National Academy of Sciences of the United States of America 201612094.

Zhang, L., Andersen, K.H., Dieckmann, U., Brännström, Å., 2015. Four types of interference competition and their impacts on the ecology and evolution of size-structured populations and communities. Journal of Theoretical Biology 380, 280–290.

Zia, A., Norton, B.G., Metcalf, S.S., Hirsch, P.D., Hannon, B.M., 2014. Spatial discounting, place attachment, and environmental concern: toward an ambit-based theory of sense of place. Journal of Environmental Psychology 40, 283–295.

Zobitz, J., Desai, A., Moore, D., Chadwick, M., 2011. A primer for data assimilation with ecological models using Markov Chain Monte Carlo (MCMC). Oecologia 167, 599–611.

Chapter 21

The Big Role of Coastal Communities and Small-Scale Fishers in Ocean Conservation

Anthony Charles

Saint Mary's University, Halifax, NS, Canada

INTRODUCTION

As has become commonplace in many marine-focused articles these days; this chapter begins by noting that the world's oceans are increasingly under threat as a result of impacts that range from fishing and ocean mining to coastal industries and land-based pollution and to climate change and urbanization (Halpern et al., 2007). A crucial implication of these threats is the risk posed to the livelihoods of people in coastal areas. This chapter is broadly about responses to these threats and risks through ocean conservation, that is, the broad variety of stewardship practices and policies that are undertaken by society (including governments as well as coastal communities and components of civil society) to maintain and improve the state of the ocean, thus supporting sustainable uses. In particular, this chapter focuses on decision making about ocean conservation, that is, governance, including the values and goals underlying decisions, who is involved in that decision making, and the institutions within which it takes place.

Clearly, there are many jurisdictional levels at which that governance can occur, from local (e.g., municipalities and communities) to national to multinational. Much attention has been paid by governments and international bodies to the national and multinational levels of conservation and management, for example, with large marine ecosystems (LMEs) (Olsen et al., 2006). At the other end of the scale, the many coastal communities around the world and their small-scale fishers—located on bays, estuaries, and small stretches of coastline—have undertaken a very large number of local initiatives in ocean conservation, often with considerable success. This chapter focuses on that local level of coastal communities and small-scale fishers, assessing the nature of the successes and the potential for increased attention in ocean conservation to the local level.

Conservation for the Anthropocene Ocean. http://dx.doi.org/10.1016/B978-0-12-805375-1.00021-0

In this analysis, the chapter draws on research by the Community Conservation Research Network (CCRN), a multiyear global initiative involving indigenous, community, academic, government, and nongovernmental partners (CCRN, 2016a). Applying a social–ecological systems approach and comparing across over 20 sites internationally, the CCRN aims to provide guidance to both local communities and policy makers on successful paths of stewardship and livelihood sustainability. This work is being done through an exploration of why and how local communities (including coastal and small-scale fishing communities) engage in environmental stewardship, how those initiatives interact with livelihoods, and how governments either help or hinder conservation efforts.

The chapter begins with a brief review of the historical evolution of ocean conservation, focusing on participatory and integrative approaches. This is followed by discussion of the role of local-level coastal communities and small-scale fishers in ocean conservation. Finally, the practical and policy implications are discussed of enhancing the national and international focus placed on small-scale fishers and coastal communities as catalysts of ocean conservation.

A BRIEF AND SELECTIVE HISTORY OF OCEAN CONSERVATION

Garcia et al. (2014) document the evolution globally of governance approaches in each of fisheries management and marine biodiversity conservation—referred to as two "streams of governance." That analysis notes that until the late 1800s or early 1900s, both these streams—one focused on managing human use of fishery resources, the other on protecting nature in the oceans—tended to involve decision making primarily of local, community-based, and/or self-regulated forms. Largely over the course of the 1900s, the governance focus shifted to top-down governmental decision making over natural resources and the environment; as this became more prominent and enforced, local governance consequently declined.

Thus the dominant paradigm of 20th century fishery management was characterized by a centralized, top-down approach, one in which governments focused on the goal of regulating and controlling the activities of fishers, and came to have a narrow view of fishers as resource exploiters aiming to catch fish as rapidly and exhaustively as possible (Charles, 1995). To a considerable extent, marine science evolved to support this approach through a similarly centralized focus on large-scale fish stock and marine environmental assessments. This governmental approach produced considerable success in developing fishery and marine science, but at the same time, these successes led to a misplaced belief that the problems of fisheries and marine management could be "solved" strictly through this centralized mode of work (Charles, 1998).

A number of major fish stock collapses toward the end of the 20th century led to a rethinking both of fisheries management and of marine science and conservation (see, e.g., Charles, 1998). Although this resulted in many changes, here two key directions are highlighted: (1) a shift to a broader "systems" perspective on the fishery, one that considers more comprehensively the various ecosystems and multiple human impacts, and (2) a move to more participatory approaches in ocean and resource management.

The first of these involves two major thrusts. First, ecosystem-based management (Pikitch et al., 2004; Charles, 2014; Long et al., 2015) has become dominant worldwide, as a wide range of local, national, and international bodies have shifted in recent decades from a narrow single-species focus to one looking at whole ecosystems. In the fishery sector, this has inspired an "ecosystem approach to fisheries" (de Young et al., 2008; FAO, 2003), in keeping with "systems thinking" and more general ideas of ecosystem-based management. Although some approaches to this remain relatively narrow, a full view (FAO, 2003, 2009) seeks to (1) broaden the conservation focus from solely about individual fish stocks to better consider the structure, dynamics, and quality of marine ecosystems and (2) link these ecosystem considerations with those relating to the structure, dynamics, and quality of the human system, including the well-being of fishers and fishing communities and related human dimensions.

Parallel to ecosystem-based management is integrated management, a related holistic approach (e.g., Charles, 2012; Forst, 2009) with a focus on planning and managing all the economic sectors that use the ocean within a specific spatial area, such as a coastal zone with fishing, aquaculture, tourism, shipping, etc. (Cicin-Sain et al., 1998). Integrated management is a vehicle to address the flaws in the typically "silo" structure of most governments and their departmental arrangements that often has fishery agencies only looking at fisheries, forestry departments only at forestry, health agencies restricting attention to health, and so on.

Integrated management recognizes that many economic sectors may be using marine natural resources, which often provide the key engine of the local or regional economy and the backbone of many individual coastal communities. Although each sector has its own importance individually, each also supports other parts of the economy, so integrated management can help sectoral policies to take into account these effects, together with broader community economics (Charles et al., 2010). Consider tourism, for example, as a key economic activity in many coastal areas. Tourists are often drawn to an area by the quality of the local environment, and cultural aspects, such as visiting small communities, fishing boats, and wharves along the seashore. Thus fisheries support tourism and contributes more to the economy than single-sector numbers would show.

The current emphasis, in many locations globally, on ecosystem-based management and integrated management is exemplified in a major push toward expanding the extent of ocean space devoted to marine protected areas (MPAs)—specific delineated areas in which additional restrictions on human uses are put in place, relative to neighboring areas (Cicin-Sain and Belfiore, 2003). These MPAs are inherently conservation oriented but can take various forms, such as "no-take" (no extraction of resources permitted) or multiple zones (each limiting human uses differently). MPAs have the potential to reflect an integrated ecosystem approach. That can be the case if they mesh well with marine eco-systems and apply to multiple ocean use sectors in a suitably integrated man-ner. MPAs are included in international agreements such as the Convention on Biological Diversity (2004), and guidelines such as those of the Food and Agriculture Organization (FAO) of the United Nations (FAO, 2011), but still to be resolved are concerns about how they interact with marine-based livelihoods (Charles et al., 2016). An extensive research literature has developed on how MPAs should be implemented to achieve multiple social and ecological goals (e.g., Ban et al., 2011; Christie, 2004; Christie and White, 2007; Jones, 2002; Pollnac et al., 2010; Pomeroy et al., 2004).

The second shift, toward greater participation in ocean and resource man-agement, represents a change from the formerly pervasive top-down paradigm toward a new "cooperative" approach of comanagement (Pinkerton, 1989). This reflects a growing consensus, in both terrestrial and marine systems, of the need for a stronger role for civil society in stewarding our environment and managing our resources, to overcome the gaps and shortcomings of conventional top-down decision making that is typically "disconnected" from people and communities. Two major problems of the latter have been (1) the waste of resources resulting from a failure to draw on the energy and talent for conservation among resource users and local people and (2) poor compliance, when rules are imposed from afar rather than developed cooperatively, within a local context.

There is now a recognition that conservation is enhanced when the people and communities dependent on resources take on some of the responsibility for managing (making decisions about) those resources. Indeed, this is an insight that arose strongly within research on the Commons—locations (such as ocean space) and resources (such as fisheries) that are typically subject to collective use and management. The sustainability linkages between local communities and local ecosystems and the process of local-level stewardship by communities over their local resources—such as water, fish, trees, or even urban resources, such as police and roads—are now well-documented (e.g., by Nobel prize lau-reate Elinor Ostrom, 1990). Hundreds of case studies have been analyzed, by Ostrom and many others, on how communities work together often in conjunc-tion with government but without top-down control to use resources wisely.

Accordingly, the modern approach of comanagement, applied in coastal and marine realms, has governments working with fishing organizations and coastal communities to develop and implement management measures. This applies specifically for fisheries (McCay and Jentoft, 1996; Pinkerton, 2009; Wilson et al., 2003) and more broadly for ecosystem based and integrated management (Kearney et al., 2007). Along with this, we are seeing an evolution of marine science and conservation from a conventional model that typically ignored resource users and communities toward much greater engagement with fishers, other resource users, and coastal communities to better understand and conserve the marine environment. In the analysis of Garcia et al. (2014), these changes reflect a partial return to the greater range of involvement that was present in the earlier days of resource management as opposed to the intervening focus on a dominant governmental presence.

The shift toward greater participation in ocean conservation and management appears strongly with respect to spatial management of oceans and, in particular, the implementation of MPAs, as discussed earlier. In MPA creation and operation, the crucial role of human dimensions (including participatory approaches to decision making as well as suitable access rights arrangements and issues of fairness and equity) is now widely recognized (Charles and Wilson, 2009; Christie, 2004; Christie and White, 2007). This has led to a growing focus on initiatives such as locally managed marine areas (LMMA)—a form of MPA that is directly created by local communities and marine user groups (Govan, 2009; Govan et al., 2008; Rocliffe et al., 2014), and which relates as well to the global movement to recognize and support indigenous and community conserved areas (ICCA—Berkes, 2009; Borrini-Feyerabend et al., 2004, 2010; Corrigan and Hay-Edie, 2013; Kothari et al., 2013).

The two key components of the evolution of ocean conservation (see Garcia et al., 2014)—broader systems perspectives (ecosystem-based management and integrated ocean management) and participatory comanagement of ocean space and resources—are intertwined and notably linked through their fundamental roots in and need for deep appreciation and understanding of human dimensions (de Young et al., 2008; Charles, 2009, 2014). Although both of the two shifts have occurred at all scales of ocean conservation, this chapter focuses on insights produced from success stories of marine ecosystem conservation carried out at the scale of local coastal and fishing communities.

COASTAL COMMUNITIES AND OCEAN CONSERVATION

Local communities and resource users, wherever they are located, are often highly motivated to protect their local environment, as their individual and collective livelihoods have close interaction with these ecosystems (Borrini-Feyerabend et al., 2010; Govan et al., 2008). Strong ecosystem services support communities in terms of sustainable jobs and social services (such as education, health) and the capacity to bounce back from shocks that arise from time to time. A community in tune with its environment can better meet the fundamental need of making a living by using the local environment while at the same time protecting the environment from negative human impacts. That motivation explains why, despite much attention by governments and international bodies to large-scale approaches, in reality much of the progress in conservation is at a local level. Given the chance, local communities and resource user bodies can develop and utilize their capacity to resolve environmental and livelihood challenges, in ways that make a difference locally (and indeed, may go further in inspiring others worldwide, as with the LMMA and ICCA initiatives noted earlier).

This approach, often referred to as community-based management and community conservation (e.g., Berkes, 2006; Kearney et al., 2007), though not universally applicable, is found in a large variety of locations globally (e.g., Nasuchon and Charles, 2010; Berkes et al., 2001). In some cases, this involves self-management by communities and resource users, in systems that may well endure despite government. Often, however, when governments are willing, local communities and/or resource user organizations share power and responsibility with government to jointly engage in decision making over the local resources and local environments on which they depend.

Community-based approaches, where they are effective, have been shown to aid resource conservation and sustainable development by (1) better utilizing traditional local knowledge and community understanding of local ecosystems; (2) empowering local resource users and communities, leading to greater acceptance of conservation measures and resulting compliance; and (3) drawing on effective community mechanisms to resolve conflicts over resource use. Thus if communities have strong local rules and institutions, their common resources can be used sustainably (cf. Hardin, 1968; Ostrom, 1990).

The successes of local communities around the world show how people work together to conserve, manage, and improve local environments and build sustainable local economies—even if many of these successes happen "under the radar" and may even be unknown to corresponding national governments. The importance of this local-level community stewardship over local ecosystems and resources is certainly true of communities in coastal areas of the world, many of which are seeing that they can engage in ocean conservation to both improve their livelihoods and protect their communities (e.g., Armitage et al., 2017; Berkes, 2004; Jentoft, 2000; Pinkerton, 2009; Weeratunge et al., 2014).

Unlike governments that are structured in departments ("silos") with separate consideration of resource use, environmental considerations, social aspects, etc., local communities are inherently integrated, reflecting a reality that their quality of life depends jointly on multiple economic activities as well as on environmental quality and social well-being. This reality is reflected in the success of many LMMAs, as noted earlier. This nature of community-based approaches makes them well-placed to draw on the major thrusts in modern ocean conservation discussed earlier—certainly the imperative for more participatory management, which is naturally desired at a local level and also the holistic approaches of ecosystem-based management and integrated management. Indeed, the benefits of local-level ocean conservation could be expanded through an increased focus of governments to applying ecosystem-based management and integrated management in small-scale ecosystems and at the local scale of the community (Charles et al., 2010; Long et al., 2016).

As an example of how coastal communities engage in environmental stewardship, consider the small community of Koh Piyak, on an island off the coast of Thailand (CCRN, 2016b). Koh Piyak's 43 households are involved in a range of resource-based livelihood activities that include fishing and coconut production; this resource use is accompanied by significant community stewardship initiatives to protect the island's resources, such as through the replanting of coral reefs (a practice fitting well with the ocean conservation trend to ecosystem-based management). In addition, the community is diversifying its livelihoods through tourism development, including a program of providing homestays and producing souvenirs. This supports both environmental stewardship and livelihood security, as it reduces (or avoids increases in) pressure on environment and resources. It is important to note that this combination, in Koh Piyak, of stewardship and ocean conservation together with innovative, culturally appropriate economic development (CCRN, 2016b) is aided by the local culture and beliefs of the community, which produce effective participation in decision making, sharing of natural resources in the community, and a sense of equity (e.g., reflected in the establishment of a community market to sell food at a fair price).

SMALL-SCALE FISHERS AND OCEAN CONSERVATION

In considering ocean conservation, the role of small-scale fishers is distinct from, yet closely related to, that of coastal communities. Over 90% of the millions of fishing people around the world are in small-scale fisheries (FAO, 2016), using small boats that stay close to shore or even harvesting along the shore, and typically viewing fishing as both a livelihood and a way of life. This close-to-shore nature means that most small-scale fishers have close ties to their local communities. The fishers care about the well-being of their community, and at the same time, small-scale fisheries play a big role as sustainable economic engines of those coastal communities and as the heart of the social fabric of the coast. This synergistic relationship is at the heart of the key role that small-scale fishers play in ocean conservation.

I had the opportunity to witness firsthand that stewardship imperative among small-scale fishers around a quarter century ago, at the time of Canada's famous cod fishery collapse (Charles, 1995). In serving on the Fisheries Resource Conservation Council, an advisory council formed by the Canadian government in response to the collapse, I heard from many small-scale fishers who, although certainly concerned about their own livelihoods, also had a deep concern for the future of their communities. The cod stocks that had collapsed, and on which these small-boat fishers had relied, were just off the coast from their communities; the future of the communities was tied together with the future of the fish, so conservation mattered (Charles et al., 2007).

Despite the importance of small-scale fisheries for achieving conservation (as well as social and economic) goals (FAO, 2016), many governments and global agencies have undervalued them. This has ranged from a basic lack of recognition to negative attitudes (e.g., viewing small-scale fisheries as out-of-date and in need of "phasing out") to detrimental policy impacts (e.g., providing greater fish allocations to industrial fleets; e.g., Pauly, 2006). Such a negative policy bias has often resulted in a worsening of the state of small-scale fisheries. In many cases, for example, government policy deliberately or inadvertently led to ownership (and location) of the fisheries becoming highly concentrated in fewer hands and in a small number of centers, with coastal communities suffering. A dramatic case of this was in British Columbia, Canada, where small communities lost their fishing boats, and their main livelihood, through such a policy favoring large-scale fisheries (e.g., Pinkerton and Edwards, 2009). Beyond the fishery, neglect of small-scale fisheries can lead to consequent negative impacts in other economic sectors, notably tourism—as small communities lose their fishing livelihoods and their fishing heritage, becoming less attractive to tourists. These negative social and economic impacts of the undervaluing of small-scale fisheries are accompanied by a loss of their conservation benefits. In particular, when small-scale fisheries are neglected relative to large-scale industrial fleets (the most heavily subsidized ones), there can be significant negative environmental impacts.

In recent years, fishery organizations around the world have pushed for better recognition and support of small-scale fisheries. Notably, the FAO of the United Nations, responsible for tackling food and poverty issues globally, responded through a global initiative leading to successful adoption of the *Voluntary Guidelines for Securing Sustainable Small-scale Fisheries in the Context of Food Security and Poverty Eradication*, commonly known as the *Small-scale Fisheries Guidelines* (FAO, 2015). These guidelines will, for years to come, provide a stronger focus on such fisheries, including an emphasis on the security of fishing rights and tenure (Charles, 2013), and notably promoting the role and responsibility of small-scale fishers in environmental stewardship.

A concrete example of the stewardship role of small-scale fishers is provided within the indigenous Nuu-chah-nulth territory on the west coast of Vancouver Island, in the Canadian province of British Columbia (CCRN, 2016b). The Nuu-chah-nulth people, currently numbering over 8000, have been living for thousands of years on land that includes coastal ecosystems and watersheds, with a society, economy, and culture that is deeply connected to their natural resources. There are multiple Nuu-chah-nulth communities in the territory, and for these communities, salmon fishing is of great importance culturally, with the Nuu-chah-nulth now building their capabilities to utilize a broad range of marine resources and developing suitable management plans to benefit their communities and to ensure sustainability. The Nuu-chah-nulth approach, such as those of many other indigenous peoples, is based on traditional management incorporating the holistic principles underlying ecosystem-based and integrated management. The resulting stewardship efforts are accompanied by higher-level policy and legal initiatives, including court cases that aim to ensure that the Canadian government recognizes traditional resource access rights, so that Nuu-chah-nulth communities can achieve their long-term livelihood potential. The lesson here is that a combination of long-standing cultural values, traditional stewardship practices, rights over local resources, and a crucial need for sustainable livelihoods, leads to practical conservation efforts (e.g., fishery management plans) as well as coordinated high-level policy engagement by the Nuu-chah-nulth communities (CCRN, 2016b).

PRACTICAL AND POLICY IMPLICATIONS

Combining the discussions so far in this chapter on (1) recent trends in conserva-tion to more systems-based and participatory approaches and (2) the dual roles of coastal communities and small-scale fishers in conservation, and although recog-nizing that each example of local conservation is geographically, administratively, and culturally distinct, there are certain emerging lessons for ocean conservation that arise from looking collectively at multiple case studies (CCRN, 2016a,b). In particular, three key ingredients seem widely applicable in underlying the success of a local-level community approach to conservation and sustainable livelihoods.

Knowledge

Local conservation can draw on local knowledge, held by those dependent on and/or living beside the ocean, and who have a strong understanding of local ecosystems. This knowledge—variously known as traditional ecological knowl-edge, indigenous knowledge, fisher knowledge, or simply "local knowledge"—can help to produce a better understanding of the marine environment and lead to more effective conservation. Indeed, the use of all sources of available knowl-edge might be seen as a prerequisite for effective local conservation efforts—as demonstrated in cases globally of ICCA, as referenced earlier in this chapter.

Participation

Small-scale local conservation initiatives can draw on participatory approaches, and thus be carried out jointly by community members and ocean users, together with governments and scientists, in a manner parallel to that of comanage-ment. This can help to build greater acceptance at the local level of the need for marine conservation, producing greater local "buy-in" that can extend to coastal resource management measures.

Institutions

Although there are typically considerable human complexities and jurisdic-tional concerns at a local (coastal) scale, particularly related to the land–sea interface, there is also a great potential to draw on existing human institutions and community support for knowledge acquisition, conservation, and manage-ment. Indeed, recognizing and drawing on strong human institutions and social cohesion within coastal communities can provide a "comparative advantage" for local-level ocean conservation, which can enhance the sustainability of con-servation efforts through time.

   This three-point rationale points to the value of community conservation in supporting livelihoods and conserving ecosystems, through participatory and cooperative initiatives at a local scale. Achieving the potential does require some concrete actions, particularly in terms of adjustments to government policy. The following four policy implications arise out of the analysis mentioned earlier.

	Policy Implication	**Analysis**
1.	Achieving the full ocean conservation potential of coastal communities and small-scale fishers requires greater attention to and mainstreaming of this level of conservation.	To fully realize the potential of coastal communities and small-scale fishers in contributing to ocean conservation, there is a need for broad policy measures, nationally and internationally, to better support and build capacity for local-level conservation and stewardship initiatives. Such initiatives must nevertheless recognize the diversity and complexity of arrangements at a local level, and corresponding variations in the stewardship role of coastal communities and small-scale fishers (e.g., Agrawal and Gibson, 1999; Béné et al., 2007; Berkes, 2004). Indeed, a fundamental determinant of success or failure of local-level ocean conservation is certainly the interest and capability of coastal communities and fisher organizations to participate in such activities. Also important are the values, objectives, attitudes, and personalities of those involved, both those at the local level and those from governmental institutions.
2.	Government policy must better connect ocean conservation and coastal communities, so that decisions made by governments about ocean space and resources fully consider effects on communities.	Although governments must always consider their responsibilities to maintain resources, they should also implement a "community screen" that ensures that community impacts are considered when making decisions about natural resources to avoid serious harm from the wrong decisions. This implies the need for a strong connection, supported by government, between sustainable economies and sustainable communities—both for the sake of the communities themselves and to maintain the stewardship function that communities often play. A similar point applies to small-scale fisheries in terms of ensuring that ocean conservation initiatives are evaluated in part through assessment of positive and negative impacts on small-scale fishers and their stewardship activities. It is necessary not only to "take into account" coastal communities and small-scale fishers in ocean conservation but also to make their stewardship role more effective through suitable policy development. An example would be to follow the lead of some governments that have developed systems of collective community rights assigned at a local level, which acknowledge the community nature of such situations, by providing secure access to resources for local communities and specifying the manner in which the community can take part in the management of those resources (Charles, 2009, 2013).

Continued

	Policy Implication	**Analysis**
3.	The relevant scientific and management agencies must adapt institutionally to new realities, which can require restructuring programs and reassigning resources to better align with communities and ocean users.	Governments must (1) consider community-focused values, as these underlie local aspirations and goals; (2) provide local communities with the "legal space" needed to make local decisions, (3) deal with multiple scales of management, connecting the local to the regional, and (4) design solutions suited to a given context, which depends on conservation needs and economic realities (Charles et al., 2010).
		There may be a need for government support in terms of building skills and organizational capacity, developing the capability to overcome trade-offs that may arise in terms of the time and effort required (between participation of this sort and immediate pursuit of livelihoods, such as fishing) and helping in the process of communities engaging with scientific agencies and governmental institutions that also manage ocean resources.
4.	Opportunities should be sought for "scaling-up" from initiatives of coastal communities and small-scale fishers to large-scale ocean management, or for "scaling down" from large-scale to local, coastal efforts.	For example, some cases of local-level ocean conservation have produced conditions in which new national or subnational management arrangements have been able to emerge, such as in Chile (Gelcich et al., 2010), where the local conservation efforts of fishers, working with scientists, led to a model of community-level MPAs that eventually entered national legislation and management systems. Similarly, the model of LMMA, described earlier, has expanded globally, reflecting an important illustration of "scaling-up."
		Even if a particular scale is being emphasized, it is important to determine mechanisms to connect to higher or lower scales—e.g., linking together a set of local initiatives so as to be relevant at a national scale, or embedding within large marine ecosystems the capability to carry out local-level efforts supportive of higher-level goals. In seeking the appropriate balance across scales, there may be social, economic, cultural, ecological, or biophysical factors that bear on the choice, as may current realities of capacity, institutions, and governance.
		The capacity for public and community involvement, and the existence of supportive institutions, may increase the efficiency of conservation efforts at any scale, whereas jurisdictional challenges and governance limitations may reduce effectiveness. Thus cost implications need to be considered, including possible economies of scale, and also efficiencies arising through devolution of authority and the social capital that may be available locally.

CONCLUSIONS

The world's oceans face many threats, with associated threats to the livelihoods that oceans support. Success stories in local-level ocean conservation, carried out by coastal communities and small-scale fishers worldwide, indicate that, while local initiatives are undoubtedly challenging at times, they can combine multiple sources of knowledge, an integrated perspective, and participatory and cooperative approaches, to increase the overall efficiency of conservation and perhaps even to overcome otherwise intractable problems.

Accordingly, this chapter makes a case for greater attention, in national and international policy, to conservation at the local level, drawing on the participatory and community-based approaches to ocean conservation found among small-scale fishers and coastal communities. Ocean conservation will not only benefit from but may fundamentally depend on these local stewardship practices. This argument is not one contrary to larger-scale initiatives—which have tended to receive much attention and funding—but rather one in favor of an appropriate balance across levels of action and decision making. This fits closely with the emerging consensus on the need for a multilevel and cross-scale approach to governance of all forms, taking into consideration social and ecological realities, corresponding costs, capacities, and institutional arrangements, and the relative benefits of engaging at each level of decision making.

Ultimately, the ocean conservation practices found among coastal communities and small-scale fishers provide an inspiring vision, one embracing the idea that protecting livelihoods and protecting the environment must go together, one that includes people and communities in the process, and one that uses community decision making to improve both conservation and community well-being.

ACKNOWLEDGMENTS

I am grateful to many colleagues, including those in the Community Conservation Research Network, the Coastal Community-University Research Alliance, and the FAO of the United Nations, for many helpful discussions and insights that supported the writing of this chapter I am also grateful to the editors of the book and to two anonymous referees for helpful comments that considerably improved the chapter. Of course, any remaining errors are the responsibility of the author. Financial support is acknowledged from the Social Sciences and Humanities Research Council of Canada and the Natural Sciences and Engineering Research Council of Canada.

REFERENCES

Agrawal, A., Gibson, C.C., 1999. Enchantment and disenchantment: the role of community in natural resource conservation. World Development 27, 629–649.

Armitage, D., Berkes, F., Charles, A., 2017. Governing the Coastal Commons: Communities, Resilience and Transformation. Earthscan, Routledge/Taylor & Francis, Oxford, UK.

Ban, N.C., Adams, V., Almany, G.R., Ban, S., Cinner, J., McCook, L.J., Mills, M., Pressey, R.L., White, A., 2011. Designing, implementing and managing marine protected areas: emerging trends and opportunities for coral reef nations. Journal of Experimental Marine Biology and Ecology 408, 21–31.

Béné, C., Macfadyen, G., Allison, E.H., 2007. Increasing the Contribution of Small-Scale Fisheries to Poverty Alleviation and Food Security. FAO Technical Paper No. 481. Food and Agriculture Organization of the United Nation, Rome, Italy. 125 p.

Berkes, F., 2009. Community conserved areas: policy issues in historic and contemporary context. Conservation Letters 2 (1), 19–24. http://dx.doi.org/10.1111/j.1755-263X.2008.00040.x.

Berkes, F., 2006. From community-based resource management to complex systems: the scale issue and marine commons. Ecology and Society 11, 1–15.

Berkes, F., 2004. Rethinking community-based conservation. Conservation Biology 18, 621–630.

Berkes, F., Mahon, R., McConney, P., Pollnac, R., Pomeroy, R., 2001. Managing Small-Scale Fisheries: Alternative Directions and Methods. International Development Research Centre, Ottawa, Canada. 320 pp.

Borrini-Feyerabend, G., Kothari, A., Alcom, J., Amaya, C., Bo, L., Campese, J., Carroll, M., et al., 2010. Strengthening What Works: Recognising and Supporting the Conservation Achievements of Indigenous Peoples and Local Communities. IUCN CEESP Briefing Note 10 https://portals. iucn.org/library/node/9672.

Borrini-Feyerabend, G., Kothari, A., Oviedo, G., 2004. Indigenous and Local Communities and Protected Areas: Towards Equity and Enhanced Conservation. IUCN, Gland, Switzerland and Cambridge, UK. xviii + 111 pp.https://portals.iucn.org/library/efiles/documents/PAG-011.pdf.

CBD (Convention on Biological Diversity), 2004. Decision Adopted by the Conference of the Parties to the Convention on Biological Diversity at its Seventh Meeting. UNEP/CBD/COP/ DEC/VII/5, 13 April 2004 www.cbd.int/doc/decisions/cop-07/cop-07-dec-05-en.pdf.

CCRN, 2016a. Community Conservation Research Network. Halifax, Canada www. CommunityConservation.Net.

CCRN, 2016b. Community Stories: Community Conservation Research Network. Halifax, Canada http://www.communityconservation.net/community-stories.

Charles, A., 2014. Human dimensions in marine ecosystem-based management (Chapter 3). In: Fogarty, M.J., McCarthy, J.J. (Eds.), Marine Ecosystem-Based Management. The Sea: Ideas and Observations on Progress in the Study of the Seas, vol. 16. Harvard University Press, Cambridge, USA. 568 p.

Charles, A., 2013. Governance of tenure in small-scale fisheries: key considerations. Land Tenure Journal 1, 9–37.

Charles, A., 2012. People, oceans and scale: governance, livelihoods and climate change adaptation in marine social-ecological systems. Current Opinion in Environmental Sustainability 4, 351–357.

Charles, A., 2009. Rights-based fisheries management: the role of use rights in managing access and harvesting. In: Cochrane, K.L., Garcia, S.M. (Eds.), A Fishery Manager's Guidebook. Wiley-Blackwell, Oxford, UK, pp. 253–282.

Charles, A., 1998. Beyond the status quo: re-thinking fishery management. In: Pitcher, T., Pauly, D., Hart, P. (Eds.), Re-inventing Fisheries Management. Kluwer, pp. 101–111.

Charles, A., 1995. The Atlantic Canadian groundfishery: roots of a collapse. Dalhousie Law Journal 18, 65–83.

Charles, A., Westlund, L., Bartley, D.M., Fletcher, W.J., Garcia, S., Govan, H., Sanders, J., 2016. Fishing livelihoods as key to marine protected areas: insights from the World Parks Congress. Aquatic Conservation: Marine and Freshwater Ecosystems 26, 165–184. http://dx.doi. org/10.1002/aqc.2648.

Charles, A., Wiber, M., Bigney, K., Curtis, D., Wilson, L., Angus, R., Kearney, J., Landry, M., Recchia, M., Saulnier, H., White, C., 2010. Integrated management: a coastal community perspective. Horizons 10, 26–34.

Charles, A., Wilson, L., 2009. Human dimensions of marine protected areas. ICES Journal of Marine Science 66, 6–15.

Charles, A., Bull, A., Kearney, J., Milley, C., 2007. Community-based fisheries in the Canadian Maritimes. In: McClanahan, T., Castilla, J.C. (Eds.), Fisheries Management: Progress Towards Sustainability. Blackwell Publishing, Oxford, UK, pp. 274–301.

Christie, P., 2004. Marine protected areas as biological successes and social failures in Southeast Asia. American Fisheries Society Symposium 42, 155–164.

Christie, P., White, A., 2007. Best practices for improved governance of coral reef marine protected areas. Coral Reefs 26, 1047–1056.

Cicin-Sain, B., Belfiore, S., 2003. Linking marine protected areas to integrated coastal and ocean management: a review of theory and practice. Ocean and Coastal Management 48, 847–868.

Cicin-Sain, B., Knecht, R.W., Jang, D., Fisk, G.W., 1998. Integrated Coastal and Ocean Management: Concepts and Practices. Island Press.

Corrigan, C., Hay-Edie, T., 2013. A Toolkit to Support Conservation by Indigenous Peoples and Local Communities: Building Capacity and Sharing Knowledge for Indigenous Peoples' and Community Conserved Territories and Areas (ICCAs). UNEP-WCMC, Cambridge, UK.

de Young, C., Charles, A., Hjort, A., 2008. Human Dimensions of the Ecosystem Approach to Fisheries: An Overview of Context, Concepts, Tools and Methods. Fisheries Technical Paper No. 489. Food and Agriculture Organization of the United Nations, Rome, Italy. 152 p.

FAO, 2016. Small-Scale Fisheries. Food and Agriculture Organization of the United Nations, Rome, Italy. http://www.fao.org/3/a-au832e.pdf.

FAO, 2015. Voluntary Guidelines for Securing Sustainable Small-Scale Fisheries in the Context of Food Security and Poverty Eradication. Food and Agriculture Organization of the United Nations, Rome, Italy.

FAO, 2011. Fisheries Management. 4. Marine Protected Areas and Fisheries. FAO Technical Guidelines for Responsible Fisheries. No. 4, Suppl. 4. FAO, Rome. 198 p.

FAO, 2009. Fisheries Management. 2. The Ecosystem Approach to Fisheries. 2.2 Human Dimensions of the Ecosystem Approach to Fisheries. FAO Technical Guidelines for Responsible Fisheries. No. 4, Suppl. 2, Add. 2. FAO, Rome. 88 p.

FAO, 2003. The Ecosystem Approach to Fisheries. Issues, Terminology, Principles, Institutional Foundations, Implementation and Outlook. FAO Fisheries Technical Paper. No. 443. Rome 71 pp.

Forst, M.F., 2009. The convergence of integrated coastal zone management and the ecosystems approach. Ocean & Coastal Management 52, 294–306.

Garcia, S.M., Rice, J., Charles, A., 2014. Governance of marine fisheries and biodiversity conservation: a history. In: Garcia, S.M., Rice, J., Charles, A. (Eds.), Governance of Marine Fisheries and Biodiversity Conservation: Interaction and Coevolution. Wiley-Blackwell, Oxford, UK, pp. 3–17. 552 p.

Gelcich, S., Hughes, T.P., Olsson, P., Folke, C., Defeo, O., Fernández, M., Foale, S., Gunderson, L.H., Rodríguez-Sickert, C., Scheffer, M., Steneck, R.S., 2010. Navigating transformations in governance of Chilean marine coastal resources. Proceedings of the National Academy of Sciences 107, 16794–16799.

Govan, H., July 25, 2009. Achieving the Potential of Locally Managed Marine Areas in the South Pacific. SPC Traditional Marine Resource Management and Knowledge Information. Bulletin.

Govan, H., Aalbersberg, W., Tawake, A., Parks, J., 2008. Locally-Managed Marine Areas: A Guide for Practitioners. The Locally-Managed Marine Area Network, Fiji.

Halpern, B.S., Selkoe, K.A., Micheli, F., Kappel, C.V., 2007. Evaluating and ranking the vulner-ability of global marine ecosystems to anthropogenic threats. Conservation Biology 21, 1301–1315.

Hardin, G., 1968. The tragedy of the commons. Science 162, 1243–1247.

Jentoft, S., 2000. The community: a missing link of fisheries management. Marine Policy 24, 53–60.

Jones, P.J.S., 2002. Marine protected area strategies: issues, divergences and the search for middle ground. Reviews in Fish Biology and Fisheries 11, 197–216.

Kearney, J., Berkes, F., Charles, A., Pinkerton, E., Wiber, M., 2007. The role of participatory governance and community-based management in integrated coastal and ocean management in Canada. Coastal Management 35, 79–104.

Kothari, A., Camill, P., Brown, J., 2013. Conservation as if people also mattered: policy and practice of community-based conservation. Conservation and Society. http://www.conservationandsociety.org/text.asp?2013/11/1/1/110937.

Long, R.D., Charles, A., Stephenson, R.L., 2016. Key principles of ecosystem-based management: the fishermen's perspective. Fish and Fisheries. http://dx.doi.org/10.1111/faf.12175.

Long, R.D., Charles, A., Stephenson, R.L., 2015. Key principles of marine ecosystem-based management. Marine Policy 57, 53–60.

McCay, B.J., Jentoft, S., 1996. From the bottom up: participatory issues in fisheries management. Society & Natural Resources 9, 237–250.

Nasuchon, N., Charles, A., 2010. Community involvement in fisheries management: experiences in the Gulf of Thailand countries. Marine Policy 34, 163–169.

Olsen, S.B., Sutinen, J.G., Juda, L., Hennessey, T.M., Grigalunas, T.A., 2006. A Handbook on Governance and Socioeconomics of Large Marine Ecosystems. Coastal Resources Center, University of Rhode Island.

Ostrom, E., 1990. Governing the Commons: The Evolution of Institutions for Collective Action. Cambridge University Press, Cambridge, UK.

Pauly, D., 2006. Major trends in small-scale marine fisheries, with emphasis on developing countries, and some implications for the social sciences. Maritime Studies 4, 7–22.

Pikitch, E.K., Santora, C., Babcock, E.A., Bakun, A., Bonfil, R., Conover, D.O., et al., 2004. Ecosystem-based fishery management. Science 305, 346–347.

Pinkerton, E., 2009. Coastal marine systems: conserving fish and sustaining community livelihoods with co-management. In: Chapin, F.S., Kofina, G.P., Folke, C. (Eds.), Principles of Ecosystem Stewardship. Springer Science and Business Media, New York, NY, USA, pp. 241–257.

Pinkerton, E.W., 1989. Cooperative Management of Local Fisheries. University of British Columbia Press, Vancouver, Canada.

Pinkerton, E., Edwards, D.N., 2009. The elephant in the room: the hidden costs of leasing individual transferable fishing quotas. Marine Policy 33, 707–713.

Pollnac, R., Christie, P., Cinner, J., Dalton, T., Daw, T., Forrester, G., Graham, N., McClanahan, T., 2010. Marine reserves as linked social-ecological systems. Proceedings of the National Academy of Sciences 107, 18262–18265.

Pomeroy, R.S., Parks, J.E., Watson, L.M., 2004. How is Your MPA Doing? A Guidebook of Natural and Social Indicators for Evaluating Marine Protected Area Management Effectiveness. IUCN, Gland, Switzerland, and Cambridge, UK. 216 pp.

Rocliffe, S., Peabody, S., Samoilys, M., Hawkins, J.P., 2014. Towards a network of locally managed marine areas (LMMAs) in the Western Indian Ocean. PLoS One 9 (7), e103000. http://dx.doi.org/10.1371/journal.pone.0103000.

Weeratunge, N., Béné, C., Siriwardane, R., Charles, A., Johnson, D., Allison, E.H., Nayak, P.K., Badjeck, M.-C., 2014. Small-scale fisheries through the wellbeing lens. Fish and Fisheries 15, 255–279.

Wilson, D.C., Nielsen, J.R., Degnbol, P. (Eds.), 2003. The Fisheries Co-management Experience: Accomplishments, Challenges and Prospects. Fish and Fisheries Series 26, Kluwer Academic Publishers, Dordrecht, The Netherlands.

Chapter 22

Innovations in Collaborative Science: Advancing Citizen Science, Crowdsourcing and Participatory Modeling to Understand and Manage Marine Social–Ecological Systems

Steven Gray[1], Steven Scyphers[2]
[1]*Michigan State University, East Lansing, MI, United States;* [2]*Northeastern University, Boston, MA, United States*

INTRODUCTION

Successful marine conservation in the Anthropocene demands science that better accounts for the interconnected relationships between people and oceans. Fortunately, the last decade has seen a vast increase in the number of interdisciplinary studies, frameworks, and models for explaining complex environmental problems as a function of intertwined social and ecological processes (Liu et al., 2007; Ostrom, 2009; Collins et al., 2011). The popularity of this approach has given rise to a new paradigm in environmental science and natural resource management focused on investigating coupled social–ecological systems (SES), or coupled human–natural systems, noting that many environmental issues cannot be well understood by relying solely on disciplinary scientific approaches alone (Binder et al., 2013). This systems approach has become commonplace in scientific literature and rests upon the notion that human societies are nested within nature and there are complex feedbacks linking humans and their environment (Berkes and Folke, 1998; Liu et al., 2007). Although definitions of SES vary slightly (see for example Redman et al., 2004), they tend to focus on understanding the complexity and dynamic nature of human–environment interactions through a systems-based perspective operationalized either theoretically or empirically.

At the same time that the SES framework has come to prominence and provided new way for thinking about environmental issues in the new era of the

Conservation for the Anthropocene Ocean. http://dx.doi.org/10.1016/B978-0-12-805375-1.00022-2
463

Anthropocene, the last decade has also seen dramatic advances in the tools used by science to understand and manage these coupled systems. Specifically, there has been a considerable increase in the number of studies that include the public or specific stakeholders as a partner in the scientific process through "citizen science" (Cooper et al., 2007; Dickinson et al., 2010; Wechsler, 2014; Shirk et al., 2012; Bonney et al., 2014). Since the inception of citizen science in the mid-1990s whereby Alan Irwin (1995) presented a call to open up science and science-policy processes to the public (Riesch and Potter, 2014), the ideas associated with citizen science have grown to encompass several approaches for including nonscientists and other relevant stakeholders in the creation, deliberation, and communication of scientific research.

Today, the study of citizen science has been so popularized, especially in environmental and conservation research, it has even been suggested that it is its own discipline due to the many impacts that citizen science has had on scientific research and society (Jordan et al., 2015). For example, citizen science has been shown to positively influence research outcomes (e.g., enhanced data coverage, resolution) and outputs (e.g., increasing research publications), environmental decision-making (e.g., action and legislation), and the individuals participating in these programs (e.g., increased personal skills, positive social relationships, and scientific literacy) (Shirk et al., 2012). The "crowd wisdom" generated by the knowledge and observations of citizens has frequently proven valuable for documenting environmental change and supporting or generating new hypotheses for empirical studies (Alessa et al., 2013). However, these outcomes are highly dependent on the degree to which the public is involved in the scientific process, and public inclusion can vary greatly based on the goals and design of the citizen-science project. For instance, a recent NSF-funded report (Bonney et al., 2009) identified three major categories of citizen-science participation, separated by the degree to which the public is included in the scientific process as: (1) *contributory projects* that are usually scientist-designed and the public is included mainly in data collection; (2) *collaborative projects* that are structured by scientists but citizens are provided opportunities to provide some input on project design and in data collection; and (3) *cocreated projects* that are more democratic partnerships where the public is actively engaged with all steps of the scientific process. In addition, Regalado (2015) identified the potential of public-initiated scientific research, in which a project is done by members of the public with professional scientists joining at a later stage. Recognizing that these categories of participation could be even further subdivided (see Shirk et al., 2012), it is reasonable to consider public participation in science as a continuum where both the emergent scientific and social outcomes of citizen-science projects depend on the context of the scientific problem and the structure of volunteer involvement.

Although the literature on citizen science describes it as an increasingly popular way to *collaboratively understand* ecosystems and *collaboratively make decisions* about natural resources, to date there is little information that

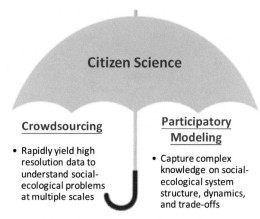

FIGURE 22.1   Crowdsourcing and participatory modeling are approaches we consider to be under the broader umbrella of citizen science. Both will play a substantial, but unique, role in understanding marine SES in the future.

specifically frames the unique and rapidly evolving challenges and opportunities for citizen science in marine systems. Here, we provide an overview of how citizen science has been applied in marine research and discuss how emerging technologies can be used in the future to improve marine science and conservation. Specifically, we use case studies to demonstrate: (1) how marine applications of citizen science can improve our understanding of social–ecological problems through rapid crowdsourcing of high resolution data; (2) how participatory modeling can capture complex and useful knowledge about the structure and function of marine SES; and (3) the role that web-based technologies may play in crowdsourcing data and harnessing collective intelligence for marine conservation in the Anthropocene (Fig. 22.1).

CROWDSOURCING ENVIRONMENTAL DATA THROUGH CITIZEN SCIENCE: TAPPING INTO THE HUMAN OBSERVING NETWORK

One key outcome routinely associated with citizen science is an improved understanding of our natural world. There is growing evidence that distributed networks of citizen scientists can contribute to scientific understanding of the structure and dynamics of ecosystems, both locally and globally (Bonney et al., 2014). In some instances, citizen-science monitoring has documented changes in ecosystems that would otherwise be prohibitively expensive or logistically impossible to characterize (Wolkovich and Cleland, 2010). Like all research approaches, citizen science has costs, limitations, and vulnerabilities that must be considered. For instance, citizen-science programs that rely on volunteered observations from multiple sources often lack standardized measures of effort. Further, large-scale programs face

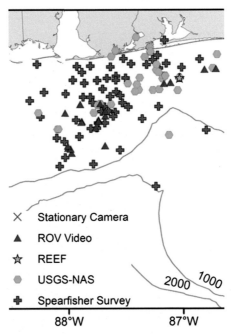

FIGURE 22.2 Map from Scyphers (2014) showing that the observations of spearfishers provided greater spatial coverage and higher resolution than traditional fisheries monitoring. Each symbol on the map indicates an observation of Indo-Pacific lionfish.

significant start-up costs and logistical hurdles. The vast data resources produced by modern web and mobile applications, such as iNaturalist.org and ebird.org, highlight the value of mainstreaming citizen science. However, as with all scientific approaches, innovations in citizen science must coincide with a consistent commitment and evolving efforts to ensure data quality (Crall et al., 2010) (Fig. 22.2).

Theorists in this area have suggested conditions under which such "swarm intelligence" approaches might be feasible, specifically within the context of generating information under "data poor" conditions. For example, crowd-based estimates are thought to be reliable when four conditions are met: (1) the study participants represent diverse opinions; (2) make judgments independent of each other and without outside influences; (3) are free of any fundamental biases that would cause them to systematically over- or underestimate a resource; and (4) truthfully report their estimates (Arlinghaus and Krause, 2013). Under these conditions, researchers can expect that any ill-informed participants are equally likely to overestimate the actual resource size as they are to underestimate it. When estimates are aggregated statistically, their contributions will therefore cancel each other out and the aggregated result will be close to the actual data (Arlinghaus and Krause, 2013). Until recently, tools that provide ways to rapidly design and test experiments to determine contexts under which crowds are

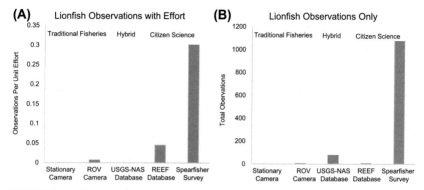

(A) Lionfish Observations with Effort

Traditional Fisheries   Hybrid   Citizen Science

Observations Per Unit Effort

Stationary Camera | ROV Camera | USGS-NAS Database | REEF Database | Spearfisher Survey

(B) Lionfish Observations Only

Traditional Fisheries   Hybrid   Citizen Science

Total Obervations

Stationary Camera | ROV Camera | USGS-NAS Database | REEF Database | Spearfisher Survey

FIGURE 22.3 Comparison of traditional fisheries and citizen-science data as observing systems for invasive lionfish in the Gulf of Mexico (Scyphers et al., 2015). Stationary cameras and remotely operated vehicles (ROV) with cameras are two common approaches for quantitatively monitoring reef fish populations. The U.S. Geological Survey's Nonindigenous Aquatic Species (USGS-NAS) database is the national repository for aquatic nonindigenous species and contains sightings data from both traditional monitoring and citizen-science sources. The REEF Database contains fish sighting data collected by volunteer divers trained as citizen scientists. The spearfisher survey represents the collective knowledge and experiences of licensed spearfishers documented through an online survey (e.g., Morris and Whitfield, 2009; Green et al., 2012).

wise (and when they are not), and under what conditions such approaches are appropriate for resource management have largely been limited due to a lack of available technologies. In the context of understanding marine ecosystems, citizen-science observations have proven valuable for documenting ecosystem change, as well as the human dimensions of conservation and management initiatives (e.g., Scyphers et al., 2014; Ward-Paige et al., 2010a). Here, the observations and collective knowledge of fishers and divers have been used to document species declines, shifts in community structure, and inform conservation initiatives (Scholz et al., 2004; Stallings, 2009). For example, the lionfish invasion across the US Atlantic coast is considered the best-documented marine invasion to date and provides an excellent illustration of the value of crowdsourcing environmental data. Over the past decade, lionfish have spread to an area estimated >3 million km², where they often occur in densities far greater than in their native habitats (Schofield, 2010; Schofield et al., 2013). Although the invasion is still relatively new, some early studies have documented alarming reductions in densities of native fish coincided with lionfish invasion (Fig. 22.3).

A recent study by Scyphers et al. (2015) compared traditional fisheries monitoring and three different sources of citizen-science data for detecting the spread of invasive Indo-Pacific lionfish: a federal database of nonindigenous species sightings (i.e., USGS-NAS), a database of fish surveys conducted by trained volunteer divers, and from a targeted survey of recreational spearfishers. They found that citizen observations documented lionfish 1–2 years earlier and more frequently than traditional reef fish monitoring programs. Citizen observations first documented lionfish in 2010, followed by rapid expansion in 2011

(+367%). Such studies reveal the value of using fishers as citizen scientists, providing a living observatory network for rapid and broad-scale ecosystem monitoring (Alessa et al., 2013).

Even though such observational approaches to citizen science has been repeatedly demonstrated effective in marine ecosystems, there are some additional challenges that must be considered. From a scientific perspective, underwater visual surveys may overestimate the abundance of large and highly mobile species if the potential for species entering and exiting the study area is not accounted for (Ward-Paige et al., 2010b). From a cost and logistics perspective, citizen science in marine ecosystems require specialized gear and skillsets (e.g., snorkeling, SCUBA) that are often costly and result in a narrower pool of potential participants. As with any research design that involves complex environmental data collected by citizens or stakeholders, the nature of the community involved and how the data are collected and reported must be considered in the design of marine citizen-science projects.

In addition to increasing the scale and scope of biological observations, citizen science can also be used to rapidly assess how well conservation and management initiatives are performing. In fisheries, management often relies on the compliance and participation of fishers for effective policies. For instance, fish captured from deep water sometimes experience internal injuries caused by barotrauma (i.e., gas expansion in swim bladder, body cavity). A management policy aimed at increasing post-release survival of discarded fish relies on the mandatory use of venting tools (i.e., hollow syringe-like devices) to release expanded gases and mitigate the effects of capture. However, the effectiveness of venting is disputed resulting in a management issue surrounded by uncertainty. A recent study by Scyphers (2014) sought to better understand the human dimensions and efficacy of this participatory management strategy. Using an online survey, anglers revealed that there is high "buy-in" among recreational fishers and considerable belief that venting is beneficial for released reef fishes. However, their work also showed that a substantial proportion of anglers, including some highly experience anglers, are not implementing the management measure effectively (Fig. 22.4). For instance, when asked to identify the ideal location for inserting the venting tool, many anglers indicated the protrusion of the fish's stomach or any adjacent major internal organ. Considering that a large proportion of anglers implementing venting techniques may be doing so improperly, the authors concluded that more attention should be directed toward angler considerations, behaviors, and the social costs of mandates to avoid losing the trust of anglers.

PARTICIPATORY MODELING TO UNDERSTAND SOCIAL–ECOLOGICAL COUPLING

In addition to crowdsourcing environmental data, citizen science has also been linked to participatory approaches that seek to answer broader questions about the complex structure and function of marine SES (Nayaki et al., 2014; Gray et al., 2017a). Given the wide range of stakeholders that use and interact with

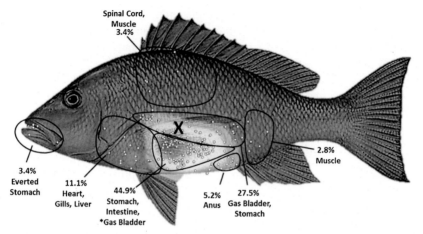

FIGURE 22.4 Figure from Scyphers et al. (2014) showing results from a survey question where recreational anglers were asked to demonstrate their knowledge of best practices for using venting tools to mitigate the consequences of barotrauma in reef fish. The overlaid anatomical regions and heatmap of responses highlight the management implications of improperly executed venting and the value of crowdsourcing data on participatory management strategies.

marine environments, knowledge about how ecosystems and social systems interact is diverse (Stier et al., 2017). In fact, it had been suggested that in order to create more sustainable management strategies, stakeholders must forge new relationships to enhance multidirectional information flows, learn from each other, and together, develop flexible ways of understanding and managing their environments (Carpenter and Gunderson, 2001). Although there is general agreement on the importance of citizen involvement within adaptive decision-making processes, it is unclear how best this should happen, what form it should take (Abelson et al., 2003; Rowe and Frewer, 2000), and how participation in environmental knowledge generation contributes to shared understanding of environmental issues. What constitutes "appropriate" venues of knowledge sharing varies within and between stakeholder groups, research design, and decision-contexts, and is highly influenced by social conditions in which decisions are made. These conditions include hierarchies like differences in values, power struggles, degree of participation of those affected, and the inherent uncertainty of complex system behavior (Biggs et al., 2011).

As a result of these issues, participatory approaches and software tools have emerged over the last decade as a way to model and collaboratively discuss the structure and function of SES with stakeholders. The justification for including stakeholders in scientific modeling practices is based on the acknowledgment that: (1) model-based reasoning is a predominant and preferred basis of environmental decision-making in contemporary environmental management; (2) public participation is an essential component to informed environmental decision-making; and (3) stakeholder groups often hold unique and complex knowledge that is useful for understanding the dynamics of SES (Gray et al., 2015).

This increased interest in participatory modeling has given rise to a range of stakeholder-centered modeling tools, practices, and guidelines that aim to provide decision support in participatory environmental planning contexts (Voinov and Bousquet, 2010; Voinov et al., 2016). However, even with this increase in tool and software development, some critics have cautioned that diversity of modeling practices does not necessarily indicate diversity in function (Jones et al., 2009) and the most significant contribution of including stakeholders in modeling is community learning, facilitated by structured knowledge sharing between citizens, scientist, and managers (Voinov and Bousquet, 2010).

Given the popularity of participatory modeling approaches in the last few years, synthesis research has begun to identify trends in both the process and products associated with participatory modeling (Gray et al., 2017b) and their implications for improved marine conservation decision-making. For example, Sandker et al. (2010) found that although there are several common tools associated with the practice of participatory modeling, including Bayesian methods, Agent-Based Modeling, and Systems Dynamic Modeling among others, the largest contribution of engaging stakeholders in modeling included stimulating cross-sector planning and facilitating discussions that helped participants to confront the drivers of environmental change and to recognize trade-offs in management strategies. The authors also found evidence of environmental decision-making outcomes, such as the coproduction of knowledge leading to management decisions as a result of the modeling process, but this was largely dependent on types of stakeholders that were included in the process. In terms of the modeling processes, case studies evaluated in their review varied in terms of the degree of participation among stakeholders, measured in terms of the amount of time that scientists, nongovernmental organizations, and stakeholders collaborated, which ranged from creating models in a single workshop to prolonged involvement with a group of stakeholders over several months (Sandker et al., 2010). Others have also suggested that different modeling techniques may be more or less appropriate to employ given variation in the community engaged in the modeling process and the types of modeling practices that are useful for a given decision-making context (Gray et al., 2015).

In terms of using participatory modeling methods for marine conservation, there has recently been a considerable increase in the number of case studies that adopt these approaches. These marine resource-based participatory modeling applications include marine spatial planning (Gray et al., 2014) and coastal hazard adaptation (Henly-Shepard et al., 2015). However, given the mandates for stakeholder involvement in policy making in the EU and in the USA (Magnuson Stevens), participatory modeling seems to be particularly useful for fisheries management to understand marine systems that span multiple ecological scales using several different system-based modeling software packages.

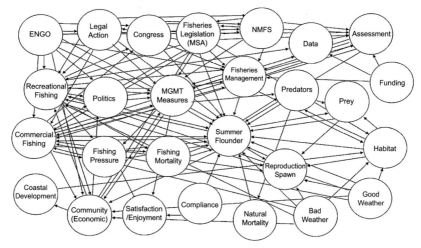

FIGURE 22.5 Representation of the mid-Atlantic coast summer flounder SES as informed by the aggregation of multiple fishery stakeholders individual "mental models."

For example, at the local scale, Webler et al. (2016) used a dialog-based group concept mapping called VCAPS (vulnerability, consequences, and adaptation planning scenarios) and System Dynamics Modeling to elicit and integrate both local and expert knowledge about vulnerability to climate change in a lobster fishery in Maine. The study involved a prolonged engagement period with ecologists, lobstermen, and other community members over a 2-year period, which included two concept mapping workshops to understand the structure of social–ecological dynamics of the fishery related to climate change impacts, and three system dynamics workshops to build a functional and parameterized model to predict how fishing effort and climate change relate to economic productivity in the fishery.

On a more regional scale, Gray et al. (2012) used fuzzy cognitive mapping with different stakeholder groups involved in the summer flounder fishery in the mid-Atlantic coast in the USA. The authors conducted a series of interviews with 35 individuals from 6 different fishery stakeholder groups, including recreational fishermen, commercial fishermen, members of the pre- and postharvest sectors, environmental NGO representatives, fisheries scientists, and fishery managers that were selected based on their involvement in the policy-making process. The goal of this study was to understand how local expertise reflects different beliefs and values, and how these more pluralistic views of marine dynamics could be integrated and used to decrease uncertainty surrounding the dynamics of the fishery. The result was a wider scenario-capable systems-based model that integrated the knowledge of all stakeholder experts involved in the fishery, thus providing a more complete view of the systems being managed (Fig. 22.5).

FUTURE DIRECTIONS OF USING CITIZEN SCIENCE TO UNDERSTAND MARINE SES

Citizen science has already proven to be very effective in expanding data collection capacity, enhancing our understanding of how human and natural ecosystems are coupled, and in improving stakeholder participation in decision-making about these systems. However, new frontiers and tools are increasingly emerging that hold considerable promise for more collaborative forms of marine science in the future. Although many of these new approaches, and specifically software tools, were not developed for use in collaborative science or with natural resource management in mind, new web-enabled platforms that capture the "wisdom of the crowds" or types of "collective intelligence" now provide marine scientists with new and powerful approaches to better understand and manage marine systems. We identify two of these emerging web-based approaches (e.g., UNU and Mental Modeler) to collaborative research and modeling, and discuss some of the trade-offs and considerations associated with the selection of different approaches.

Innovations in Web-Based Collective Intelligence

While still in the early stages of development, new web-based applications such as UNU (www.unu.ai) (Rosenberg, 2015) provide exciting and novel approaches to harnessing collective intelligence of marine stakeholders. Such "swarm intelligence" platforms hold great promise for (1) identifying value-based judgments regarding the types of scientific research that should be conducted; and (2) providing new ways to estimate and forecast resource assessments by engaging groups of knowledgeable stakeholders that interact with marine systems to share their knowledge and opinions. The design of the UNU interface is modeled after "biological swarms," enabling groups of online users to individually pose questions (selected by a moderator) and, as a group, answer these questions, make decisions, and resolve dilemmas by working together in unified online dynamic system (Rosenberg, 2015). The platform is synchronous, meaning users can explore decision spaces together, and the software structures these online social groups of users through a process of "social swarming" in real time intended to promote group convergence on a preferred solution in a matter of seconds. Fig. 22.6A and B show current testing of the software with online user groups, where a question is posed about whether more resources should be devoted to the colonization of space. The crowd collectively settled into the answer of "Yes" provided with alternative options "No," "Maybe," or "Bad Question." To arrive at the answer, all logged-in users use individual horseshoe shaped magnets to pull a circle toward one of several predefined answers (which can themselves be generated by the group).

While not currently used by natural resource users, Fig. 22.6C and D show hypothetical conditions of software use with groups of scientists, commercial fishermen, recreational fishermen or combinations of these, or other knowledgeable

(A)   **(B)**

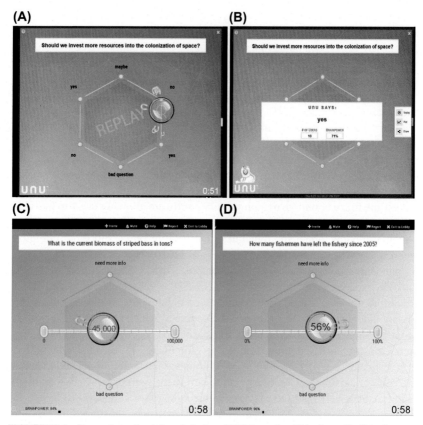

FIGURE 22.6 Screenshots of web-based platform UNU (unu.ai), which allows distributed groups of users to log in and collaboratively answer questions using a "swarm intelligence" approach. Panels A and B show actual data from crowdsourcing an answer to the value-based question of increasing resources for the colonization of space research. Panels C and D show hypothetical data in a marine science context asking fisheries stakeholders (e.g., groups of scientists, commercial or recreational fishermen, or both) to collaboratively estimate the current biomass in tons of recreationally and commercially important Atlantic species of striped bass (C) in order to estimate true stock estimates and to estimate the percentage of fishermen that have left the fishery since 2005 (D).

groups to: (1) estimate current biomass of economically important fish stocks; and (2) estimate social changes such as the percentage of fishermen that have left the fishery since 2005. Tools like UNU may provide clear mechanisms to test often hypothesized, but empirically untested ideas using "wisdom of the crowd." For instance, such approaches could have the potential to generate independent estimates of population sizes for natural resources (e.g. stock estimates) which can be aggregated to approximate true sizes (Arlinghaus and Krause, 2013). While previous researchers have suggested averaging individual resource estimates collected through surveys as one of the most straightforward ways to collect such crowdsourced data (Arlinghaus and Krause, 2013), relatively few studies have attempted to compare averaged resource estimates to scientifically or empirically developed resource assessments, with some recent exceptions (see Predavec et al., 2016). It has been noted that wisdom of the crowd, collective cognition, and swarm intelligence all essentially refer to a process of individuals independently acquiring information and processing this information through social interaction to produce a solution to a cognitive problem that cannot be arrived at by any single individual (Krause et al., 2010).

Other new web-based technologies that take a unique approach to collaborative science include Mental Modeler (www.mentalmodeler.org) and DESIM (descriptive executable simulation modeling), platforms which provide scientists new ways of "crowdsourcing" mental models of groups of natural resource stakeholders asynchronously. Mental Modeler, developed by Gray et al. (2013) is an online fuzzy cognitive mapping (FCM) software allowing users to individually or collaboratively create dynamic representations of their belief systems, or mental models, about the dynamics of natural resource and other complex systems. DESIM, developed by Pfaff et al. (2016) is a complementary, but independent, software package that decomposes the FCM generated in Mental Modeler into pairwise comparisons (Fig. 22.7) which can then be converted into online survey questions and administered to large groups of users online to validate model structure and evaluate the degree to which complex understanding of resource systems are shared across local and scientific experts. Using interviews or web-based approaches to develop FCMs, DESIM asks the online participants to agree/disagree with causal connections in the model and to compare pairs of existing connections with regard to their strength. Analytical hierarchy process (AHP) is used to compute the strength of connections, based on all pairwise comparisons by all online study participants, thus providing a very robust, "crowdsourced" FCM model. Such approaches allow the complex structure of SES to be defined along with the areas of uncertainty of the dynamics of SES. Additionally, because FCMs are based on graph theory and matrix algebra, these platforms can be used to generate environmental and social change scenarios based on crowd knowledge to understand how these complex systems may react to future or hypothetical changes or perturbations.

Fig. 22.7 shows a hypothetical FCM collected through interviews with several coastal experts about the relationship between oil and gas operations, unexpected oil spills and their structural relation and influence on ecosystem, and economic dynamics of a coastal area. Once individual or group mental models are generated with Mental Modeler through interviews of focus groups, DESIM allows pairwise relationships to be converted into individual survey questions, which can be administered to large groups of stakeholders online to validate or evaluate the degree to which mental models about marine SES are shared across marine stakeholder groups. Further, the final crowdsourced model reconstructed through survey responses can be subjected to scenario analysis and used to estimate how qualitative changes in system states are linked with social or ecological conditions under different policy, social, or natural conditions.

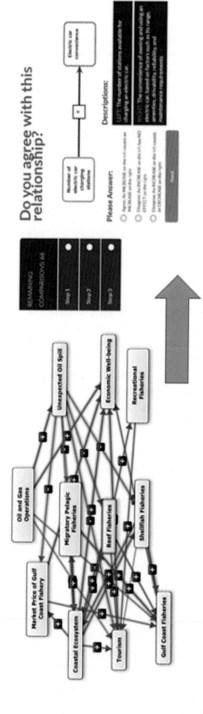

FIGURE 22.7   FCM created in Mental Modeler can be translated into a series of survey questions in DESIM which can be translated back into a "crowdsourced" mental model to estimate structure and dynamics of marine systems.

Trade-Offs in Participatory Modeling Approaches: Selecting the Right Tool for the Job

As more participatory approaches to collaborative modeling continue to become more mainstream with technological advances (Voinov et al., 2016), it is important to select participatory modeling methods based on the community involved in the modeling process, research questions, or management goals, and how each tool differs across dimensions (Gray et al., 2017b). Although the theory behind each of these software-based or web-enabled tools continues to develop with new methodological and technological advances, the strengths and weaknesses of different approaches should be taken into consideration when designing collaborative science projects.

Certain participatory modeling methods may be more or less amenable to different types of marine stakeholders involved in the modeling process based on the amount of training required to create and analyze a model or to provide data points for an assessment. Although narrative scenario analysis and qualitative concept mapping lend themselves to use across a wider range of communities because they are more flexible than semiquantitative approaches, the output of these models is often not dynamic, thus limiting their ability to be used to evaluate competing system states through post hoc analyses (Gray et al., 2015, 2017b). Additionally, although to varying degrees most methods allow stakeholders and scientists to define the concepts, components, or variables that constitute the state space of the system modeled, some methods are more flexible in terms of the types of relationships that can be defined between variables. FCM and Agent-Based Modeling, for example, can represent feedback relationships between variables, whereas Bayesian belief network relationships are unidirectional. Although to some extent, all SES modeled through these efforts are defined in terms of time and space, the degree to which model outputs can be interpreted in spatial or temporal units by stakeholders varies and thus may influence analytical abilities to draw meaningful conclusions that facilitate management action. When considered together on a spectrum, as tools transition from more flexible and qualitative to more parameterized and semiquantitative, ease of stakeholder use decreases while the ability to explicitly evaluate competing system states increases. Further, although semiquantitative approaches may provide a wide range of opportunities for post hoc analysis, they may limit the degree to which stakeholder values and knowledge are integrated into model-based assessments.

CONCLUSION

Although innovations in citizen science have tested new ways for researchers, managers, and society to *collaboratively understand* and *collaboratively make decisions* about ecosystems natural resources, how citizen and other participatory forms of science will support marine conservation in the future is unclear. However, given new innovations and a strong history of success, marine scientists will likely have access to a diverse toolkit allowing them to harness citizens' knowledge and participation in new ways in order to: (1) prioritize the scientific questions that are asked about marine systems through web-based technologies (Rosenberg, 2015); (2) understand the complex nature and dynamics of these systems based on integrating a range of local expert knowledge (Gray et al., 2012); (3) use stakeholders to collect empirical data about changes in these systems (Scyphers et al., 2015); and (4) adaptively evaluate whether management decisions having desired impacts to be considered as new research questions are developed (Scyphers et al., 2013). Such collaborative science approaches are promising and will only continue to expand, transitioning the nature of knowledge by diverse stakeholders away from being a management liability that impedes conservation action (Biggs et al., 2011) into strength. Citizens, scientists, and managers collectively seek to define, understand, and address modern marine problems presented by the complex and dynamic nature of the marine environment. For both science and conservation, collaboration and inclusive participation will be critical for understanding and responding to the novel and rapid changes anticipated to occur in the Anthropocene.

REFERENCES

Abelson, J., Forest, P.G., Eyles, J., Smith, P., Martin, E., Gauvin, F.P., 2003. Deliberations about deliberative methods: issues in the design and evaluation of public participation processes. Social Science & Medicine 57 (2), 239–251.

Alessa, L., Kliskey, A., Myers, M., et al., 2013. Community Based Observing Networks (CBONs) for Arctic Adaptation and Security (White paper prepared for Arctic Observing Summit, Vancouver, BC).

Arlinghaus, R., Krause, J., 2013. Wisdom of the crowd and natural resource management. Trends in Ecology & Evolution 28 (1), 8–11.

Berkes, F., Folke, C. (Eds.), 1998. Linking Social and Ecological Systems: Management Practices and Social Mechanisms for Building Resilience. Cambridge University Press, Cambridge, UK.

Biggs, D., Abel, N., Knight, A.T., Leitch, A., Langston, A., Ban, N.C., 2011. The implementation crisis in conservation planning: could "mental models" help? Conservation Letters 4, 169–183.

Binder, C.R., Hinkel, J., Bots, P.W., Pahl-Wostl, C., 2013. Comparison of frameworks for analyzing social-ecological systems. Ecology and Society 18 (4), 26.

Bonney, R., Ballard, H., Jordan, R., McCallie, E., Phillips, T., Shirk, J., Wilderman, C.C., 2009. Public Participation in Scientific Research: Defining the Field and Assessing Its Potential for Informal Science Education. A CAISE Inquiry Group Report (Online Submission).

Bonney, R., Shirk, J.L., Phillips, T.B., Wiggins, A., Ballard, H.L., Miller-Rushing, A.J., Parrish, J.K., 2014. Next steps for citizen science. Science 343 (6178), 1436–1437.

Carpenter, S.R., Gunderson, L.H., 2001. Coping with collapse: ecological and social dynamics in ecosystem management: like flight simulators that train would-be aviators, simple models can be used to evoke people's adaptive, forward-thinking behavior, aimed in this instance at sustainability of human–natural systems. BioScience 51, 451–457.

Collins, S.L., Carpenter, S.R., Swinton, S.M., Orenstein, D.E., Childers, D.L., Gragson, T.L., Knapp, A.K., 2011. An integrated conceptual framework for long-term social–ecological research. Frontiers in Ecology and the Environment 9 (6), 351–357.

Cooper, C.B., Dickinson, J., Phillips, T., Bonney, R., 2007. Citizen science as a tool for conservation in residential ecosystems. Ecology and Society 12 (2), 11.

Crall, A.W., Newman, G.J., Jarnevich, C.S., Stohlgren, T.J., Waller, D.M., Graham, J., 2010. Improving and integrating data on invasive species collected by citizen scientists. Biological Invasions 12 (10), 3419–3428.

Dickinson, J.L., Zuckerberg, B., Bonter, D.N., 2010. Citizen science as an ecological research tool: challenges and benefits. Annual Review of Ecology, Evolution and Systematics 41, 149–172.

Gray, S., Chan, A., Clark, D., Jordan, R.C., 2012. Modeling the integration of stakeholder knowledge in social-ecological system decision-making: benefits and limitations to knowledge diversity. Ecological Modeling 229, 88–96.

Gray, S.A., Gray, S., Cox, L.J., Henly-Shepard, S., January, 2013. Mental modeler: a fuzzy-logic cognitive mapping modeling tool for adaptive environmental management. In: 2013 46th Hawaii International Conference on System Sciences (HICSS). IEEE, pp. 965–973.

Gray, S.R.J., Gagnon, A.S., Gray, S.A., O'Dwyer, B., O'Mahony, C., Muir, D., Devoye, R.J.N., Falaleeva, M., Gault, J., 2014. Are coastal managers detecting the problem? Assessing stakeholder perception of climate vulnerability using Fuzzy Cognitive Mapping. Ocean & Coastal Management 94, 74–89.

Gray, S.A., Gray, S., De Kok, J.L., Helfgott, A.E.R., O'Dwyer, B., Jordan, R., Nyaki1, A., 2015. Using fuzzy cognitive mapping as a participatory approach to analyze change, preferred states, and perceived resilience of social-ecological systems. Ecology and Society 20 (2), 11.

Gray, S., Jordan, R.C., Crall, A., Newman, G., Hmelo-Silver, C., Huang, J., Novak, W., Mellor, D., Frensley, T., Prysby, M., Singer, A., 2017a. Combining participatory modelling and citizen science to support volunteer conservation action. Biological Conservation (in press).

Gray, S., Voinov, A., Paolisso, M., Jordan, R.C., Todd BenDor, P., Glynn, B., Hedelin, K., Hubacek, J., Introne, N., Kolagani1, B., Laursen, C., Prell, L., Schmitt-Olabisi, A., Singer, E., Sterling, M., Zellner, 2017b. Purpose, Processes, Partnerships, and Products: 4Ps to advance participatory socio-environmental modeling. Ecological Applications (in press).

Green, S.J., Akins, J.L., Maljkovic, A., Côté, I.M., 2012. Invasive lionfish drive Atlantic coral reef fish declines. PLoS One 7, e32596.

Henly-Shepard, S., Gray, S., Cox, L., 2015. Facilitating community adaptation through participatory modeling and social learning. Environmental Science and Policy 45, 109–122.

Irwin, A., 1995. Citizen Science, vol. 136. Routledge, London.

Jones, N.A., Perez, P., Measham, T.G., Kelly, G.J., d'Aquino, P., Daniell, K.A., Ferrand, N., 2009. Evaluating participatory modeling: developing a framework for cross-case analysis. Environmental Management 44 (6), 1180–1195.

Jordan, R., Crall, A., Gray, S., Phillips, T., Mellor, D., 2015. Citizen science as a distinct field of inquiry. BioScience 65 (2), 208–211.

Krause, J., Ruxton, G.D., Krause, S., 2010. Swarm intelligence in animals and humans. Trends in Ecology & Evolution 25 (1), 28–34.

Liu, J., Dietz, T., Carpenter, S.R., et al., 2007. Complexity of coupled human and natural systems. Science 317, 1513–1516.

Morris Jr., J.A., Whitfield, P., 2009. Biology, ecology, control and management of the invasive Indo-Pacific lionfish: an updated integrated assessment. NOAA Technical Memorandum NOS NCCOS 99 57 pp.

Nayaki, A., Gray, S., Lepczyk, J., Skibins, D.R., 2014. Understanding the hidden drivers and local-scale dynamics of the bushmeat trade through participatory modeling. Conservation Biology 28 (5), 1403–1414.

Ostrom, E., 2009. A general framework for analyzing sustainability of social-ecological systems. Science 325.

Pfaff, M.S., Drury, J.L., Klein, G.L., September 2016. Modeling knowledge using a crowd of experts. In: Proceedings of the Human Factors and Ergonomics Society Annual Meeting, vol. 60, No. 1. SAGE Publications, pp. 183–187.

Predavec, M., Lunney, D., Hope, B., Stalenberg, E., Shannon, I., Crowther, M.S., Miller, I., 2016. The contribution of community wisdom to conservation ecology. Conservation Biology 30 (3), 496–505.

Redman, C.L., Grove, J.M., Kuby, L.H., 2004. Integrating social science into the Long-Term Ecological (LTER) network: social dimensions of ecological change and ecological dimensions of social change. Ecosystems 7 (2), 161–171.

Regalado, C., 2015. Promoting playfulness in publicly initiated scientific research: for and beyond times of crisis. International Journal of Play 4 (3), 275–284.

Riesch, H., Potter, C., 2014. Citizen science as seen by scientists: methodological, epistemological and ethical dimensions. Public Understanding of Science 23 (1), 107–120.

Rosenberg, L.B., September 2015. Human swarming, a real-time method for parallel distributed intelligence. In: Swarm/Human Blended Intelligence Workshop (SHBI). IEEE, pp. 1–7.

Rowe, G., Frewer, L.J., 2000. Public participation methods: a framework for evaluation. Science, Technology & Human Values 25 (1), 3–29.

Sandker, M., Campbell, B.M., Ruiz Perez, M., Sayer, J.A., Cowling, R.M., Kassa, H., Knight, A., 2010. The role of participatory modeling in landscape approaches to reconcile conservation and development. Ecology and Society 15 (2).

Schofield, P.J., 2010. Update on geographic spread of invasive lionfishes (*Pterois volitans* [Linnaeus, 1758] and *P. miles* [Bennett, 1828]) in the western north Atlantic Ocean, Caribbean Sea and Gulf of Mexico. Aquatic Invasions 5, S117–S122.

Schofield, P.J., Morris Jr., J.A., Langston, J.N., Fuller, P.F., 2013. *Pterois volitans/miles* Factsheet. USGS Nonindigenous Aquatic Species Database. Gainesville, FL. http://nas.er.usgs.gov/taxgroup/fish/lionfishdistribution.aspx.

Scholz, A., Bonzon, K., Fujita, R., et al., 2004. Participatory socioeconomic analysis: drawing on fishermen's knowledge for marine protected area planning in California. Marine Policy 28, 335–349.

Scyphers, S.B., Fodrie, F.J., Hernandez, F.J., Powers, S.P., Shipp, R.L., 2013. Venting and reef fish survival: perceptions and participation rates among recreational anglers in the northern Gulf of Mexico. North American Journal of Fisheries Management 33, 1071–1078.

Scyphers, S.B., Powers, S.P., Akins, J.L., et al., 2015. The role of citizens in detecting and responding to a rapid marine invasion. Conservation Letters 8, 242–250.

Shirk, J.L., Ballard, H.L., Wilderman, C.C., Phillips, T., Wiggins, A., Jordan, R., Bonney, R., 2012. Public participation in scientific research: a framework for deliberate design. Ecology and Society 17 (2), 29.

Stallings, C.D., 2009. Fishery-independent data reveal negative effect of human population density on Caribbean predatory fish communities. PLoS One 4, e5333.

Stier, A., Samhouri, J., Gray, S., Martone, R., Mach, M., Halpern, B., Kappel, C., Scarborough, C., Levin, P., 2017. Integrating expert opinion into food web conservation and management. Conservation Letters 10 (1), 67–76.

Voinov, A., Bousquet, F., 2010. Modelling with stakeholders. Environmental Modelling & Software 25 (11), 1268–1281.

Voinov, A., Kolagani, N., McCall, M.K., Glynn, P.D., Kragt, M.E., Ostermann, F.O., Ramu, P., 2016. Modelling with stakeholders–next generation. Environmental Modelling & Software 77, 196–220.

Ward-Paige, C.A., Mora, C., Lotze, H.K., et al., 2010a. Large-scale absence of sharks on reefs in the greater-Caribbean: a footprint of human pressures. PLoS One 5, e11968.

Ward-Paige, C.A., Flemming, J.M., Lotze, H.K., 2010b. Overestimating fish counts by non instantaneous visual censuses: consequences for population and community descriptions. PLoS One 5, e11722.

Webler, T., Stancioff, E., Goble, R., Whitehead, J., 2016. Participatory modeling and community dialog about vulnerability of lobster fishing to climate change. In: Gray, S., Paolisso, M., Jordan, R.C., Gray, S. (Eds.), Environmental Modeling with Stakeholders: Theory, Methods and Applications. Springer Publishing, New York City.

Wechsler, D., 2014. Crowdsourcing as a method of transdisciplinary research—tapping the full potential of participants. Futures 60, 14–22.

Wolkovich, E.M., Cleland, E.E., 2010. The phenology of plant invasions: a community ecology perspective. Frontiers in Ecology and the Environment 9, 287–294.

Chapter 23

Looking Forward: Interconnectedness in the Anthropocene Ocean

Melissa R. Poe[1], Phillip Levin[2]

[1]University of Washington, Seattle, WA, United States; [2]University of Washington, School of Environment and Forest Sciences, Seattle, WA, United States

OUR SHARED OCEAN CHALLENGES

The stakes are high in this time of unprecedented changes in the Anthropocene ocean. Threats—many already manifest—range from fisheries collapse to species extinctions, invasions, and shifts in distribution; from habitat degradation to altered ecosystem structure and function; from increased storm exposure and rising sea levels to coastal erosion and inhospitable ocean chemistry. Not only do we see and foresee wide-ranging ecological impacts, changes in the ocean, and the underlying drivers of change also create social injustices. Human rights abuses, human trafficking, and forced labor in fisheries are widely reported. Moreover, poverty, food insecurity, and the alienation of resource rights continue to put the most vulnerable people at risk. The stakes are even more complex and challenging when considered together as cumulative impacts from a multitude of social and ecological stressors. For optimists, heck even for realists, reckoning with these risks is downright sobering. And what is most sobering? These transformations result primarily from our own activities.

We are faced not simply with *how to survive* the existing and emergent threats, but crucially, *how shall we create a caring and meaningful future* for ourselves and our coinhabitants of the planet earth. And if we are to take responsibility for such a future, how might we direct and scale our actions effectively? This is the shared challenge of the Anthropocene.

What is different about ocean conservation in the Anthropocene? Observations and predictions of social–ecological change reflect more intense, expansive, and frequent impacts to people and nature. This new epoch is marked by evidence that earth systems have shifted into a "no-analogue state" (Crutzen and Steffen, 2003; Hamilton et al., 2015). While many of the reasonably well-understood

Conservation for the Anthropocene Ocean. http://dx.doi.org/10.1016/B978-0-12-805375-1.00023-4

and age-old earth system dynamics continue, the Anthropocene ocean presents different thresholds of resilience, and tipping points with increasing social and ecological uncertainties (Chapters 2 and 13).

Epochal changes (e.g., from the Holocene to the Anthropocene) are marked by the crossing of thresholds when a distinctive tipping point ushers in a new epoch. Geologists and paleontologists look at fossil records and sediments layered in geological timescales, or strata, to interpolate environmental changes between points in time. These records typically show mass extinctions, changes in species diversity, and geographic distribution (Holland, 2016). In social systems, the transition to the Anthropocene has been marked by transformations in economic, cultural, and political systems: influencing modes and means of production, altering access to the benefits and services (and exposure to waste and disservices) of nature, and changing people's relationships to nature and other people. What are the social parallels to crossing ecological thresholds such as species extinctions and biodiversity losses? Community economic systems (e.g., small scale fisheries and subsistence-trade networks) in many cases have become subsumed into capitalist ones; many local knowledge and language systems have been lost to cosmopolitan (i.e., Western) science, resulting in lost cultural diversity and understanding; and people have been displaced from their homelands as they entered the wage economy or as resources became privatized. And this list goes on.

Anthropogenic drivers of *global* climate change are largely mismatched with the scale of *local* impacts and burdens to communities and places. These scale mismatches are as much spatial (global–local) as they are temporal (past–future). Ironically, many local communities with marine-based livelihoods are burdened by both ends of a globalization–climate change squeeze. On one hand, communities are directly affected by changes to ecosystems that negatively impact their ways of life (e.g., loss of marine subsistence foods, or displacement from sea-level rise). On the other hand, these communities are excluded from the goods and benefits of global consumption driving climate change in the first place (e.g., large-scale extraction and development, often driven by consumers in developed countries; see Chapter 5). In many cases, too, mitigation of climate impacts burdens communities already most acutely vulnerable (e.g., local harvest restrictions imposed on those facing food shortages with few available alternatives, etc., see Wassilie and Poe, 2015).

Just as the ecological degradation from global carbon emissions and climate change produce cross-scalar impacts, economic globalization has also been a scale-crossing force in social–ecological systems characterizing "modernity" (Giddens, 2013). An inherently scale-based concept, globalization (and earlier "world systems" theories, Hopkins and Wallerstein, 1996) shows how these processes have worked hand in glove with ecological effects associated with the Anthropocene. Indeed, some mark the inception of the epoch with the onset of the industrial revolution and its earth-transforming, fossil fuel–driven technological development in the Western world (which is also, arguably facilitated by colonialism and imperialism of emerging nation-states, Hardt and Negri, 2000). The "Capitalocene" has been suggested as an alternative epochal title given how closely tied the Anthropocene is to the advance and effects of global capitalist resource consumption and social restructuring (Moore, 2016).

Yet, the primacy given to the historical juncture of European agro-industrial-driven changes to ecosystems risks erasing the landscape-scale modifications by First Peoples that have been underway for millennia. Indeed, biophysical forms and ecological functions have long been shaped by humans (e.g., clam gardens, high-grading salmon, and estuarine root gardens seen in Chapter 9; and see Balée, 2006 for a broader discussion of anthropogenic traces of historical ecology). Moreover, many coastal communities (in particular, indigenous communities) have already long endured ecological and cultural disruptions from colonialism and cumulative impacts of capitalism (see Chapter 18; see Whyte, 2016). Thus, when we speak in resilience terms of "bouncing back" (e.g., Davoudi, 2012), especially regarding indigenous communities, we must ask: what exactly is the "state" to bounce back and return to?

Conservation in the Anthropocene shall scarcely be status quo. Varied exposures, risks, and responses call for new and diverse knowledge and modes of action to ensure possible future resilience. Uncertain, unprecedented, and uneven threats will require new sorts of collaborations and participation in our efforts to define, understand, predict, and tackle collective conservation and justice challenges. As we work toward a livable future, our solutions will likely be far different than the sustainability references points based on systems of the past. A central challenge facing ocean conservation in the Anthropocene is to develop practical means for addressing the interdependence of ecosystems and human well-being, advancing the fundamental interdisciplinary science that underlies conservation practice, and implementing this science in decisions to manage, preserve, and restore ocean ecosystems.

CONSERVATION PRAXIS AND THE INTERPLAY OF THEORY AND PRACTICE

In this book we build upon case studies of ocean conservation in diverse situations around the world. These were selected for their effectiveness in orienting the reader to a range of principles and experiences of conservation in the context of the Anthropocene. We frame the challenges in the first section of the book, and then in the second section, we focus the reader on a variety of *principles* for science in the Anthropocene. Among the lessons highlighted are the importance of thinking across disciplinary silos, bringing together multiple streams of knowledge, including diverse biophysical and social sciences as well as citizen science and indigenous knowledges. Topics include how we think about the taxonomies of species (e.g., Chapter 10), human communities (e.g., Chapter 7), and the flows and access of goods and services (e.g., Chapters 11 and 12). Then, in the third section, the chapters illustrate many of these principles in action through a series of case studies on conservation *practices*. The case studies present a breadth of lessons from experiences with participatory science and decision-making, marine protected areas, sustainable development planning, economic trade-off assessments, and human rights implications of fisheries management.

Focusing on conservation principles coupled with conservation practices brings to light important lessons about *conservation praxis*. Praxis, in this case, refers to the adaptive learning and values that come from the interplay of applying theories (i.e., principles) to actions (i.e., practices) in conservation around the world. This interplay allows us to explore important overarching and productive themes running through this book: methods and processes of *knowledge and understanding*, as well as *management and decision-making*. We endeavor to better understand the world, and to act. Conservation praxis informs thinking, doing, learning, reflecting, and changing. Being adaptive with emergent and experiential wisdoms at the intersections of our held theories and in situ experiences of conservation will be important to our future resilience.

Knowledge and Understanding

Conservation began as an applied science field focused almost entirely on biology and related disciplines (Kareiva and Marvier, 2012; also noted in Chapter 6), but an increasing recognition of the integrated nature of social and ecological systems has shifted the field toward more interdisciplinary and transdisciplinary approaches (Levin et al., 2016). Indeed, the uncertainties and multiple (shared and competing) values of an increasingly complex future require synthesizing and producing ever richer and broader interdisciplinary and transdisciplinary understandings of the ocean and its people. Conservation sciences now include economics, anthropology, political science, geography, sociology, psychology, and other social sciences (Bennett et al., 2016; Mascia et al., 2003); these are added to or combined with more conventional modes of inquiry such as biology, ecology, oceanography, botany, genetics, physiology, evolutionary ecology, mathematical modeling, and the like. Our volume is committed to applying these diverse tools to ocean conservation. Consolidating these diverse streams of knowledge may not be easy, but the complexity of challenges we now face requires it, not only for improved understanding, but also for effective and creative solutions.

Toward this end, several chapters show how modeling approaches can be effective tools in consolidating the information that might otherwise be empirically challenging. Plagányi and Fulton (Chapter 20) give a modeling case study in the Torres Strait, where conventional biological and economic streams of information were expanded to include sociocultural dimensions, such as equity in a commercial fishery. Lewis et al. (Chapter 18) use bioeconomic models to illustrate the feedback between fishing-related environmental degradation and human-rights abuses. Inclusion of social information into classic bioeconomic models has enabled trade-off analyses that consider the "triple bottom line" (Halpern et al., 2013). This triple-bottom-line trade-off analysis is also demonstrated by Voss et al. (Chapter 17) in their study on cod recovery in the Baltic Sea: the authors bring ecological, economic, and social endpoints to bear on management options for cod, herring, and sprat that could better balance ecological conservation and economic profitability goals with equity goals across different fishing sectors and places.

Aside from simulated futures, other tools for synthesizing information across disciplines are illustrated in Chapter 16, through the use of ecosystem services valuation, based in community-based science priorities for sustainability goals, a hallmark of the Natural Capital Project. Another example includes Gray and Scyphers' Chapter 22 on public participatory and citizen science approaches. Coproduction of knowledge through participatory methods and citizen science is complementary to the broader field of local knowledge.

Local knowledges, and more specifically, traditional knowledges, (collectively TLKs), provide valuable ecological information based on long-term interactions with the environment, often in specific places and transferred over multiple generations (Berkes et al., 2000). TLKs are based in ecological observations, but also reflect cultural cognitive processes of people in particular places. Repeat, frequent, and intensive interactions and experimentation that form TLK not only improve understanding of community ecologies, but these local (folk) sciences can also contribute a deeper understanding of people–environment relationships.

The future is transdisciplinary. The watershed of coupled human and biophysical sciences has come at a time when scientists, policy makers, and society are seeking improvements in communication for actionable research. This includes the need to bring together various bodies of knowledge and social actors toward collaborative solutions (e.g., through the use of ecosystem services valuation, based in community-based sustainability priorities discussed in Arkema and Ruckleshaus). Transdisciplinary and participatory approaches help ensure that conservation science and practices are relevant to communities and decision makers; these tools not only build the most relevant and best available science, they ultimately assist with the difficult task of evaluating trade-offs and informing policy.

Science and policy integration are components of adaptive management for ocean sustainability. In her chapter on principles for interdisciplinary conservation, Leslie (Chapter 6) outlines the following four principles: attention and openness when defining the coupled systems; commitment to useful, solutions-oriented scholarship; mindful engagement with collaborators, including community members; and finally, humility. Likewise, in their chapter on bridging the science–policy interface, Sullivan et al. (Chapter 1) emphasize the importance of two-way dialogues between conservation scientists and society; this type of dialogue, they point out, often results in more trusted and responsive information, which in turn, motivates adaptive management.

Management and Decision-Making

Adaptive management and decision-making are processes within the larger domain of ocean governance: the institutions, processes, and power dynamics that shape how we act.

Taking action increasingly requires participation from multiple stakeholders. The opportunity here is to democratize conservation. This entails a more inclusive integration of multiple knowledges, diverse values, participatory decision-making, and negotiation. Several chapters explore questions of governance, reviewing top-down approaches as well as local scale political action. One example includes a community-based stewardship effort to replant the coral reefs near Koh Piyak, Thailand described by Charles in Chapter 21. These examples remind us that conservation governance systems and actors are complex, and can be formal (i.e., state actors) and informal, and scales of action need to fit conservation challenges (see also Chapter 8).

Multi-actor participatory processes include knowledge sharing (see Chapter 22) and negotiation (see Chapter 14). In the former, participants add to the body of knowledge information for each decision context. In the latter, stakeholder processes become a space for negotiating political agendas and diverse objectives, including what kinds of information will be used, which is ultimately a process subject to power dynamics. In all cases, power dynamics are at play in the decisions we make to conserve and manage the ocean (see also Chapter 15 on cooperation and conflict by Basurto et al.).

Attention to governance matters because governance constitutes the context and process through which policies are established and implemented, and thus governance shapes how marine resources are used and protected, including where, how, and by whom.

Food, for example, is among the important resources and ecosystem services provided by the ocean. But access to ocean foods is affected by policy. Koehn et al. (Chapter 4) raise the prospects of policy interventions to improve how fisheries and aquaculture might feed a growing global population. Here they note that food security and health are rarely the objectives of fisheries policies, which instead tend to orient around maximizing economic gains. Their chapter explores: what might food-maximizing alternatives to fisheries management look like and what steps are to be taken to get there? Food security, and importantly, food sovereignty are again taken up in Chapter 19 by Satterfield et al. The authors describe the impacts of first colonial policies, and now climate change to indigenous food sovereignty in British Columbia. Not only do local initiatives to restore marine food practices promise greater nutritional benefits, but these also restore cultural–ecological practices of First Nations that are fundamental to their identities, ways of life, and rights to traditional territories.

The question of policy effects on resource rights and fisheries privatization is raised by Donkersloot and Carothers (Chapter 12). Catch shares policies in their Alaska case study have negatively impacted fishing communities by deepening inequalities, hampering new participants, and alienating local communities from access to resources, while also failing to curtail bycatch or consistently meet other conservation goals. Instead, Donkersloot and Carothers describe promising new community arrangements such as community development quota systems that better account for the distribution of benefits and burdens across fisheries communities and actors, and offer new hope for socially and ecologically sustainable fisheries. Their intervention orients us toward alternative systems for management during uncertain periods of ocean and social change.

THINKING AND ACTING IN AN UNCERTAIN FUTURE

Applying more holistic approaches to ocean management, with sensitivity to coupled social–ecological health, will require that we work together. In many ways, this can be accomplished through renewed commitments to participatory and ecosystems-based thinking and management for our shared social–ecological communities.

The challenge for social–ecological systems is detecting tipping points before it is too late, particularly in cases where interventions could alter outcomes. All too often, losses are identified in hindsight and ocean systems are particularly difficult to see (see Chapter 3), or may be muddied by a myriad of social–ecological changes (e.g., Morlein and Carothers, 2012) or by effects that are obscured by smoothing of detectable patterns at different scales (Barnosky et al., 2012). When changes have gone unnoticed, it is often the case that baselines have shifted. Novel systems become normalized (Pauly, 1995). The "new normal" in many local and regional contexts are more homogenous and vulnerable than social–ecological systems of the past.

Conservation practice during intensifying times of ocean uncertainties depends on being able to identify early warning signs and mechanisms of change in order to act proactively to reduce threats (Barnosky et al., 2012). Scientists are exploring methods for identifying new early warning indicators toward that end. For example, Dulvy and Kindsvater argue for lowering the bar for acceptable statistical error used in hypothesis testing to give timely precautionary warnings of approaching species extinction. Halpern (Chapter 13) presses for resource management approaches that explicitly account for uncertainty rather than remove it. We need to know a *little*, but maybe not everything, to minimize the chances of crossing a tipping point, Halpern argues.

The urgency to pin down the information on how and when things will change, however, can cloud science and action that instead must focus on understanding and supporting resilient social and ecological systems and adaptation. The more incomplete our understanding of the changes advancing in the Anthropocene, the easier it is to simplify our responses to them. This point can be seen through (Chapter 10) on species now out of place in the Anthropocene.

We still know very little about non-native species, with tendencies to catego-rize these in dichotomies of "good" or "bad" without regard to the processes, functions, and even impacts of changes to ecological communities. The same critique can be made about our understanding of other changing phenomena in the Anthropocene. Cote pushes us, then, to attend to the empirics of change, and to reexamine our held beliefs of what belongs and what does not, especially under conditions of increasing complexity. We also have to accept that some unpredictable and nonlinear changes may be beyond our control (Head, 2016).

The futures of people and nature are entwined. Convergences have the potential to inspire novel conservation actions. Indeed, when we truly embrace a concept of social–ecological systems, the suite of conservation tactics also becomes diverse. Familiar goals such as protection of biological diversity and habitats are also now augmented by conservation goals oriented toward human well-being. Ecosystem services and human well-being are increasingly framed as shared endpoints of conservation (Breslow et al., 2016; Hicks et al., 2016). By giving more attention to "improving how humans relate to each other and with the material and cultural world" Basurto et al. (Chapter 15) suggest: we can "create better solutions to conservation problems we face in common."

We conclude by asking: what if social–ecological well-being is conceived as *means to an end*, rather than simply an end itself?

After all, "well-being" is an action word. Being well. Being. Action in the Anthropocene is as much about how we envision and learn in our uncertain future(s), as it is about the practices and processes through which we ensure our collective continuance (e.g., Whyte, 2013). These practices take shape in rela-tionship to one another, connected to other beings and their multiple complex systems: *interconnected*.

INTERCONNECTEDNESS

The image on the front cover of our book is a Haida ceremonial canoe and replica of Bill Reid's original canoe carved from 450 year old cedar tree. Both canoes are located at the Haida Heritage Center, near the village of Skidegate in the islands of Haida Gwaii, British Columbia, Canada, where we collaborate and where this photograph was taken. We chose this image in part for its obvious depiction of the dynamic connections that people have with the ocean. The canoe car-ries people out to sea where they journey, fish, share, negotiate, observe, pray, recreate, care for ocean creatures, and restore cultural practices. The image also inspires the principle we put forth to conclude this collection of chapters on the challenges and actions of ocean conservation in the Anthropocene: intercon-nectedness. In their guiding ethics of the marine relationship, the Haida have identified *Gina waadluxan gud ad kwaagiida* as a key principle, "everything depends on everything else." This vision considers "the relationships between species and habitats, and accounts for short-term, long-term and cumulative effects of human activities on the environment. Interrelationships are accounted for across spatial and temporal scales and across agencies and jurisdictions" (Council of Haida Nation, Haida Marine Plan).

In this book, and going forward, we promote this vision of interconnectedness as an underlying principle and approach for interdisciplinary conservation. Interconnectedness orients us toward our responsibilities: the responsibilities we have with nature and with one another as we make decisions that help us adapt and thrive in just, honorable, and sustainable ways.

REFERENCES

Balée, W., 2006. The research program of historical ecology. Annual Review of Anthropology 35, 75–98.

Barnosky, A.D., Hadly, E.A., Bascompte, J., Berlow, E.L., Brown, J.H., Fortelius, M., Wayne, M.G., et al., 2012. Approaching a state shift in Earth's biosphere. Nature 486 (7401), 52–58.

Bennett, N.J., Roth, R., Klain, S.C., Chan, K., Christie, P., Clark, D.A., Cullman, G., Curran, D., Durbin, T.J., Epstein, G., Greenberg, A., 2016. Conservation social science: understanding and integrating human dimensions to improve conservation. Biological Conservation 204, 93–108.

Berkes, F., Colding, J., Folke, C., 2000. Rediscovery of traditional ecological knowledge as adaptive management. Ecological Applications 10, 1251–1262.

Breslow, S.J., Sojka, B., Barnea, R., Basurto, X., Carothers, C., Charnley, S., Coulthard, S., Dolšak, N., Donatuto, J., García-Quijano, C., Hicks, C.C., Levine, A., Mascia, M.B., Norman, K., Poe, M.R., Satterfield, T., St. Martin, K., Levin, P.S., 2016. Conceptualizing and operationalizing human wellbeing for ecosystem assessment and management. Environmental Science & Policy 66, 250–259.

Crutzen, P.J., Steffen, W., 2003. How long have we been in the Anthropocene era? Climatic Change 61 (3), 251–257.

Davoudi, S., 2012. Resilience: a bridging concept or a dead end? Planning Theory & Practice 13 (2), 299–333.

Giddens, A., 2013. The Consequences of Modernity. John Wiley & Sons.

Hamilton, C., Bonneuil, C., Gemenne, F., 2015. Thinking the Anthropocene. The Anthropocene and the Global Environmental Crisis: Rethinking Modernity in a New Epoch, pp. 1–14.

Halpern, B.S., Klein, C.J., Brown, C.J., Beger, M., Grantham, H.S., Mangubhai, S., Ruckelshaus, M., Tulloch, V.J., Watts, M., White, C., Possingham, H.P., 2013. Achieving the triple bottom line in the face of inherent trade-offs among social equity, economic return, and conservation. Proceedings of the National Academy of Sciences of the United States of America 110 (15), 6229–6234.

Hardt, M., Negri, A., 2000. Empire. Harvard.

Head, L., 2016. Hope and Grief in the Anthropocene: Re-conceptualising Human–Nature Relations. Routledge.

Hicks, C., Levine, A., Agrawal, A., Basurto, X., Breslow, S., Carothers, C., Charnley, S., Coulthard, S., Dolsak, N., Donatuto, J., Garcia Quijano, C., Mascia, M.B., Norman, K., Poe, M., Satterfield, T., St. Martin, K., Levin, P.S., 2016. Engage key social concepts for sustainability. Science April 1, 352 (6281).

Holland, S.M., 2016. Ecological disruption precedes mass extinction. Proceedings of the National Academy of Sciences of the United States of America 201608630.

Hopkins, T.K., Wallerstein, I.M., 1996. The Age of Transition: Trajectory of the World System, 1945–2025. Palgrave Macmillan.

Kareiva, P., Marvier, M., 2012. What is conservation science? BioScience 62 (11), 962–969.

Levin, P.S., Breslow, S.J., Harvey, C.J., Norman, K.C., Poe, M.R., Williams, G.D., Plummer, M.L., 2016. Conceptualization of social-ecological systems of the California current: an examination of interdisciplinary science supporting ecosystem-based management. Coastal Management 44 (5), 397–408.

Mascia, M.B., Brosius, J.P., Dobson, T.A., Forbes, B.C., Horowitz, L., McKean, M.A., Turner, N.J., 2003. Conservation and the social sciences. Conservation Biology 17 (3), 649–650.

Moerlein, K., Carothers, C., 2012. Total environment of change: impacts of climate change and social transitions on subsistence fisheries in northwest Alaska. Ecology and Society 17 (1).

Moore, J. (Ed.), 2016. Anthropocene or Capitalocene? Nature, History, and the Crisis of Capitalism. PM Press.

Pauly, D., 1995. Anecdotes and the shifting baseline syndrome of fisheries. Trends in Ecology and Evolution 10 (10), 430.

Wassilie, K., Poe, M., 2015. Pacific Walrus and Coastal Alaska Native Subsistence Hunting: Considering Vulnerabilities From Ocean Acidification. Earthzine.

Whyte, K., September 8, 2016. Our Ancestors' Dystopia Now: Indigenous Conservation and the Anthropocene. Routledge Companion to the Environmental Humanities, Forthcoming. Available at SSRN: https://ssrn.com/abstract=2770047.

Whyte, K.P., 2013. Justice forward: tribes, climate adaptation and responsibility. Climatic Change 120 (3), 517–530.

Index

Printed in the United States
By Bookmasters